Redox Active Molecules in Cancer Treatments

Redox Active Molecules in Cancer Treatments

Editors

Višnja Stepanić
Marta Kučerová-Chlupáčová

MDPI • Basel • Beijing • Wuhan • Barcelona • Belgrade • Manchester • Tokyo • Cluj • Tianjin

Editors
Višnja Stepanić
Laboratory for Machine Learning and Knowledge Representation
Ruđer Bošković Institute
Zagreb
Croatia

Marta Kučerová-Chlupáčová
Department of Pharmaceutical Chemistry and Pharmaceutical Analysis
Charles University
Hradec Králové
Czech Republic

Editorial Office
MDPI
St. Alban-Anlage 66
4052 Basel, Switzerland

This is a reprint of articles from the Special Issue published online in the open access journal *Molecules* (ISSN 1420-3049) (available at: www.mdpi.com/journal/molecules/special_issues/Redox_Active_Cancer).

For citation purposes, cite each article independently as indicated on the article page online and as indicated below:

LastName, A.A.; LastName, B.B.; LastName, C.C. Article Title. *Journal Name* **Year**, *Volume Number*, Page Range.

ISBN 978-3-0365-6731-0 (Hbk)
ISBN 978-3-0365-6730-3 (PDF)

Contents

About the Editors

Višnja Stepanić

Višnja Stepanić, Ph.D. received her doctorate from the Faculty of Science, University of Zagreb, Croatia. She worked in the pharmaceutical industry at PLIVA and GSK and now works in the Machine Learning and Knowledge Representation Laboratory at the Rudjer Bošković Institute in Zagreb, Croatia. Dr. Stepanić has participated in numerous research projects mainly in the fields of medicinal chemistry and chemical biology. In her interdisciplinary collaborations, she focuses on the analysis of modes of action of compounds with mainly redox activities, using molecular modelling (DFT calculations, molecular docking, and dynamics) and big data analysis (pattern recognition, QSAR). She has been a member of several COST actions, scientific committees of several international scientific meetings, and international panels. She is a member of the Committee of the Medicinal and Pharmaceutical Chemistry Section of the Croatian Chemical Society, a member organization of the European Federation of Medicinal Chemistry and Chemical Biology (EFMC).

Marta Kučerová-Chlupáčová

Dr. Marta Kučerová, Ph.D. obtained her master's degree in Pharmacy as well as her Ph.D. in Pharmaceutical Chemistry at Charles University, Faculty of Pharmacy in Hradec Králové (Czech Republic). She works as an assistant professor in the Department of Pharmaceutical Chemistry and Pharmaceutical Analysis of the Faculty of Pharmacy in Hradec Králové. She was employed in different research projects. She spent two research internships in the group of Prof. Dr. Michael Gütschow at the University of Bonn (Germany). In her research, she focuses on the synthesis of potential drugs, especially from the group of nitrogen-containing heterocycles, in silico drug design, and medicinal chemistry of antimicrobial and anticancer compounds.

She is a member of the State Examination Committee and the State Rigorosum Examination Committee for Pharmaceutical Chemistry at the faculty and a member of the committee of the Section of Synthetic Drugs in the Czech Pharmaceutical Society, which is an adhering organization of the European Federation for Medicinal chemistry and Chemical biology (EFMC). Marta Kučerová holds the function of the chair of the Editorial Committee of the Faculty of Pharmacy in Hradec Králové and is a member of the Board for Equal Opportunities at Charles University (Czech Republic).

molecules

Editorial

Redox Active Molecules in Cancer Treatments

Višnja Stepanić [1,*] and Marta Kučerová-Chlupáčová [2,*]

1 Laboratory for Machine Learning and Knowledge Representation, Ruđer Bošković Institute, Bijenička 54, 10000 Zagreb, Croatia
2 Department of Pharmaceutical Chemistry and Pharmaceutical Analysis, Faculty of Pharmacy in Hradec Králové, Charles University, Ak. Heyrovského 1203/8, 500 05 Hradec Králové, Czech Republic
* Correspondence: visnja.stepanic@irb.hr (V.S.); kucerom@faf.cuni.cz (M.K.-C.); Tel.: +385-1-457-1356 (V.S.); +420-495-067-372 (M.K.-C.)

Cancer is one of the leading causes of death worldwide, with nearly 10 million deaths in 2020 [1]. Redox active molecules in the diet, dietary supplements, or in approved drug preparations are used to prevent and treat cancer.

The main objective of this Special Issue, "Redox Active Molecules in Cancer Treatments", in the journal *Molecules* is to present the results of in vitro, in vivo, and/or in silico studies on the biological effects and activities of anti- and pro-oxidant molecules observed in original research studies or collected and discussed in review articles. This goal is achieved by compiling seventeen articles. They present antioxidative or targeted oxidative effects of miscellaneous small-molecular-weight compounds or proteins against a variety of cancer types:

- An endogenous compound—melatonin [2].
- Natural plant compounds (naringenin [3], papaverine [4], polyphenols isolated from *Myrciaria trunciflora* [5] or *Anneslea fragrans* [6], and seed-derived peptides [7]), natural compounds also found in animals (melatonin [2,8]), and peptides as well as proteins from Jellyfish venom [9].
- Synthetic compounds, i.e., alkyl thiols [10], dimethyl sulfoxide [11], metformin and S63845 [12], the ruthenium complex $[Ru(Phen)_3]^{2+}$ [13], and copper-based compounds—Casiopeinas [14].
- Different formulations, i.e., peptide fractions from germinated soybeans conjugated to Fe_3O_4 nanoparticles [15] and astaxanthin microparticles in combination with pentoxifylline [16].
- Proteins (aquaporins [17]) and nuclear factor erythroid-2-related factor 2 (NRF2) [8].

The studies explored diverse anticancer mechanisms of action of redox-active molecules in association with specific signaling pathways by using in vitro and in vivo methods. Some studies investigated the use of redox-active compounds to alleviate radiation-induced fibrosis, which is a side-effect of radiotherapy [16], or to detect oxygen in vitro and in vivo [13]. Most studies examined the effect of the tested compounds on cancer cell viability/proliferation assays [2–4,6,11,12] and/or analyses of reactive oxygen species concentrations [2,3,6,11,15,16]. Some other studies used in vitro assays such as cell cycle analyses [2–4,9], DNA fragmentation assays [3,9], analyses of the expression of apoptosis-related proteins and/or genes [9,11,12], etc. The two included studies are based on the application of state-of-the-art chemoinformatic analysis and modeling approaches—molecular docking and molecular dynamics [7,18].

The whole series of thirteen experimental investigations and one computational study is accompanied by three review articles focusing on aquaporins as redox regulators in breast cancer [17], natural compounds affecting ferroptosis [18], and modulation of NRF2 expression at the mRNA and protein levels [8].

We hope that readers will enjoy the book and glean interesting and useful information from the particular studies.

Citation: Stepanić, V.; Kučerová-Chlupáčová, M. Redox Active Molecules in Cancer Treatments. *Molecules* **2023**, *28*, 1485. https://doi.org/10.3390/molecules28031485

Received: 29 January 2023
Accepted: 31 January 2023
Published: 3 February 2023

Funding: This research received no external funding.

Acknowledgments: We would like to thank all of the authors who contributed to this Special Issue.

Conflicts of Interest: The authors declare no conflict of interest.

References

1. World-Health-Organization. Cancer. Available online: https://www.who.int/news-room/fact-sheets/detail/cancer (accessed on 11 January 2023).
2. Chok, K.C.; Koh, R.Y.; Ng, M.G.; Ng, P.Y.; Chye, S.M. Melatonin Induces Autophagy via Reactive Oxygen Species-Mediated Endoplasmic Reticulum Stress Pathway in Colorectal Cancer Cells. *Molecules* **2021**, *26*, 17. [CrossRef] [PubMed]
3. Lee, C.W.; Huang, C.C.Y.; Chi, M.C.; Lee, K.H.; Peng, K.T.; Fang, M.L.; Chiang, Y.C.; Liu, J.F. Naringenin Induces ROS-Mediated ER Stress, Autophagy, and Apoptosis in Human Osteosarcoma Cell Lines. *Molecules* **2022**, *27*, 16. [CrossRef] [PubMed]
4. Gomes, D.A.; Joubert, A.M.; Visagie, M.H. In Vitro Effects of Papaverine on Cell Proliferation, Reactive Oxygen Species, and Cell Cycle Progression in Cancer Cells. *Molecules* **2021**, *26*, 19. [CrossRef] [PubMed]
5. Augusti, P.R.; Quatrin, A.; Mello, R.; Bochi, V.C.; Rodrigues, E.; Prazeres, I.D.; Macedo, A.C.; Oliveira-Alves, S.C.; Emanuelli, T.; Bronze, M.R.; et al. Antiproliferative Effect of Colonic Fermented Phenolic Compounds from Jaboticaba (Myrciaria trunciflora) Fruit Peel in a 3D Cell Model of Colorectal Cancer. *Molecules* **2021**, *26*, 13. [CrossRef] [PubMed]
6. He, S.Y.; Cui, X.Y.; Khan, A.; Liu, Y.P.; Wang, Y.D.; Cui, Q.M.; Zhao, T.R.; Cao, J.X.; Cheng, G.G. Activity Guided Isolation of Phenolic Compositions from Anneslea fragrans Wall. and Their Cytoprotective Effect against Hydrogen Peroxide Induced Oxidative Stress in HepG2 Cells. *Molecules* **2021**, *26*, 14. [CrossRef] [PubMed]
7. Chai, T.T.; Koh, J.A.; Wong, C.C.C.; Sabri, M.Z.; Wong, F.C. Computational Screening for the Anticancer Potential of Seed-Derived Antioxidant Peptides: A Cheminformatic Approach. *Molecules* **2021**, *26*, 21. [CrossRef] [PubMed]
8. Aliyev, A.T.; Panieri, E.; Stepanić, V.; Gurer-Orhan, H.; Saso, L. Involvement of NRF2 in Breast Cancer and Possible Therapeutical Role of Polyphenols and Melatonin. *Molecules* **2021**, *26*, 18. [CrossRef]
9. Tawfik, M.M.; Eissa, N.; Althobaiti, F.; Fayad, E.; Abu Almaaty, A.H. Nomad Jellyfish Rhopilema nomadica Venom Induces Apoptotic Cell Death and Cell Cycle Arrest in Human Hepatocellular Carcinoma HepG2 Cells. *Molecules* **2021**, *26*, 14. [CrossRef] [PubMed]
10. Heymans, V.; Kunath, S.; Hajieva, P.; Moosmann, B. Cell Culture Characterization of Prooxidative Chain-Transfer Agents as Novel Cytostatic Drugs. *Molecules* **2021**, *26*, 12. [CrossRef] [PubMed]
11. Sangweni, N.F.; Dludla, P.V.; Chellan, N.; Mabasa, L.; Sharma, J.R.; Johnson, R. The Implication of Low Dose Dimethyl Sulfoxide on Mitochondrial Function and Oxidative Damage in Cultured Cardiac and Cancer Cells. *Molecules* **2021**, *26*, 15. [CrossRef] [PubMed]
12. Valiuliene, G.; Vitkeviciene, A.; Skliute, G.; Borutinskaite, V.; Navakauskiene, R. Pharmaceutical Drug Metformin and MCL1 Inhibitor S63845 Exhibit Anticancer Activity in Myeloid Leukemia Cells via Redox Remodeling. *Molecules* **2021**, *26*, 13. [CrossRef] [PubMed]
13. Huntosova, V.; Horvath, D.; Seliga, R.; Wagnieres, G. Influence of Oxidative Stress on Time-Resolved Oxygen Detection by Ru(Phen)(3) (2+) In Vivo and In Vitro. *Molecules* **2021**, *26*, 24. [CrossRef]
14. Ramirez-Palma, L.G.; Espinoza-Guillen, A.; Nieto-Camacho, F.; Lopez-Guerra, A.E.; Gomez-Vidales, V.; Cortes-Guzman, F.; Ruiz-Azuara, L. Intermediate Detection in the Casiopeina-Cysteine Interaction Ending in the Disulfide Bond Formation and Copper Reduction. *Molecules* **2021**, *26*, 12. [CrossRef]
15. Augusto-Jimenez, Y.E.; Gonzalez-Montoya, M.; Naranjo-Feliciano, D.; Uribe-Ramirez, D.; Cristiani-Urbina, E.; Diaz-Aguila, C.; Yee-Madeira, H.; Mora-Escobedo, R. Antioxidant Activity of Bioactive Peptide Fractions from Germinated Soybeans Conjugated to Fe_3O_4 Nanoparticles by the Ugi Multicomponent Reaction. *Molecules* **2021**, *26*, 15. [CrossRef] [PubMed]
16. Binatti, E.; Zoccatelli, G.; Zanoni, F.; Dona, G.; Mainente, F.; Chignola, R. Effects of Combination Treatments with Astaxanthin-Loaded Microparticles and Pentoxifylline on Intracellular ROS and Radiosensitivity of J774A.1 Macrophages. *Molecules* **2021**, *26*, 11. [CrossRef] [PubMed]
17. Milković, L.; Čipak Gašparović, A. AQP3 and AQP5-Potential Regulators of Redox Status in Breast Cancer. *Molecules* **2021**, *26*, 14. [CrossRef] [PubMed]
18. Stepanić, V.; Kučerová-Chlupáčová, M. Review and Chemoinformatic Analysis of Ferroptosis Modulators with a Focus on Natural Plant Products. *Molecules* **2023**, *28*, 475. [CrossRef] [PubMed]

molecules

Article

Melatonin Induces Autophagy via Reactive Oxygen Species-Mediated Endoplasmic Reticulum Stress Pathway in Colorectal Cancer Cells

Kian Chung Chok [1], Rhun Yian Koh [2], Ming Guan Ng [1], Pei Ying Ng [3] and Soi Moi Chye [2,*]

1 School of Health Science, International Medical University, Kuala Lumpur 57000, Malaysia; chok.kianchung@student.imu.edu.my (K.C.C.); NG.MINGGUAN@student.imu.edu.my (M.G.N.)
2 Division of Biomedical Science and Biotechnology, School of Health Science, International Medical University, Kuala Lumpur 57000, Malaysia; rhunyian_koh@imu.edu.my
3 School of Postgraduate, International Medical University, Kuala Lumpur 57000, Malaysia; NG.PEIYING@student.imu.edu.my
* Correspondence: chye_soimoi@imu.edu.my; Tel.: +60-327317220

Abstract: Even though an increasing number of anticancer treatments have been discovered, the mortality rates of colorectal cancer (CRC) have still been high in the past few years. It has been discovered that melatonin has pro-apoptotic properties and counteracts inflammation, proliferation, angiogenesis, cell invasion, and cell migration. In previous studies, melatonin has been shown to have an anticancer effect in multiple tumors, including CRC, but the underlying mechanisms of melatonin action on CRC have not been fully explored. Thus, in this study, we investigated the role of autophagy pathways in CRC cells treated with melatonin. In vitro CRC cell models, HT-29, SW48, and Caco-2, were treated with melatonin. CRC cell death, oxidative stress, and autophagic vacuoles formation were induced by melatonin in a dose-dependent manner. Several autophagy pathways were examined, including the endoplasmic reticulum (ER) stress, 5′–adenosine monophosphate-activated protein kinase (AMPK), phosphoinositide 3-kinase (PI3K), serine/threonine-specific protein kinase (Akt), and mammalian target of rapamycin (mTOR) signaling pathways. Our results showed that melatonin significantly induced autophagy via the ER stress pathway in CRC cells. In conclusion, melatonin demonstrated a potential as an anticancer drug for CRC.

Keywords: melatonin; autophagy; colorectal cancer cells; reactive oxygen species; endoplasmic reticulum stress

Citation: Chok, K.C.; Koh, R.Y.; Ng, M.G.; Ng, P.Y.; Chye, S.M. Melatonin Induces Autophagy via Reactive Oxygen Species-Mediated Endoplasmic Reticulum Stress Pathway in Colorectal Cancer Cells. *Molecules* **2021**, *26*, 5038. https://doi.org/10.3390/molecules26165038

Academic Editors: Višnja Stepanić, Marta Kučerová-Chlupáčová and Simona Rapposelli

Received: 16 March 2021
Accepted: 29 July 2021
Published: 20 August 2021

1. Introduction

Colorectal cancer (CRC) ranks third in incidence and second in cause of cancer death worldwide, with more than 1.9 million new cases and 935,000 deaths in the year 2020 [1]. The number of deaths caused by CRC worldwide is predicted to increase to 2.5 million in the year 2035 [2]. The 5-year survival rate of CRC is around 90%, but it drops to around 10% for metastatic CRC [3]. The high mortality rates of CRC are due to the fact that 25% of CRC cases have metastases at the time of diagnosis and up to 50% of patients develop them after diagnosis [4].

Although new drugs against CRC have been developed in recent years, chemotherapy remains the mainstream of the first-line treatment [5]. Chemotherapy for CRC has undesirable side effects, which induce death of normal cells and reduce patients' quality of life [6,7]. Hence, drugs targeting CRC effectively and without normal cell toxicity are urgently required to improve the patients' quality of life. In previous studies, melatonin alleviated the undesirable side effects and enhanced the anticancer effects when used as an adjuvant therapy in existing anticancer treatment [8–10]. Studies proved that melatonin inhibits cell proliferation and induces apoptosis in breast cancer [11], induces anti-angiogenic effects in liver cancer [12], and inhibits invasion and migration in ovarian cancer [13]. Melatonin has shown promising anticancer effects by anti-inflammation and antioxidation in cancer cells [9,14]. Our previous publication reported that melatonin prevented oxidative stress-induced mitochondrial dysfunction in high glucose-treated Schwann cells, proving the antioxidant properties of melatonin [15]. Melatonin also increases DNA repair capacity in oxidant-challenged cells by preventing the initiation and progression of cancer directly as a free radical scavenger and indirectly as an antioxidant enzyme regulator [16]. Combining antioxidant vitamins A and E with irinotecan can increase its cytotoxic effect. However, when melatonin was combined with irinotecan, no significant increment of cytotoxic effect was reported [17], although melatonin is a well-documented antioxidant, and it is also a conditional prooxidant, depending on the dose, cell types, and drug–drug interactions. Several studies reported melatonin induces reactive oxygen species (ROS) accumulation in cancer cell, which leads to cancer cell death [18,19]. Hence, our study investigated the oxidation properties of melatonin in CRC cells.

Autophagy plays a critical role in cancer and its function in cancer is controversial [20]. Although autophagy is a stress-related cell adaptation mechanism to avoid cell death, occasionally, it may be an alternative death pathway called autophagic cell death [21]. In HCT116 colon cancer cells and DU145 prostate cancer cells, endoplasmic reticulum (ER)-stress-induced autophagy removes polyubiquitinated protein aggregates and reduces cellular vacuolization [22]. A study of sphingolipid homeostasis revealed that the ER-stress-induced autophagy together with the Akt pathway activation in a protein kinase R (PKR)-like endoplasmic reticulum kinase (PERK)-dependent manner counteracts the ER-stress-induced apoptotic signaling [23]. In contrast, bis(dehydroxy)curcumin induces apoptosis-independent autophagic cell death in cancer [24]. Hence, the interplay between apoptosis and autophagy is tissue- and condition-dependent. A deeper look into the role of autophagy in CRC treated with melatonin is required to further support our previous work and the use of melatonin in anticancer treatment [14].

It would be premature to use melatonin to treat CRC because the mechanisms underlying its anticarcinogenic properties need to be investigated before melatonin is used as a standalone anticancer drug or as an adjuvant to other anticancer drugs. In the present work, we investigated the underlying anticancer mechanism of melatonin using HT-29, SW48, and Caco-2 CRC cells; the mechanism included oxidative stress, ER stress, AMPK, and PI3K/Akt/mTOR pathways.

2. Results

2.1. The Effects of Melatonin and Tauroursodeoxycholic Acid (TUDCA) on Cell Viability of CRC Cells

The HT-29, SW48, and Caco-2 cells were treated for 72 h with various concentrations of melatonin (0.5, 1.0, 1.5, 2.0, and 2.5 mM) and with vehicle control (98% ethanol, as (melatonin was dissolved in ethanol to a final concentration of 0.5%). The cell viabilities of HT-29, SW48, and Caco-2 were quantified with 3-(4,5-dimethylthiazol-2-yl)-2,5-diphenyltetrazolium bromide (MTT) assay (Figure 1A). The vehicle control and 0.5 mM melatonin had little inhibitory effect on the viability of the HT-29 cells. Melatonin alone significantly reduced the viability of the HT-29 cells in a dose-dependent manner (by 87, 77, 60, and 53% with 1.0, 1.5, 2.0, and 2.5 mM melatonin treatment, respectively). It also significantly reduced the viability of the SW48 cells (by 78, 58, and 48% with 1.5, 2.0, and 2.5 mM melatonin, respectively) and of the Caco-2 cells (68 and 73% with 2.0, and 2.5 mM melatonin, respectively).

In order to understand the role of ER stress in cell death, TUDCA, an ER stress inhibitor, was used in this study [25]. Trypan blue cell viability assay was performed for the HT-29, SW48, and Caco-2 cells treated with varying melatonin concentrations (0.5, 1.0, 1.5, 2.0, and 2.5 mM), as well as with their combinations with 0.1 mM TUDCA and 1 mM TUDCA. Melatonin alone induced significant cell death in all CRC cell lines, whereas TUDCA reversed melatonin-induced cytotoxic effects on all CRC cell lines; the potency of cytotoxic effects being melatonin alone > melatonin with 0.1 mM TUDCA > melatonin with 1 mM TUDCA (Figure 1C,D).

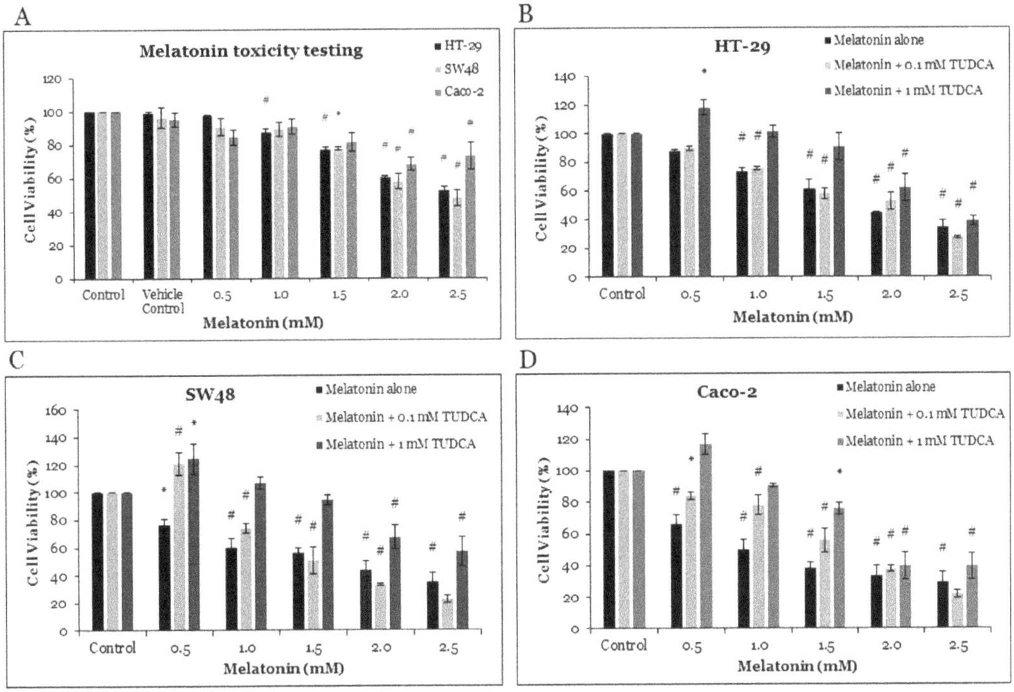

Figure 1. Cell viability assay: (**A**) HT-29, SW48, and Caco-2 were treated with various concentrations of melatonin for 72 h. (**B–D**) HT-29, SW48, and Caco-2 were treated with various concentrations of melatonin alone or together with 0.1 mM TUDCA or with 1 mM TUDCA. Means ± standard error of mean (SEM) of three separate experiments are shown. Two-way analysis of variance (ANOVA), followed by Dunnett's post hoc test, was performed to compare each treatment group and control; statistically significantly different from control: *, $p < 0.05$; #, $p < 0.01$.

2.2. *Melatonin Induces Oxidative Stress in HT-29 Cells*

We performed dichloro-dihydro-fluorescein diacetate (DCFH-DA) assay to quantify intracellular oxidative stress in the HT-29 cells after melatonin treatment (0.5, 1.0, 1.5, 2.0, and 2.5 mM) for 12, 24, and 48 h. Generally, the higher concentrations of melatonin increased ROS production in the HT-29 cells. At the 24 h time point, ROS levels in 2.0 and 2.5 mM melatonin-treated HT-29 cells were significantly elevated 2.99 and 4.22 times, respectively, as compared with the control. Thereafter, the ROS levels at 48 h post-treatment (significantly increased by ≥1.5 mM melatonin) were similar to the ROS levels at 12 h post-treatment (significantly increased by ≥1.0 mM melatonin) (Figure 2).

2.3. *Melatonin Treatment Induces Autophagic Vacuoles Formation in HT-29 Cells*

As autophagic cells form vacuoles, acridine orange forms aggregates that emit bright red fluorescence in acidic vesicles but emit green fluorescence in the nucleus and cytoplasm of the cells [26]. To further understand the mechanism of the HT-29 cell death induced by melatonin, fluorescent staining with acridine orange was done on the HT-29 cells after treatment with melatonin for 72 h (Figure 3A–F). The percentual ratios of bright red fluorescent cells to green fluorescent cells were evaluated. With the increasing concentration of melatonin, the percentage of autophagic vacuoles-positive cells increased, suggesting a dose-dependent mechanism of autophagy activation induced by the melatonin treatment. The proportions of autophagic vacuoles-positive cells after melatonin treatments were significantly different from those in control cells (22.7, 28.5, and 60.8% in case of 1.5, 2.0, and 2.5 mM melatonin, respectively) (Figure 3G).

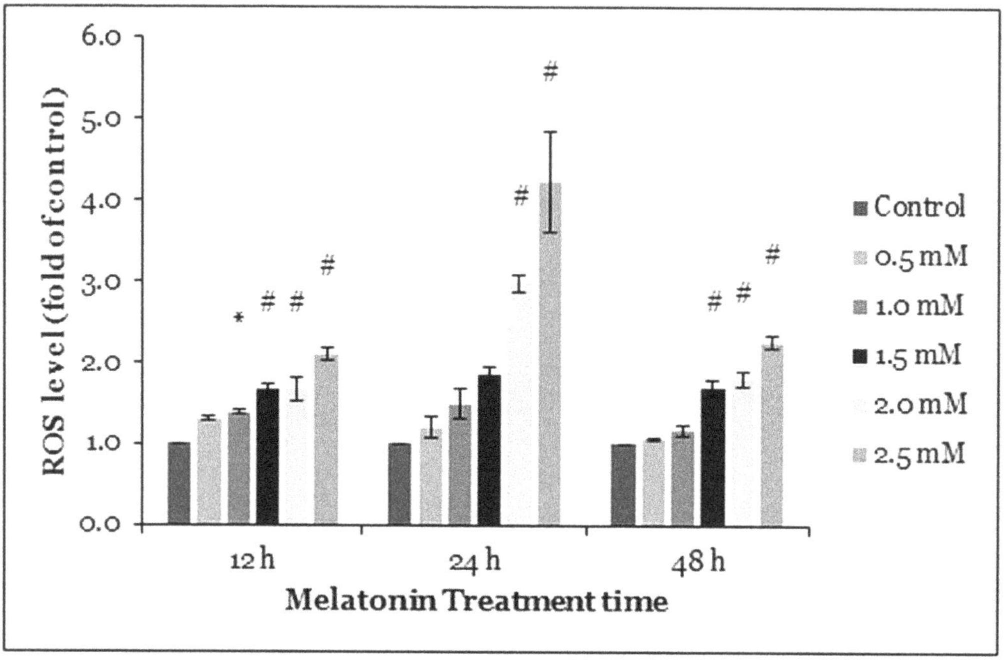

Figure 2. ROS generation in HT-29 cells after treated with various concentrations of melatonin for 72 h. The ROS generation was measured by DCFH-DA staining. Each bar represents mean ± SEM of three independent experiments. One-way ANOVA, followed by Dunnett's post hoc test, was performed to compare each treatment group with the control. Statistically significantly different from the control: *, $p < 0.05$; #, $p < 0.01$.

2.4. Melatonin Alters ER Stress-Related Protein Expressions in HT-29 Cells

The PERK and serine/threonine-protein kinase/endoribonuclease (IRE1) activations depend on the binding immunoglobulin protein (BiP) activity. BiP is a sensor for unfolded proteins in the ER. When unfolded proteins accumulate during stress, BiP binds them, leaving PERK and IRE1 free to be activated. The activation of IRE1 could increase AMPK pathway activation and mammalian ortholog of the yeast autophagy-related gene 6 (Beclin-1) expression, leading to autophagy activation. The activation of PERK could induce autophagy-related proteins (ATGs) gene expression via regulation of C/EBP-homologous protein 10 (CHOP) transcription factors [27]. The eukaryotic translation initiation factor 2 subunit alpha (eIF2α) activation by PERK is also crucial for the ATGs protein expression and autophagosome formation [28]. Protein disulfide isomerase (PDI) is the PERK activator and the inactivated PDI reduces PERK signaling [29]. The expression level of calnexin increases in response to ER stress-causing stressors [30].

To find out whether ER stress is affected in the HT-29 cells after treatment with melatonin, Western blotting was performed to study the protein expressions of ER stress-related proteins. Results suggested increased expressions of ER stress-related proteins BiP, IRE1-α, PDI, PERK, CHOP, and calnexin and an increased p-e-IF2α/e-IF2α ratio. Under the 2.0 and 2.5 mM melatonin treatments, the levels of BiP and CHOP were significantly increased, and under the 2.5 mM melatonin treatment, the p-e-IF2α/e-IF2α protein ratio was also statistically increased (Figure 4). The results indicated that melatonin induced the ER stress pathway activation in the HT-29 cells.

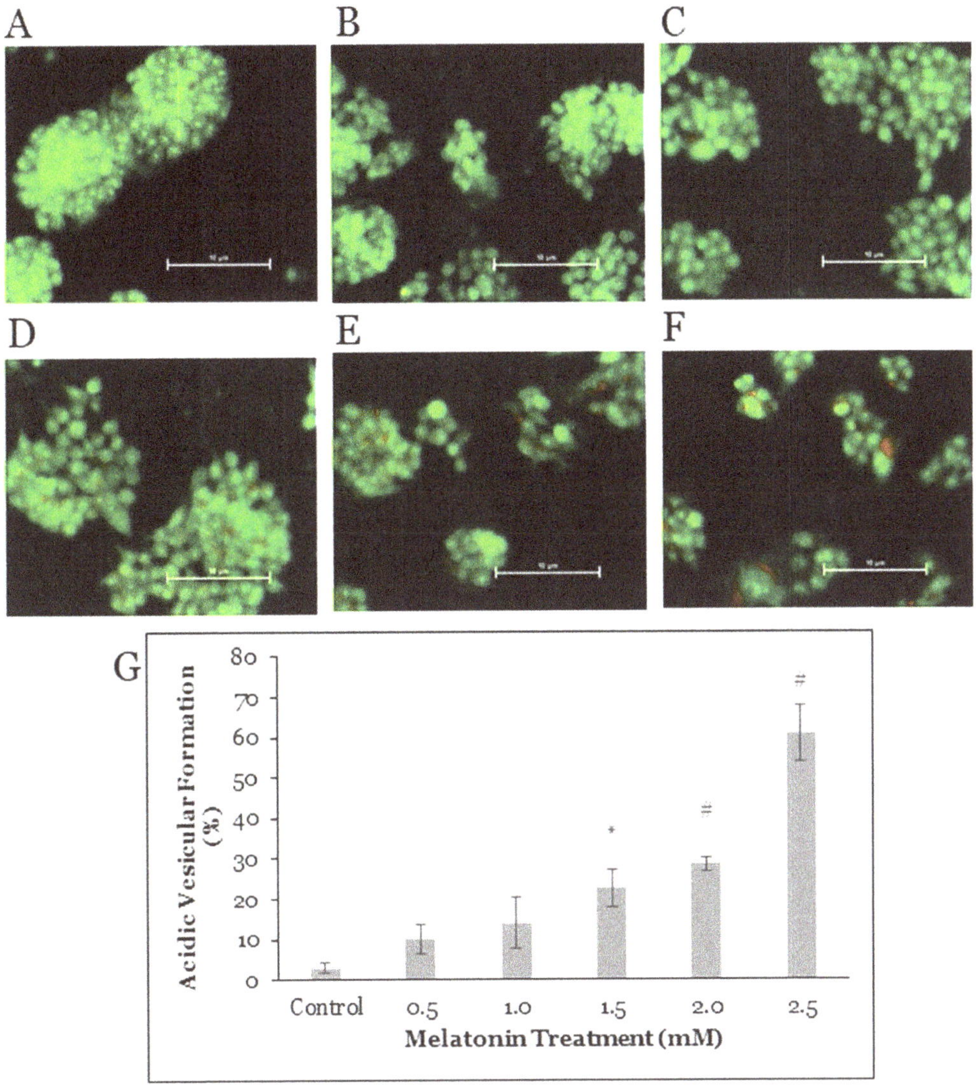

Figure 3. Acridine orange fluorescent staining in HT-29 cells. The HT-29 cells were treated with various concentrations of melatonin for 72 h: (**A**) control, (**B**) 0.5 mM, (**C**) 1.0 mM, (**D**) 1.5 mM, (**E**) 2.0 mM, and (**F**) 2.5 mM. Cells were stained with 5 μg/mL of the acridine orange dye for 20 min, then visualized under a fluorescence microscope with 200× magnification. (**G**) The bar graph represents the means ± SEM of three separate experiments. One-way ANOVA, followed by Dunnett's post hoc test, was performed to compare each treatment group with the control. Statistically significantly different from control: *, $p < 0.05$; #, $p < 0.01$.

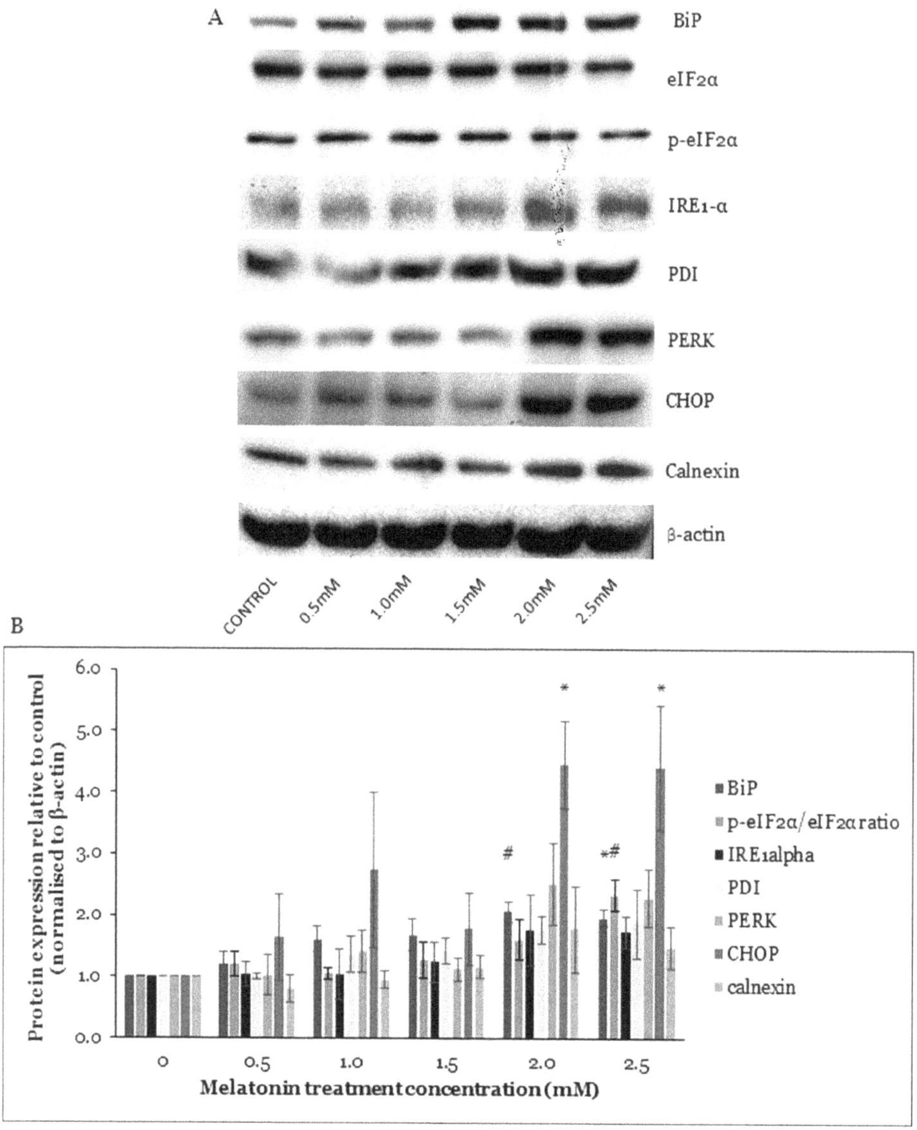

Figure 4. Western blot analysis of the ER stress in HT-29 cells after treated with various concentrations of melatonin for 72 h. (**A**) Expression levels of ER stress-related proteins were measured by Western blot analysis. Beta-actin (β-actin) served as a loading control. (**B**) Bar chart of the protein expressions; each bar represents a mean ± SEM of three independent experiments. One-way ANOVA, followed by Dunnett's post hoc test, was performed to compare each treatment group with control; statistically significantly different from control: *, $p < 0.05$; #, $p < 0.01$.

2.5. Melatonin Induces Changes in the AMPK Pathway

AMPK is a crucial cellular energy sensor protein and is activated by a low energy state in the cell. Phosphorylation of AMPK inhibits energy-consuming activities and promotes energy production under metabolic stress [31]. AMPK contributes to autophagosome maturation and lysosomal fusion [32]. Furthermore, the activation of AMPK downregulates the mTOR pathway to induce cell death [33].

To elucidate whether melatonin can affect the AMPK pathway in the HT-29 cells, Western blot analysis was carried out. Results showed an alteration in the level of the AMPK pathway proteins. The expression levels of the AMPKα, p-AMPKα, and p-AMPKβ increased with increasing concentration of melatonin. However, these changes were not statistically significant (Figure 5).

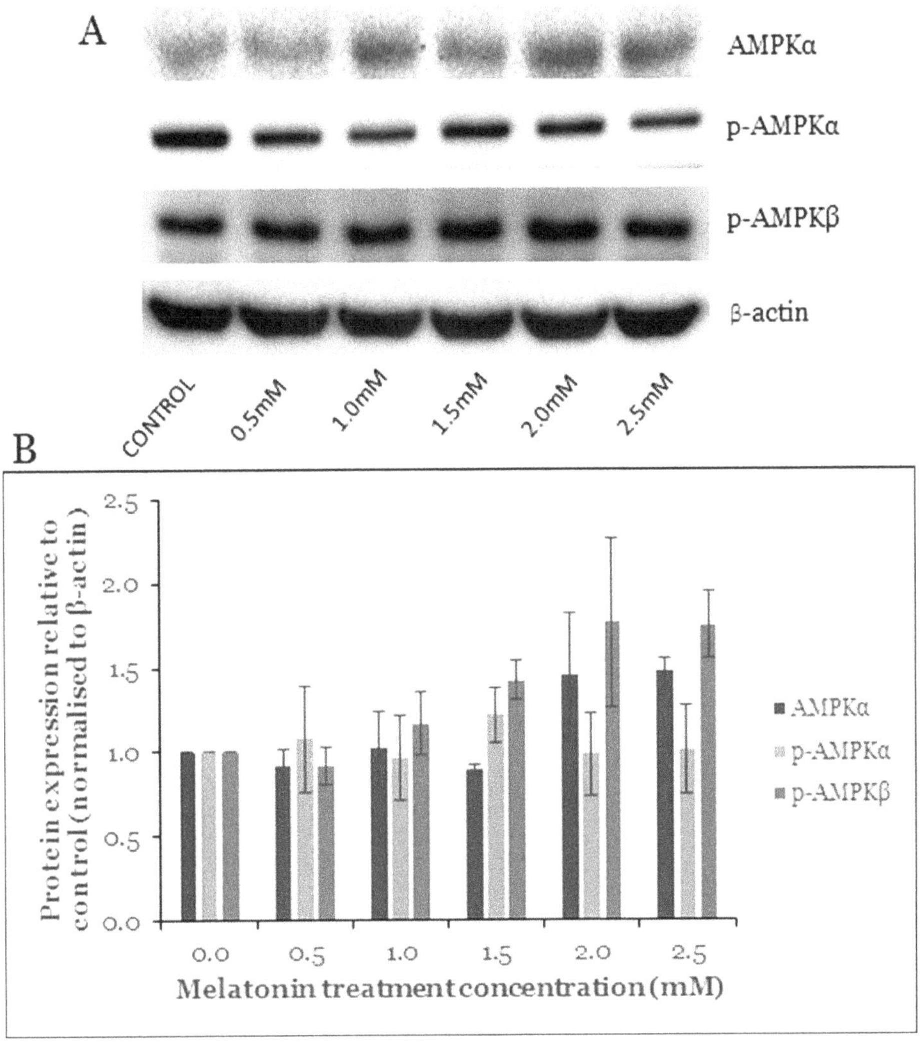

Figure 5. Western blot analysis of the AMPK pathway in HT-29 cells after treated with various concentrations of melatonin for 72 h. (**A**) Expression levels of AMPK proteins were assessed by western blot analysis. (**B**) Bar chart of the protein expressions; each bar represents a mean ± SEM of three independent experiments. One-way ANOVA, followed by Dunnett's post hoc test, was performed to compare each treatment group with control.

2.6. Melatonin Induces Changes in the PI3K/Akt/mTOR Pathway

The PI3K/Akt/mTOR pathway plays a critical role in controlling cell survival, proliferation, apoptosis, and autophagy [11]. Akt is directly activated by PI3K and is a major effector of PI3K involved in cancer growth. Akt signaling leads to increased cellular growth and survival. One of the major effectors downstream of Akt is mTOR [34]. In our study, the PI3K/Akt/mTOR pathway protein levels were studied after the HT-29 cells were treated with melatonin. Results showed changes in the levels of the PI3K/Akt, mTOR, and phosphorylated mammalian target of rapamycin (p-mTOR) proteins (Figure 6A). However, the changes in levels of the PI3K, phosphorylated phosphatidylinositol 3 kinase (p-PI3K), Akt, mTOR, and p-mTOR were not statistically significant (Figure 6B).

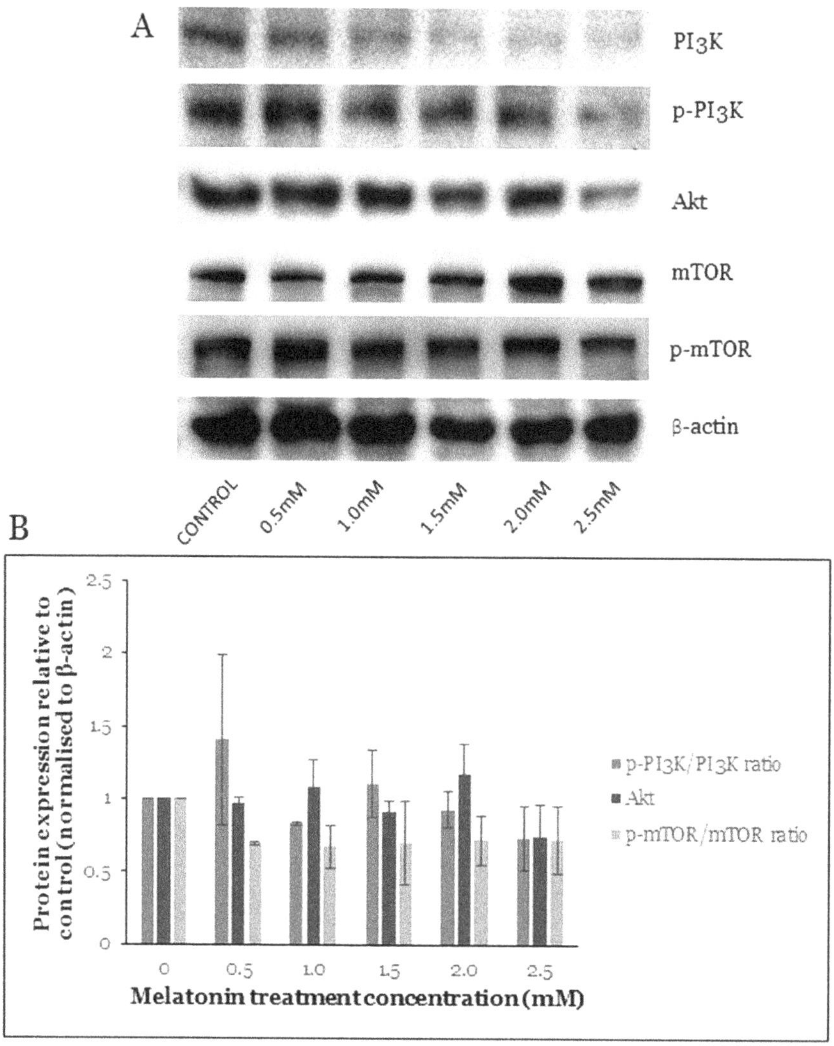

Figure 6. Western blot analysis of the PI3K/Akt/mTOR pathway in HT-29 cells after treated with various concentrations of melatonin for 72 h. (**A**) Expression levels of PI3K/Akt/mTOR proteins were assessed by Western blot analysis. (**B**) Bar chart of protein expressions; each bar represents a mean ± SEM of three independent experiments. One-way ANOVA, followed by Dunnett's post hoc test, was performed to compare each treatment group with control.

2.7. Melatonin Alters the Level of Autophagy Proteins

The formation of an autophagosome is a stepwise process, which involves initiation, nucleation, elongation, maturation, and degradation. The nucleation depends on Beclin-1 activation and its formation of complexes with other autophagy-related proteins. The following elongation involves Atg5 and the mammalian homologue of yeast ATG8 (LC3 protein family) [35]. Beclin-1, Atg5, and LC3 are commonly used autophagy markers for CRC study [36].

Western blot analysis of the HT-29 cells treated with various melatonin concentrations showed that protein levels of the Beclin-1 and Atg5 were increased but not statistically significantly (Figure 7B). LC3-I (molecular weight of 16) is a cytoplasmic form processed into LC3-II (molecular weight of 14), which is autophagosome-membrane-bound. Hence the amount of LC3-II is correlated with the extent of autophagosome formation. In our study, after 72 h treatment of the HT-29 cells with melatonin, the LC3-II/LC3-I ratio was significantly increased at melatonin concentrations of 1.5 mM and 2.5 mM.

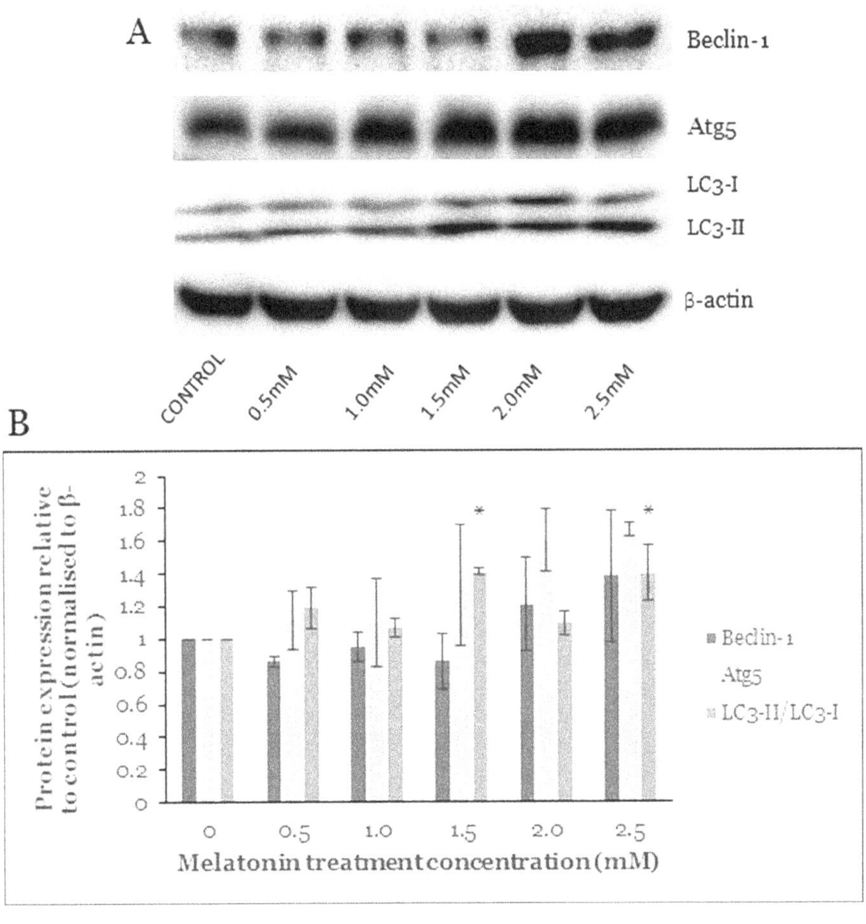

Figure 7. Western blot analysis of the melatonin-treated HT-29 cells. Effects of melatonin on autophagy pathway protein expressions in HT-29 cells after treated with various concentrations of melatonin for 72 h. (**A**) Expression levels of autophagy proteins were assessed by western blot analysis. (**B**) Bar chart of proteins expression; each bar represents a mean ± SEM of three independent experiments. One-way ANOVA was performed to compare each treatment group with control; statistically significantly different from control: *, $p < 0.05$.

3. Discussion

The past decade has seen an enormous interest in melatonin due to its therapeutic potential in a variety of diseases, such as diabetes [37], intervertebral disc degeneration [38], cardiac anomalies [39], colitis [40], and cancer [41]. Melatonin induces both apoptosis and autophagy in Hodgkin lymphoma cells [42]. It also promotes synergistic cytotoxic effects with the chemotherapeutic drug sorafenib in hepatoma cancer cells [43]. Interestingly, autophagic vacuolization was reported to be necessary for the completion of apoptosis in another study [44]. Therefore, the study of cancer cell death should include both apoptosis and autophagy cell death mechanisms to gain a better understanding of pharmacodynamics of a particular drug. In the present study, we investigated the activation of autophagy in melatonin-treated HT-29, SW48, and Caco-2 cells, further supporting our previous work [14].

MTT cell viability assay was used to evaluate cytotoxic properties of melatonin. This method is a well-established colorimetric method for assessing cell viability and proliferation. After 72 h treatment, melatonin significantly inhibited the viability of the HT-29, SW48, and Caco-2 cells in a dose-dependent manner. It has been proven in our previous study that 72 h is the most effective time for killing cancer cells [14]. Moreover, Huang et al. (2020) also demonstrated that melatonin inhibited the survival of human gastric cancer cells under ER stress after 72 h of treatment [45] Using trypan blue cell viability assay, we found that the half-maximal inhibitory concentration (IC50) of melatonin for the HT-29, SW48, and Caco-2 cells were 1.89, 1.93, and 1.77 mM, respectively. Hence, treating these cells with melatonin concentration higher than 1 mM is necessary to achieve IC50, which is relevant for the potential of melatonin as a single anticancer agent. Moreover, only HT-29 showed a significant decrement of cell viability under 1.0 mM melatonin treatment, whereas SW48 and Caco-2 showed statistically significant decrement of cell viability at 1.5 mM and 2.0 mM, respectively. Our results are in agreement with those of Guangyu et al. (2021), who found that melatonin (0.1–2.0 mM) significantly inhibited CRC cell viability in a time- and dose-dependent manner [46]. Moreover, Farriol et al. proved that treatment with 3 mM melatonin could achieve the highest cell death (47%) in the CT-26 murine CRC cell line [47]. Generally, the trypan blue assay detected lower thyroid cancer-cell viability than the MTT assay, and the two assays were highly correlated ($r = 0.99$, $p < 0.001$) in the measurement of thyroid cancer-cell viability after melatonin treatment [48].

TUDCA is an effective ER stress inhibitor, which reduces ER stress-associated protein expression and ER stress-mediated cell death [49]. Hence, we selected TUDCA to assess the effects of melatonin-induced ER stress-mediated cell death. TUDCA attenuated the melatonin-mediated cell death in all CRC cell lines and TUDCA attenuated it stronger at 1 mM than at 0.1 mM. The cytotoxic effects of melatonin were in this descending order: melatonin alone > melatonin with 0.1 mM TUDCA > melatonin with 1 mM TUDCA (Figure 1B–D). Thus, we conclude that melatonin-mediated CRC cell death is dependent on ER stress.

Autophagy is closely related to oxidative stress in which redox signaling regulates the autophagy activities and autophagy regulates oxidative stress levels with the participation of mitochondria and the activation of mitophagy [50]. Research found that accumulation of ROS, membrane lipid oxidation, and loss of plasma membrane integrity are the main causes of autophagy. Catalase, the leading ROS scavenger, is selectively degraded in the autophagy process, leading to abnormal ROS accumulation. Caspases directly cause catalase degradation and ROS accumulation, which can be prevented by autophagy inhibitors [51]. Melatonin was reported to induce cancer cell death through a calmodulin-dependent ROS production [18]. We speculate that melatonin might induce CRC cell death via a mechanism similar to that of prooxidants. Apart from that, several anticancer therapeutic agents have been developed that promote oxidation in cancer cells. For instance, curcumin induced ROS accumulation, ER stress upregulation, and vacuolated cell death, which was also reported in our previous research about the HT-29 cells treated with melatonin [52]. Delicaflavone, a novel anticancer agent, a biflavonoid from *Selaginella doederleinii Hieron*, induced ROS accumulation and inhibited the PI3K/Akt/mTOR signaling pathway, demonstrating the importance of studying ROS and the PI3K/Akt/mTOR pathway in the context of cancer [53].

The formation of autophagic vacuoles is a well-established feature of autophagic cells [26,54]. Acridine orange forms aggregates that emit bright red fluorescence in acidic vesicular organelles and green fluorescence in the cytoplasm and nucleus [26]. Acridine orange staining is a quick, accessible, and reliable method to determine the amount of acidic vesicular organelles, which increases upon autophagy induction, and ratiometric analysis of acridine orange staining is a well-established method in the study of autophagy. Lysosomal degradation of unwanted organelles is the last step in the autophagy, an indication of autophagy activation and prerequisite for autophagic cell death [45,55,56]. Based on our data, melatonin induces autophagy and autophagic vacuoles formation in the HT-29 cells in a dose-dependent manner. The percentage of autophagic vacuoles-positive cells was significantly different from control (22.7, 28.5, and 60.8% after 1.5, 2.0, and 2.5 mM melatonin treatment, respectively).

A variety of physiological and pathological conditions can lead to the accumulation of misfolded proteins in the ER, leading to ER stress. The unfolded protein response (UPR) attenuates the ER stress and re-establishes protein homeostasis. The UPR-signaling pathway is overexpressed in various types of tumors and plays a key role in tumor growth, adaptation, and resistance to cancer treatment. However, UPR activation promotes cell death during prolonged ER stress [28]. A previous study reported that melatonin activated autophagy by controlling ER stress in gastric cancer cells [57]. Similarly as in our study, the ER stress-related genes expressions were increased after melatonin treatment, indicating that autophagy was induced by ER stress. Calnexin is a calcium-binding protein that keeps newly synthesized glycoproteins inside the ER to ensure proper protein folding and to ensure protein quality [58]. In our study, BiP and CHOP protein levels were significantly increased. Increased phosphorylation of eukaryotic translation initiation factor 2 subunit alpha (p-eIF2α) was reported after melatonin treatment, which may have been related to increased ER stress. ER associates with early autophagic structures called isolation membranes (IMs) [28]. The promoted release of BiP and activation of the eIF2α/CHOP pathway could induce autophagy [59], in agreement with our Western blot results that suggested upregulation of BiP, p-eIF2α/eIF2α, and CHOP expressions. The PERK/CHOP pathway could induce ATGs protein expressions followed by autophagosome formation [27]. We observed an increase in PERK (not statistically significant, $p = 0.066$ and 0.133 at 2.0 and 2.5 mM melatonin, respectively) and CHOP (statistically significant, $p = 0.040$ and 0.043 at 2.0 and 2.5 mM melatonin, respectively); therefore, the PERK/CHOP pathway could potentially contribute to melatonin-mediated CRC cell death via autophagy.

AMP-activated protein kinase (AMPK) is an important energy sensor and is activated by low energy availability in the cell [60]. In addition, AMPK activates tuberous sclerosis complex 2 (TSC2), which in turn suppresses mTOR complex 1 (mTORC1), thus promoting autophagosome formation [61]. A study reported that the AMPK contributes to autophagosome maturation and autophagosome–lysosome fusion [32]. Combined treatment using melatonin and 5-fluorouracil in CRC stem cells increased the phosphorylation of AMPK but decreased the phosphorylation of mTOR [62]. However, we observed only a statistically insignificant increase in AMPKα and p-AMPKβ, whereas p-AMPKα remained unchanged for all melatonin treatment concentrations.

The PI3K/Akt pathway plays a major role in regulating cell survival, proliferation, and apoptotic cell death. Melatonin was shown to inhibit breast cancer cell proliferation via the PI3K/Akt pathway [11]. The co-treatment of melatonin with ER stress inducers promoted melanoma cell death by suppressing the PI3K/Akt pathway [63]. The inactivated mTOR promotes autophagy, and the mTOR also serves as a central cell-growth regulator that integrates growth factors and nutrient signals. The AMPK and the PI3K/Akt pathways converge on mTOR with opposing regulatory effects; AMPK regulates mTOR negatively while PI3K/Akt positively [64]. In our study, we observed a decrease in the p-PI3K/p-PI3K ratio, Akt, and p-mTOR/mTOR protein ratio but not statistically significant.

The autophagy pathway requires five steps: initiation, nucleation, elongation, maturation, and degradation with the involvement of numerous Atg and non-Atg proteins. Beclin-1, Atg5, and LC3-II are the commonly used indicators for autophagy [36]. Based on our study, the upregulation of Beclin-1, Atg5, and LC3-II/LC3-I ratio were observed in HT-29 cells after melatonin treatment for 72 h. Beclin-1 involves every major step in autophagic pathways, from autophagosome formation to the maturation of the autophagosome/endosome [35]. In a previous study, ER stress inducers activated the Beclin-1, Atg5, and LC3-II proteins and induced cell death in colon and prostate cancer. The ER stress inducers caused cancer cell death in an Atg5-dependent manner, in accordance with our results [22]. A betulinic acid (BA) analogue, 2c, caused autophagic cell death in CRC cells via upregulation of Beclin-1, Atg5, and LC3 [26]. A similar autophagic cell death mechanism was also observed in A375 and CHL-1 cell lines, ophiobolin A increases ROS accumulation and LC3-II protein expression to induce autophagic cell death in human melanoma cells [65].

Autophagy is a complicated and dynamic intracellular mechanism, which has multiple variations, such as chaperone-mediated autophagy, microautophagy, and macroautophagy. Our study focused on the macroautophagy, investigation of the other variations of autophagy could potentially provide an extensive understanding on how melatonin induces autophagic cell death in CRC. Moreover, there are also canonical and non-canonical autophagy pathways that require further investigation [64]. A comprehensive mapping of the autophagy network could provide us with an anticancer target and even previously unknown molecular targets for various diseases.

4. Materials and Methods

4.1. Materials

Dulbecco's modified Eagle medium (DMEM), fetal bovine serum (FBS), MTT, melatonin, and acridine orange were purchased from Sigma Aldrich (Saint Louis, MO, USA). TUDCA was purchased from Merck Millipore (Burlington, MA, USA). Trypsin–ethylenediaminetetraacetic acid (EDTA) was purchased from Gibco (Gibco, Loughborough, UK). Dimethyl sulfoxide (DMSO) and bovine serum albumin (BSA) were purchased from Nacalai Tesque (Kyto, Japan). Primary antibodies against β-actin, PI3K, p-PI3K, Akt, AMPKα23a, p-AMPKα, AMPKβ, autophagy-related protein Atg5, Beclin-1, BiP, calnexin, eIF2α, p-eIF2α, IRE1α, PDI, microtubule-associated protein 1 light chain 3 (LC-3), mTOR, p-mTOR, PERK, and CHOP and horseradish peroxidase-conjugated secondary antibodies were purchased from Cell Signalling Technology (Danvers, MA, USA). SuperSignal^{TM} West Femto Maximum Sensitivity Substrate was purchased from Thermo Fisher Scientific (Waltham, MA, USA). Melatonin was dissolved in 98% absolute ethanol; the stock concentration of melatonin was maintained at 500 mM for every experiment. The final volume of ethanol in cell culture was maintained below 0.5%. TUDCA was dissolved in phosphate-buffered saline (PBS), the stock concentration of TUDCA was maintained at 100 mM for every experiment.

4.2. Cell Culture

HT-29, SW48, and Caco-2 were obtained from the American Type Culture Collection (ATCC). The cells were cultured in DMEM and supplemented with 10% FBS at 37 °C in an incubator with 5% CO_2. Cells were trypsinized whenever the confluency of cells reach 70%. Cells collection was carried out by rinsing the cells with PBS Biobasic (Toronto, ON, Canada) followed by the addition of trypsin-EDTA to detach the cells. The action of trypsin was later neutralized with DMEM, and cells were harvested by centrifugation at 1500 rpm for 5 min. The cells were then sub-cultured into new cell culture T-25 flasks or plated for assays.

4.3. MTT Cytotoxicity Assay

The cytotoxicity of melatonin was measured in the HT-29, SW48, and Caco-2 cells by the MTT colorimetric assay. The CRC cells were seeded at 5000 cells (in 100 μL culture medium) per well in 96-well plates and incubated for 24 h. Subsequently, the cells were treated with melatonin at various concentrations (0, 0.5, 1.0, 1.5, 2.0, and 2.5 mM). After an incubation for 72 h, the cells were incubated with 20 μL of MTT (5 mg/mL) solution for 4 h. Following medium removal, 100 μL of DMSO was added to each well and plates were gently shaken. The optical absorbance was measured at 570 nm, with 630 nm as reference wavelength, using Spectra Max3 Molecular Devices (San Jose, CA, USA). The absorbance of cells without treatment was regarded as 100% cell survival. Each experiment was repeated at least three times, beginning from the cell seeding to data analysis.

4.4. Trypan Blue Cell Viability Assay

The cytotoxicity of melatonin was measured in the HT-29, SW48, and Caco-2 cells by trypan blue dye exclusion assay. The CRC cells were seeded at 3.0×10^4 cells (in 5000 μL culture medium) per well in 60 mm^2 dishes and incubated for 24 h. Then, the cells were treated with various melatonin concentrations (0, 0.5, 1.0, 1.5, 2.0, and 2.5 mM), various melatonin concentrations + 0.1 mM TUDCA, and various melatonin concentrations + 1 mM TUDCA. After incubation for 72 h, the cells were trypsinized and collected in 15 mL tubes. The cells were stained with trypan blue for cell count using a hemocytometer and light microscope. Each experiment was repeated at least three times, beginning from the cell seeding to data analysis.

4.5. ROS Level Analysis

The HT-29 cells were cultured in 96-well plates, then treated with or without melatonin (0.5, 1.0, 1.5, 2.0, and 2.5 mM) for 12, 24, and 48 h. The cells were stained with 10 μM of DCFH-DA for 30 min. The cells were then examined under a microplate reader Spectra Max3 Molecular Devices (San Jose, CA, USA). Fluorescence intensity in the cells was detected at an excitation wavelength of 485 nm and an emission wavelength of 535 nm. Each experiment was repeated at least three times, beginning from the cell seeding to data analysis.

4.6. Fluorescence Staining

Lysosomal activation was detected by acridine orange fluorescent staining and microscopy. The HT-29 cells (1.0×10^5) were seeded onto a 60 mm^2 dish for 24 h (37 °C, 5% CO_2). The cells were then treated with melatonin at various concentrations (0, 0.5, 1.0, 1.5, 2.0, and 2.5 mM). After 72 h, the cells were washed with 200 μL of PBS and stained with Acridine orange (5 μg/mL) for 15–20 min. The cells were observed under an inverted fluorescence microscope Nikon Eclipse Ti (Tokyo, Japan). The activation of lysosomes was examined, and photomicrographs were taken using an attached camera. Each experiment was repeated at least three times, beginning from the cell seeding to data analysis.

4.7. Western Blot Analysis

Lysates of the cells exposed to melatonin for 72 h were harvested using DTT lysis buffer (62.5 mM Tris, 2% *w/v* SDS, 10% glycerol, pH 6.8, 100 mM DTT). The protein concentration was determined using Quick Start™ Bradford Protein Assay (Bio-Rad, Hercules, CA, USA). From each sample, 15 μL of proteins were separated by electrophoresis, on 7.5, 10, and 12.5% SDS-PAGE gels, based on the molecular weight of proteins. After the electro-transfer of proteins onto PVDF membranes, the membranes were blocked with 3% BSA for 1 h at room temperature. The membrane was then incubated with primary antibodies at a dilution of 1:1000, overnight at 4 °C. After three washes with TBST (0.1% Tween-20 in Tris-HCl buffered saline, TBS), the membranes were incubated with horseradish peroxidase-conjugated secondary antibodies at a dilution of 1:10,000, for 1 h at room temperature. After washing with TBST and TBS, proteins were visualized by SuperSignal™ West Femto

Maximum Sensitivity Substrate. Densitometric analysis was performed using ImageLab, Version 6.1.0, build 7 Standard Edition (Bio-Rad, Hercules, CA, USA) to scan the signals. Western blot assay results reported here are representative of at least three independent experiments (repeated from the cell seeding to data analysis).

4.8. Statistical Analysis

All experiments were performed in biological triplicates and values given are representative for at least three independent experiments. Each result of the experiments was expressed as mean ± standard error of mean (SEM). The results were statistically analyzed by one-way or two-way ANOVA followed by Dunnett's post hoc test using IBM SPSS Statistics, Version 23. Values $p < 0.05$ or $p < 0.01$ were considered as statistically significant.

5. Conclusions

In conclusion, our results indicate that melatonin induces an anticancer effect in the HT-29, SW48, and Caco-2 CRC cells via oxidative stress-mediated autophagy. Melatonin-induced autophagy involves the ER stress signaling pathway. With the effects of these pathways, stepwise autophagosome formation was confirmed via Beclin-1 (nucleation), Atg5 (elongation), and LC3 (elongation). Our study provided an insight into the underlying mechanism of melatonin-induced autophagy and its pro-oxidative properties in CRC cell.

Author Contributions: K.C.C. and M.G.N. were involved in the performance of experiments, data analysis and interpretation. P.Y.N. and R.Y.K. were involved in the conception and design of the study and supervision. R.Y.K. and K.C.C. were involved in the critical revision of the manuscript. K.C.C. was involved in drafting of the manuscript. S.M.C. obtained financial support, provided critical revision, and approved the final manuscript. All authors have read and agreed to the published version of the manuscript.

Funding: This research was funded by International Medical University, grant numbers are BMSc I-2020 (08), BMSc I-2019 (02) and BMSc I-2018 (09).

Institutional Review Board Statement: Not applicable.

Informed Consent Statement: Not applicable.

Data Availability Statement: Not applicable.

Conflicts of Interest: The authors declare no conflict of interest.

Sample Availability: Samples of the compounds are not available from the authors.

References

1. Sung, H.; Ferlay, J.; Siegel, R.L.; Laversanne, M.; Soerjomataram, I.; Jemal, A.; Bray, F. Global cancer statistics 2020: GLOBOCAN estimates of incidence and mortality worldwide for 36 cancers in 185 countries. *CA Cancer J. Clin.* **2021**, *71*, 209–249. [CrossRef]
2. Araghi, M.; Soerjomataram, I.; Jenkins, M.; Brierley, J.; Morris, E.; Bray, F.; Arnold, M. Global trends in colorectal cancer mortality: Projections to the year 2035. *Int. J. Cancer* **2019**, *144*, 2992–3000. [CrossRef]
3. Siegel, R.L.; Miller, K.D.; Jemal, A. Cancer statistics, 2019. *CA Cancer J. Clin.* **2019**, *69*, 7–34. [CrossRef] [PubMed]
4. Van Cutsem, E.; Cervantes, A.; Nordlinger, B.; Arnold, D.; ESMO Guidelines Working Group. Metastatic colorectal cancer: ESMO Clinical Practice Guidelines for diagnosis, treatment and follow-up. *Ann. Oncol.* **2014**, *25*, iii1–iii9. [CrossRef]
5. Xie, Y.-H.; Chen, Y.-X.; Fang, J.-Y. Comprehensive review of targeted therapy for colorectal cancer. *Signal Transduct. Target. Ther.* **2020**, *5*, 1–30. [CrossRef] [PubMed]
6. Garg, M.; Lincz, L.; Adler, K.; Scorgie, F.; Ackland, S.P.; Sakoff, J.A. Predicting 5-fluorouracil toxicity in colorectal cancer patients from peripheral blood cell telomere length: A multivariate analysis. *Br. J. Cancer* **2012**, *107*, 1525–1533. [CrossRef] [PubMed]
7. Pearce, A.; Haas, M.; Viney, R.; Pearson, S.-A.; Haywood, P.; Brown, C.; Ward, R. Incidence and severity of self-reported chemotherapy side effects in routine care: A prospective cohort study. *PLoS ONE* **2017**, *12*, e0184360. [CrossRef]
8. Plaimee, P.; Weerapreeyakul, N.; Barusrux, S.; Johns, N.P. Melatonin potentiates cisplatin-induced apoptosis and cell cycle arrest in human lung adenocarcinoma cells. *Cell Prolif.* **2015**, *48*, 67–77. [CrossRef]
9. Gao, Y.; Xiao, X.; Zhang, C.; Yu, W.; Guo, W.; Zhang, Z.; Li, Z.; Feng, X.; Hao, J.; Zhang, K.; et al. Melatonin synergizes the chemotherapeutic effect of 5-fluorouracil in colon cancer by suppressing PI3K/AKT and NF-κB/iNOS signaling pathways. *J. Pineal Res.* **2016**, *62*, e12380. [CrossRef] [PubMed]

10. Lu, Y.-X.; Chen, D.-L.; Wang, D.-S.; Chen, L.-Z.; Mo, H.-Y.; Sheng, H.; Bai, L.; Wu, Q.-N.; Yu, H.-E.; Xie, D.; et al. Melatonin enhances sensitivity to fluorouracil in oesophageal squamous cell carcinoma through inhibition of Erk and Akt pathway. *Cell Death Dis.* **2016**, *7*, e2432. [CrossRef] [PubMed]
11. Wang, J.; Xiao, X.; Zhang, Y.; Shi, D.; Chen, W.; Fu, L.; Liu, L.; Xie, F.; Kang, T.; Huang, W.; et al. Simultaneous modulation of COX-2, p300, Akt, and Apaf-1 signaling by melatonin to inhibit proliferation and induce apoptosis in breast cancer cells. *J. Pineal Res.* **2012**, *53*, 77–90. [CrossRef]
12. Carbajo-Pescador, S.; Ordoñez, R.; Benet, M.; Jover, R.; Palomo, A.G.; Mauriz, J.L.; González-Gallego, J. Inhibition of VEGF expression through blockade of Hif1α and STAT3 signalling mediates the anti-angiogenic effect of melatonin in HepG2 liver cancer cells. *Br. J. Cancer* **2013**, *109*, 83–91. [CrossRef]
13. Akbar, M.A.; Movassaghpour, A.A.; Ghanbari, H.; Kheirandish, M.; Maroufi, N.F.; Rahbarghazi, R.; Nouri, M.; Samadi, N. The potential therapeutic effect of melatonin on human ovarian cancer by inhibition of invasion and migration of cancer stem cells. *Sci. Rep.* **2017**, *7*, 1–11. [CrossRef]
14. Kasi, R.; Yeo, P.L.; Yen, N.K.; Koh, R.Y.; Ponnudurai, G.; Tiong, Y.L.; Chye, S.M.; Ng, K.Y. Melatonin Induces Apoptosis and Inhibits the Proliferation of Cancer Cells via Reactive Oxygen Species-mediated MAPK and mTOR Pathways. *Clin. Cancer Drugs* **2020**, *7*, 44–56. [CrossRef]
15. Tiong, Y.L.; Ng, K.Y.; Koh, R.Y.; Ponnudurai, G.; Chye, S.M. Melatonin Prevents Oxidative Stress-Induced Mitochondrial Dysfunction and Apoptosis in High Glucose-Treated Schwann Cells via Upregulation of Bcl2, NF-κB, mTOR, Wnt Signalling Pathways. *Antioxidants* **2019**, *8*, 198. [CrossRef]
16. Liu, R.; Fu, A.; Hoffman, A.E.; Zheng, T.; Zhu, Y. Melatonin enhances DNA repair capacity possibly by affecting genes involved in DNA damage responsive pathways. *BMC Cell Biol.* **2013**, *14*, 1. [CrossRef]
17. Kontek, R.; Jakubczak, M.; Matlawska-Wasowska, K. The antioxidants, vitamin A and E but not vitamin C and melatonin enhance the proapoptotic effects of irinotecan in cancer cells in vitro. *Toxicol. In Vitro* **2014**, *28*, 282–291. [CrossRef] [PubMed]
18. Zhang, H.-M.; Zhang, Y. Melatonin: A well-documented antioxidant with conditional pro-oxidant actions. *J. Pineal Res.* **2014**, *57*, 131–146. [CrossRef]
19. Bejarano, I.; Espino, J.; Barriga, C.; Reiter, R.J.; Pariente, J.A.; Rodríguez, A.B. Pro-Oxidant Effect of Melatonin in Tumour Leucocytes: Relation with its Cytotoxic and Pro-Apoptotic Effects. *Basic Clin. Pharmacol. Toxicol.* **2010**, *108*, 14–20. [CrossRef]
20. Burada, F. Autophagy in colorectal cancer: An important switch from physiology to pathology. *World J. Gastrointest. Oncol.* **2015**, *7*, 271–284. [CrossRef] [PubMed]
21. Maiuri, M.C.; Zalckvar, E.; Kimchi, A.; Kroemer, G. Self-eating and self-killing: Crosstalk between autophagy and apoptosis. *Nat. Rev. Mol. Cell Biol.* **2007**, *8*, 741–752. [CrossRef]
22. Ding, W.-X.; Ni, H.-M.; Gao, W.; Hou, Y.-F.; Melan, M.; Chen, X.; Stolz, D.B.; Shao, Z.-M.; Yin, X.-M. Differential Effects of Endoplasmic Reticulum Stress-induced Autophagy on Cell Survival. *J. Biol. Chem.* **2007**, *282*, 4702–4710. [CrossRef]
23. Lepine, S.; Allegood, J.C.; Park, M.; Dent, P.W.; Milstien, S.; Spiegel, S. Sphingosine-1-phosphate phosphohydrolase-1 regulates ER stress-induced autophagy. *Cell Death Differ.* **2010**, *18*, 350–361. [CrossRef]
24. Basile, V.; Belluti, S.; Ferrari, E.; Gozzoli, C.; Ganassi, S.; Quaglino, D.; Saladini, M.; Imbriano, C. bis-Dehydroxy-Curcumin Triggers Mitochondrial-Associated Cell Death in Human Colon Cancer Cells through ER-Stress Induced Autophagy. *PLoS ONE* **2013**, *8*, e53664. [CrossRef]
25. Feng, Q.; Wang, H.; Pang, J.; Ji, L.; Han, J.; Wang, Y.; Qi, X.; Liu, Z.; Lu, L. Prevention of Wogonin on Colorectal Cancer Tumorigenesis by Regulating p53 Nuclear Translocation. *Front. Pharmacol.* **2018**, *9*, 9. [CrossRef]
26. Dutta, D.; Chakraborty, B.; Sarkar, A.; Chowdhury, C.; Das, P. A potent betulinic acid analogue ascertains an antagonistic mechanism between autophagy and proteasomal degradation pathway in HT-29 cells. *BMC Cancer* **2016**, *16*, 23. [CrossRef] [PubMed]
27. Nagelkerke, A.; Sweep, F.C.; Geurts-Moespot, A.; Bussink, J.; Span, P.N. Therapeutic targeting of autophagy in cancer. Part I: Molecular pathways controlling autophagy. *Semin. Cancer Biol.* **2015**, *31*, 89–98. [CrossRef] [PubMed]
28. Fernández, A.; Ordóñez, R.; Reiter, R.J.; González-Gallego, J.; Mauriz, J.L. Melatonin and endoplasmic reticulum stress: Rela-tion to autophagy and apoptosis. *J. Pineal Res.* **2015**, *59*, 292–307. [CrossRef]
29. Kranz, P.; Neumann, F.; Wolf, A.; Classen, F.; Pompsch, M.; Ocklenburg, T.; Baumann, J.; Janke, K.; Baumann, M.; Goepelt, K.; et al. PDI is an essential redox-sensitive activator of PERK during the unfolded protein response (UPR). *Cell Death Dis.* **2017**, *8*, e2986. [CrossRef] [PubMed]
30. Bergeron, J.J.; Brenner, M.B.; Thomas, D.; Williams, D.B. Calnexin: A membrane-bound chaperone of the endoplasmic reticulum. *Trends Biochem. Sci.* **1994**, *19*, 124–128. [CrossRef]
31. Carling, D. The AMP-activated protein kinase cascade—A unifying system for energy control. *Trends Biochem. Sci.* **2004**, *29*, 18–24. [CrossRef]
32. Jang, M.; Park, R.; Kim, H.; Namkoong, S.; Jo, D.; Huh, Y.H.; Jang, I.-S.; Lee, J.I.; Park, J. AMPK contributes to autophagosome maturation and lysosomal fusion. *Sci. Rep.* **2018**, *8*, 1–10. [CrossRef]
33. Liu, J.; Song, N.; Huang, Y.; Chen, Y. Irisin inhibits pancreatic cancer cell growth via the AMPK-mTOR pathway. *Sci. Rep.* **2018**, *8*, 1–10. [CrossRef] [PubMed]
34. Engelman, J.A. Targeting PI3K signalling in cancer: Opportunities, challenges and limitations. *Nat. Rev. Cancer* **2009**, *9*, 550–562. [CrossRef]

35. Kang, R.; Zeh, H.J.; Lotze, M.T.; Tang, D. The Beclin 1 network regulates autophagy and apoptosis. *Cell Death Differ.* **2011**, *18*, 571–580. [CrossRef]
36. Xie, Q.; Liu, Y.; Li, X. The interaction mechanism between autophagy and apoptosis in colon cancer. *Transl. Oncol.* **2020**, *13*, 100871. [CrossRef] [PubMed]
37. Ruiz, L.; Gurlo, T.; Ravier, M.; Wojtusciszyn, A.; Mathieu, J.; Brown, M.R.; Broca, C.; Bertrand, G.; Butler, P.C.; Matveyenko, A.V.; et al. Proteasomal degradation of the histone acetyl transferase p300 contributes to beta-cell injury in a diabetes environment. *Cell Death Dis.* **2018**, *9*, 1–12. [CrossRef]
38. Chen, F.; Jiang, G.; Liu, H.; Li, Z.; Pei, Y.; Wang, H.; Pan, H.; Cui, H.; Long, J.; Wang, J.; et al. Melatonin alleviates intervertebral disc degeneration by disrupting the IL-1β/NF-κB-NLRP3 inflammasome positive feedback loop. *Bone Res.* **2020**, *8*, 1–13. [CrossRef]
39. Wang, S.; Wang, L.; Qin, X.; Turdi, S.; Sun, D.; Culver, B.; Reiter, R.J.; Wang, X.; Zhou, H.; Ren, J. ALDH2 contributes to melatonin-induced protection against APP/PS1 mutation-prompted cardiac anomalies through cGAS-STING-TBK1-mediated regulation of mitophagy. *Signal Transduct. Target. Ther.* **2020**, *5*, 1–13. [CrossRef]
40. Kim, S.W.; Kim, S.; Son, M.; Cheon, J.H.; Park, Y.S. Melatonin controls microbiota in colitis by goblet cell differentiation and antimicrobial peptide production through Toll-like receptor 4 signalling. *Sci. Rep.* **2020**, *10*, 1–10. [CrossRef]
41. Talib, W.H. Melatonin and Cancer Hallmarks. *Molecules* **2018**, *23*, 518. [CrossRef]
42. Yan, G.; Lei, H.; He, M.; Gong, R.; Wang, Y.; He, X.; Li, G.; Pang, P.; Li, X.; Yu, S.; et al. Melatonin triggers autophagic cell death by regulating RORC in Hodgkin lymphoma. *Biomed. Pharmacother.* **2020**, *123*, 109811. [CrossRef]
43. Liu, Y.; Liu, J.; Cheng, L.; Fan, L.; Wang, F.; Yu, H.; Liu, Q.; Li, Y.; Bu, L.; Li, X.; et al. Melatonin increases the anti-tumor ef-fects of sorafenib on human hepatoma cell lines via down-regulating autophagy. *Int. J. Clin. Exp. Med.* **2017**, *10*, 14109–14120.
44. Polo, R.-A.G.; Boya, P.; Pauleau, A.-L.; Jalil, A.; Larochette, N.; Souquère, S.; Eskelinen, E.-L.; Pierron, G.; Saftig, P.; Kroemer, G. The apoptosis/autophagy paradox: Autophagic vacuolization before apoptotic death. *J. Cell Sci.* **2005**, *118*, 3091–3102. [CrossRef] [PubMed]
45. Huang, Y.; Yuan, K.; Tang, M.; Yue, J.; Bao, L.; Wu, S.; Zhang, Y.; Li, Y.; Wang, Y.; Ou, X.; et al. Melatonin inhibiting the survival of human gastric cancer cells under ER stress involving autophagy and Ras-Raf-MAPK signalling. *J. Cell. Mol. Med.* **2021**, *25*, 1480–1492. [CrossRef] [PubMed]
46. Ji, G.; Zhou, W.; Li, X.; Du, J.; Li, X.; Hao, H. Melatonin inhibits proliferation and viability and promotes apoptosis in colorectal cancer cells via upregulation of the microRNA-34a/449a cluster. *Mol. Med. Rep.* **2021**, *23*, 1. [CrossRef]
47. Farriol, M.; Venereo, Y.; Orta, X.; Castellanos, J.M.; Segovia-Silvestre, T. In vitro effects of melatonin on cell prolifera-tion in a colon adenocarcinoma line. *J. Appl. Toxicol.* **2000**, *20*, 21–24. [CrossRef]
48. Ghorbani-Anarkooli, M.; Dabirian, S.; Moladoust, H.; Zendedel, A.; Bahadori, M.H. Comparison of MTT, trypan blue, and clonogenic assay, to determine the viability in human anaplastic thyroid cancer cell line. *Tehran Univ. Med. J.* **2019**, *77*, 26–32.
49. Yoon, Y.M.; Lee, J.H.; Yun, S.P.; Han, Y.-S.; Yun, C.W.; Lee, H.J.; Noh, H.; Lee, S.-J.; Han, H.J.; Lee, S.H. Tauroursodeoxycholic acid reduces ER stress by regulating of Akt-dependent cellular prion protein. *Sci. Rep.* **2016**, *6*, 39838. [CrossRef]
50. Lee, J.; Giordano, S.; Zhang, J. Autophagy, mitochondria and oxidative stress: Cross-talk and redox signalling. *Biochem. J.* **2011**, *441*, 523–540. [CrossRef] [PubMed]
51. Yu, L.; Wan, F.; Dutta, S.; Welsh, S.; Liu, Z.; Freundt, E.; Baehrecke, E.H.; Lenardo, M. Autophagic programmed cell death by selective catalase degradation. *Proc. Natl. Acad. Sci. USA* **2006**, *103*, 4952–4957. [CrossRef] [PubMed]
52. Lee, W.-J.; Chien, M.-H.; Chow, J.-M.; Chang, J.-L.; Wen, Y.-C.; Lin, Y.-W.; Cheng, C.-W.; Lai, G.-M.; Hsiao, M.; Lee, L.-M. Nonautophagic cytoplasmic vacuolation death induction in human PC-3M prostate cancer by curcumin through reactive oxygen species -mediated endoplasmic reticulum stress. *Sci. Rep.* **2015**, *5*, srep10420. [CrossRef]
53. Yao, W.; Lin, Z.; Shi, P.; Chen, B.; Wang, G.; Huang, J.; Sui, Y.; Liu, Q.; Li, S.; Lin, X.; et al. Delicaflavone induces ROS-mediated apoptosis and inhibits PI3K/AKT/mTOR and Ras/MEK/Erk signaling pathways in colorectal cancer cells. *Biochem. Pharmacol.* **2020**, *171*, 113680. [CrossRef]
54. Paglin, S.; Hollister, T.; Delohery, T.; Hackett, N.; McMahill, M.; Sphicas, E.; Domingo, D.; Yahalom, J. A novel response of cancer cells to radiation involves autophagy and formation of acidic vesicles. *Cancer Res.* **2001**, *61*, 439–444.
55. Thomé, M.P.; Chiela, E.; Villodre, E.S.; Migliavaca, C.B.; Onzi, G.R.; Felipe, K.B.; Lenz, G. Ratiometric analysis of acridine orange staining in the study of acidic organelles and autophagy. *J. Cell Sci.* **2016**, *129*, 4622–4632. [CrossRef] [PubMed]
56. Kim, J.Y.; Cho, T.J.; Woo, B.H.; Choi, K.U.; Lee, C.H.; Ryu, M.H.; Park, H.R. Curcumin-induced autophagy contributes to the decreased survival of oral cancer cells. *Arch. Oral Biol.* **2012**, *57*, 1018–1025. [CrossRef] [PubMed]
57. Williams, D.B. Beyond lectins: The calnexin/calreticulin chaperone system of the endoplasmic reticulum. *J. Cell Sci.* **2006**, *119*, 615–623. [CrossRef]
58. Hayashi-Nishino, M.; Fujita, N.; Noda, T.; Yamaguchi, A.; Yoshimori, T.; Yamamoto, A. A subdomain of the endoplasmic reticulum forms a cradle for autophagosome formation. *Nat. Cell Biol.* **2009**, *11*, 1433–1437. [CrossRef] [PubMed]
59. Hardie, D.G.; Ross, F.A.; Hawley, S.A. AMPK: A nutrient and energy sensor that maintains energy homeostasis. *Nat. Rev. Mol. Cell Biol.* **2012**, *13*, 251–262. [CrossRef] [PubMed]
60. Tripathi, D.; Chowdhury, R.; Trudel, L.J.; Tee, A.; Slack, R.S.; Walker, C.L.; Wogan, G.N. Reactive nitrogen species regulate autophagy through ATM-AMPK-TSC2-mediated suppression of mTORC1. *Proc. Natl. Acad. Sci. USA* **2013**, *110*, E2950–E2957. [CrossRef] [PubMed]

61. Lee, J.H.; Yun, C.W.; Han, Y.-S.; Kim, S.; Jeong, D.; Kwon, H.Y.; Kim, H.; Baek, M.-J.; Lee, S.H. Melatonin and 5-fluorouracil co-suppress colon cancer stem cells by regulating cellular prion protein-Oct4 axis. *J. Pineal Res.* **2018**, *65*, e12519. [CrossRef]
62. Kim, H.S.; Kim, T.-J.; Yoo, Y.-M. Melatonin Combined with Endoplasmic Reticulum Stress Induces Cell Death via the PI3K/Akt/mTOR Pathway in B16F10 Melanoma Cells. *PLoS ONE* **2014**, *9*, e92627. [CrossRef] [PubMed]
63. Hahn-Windgassen, A.; Nogueira, V.; Chen, C.-C.; Skeen, J.E.; Sonenberg, N.; Hay, N. Akt Activates the Mammalian Target of Rapamycin by Regulating Cellular ATP Level and AMPK Activity. *J. Biol. Chem.* **2005**, *280*, 32081–32089. [CrossRef] [PubMed]
64. Codogno, P.; Mehrpour, M.; Proikas-Cezanne, T. Canonical and non-canonical autophagy: Variations on a common theme of self-eating? *Nat. Rev. Mol. Cell Biol.* **2011**, *13*, 7–12. [CrossRef]
65. Rodolfo, C.; Rocco, M.; Cattaneo, L.; Tartaglia, M.; Sassi, M.; Aducci, P.; Scaloni, A.; Camoni, L.; Marra, M. Ophiobolin A Induces Autophagy and Activates the Mitochondrial Pathway of Apoptosis in Human Melanoma Cells. *PLoS ONE* **2016**, *11*, e0167672. [CrossRef] [PubMed]

molecules

Article

Naringenin Induces ROS-Mediated ER Stress, Autophagy, and Apoptosis in Human Osteosarcoma Cell Lines

Chiang-Wen Lee [1,2,3], Cathy Chia-Yu Huang [4], Miao-Ching Chi [5], Kuan-Han Lee [6,7], Kuo-Ti Peng [2], Mei-Ling Fang [8,9], Yao-Chang Chiang [1,*] and Ju-Fang Liu [10,11,*]

1 Department of Nursing, Division of Basic Medical Sciences, Chronic Diseases and Health Promotion Research Center and Research Center for Chinese Herbal Medicine, Chang Gung University of Science and Technology, Puzi City 61363, Taiwan; cwlee@gw.cgust.edu.tw
2 Department of Orthopaedic Surgery, Chang Gung Memorial Hospital, Puzi City 61363, Taiwan; mr3497@cgmh.org.tw
3 Department of Safety Health and Environmental Engineering, Ming Chi University of Technology, New Taipei City 243303, Taiwan
4 Department of Life Sciences, National Central University, Taoyuan City 320317, Taiwan; cathy80787@gmail.com
5 Department of Respiratory Care, Chang Gung University of Science and Technology, Puzi City 613, Taiwan; mcchi@gw.cgust.edu.tw
6 Division of Pulmonary and Critical Care Medicine, Chang Gung Memorial Hospital, Kaohsiung 833, Taiwan; anti0822@hotmail.com
7 Department of Pharmacy, Chia Nan University of Pharmacy and Science, Tainan 71710, Taiwan
8 Center for Environmental Toxin and Emerging-Contaminant Research, Cheng Shiu University, Kaohsiung 833, Taiwan; k6764@gcloud.csu.edu.tw
9 Super Micro Research and Technology Center, Cheng Shiu University, Kaohsiung 833, Taiwan
10 Department of Medical Research, China Medical University Hospital, China Medical University, Taichung 40402, Taiwan
11 School of Oral Hygiene, College of Oral Medicine, Taipei Medical University, Taipei 11031, Taiwan
* Correspondence: yaochang.chiang@gmail.com (Y.-C.C.); jufangliu@tmu.edu.tw (J.-F.L.)

Citation: Lee, C.-W.; Huang, C.C.-Y.; Chi, M.-C.; Lee, K.-H.; Peng, K.-T.; Fang, M.-L.; Chiang, Y.-C.; Liu, J.-F. Naringenin Induces ROS-Mediated ER Stress, Autophagy, and Apoptosis in Human Osteosarcoma Cell Lines. *Molecules* **2022**, *27*, 373. https://doi.org/10.3390/molecules27020373

Academic Editors: Višnja Stepanić and Marta Kučerová-Chlupáčová

Received: 25 September 2021
Accepted: 28 December 2021
Published: 7 January 2022

Abstract: Osteosarcoma, a primary bone tumor, responds poorly to chemotherapy and radiation therapy in children and young adults; hence, as the basis for an alternative treatment, this study investigated the cytotoxic and antiproliferative effects of naringenin on osteosarcoma cell lines, HOS and U2OS, by using cell counting kit-8 and colony formation assays. DNA fragmentation and the increase in the G2/M phase in HOS and U2OS cells upon treatment with various naringenin concentrations were determined by using the terminal deoxynucleotidyl transferase-mediated dUTP nick-end labeling assay and Annexin V/propidium iodide double staining, respectively. Flow cytometry was performed, and 2′,7′-dichlorodihydrofluorescein diacetate, JC-1, and Fluo-4 AM ester probes were examined for reactive oxygen species (ROS) generation, mitochondrial membrane potential, and intracellular calcium levels, respectively. Caspase activation, cell cycle, cytosolic and mitochondrial, and autophagy-related proteins were determined using western blotting. The results indicated that naringenin significantly inhibited viability and proliferation of osteosarcoma cells in a dose-dependent manner. In addition, naringenin induced cell cycle arrest in osteosarcoma cells by inhibiting cyclin B1 and cyclin-dependent kinase 1 expression and upregulating p21 expression. Furthermore, naringenin significantly inhibited the growth of osteosarcoma cells by increasing the intracellular ROS level. Naringenin induced endoplasmic reticulum (ER) stress-mediated apoptosis through the upregulation of ER stress markers, GRP78 and GRP94. Naringenin caused acidic vesicular organelle formation and increased autophagolysosomes, microtubule-associated protein-light chain 3-II protein levels, and autophagy. The findings suggest that the induction of cell apoptosis, cell cycle arrest, and autophagy by naringenin through mitochondrial dysfunction, ROS production, and ER stress signaling pathways contribute to the antiproliferative effect of naringenin on osteosarcoma cells.

Keywords: osteosarcoma; naringenin; ROS; ER stress; autophagy; apoptosis

1. Introduction

Cancer is a public health problem, with high mortality and disability rates worldwide [1]. According to global cancer statistics, 19.3 million new cancer-related cases and 10 million cancer-related deaths were reported in 2020 [2]. Osteosarcoma is the most prevalent primary bone cancer in children aged 10 to 15 years and young adults [3,4]. The 5-year survival rate of patients with localized osteosarcoma is 65–75% [5]; however, the prognosis of patients with relapse and metastasis is poor, with a 5-year survival rate of 10–20% [6,7]. Currently, treatment for osteosarcoma mainly involves surgical removal, neoadjuvant chemotherapy, and radiation therapy [8]. However, the chemotherapeutic agents currently used cause severe side effects in the majority of patients. Moreover, resistance to chemotherapeutic agents is another challenge in the treatment of osteosarcoma. Hence, new drugs with both low toxicity and high efficacy are urgently required.

Natural products have been widely used as anticancer therapeutics. Sixty percent of drugs approved by the Food and Drug Administration from 1984 to 1994 were isolated from natural sources, especially plants [9,10]. Naringenin [(2*S*)-4′,5,7-trihydroxyflavan-4-one] is present in various herbs and fruits, especially citrus plants [11]. Naringenin possesses antimicrobial, antioxidative, and anticancer properties [12–14]. Moreover, naringenin induces cytotoxicity in different types of cancer cells [15]. In 2019, Zhao et al. reported that naringenin (250 μM; 24 h) suppressed the migration of breast cancer cells by arresting the cell cycle at the G0/G1 phase [16]. Furthermore, Song et al. indicated that naringenin (200 μM; 24 h) caused colon cancer apoptosis through p38-dependent ATF3 activation [17]. Naringenin (500 μM; 24 h) enhanced TRAIL-induced apoptosis through the induction of DR5 expression in human A549 cells [18]. However, the effect of naringenin on osteosarcoma remains unclear.

Autophagy, which refers to the intracellular degradation of cytoplasmic materials caused by vacuoles or lysosomes in eukaryotic cells, eliminates and recycles damaged proteins to prolong the lifespan of cells [19]. It is a crucial homeostasis and cell survival mechanism that responds to environmental stresses such as starvation or pathogen infection [20]. Recent accumulating evidence indicates that autophagy also occurs under pathological conditions, such as in neurodegenerative disease or tumor development [21]. Specifically, autophagy is believed to play an important role in tumor development [22]. During the early stages of tumor formation, autophagy functions as a tumor suppressor, and autophagic activity is often impaired in cancer cells. Many anticancer drugs which lead to apoptosis can also induce autophagy-related cell death in cancer cell lines [23]. In osteosarcoma, autophagy is deregulated and functions as a protumoral or antitumoral process to suppress carcinogenesis and support the growth of established tumors [19].

This study investigated the effects of and molecular mechanisms underlying naringenin-induced autophagy and apoptosis and the interaction between autophagy and apoptosis in osteosarcoma cells. The findings of this study can provide the proof-of-concept for evaluating naringenin as an antiosteosarcoma agent.

2. Materials and Methods

2.1. Material

Primary antibodies for Bak (GTX100063), Bax (GTX109683), Bcl2 (GTX100064), Bcl-xL (GTX105661), GRP78 (GTX113340), GRP94 (GTX103203), PARP (GTX100573), calpain I (GTX102340), calpain II (GTX102499), cytochrome c (GTX108585), beclin1 (GTX134209), p53 (GTX70214), p62 (GTX102361), LC3B (GTX127375), ATG5 (GTX102360) and a voltage-dependent anion channel (VDAC; GTX104745) were purchased from GeneTex International Corporation (Hsinchu City, Taiwan). Caspase-3 (19677-1-AP) and caspase-9 (10380-1-AP) were purchased from Proteintech Group Inc. (Rosemont, IL, USA). CDK1 (MN ABE1403) and cyclin B (MM05373) was purchased from Merck KGaA, Darmstadt, Germany. Anti-mouse and anti-rabbit IgG-conjugated horseradish peroxidase, as well as rabbit polyclonal antibodies specific for β-actin (cat. no. SI-A5441; Sigma-Aldrich; Merck KGaA, Darmstadt,

Germany) were used in study. All other chemicals were obtained from Sigma-Aldrich (St. Louis, MO, USA).

2.2. Cell Line and Cell Culture

The human osteosarcoma cell lines (U2OS and HOS) and osteoblast cell line (hFOB 1.19) were purchased from the American Type Culture Collection (ATCC; Manassas, VA, USA). The hFOB1.19 osteoblast cells were cultured in a DMEM/F12 medium supplemented with 10% fetal bovine serum (FBS), 2.5 mM L-glutamine, 0.3 mg/mL G418 and 100 units/mL penicillin/streptomycin. Cells were incubated in an atmosphere of 5% CO_2 at 34 °C and subcultures were changed every 48 h.

The HOS cells were maintained in an Eagle's Minimum Essential medium supplemented with 10% FBS and 100 units/mL penicillin/streptomycin. The U2OS cells were cultured in a McCoy's 5A medium supplemented with 10% FBS and 100 units/mL penicillin/streptomycin. (Invitrogen; Thermo Fisher Scientific, Inc., Waltham, MA, USA). Cells were incubated in an atmosphere of 5% CO_2 at 37 °C and subcultures were changed every 48 h.

2.3. Cell Viability/Proliferation Assay

Cells were seeded (6×10^3) in triplicate in a 96-well plate, treated with naringenin at various concentrations (100, 250, and 500 µM), and incubated at 37 °C for 24 h under 5% CO_2. After incubation, the cell counting kit-8 (CCK-8) assay (10 µL; Sigma-Aldrich, St. Louis, MO, USA) solution was added into each well, and the plate was incubated under the same conditions for 2–5 h. A microplate reader was used to measure absorbance at 450 nm (Bio-Tek Instruments, Winooski, VT, USA).

2.4. Colony Formation Assay

HOS and U2OS cells (5×10^4) were seeded in six-well plates and treated with naringenin at the indicated concentrations (100, 250, and 500 µM) in a medium without 10% FBS for 24 h. Untreated cells were used as the control. The culture medium containing 10% FBS was replaced after 24 h of treatment and changed every 2 days without additional naringenin treatment. After incubation for 7 days, visible colonies were washed with PBS, fixed with 4% paraformaldehyde for 15 min, and stained with 0.25% crystal violet for 15 min. The images of colonies were captured through scanning. Subsequently, the plates were washed three times with double-distilled water and then with 33 % (v/v) acetic acid, and this was followed by the measurement of absorbance at 550 nm. The colony formation assay was repeated three times in duplicate wells.

2.5. Cytosolic and Mitochondrial Protein Extraction

Cytosolic and mitochondrial proteins were extracted from untreated controls and cells treated with naringenin at various concentrations (100, 250, and 500 µM) for 8 h by using the Mitochondria/Cytosol Fractionation Kit (Cat#K256-25; BioVision Inc., Milpitas, CA, USA) according to the manufacturer's protocol. Cells were collected, washed, and centrifuged for 10 min at 4 °C at 1000 rpm. Cells were then resuspended in the cytosol extraction buffer mix, incubated on ice for 10 min, and repeatedly passed through a 25-gauge needle. The homogenate mix was centrifuged for 10 min at 4 °C at 3000× g rpm. The supernatant was collected and centrifuged for 30 min at 4 °C at 15,000× g rpm. The supernatant was used as the cytosolic fraction. The pellet was resuspended in a mitochondrial extraction buffer mix and used as the mitochondrial fraction.

2.6. Western Blot Analysis

HOS and U2OS cells were treated with naringenin at various concentrations (100, 250, and 500 µM) for 8 h. After treatment, total protein was harvested and lysed in RIPA lysis buffer containing protease inhibitors. Protein concentrations were determined using the bicinchoninic acid assay kit (Sigma-Aldrich, St. Louis, MO, USA). Proteins were separated through 8–15% sodium dodecyl sulfate–polyacrylamide gel electrophoresis and transferred onto polyvinylidene fluoride membranes. The membranes were blocked with a TBST buffer containing 4% BSA for 1 h at room temperature and subsequently incubated with primary antibodies (at a dilution of 1:1000) at 4 °C overnight. The membranes were washed three times with the TBST buffer and then incubated with peroxidase-conjugated secondary antibodies (at a dilution of 1:10,000) for 1 h at room temperature. The blots were visualized using an enhanced chemiluminescence system (EMD Millipore, Billerica, MA, USA) with a UVP BioImaging System (Upland, CA, USA). Each experiment was repeated at least three times.

2.7. DAPI Staining

HOS and U2OS cells were treated with naringenin at various concentrations (100, 250, and 500 μM) for 24 h. After treatment, cells were washed with PBS, fixed in a 3.7% formaldehyde solution for 15 min, permeabilized with 0.1% Triton X-100 for 5 min, and stained with DAPI (1 μg/mL) for 5 min. All samples were examined and photographed using a Nikon Eclipse Ti inverted fluorescence microscope (software version 5.02.01).

2.8. Terminal Deoxynucleotidyl Transferase-Mediated dUTP Nick-End Labeling Assay

Apoptotic cells were quantified using the terminal deoxynucleotidyl transferase-mediated dUTP nick-end labeling (TUNEL) assay, which examines breaks in DNA strands caused during cell apoptosis by using the BD APO-DIRECT kit (BD Biosciences, San Jose, CA, USA; cat. no. 556381). HOS and U2OS cells (2×10^6) were treated with naringenin at various concentrations (100, 250, and 500 μM) for 24 h. After treatment, cells were collected and centrifuged at 950× *g* for 10 min at 4°C. Cells were fixed with 1% paraformaldehyde for 30 min on ice and washed with PBS twice. After the removal of the fixative, 0.5 mL of ethanol was added, and the mixture was incubated at −20 °C for 4 h. Subsequently, ethanol was removed through centrifugation and cellular DNA was obtained. The cellular DNA was stained with TUNEL solution (3 ng/mL TdT enzyme and 0.04 nmol FITC dUTP) at 37 °C for 1 h. After incubation with TUNEL solution, cells were washed with a rinse buffer (1 mL; BD APO-DIRECT kit; BD Biosciences; cat. no. 556381) and centrifuged at 1425× *g* for 10 min at 4 °C. The fluorescein-labeled DNA strand was detected and quantified using the BD Accuri C5 flow cytometer and BD Accuri C6 software (version 1.0.264.21, BD Biosciences).

2.9. Annexin V and Propidium Iodide Staining

The Annexin V/propidium iodide (PI) double staining assay was performed to examine cell apoptosis by using the Annexin V/PI detection kit (cat. no. PF00005; Proteintech Group, Inc., Rosemont, IL, USA). HOS and U2OS cells were treated with naringenin at various concentrations (100, 250, and 500 μM) for 24 h. After treatment, cells were collected and washed with PBS twice and then resuspended in a staining buffer containing PI and Annexin V–FITC at room temperature for 30 min; cells were placed in the dark prior to flow cytometry. Cells were analyzed using the BD Accuri C5 flow cytometer and BD Accuri C6 software (version 1.0.264.21, BD Biosciences).

2.10. Cell Cycle Analysis Using PI Staining

Apoptotic cells were quantified by examining the cell cycle. HOS and U2OS cells were treated with naringenin at various concentrations (100, 250, and 500 μM) for 24 h and collected through centrifugation (10 min at 950× *g*). Ice-cold ethanol was added to 0.5 mL of cell suspension and the mixture was then incubated at −20 °C for 4 h. Ethanol was removed through centrifugation (15 min at 1425× *g*) and cells were stained with a PI solution (0.1% Triton-X 100, 100 μg/mL of DNase-free RNase A, and 10 μg/mL of PI in PBS). After staining, cells were analyzed using the BD Accuri C5 flow cytometer and BD Accuri C6 software (version 1.0.264.21, BD Biosciences).

2.11. Intracellular ROS Production, Ca^{2+} Concentration, and Mitochondrial Mmembrane Potential

ROS generation was determined using the fluorogenic probe 2′,7′-dichlorodihydrofluorescein (H_2DCFDA; Thermo Fisher; Waltham, MA, USA). The mitochondrial membrane potential (MMP) was determined using the cationic JC-1 dye (BD Biosciences). The Ca^{2+} concentration was measured using the Ca^{2+}-sensitive fluorescent probe Fluo-4 AM (Thermo Fisher). Cells (5×10^5) were plated in six-well plates, grown to confluence, and treated with naringenin at various concentrations (100, 250, and 500 μM) for indicated times. After incubation, cells were stained with H_2DCFDA (10 μM), Fluo-4 AM (3 μg/mL), and JC-1 (5 μg/mL) to determine ROS production, Ca^{2+} levels, and MMP, respectively. NAC and DPI were used as ROS inhibitors, and BAPTA-AM was used to control the intracellular Ca^{2+} level. Cells were determined using the BD Accuri C5 flow cytometer and BD Accuri C6 software (version 1.0.264.21, BD Biosciences).

2.12. Small Interfering RNA Transfection

Small interfering RNAs (siRNAs) against ATG5 (sense: 5′-GUGAGAUAUGGUUUGA AUAdTdT-3′ and antisense: 3′-UAUUCAAACCAUAUCUCACdTdT-5′), Beclin 1 (sense: 5′-GUUUGGAGAUCUUAGAGCAdTdT-3′ and antisense: 3′-UGCUCUAAGAUCUCCAAAC dTdT-5′) and a nonspecific scrambled siRNA were purchased from Sigma-Aldrich (Merck KGaA). HOS and U2OS cells (5×10^5) were seeded in six-well plates and transfected with Lipofectamine 3000 (Invitrogen, Rockville, MD, USA) mixed with a serum-free medium containing ATG5 siRNA, Beclin 1 siRNA, and scrambled siRNA. Transfection was performed under 5%CO_2 at 37 °C for 24 h.

2.13. Autophagy Assay

Cell autophagy was determined using DAPgreen (Dojindo Molecular Technologies, Inc., Kumamoto, Japan) and acridine orange (AO; Sigma-Aldrich; Merck KgaA) according to the manufacturers' instructions. HOS and U2OS cells (3×10^5) were seeded in six-well plates overnight. Subsequently, cells were washed with PBS and incubated with DAPgreen and AO for autophagy detection at 37 °C for 30 min. After cells were washed twice with PBS, they were treated with the indicated concentrations of naringenin (100, 250, and 500 μM) for 24 h. After treatment, the image of the cells was obtained under a 200× microscope to determine the visual intensity of green fluorescence (DAPgreen) and orange/green fluorescence (AO; autophagy) using a Nikon ECLIPSE Ti and NIS-Elements AR microscope (software version 5.02.01).

2.14. Caspase Activity Assay

The caspase activity assay is based on the ability of active enzymes to cleave chromophores from the enzyme substrate Ac-DEVD-pNA (caspase-3; cat. #1008) or Ac-LEHD-pNA (caspase-9; cat. #1076; BioVision Inc., Milpitas, CA, USA). To examine the activity of caspase-3 and -9, HOS and U2OS cell lysates were prepared and incubated with caspase-3 and -9 substrates. Immunocomplexes were incubated with the peptide substrate in the assay buffer (100 mM NaCl, 50 mM 4-(2-hydroxyethyl)-1-piperazine ethanesulfonic acid, 10 mM dithiothreitol, 1 mM EDTA, 10% glycerol, and 0.1% CHAPS; pH 7.4) at 37 °C for 6 h. The release of *p*-nitroaniline was monitored using an enzyme-linked immunosorbent assay reader at 405 nm. The results are presented as the percentage change in activity compared with the untreated control.

2.15. Transmission Electron Microscopy

HOS and U2OS cells (1×10^5) were treated with naringenin as previously described. Cells were fixed for 10 min in 50% Karnovsky's fixative. Cells were collected and centrifuged at 1500× *g* for 5 min. The pellet was washed and stored in 70% Karnovsky's fixative at 4 °C until embedding and then analyzed by means of a transmission electron microscope. The sections were observed under a JEOL JEM-1400 electron microscope (Tokyo, Japan).

2.16. Statistical Analysis

All results were analyzed using the GraphPad Prism program (GraphPad, San Diego, CA, USA). Quantified results were expressed as the mean ± standard deviation (SD) and analyzed with one-way ANOVA followed by Fisher's least significant difference (LSD) post-hoc test. For all results, $p < 0.05$ was considered a significant difference.

3. Results

3.1. Induction of Cell Apoptosis in Human Osteosarcoma Cells by Naringenin

Naringenin contains the skeleton structure of flavanone and three hydroxy groups at 4′, 5′, and 7′ carbons ((2*S*)-4′,5′,7′-trihydroxyflavan-4-one; Figure 1A). To determine the effect of naringenin on human osteosarcoma, HOS and U2OS cells were treated with varying concentrations of naringenin for 24 h. The results indicated that naringenin reduced the viability of osteosarcoma cells but not normal hFOB 1.19 cells in a concentration-dependent manner (Figure 1B–D), indicating that naringenin selectively inhibited the growth of osteosarcoma cells but exhibited less cytotoxicity in normal human bone cells. In addition, the antiproliferative effect of naringenin on HOS and U2OS cells was evaluated through a colony formation assay. The results demonstrated that compared with no treatment, naringenin treatment significantly reduced the number of colonies in a dose-dependent manner (Figure 1E,F). Cell apoptosis caused by naringenin was examined using with DAPI staining. The results indicated that naringenin increased DNA condensation and morphological changes in HOS and U2OS cells (Figure 1G).

To explore whether naringenin inhibited the viability and proliferation of HOS and U2OS cells by inducing apoptosis, apoptotic cells were detected by performing Annexin V/PI double-labeling and the TUNEL assay. Treatment with varying concentrations of naringenin significantly promoted the apoptosis of osteosarcoma cells in a dose-dependent manner (Figure 2A,B). Similar results were obtained in the TUNEL assay (Figure 2C,D). To investigate the inhibitory effects of naringenin on the proliferation of HOS and U2OS cells, we examined cell cycle distribution after 24 h treatment with naringenin. The percentages of naringenin-treated osteosarcoma cells in the G2/M phase were significantly higher than those of control cells (Figure 2E). To verify this change, the levels of G2/M phase regulatory proteins (CDK1 and cyclin B) were examined by western blotting. The results indicated that compared with control cells, naringenin-treated cells exhibited decreased CDK1 and cyclin B levels and upregulated p21 expression (Figure 2F). These findings indicated that naringenin induced cell apoptosis and prolonged G2/M arrest in human osteosarcoma cells.

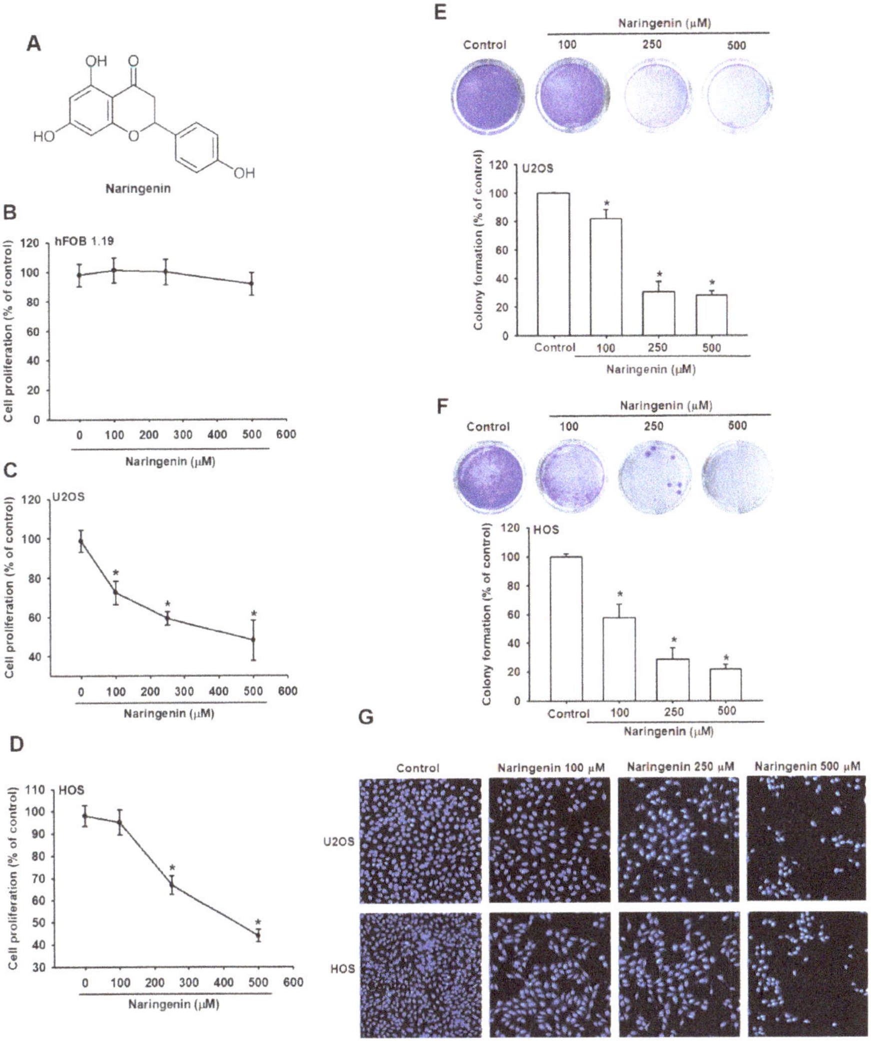

Figure 1. Naringenin inhibits the proliferation of human osteosarcoma cells. (**A**) Structure of naringenin. (**B–D**) hFOB 1.19, HOS, and U2OS cells were treated with the indicated concentration of naringenin for 24 h, and cell proliferation was determined using the CCK-8 assay. (**E**,**F**) Colony formation assay was performed in HOS and U2OS cells treated with naringenin at various concentrations (100, 250, and 500 μM) for 7 days. Quantitative results indicated the number of colonies per group. (**G**) After cells were incubated with various concentrations of naringenin for 24 h, the nucleus morphology was determined through 4′,6-diamidino-2-phenylindole staining, and cells were photographed. Magnification, ×200. Results are expressed as the mean ± SD of four independent experiments. * $p < 0.05$ compared with control.

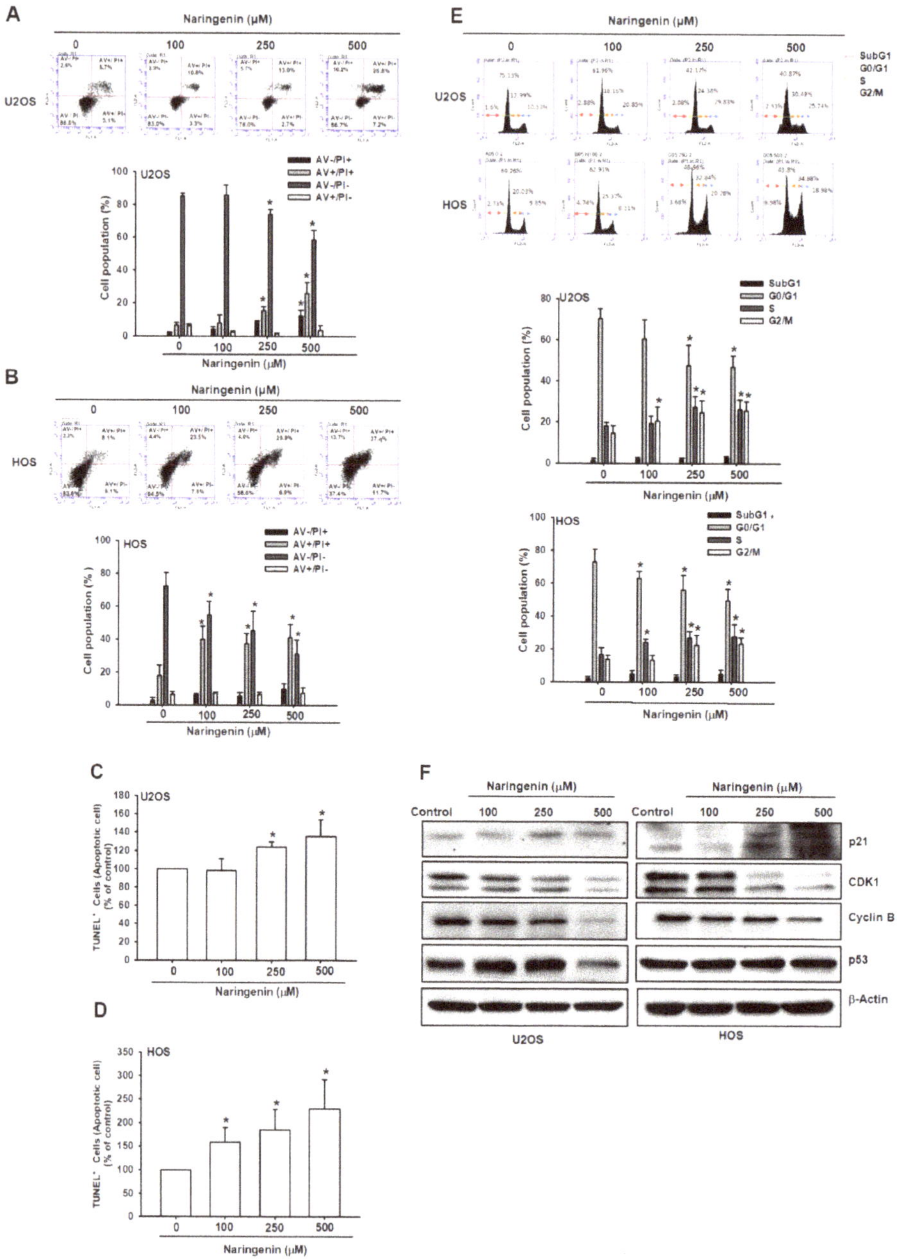

Figure 2. Naringenin induced apoptosis and cell cycle arrest at the G2/M phase in human osteosarcoma cells. (**A–D**) Cells were treated with naringenin at various concentrations (100, 250, and 500 μM) for 24 h. Apoptotic cells were determined using with Annexin V/PI and TUNEL staining. (**E**) Flow cytometry was performed to determine the cell cycle distribution of HOS and U2OS cells treated with naringenin at various concentrations (100, 250, and 500 μM) for 8 h. (**F**) Levels of p21, CDK1, cyclin B, and p53 proteins in naringenin-treated cells for 8 h were detected through western blotting. Results are expressed as the mean ± SD of four independent experiments. * $p < 0.05$ compared with control.

3.2. Induction of ROS-Mediated ER Stress by Naringenin in Osteosarcoma Cells

ROS plays a crucial role in the apoptotic process in many cell types [24]. ER stress is induced by the accumulation of ROS, leading to mitochondrial dysfunction and apoptosis [25]. As shown in Figure 3A,B, naringenin induced ROS production and accumulation. Pretreating cells with the antioxidant NAC and the inhibitor of the flavoprotein-dependent oxidase DPI reduced ROS production and naringenin-induced cell apoptosis in osteosarcoma cells (Figure 3C,D).

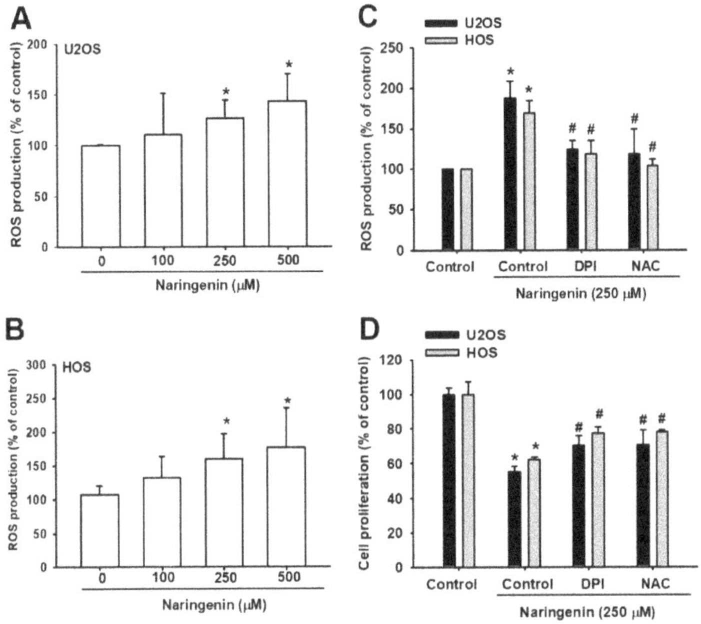

Figure 3. Naringenin induces reactive oxygen species (ROS) generation in human osteosarcoma cells. (**A**,**B**) Cells were treated with naringenin (100, 250, and 500 μM) for 1 h and stained with H_2DCFDA. The percentage of ROS production was determined by flow cytometry. (**C**) Cells were treated as described in A-B. ROS inhibitors inhibited ROS production. (**D**) After cells were pretreated with a ROS inhibitor for 1 h, followed by stimulation with naringenin for 24 h, cell viability was examined by using the CCK-8 assay. Results are expressed as the mean ± SD of three independent experiments. * $p < 0.05$ compared with the control group; # $p < 0.05$ compared with the naringenin-treated group.

To determine whether naringenin induced apoptosis by triggering ER stress, we first examined its effect on the mobilization of Ca^{2+}. The Ca^{2+} level was significantly increased in naringenin-treated cells compared with control cells (Figure 4A,B). The results revealed that aberrant Ca^{2+} imbalance increased in naringenin-treated cells in a dose-dependent manner. We next determined whether ER stress-associated proteins, glucose-regulated proteins (GRP78 and GRP 94), and calpain proteins (calpain I and calpain II) are induced by naringenin in osteosarcoma cells. As shown in Figure 4C, naringenin increased the expression of GRP78, GRP94, and calpain I and II in a dose-dependent manner. Moreover, cells pretreated with BAPTA-AM, a Ca^{2+} cell-permeable chelator, markedly reduced naringenin-mediated cell apoptosis (Figure 4D). In addition, naringenin-induced Ca^{2+} release was inhibited by ROS inhibitors (DPI or NAC; Figure 4E). These findings indicated that naringenin induced cell apoptosis in osteosarcoma cells through ROS production and caused ER stress.

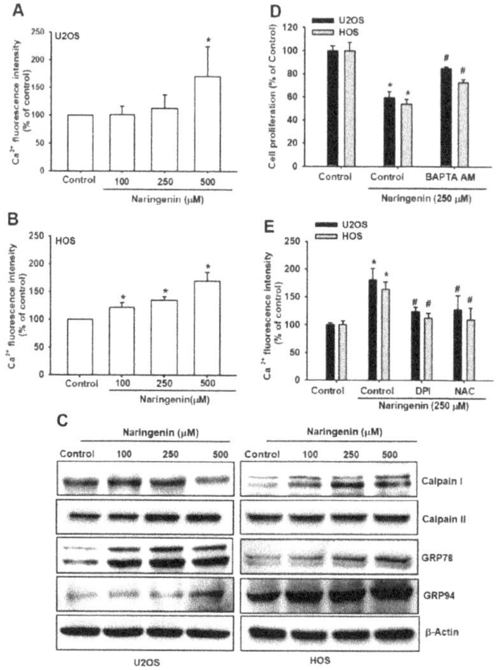

Figure 4. Naringenin triggers Ca^{2+} release and causes endoplasmic reticulum (ER) stress in human osteosarcoma cells. (**A**,**B**) Cells were treated with naringenin (100, 250 and 500 μM) for 5 h and stained with Fluo-4 AM. Ca^{2+} expression was examined by using flow cytometry. (**C**) After cells were treated with varying concentrations of naringenin for 8 h, the protein expression of GRP78, GRP94, calpain I, and calpain II was examined using with western blotting, and β-actin was used as the internal control. (**D**) After cells were pretreated with BAPTA-AM for 1 h, followed by stimulation with naringenin for 24 h, cell viability was examined using the CCK-8 assay. (**E**) After cells were pretreated with a ROS inhibitor for 1 h, followed by stimulation with naringenin for 5 h, Ca^{2+} expression was examined by using flow cytometry. Results are expressed as the mean ± SD of four independent experiments. * $p < 0.05$ compared with the control group. # $p < 0.05$ compared with the naringenin-treated group.

3.3. *Involement of Cell Apoptosis in Naringenin-Induced Mitochondrial Dysfunction in Human Osteosarcoma*

ROS production and accumulation in mitochondria reduce MMP, thereby triggering the mitochondrial apoptotic pathway [26]. To examine the effect of naringenin on HOS and U2OS cells, JC-1 staining was performed to determine the fluorescence ratio of orange and green fluorescence between normal and unhealthy mitochondria. Orange fluorescence disappeared in cells when observed under the 200× microscope, indicating mitochondrial dysfunction (Figure 5A). To verify the loss of MMP, HOS and U2OS cells were analyzed using with flow cytometry, and the results revealed a shift in MMP in naringenin-treated cells (Figure 5B,C). The findings indicated the loss of MMP in cells treated with various naringenin concentrations. Mitochondrial and cytosolic proteins, cytochrome c, and proapoptotic/antiapoptotic proteins were examined in cells treated with naringenin. The release of cytochrome c due to mitochondrial dysfunction was upregulated by protein–protein interactions between Bcl-2 proteins (Figure 5D,E). Proapoptotic proteins, Bak and Bax, and cytochrome c released in the cytosol were upregulated with the increasing concentration of naringenin (Figure 5D,E). In the mitochondrial pathway of apoptosis, the downstream signaling of caspases was activated after treatment with naringenin. To determine the primary mediators of apoptosis, caspase-3 and caspase-9 were examined in naringenin-treated HOS and U2OS cells. The results revealed the upregulation of caspases, including that of the PARP cleavage (Figure 5F). To determine the intrinsic pathway of naringenin-induced apoptosis, caspase-3 and caspase-9 activities were examined separately (Figure 5G,H). The activities of both caspases were associated with naringenin-induced cell apoptosis. Pretreatment of cells with a caspase-3 inhibitor (z-DEVD-FMK) or caspase-9 inhibitor (z-LEHD-FMK) inhibited naringenin-induced cell apoptosis (Figure 5I). Thus, these data demonstrated that naringenin induced mitochondrial dysfunction and the subsequent release of cytochrome c and activation of caspases-9 and caspases-3 in human osteosarcoma cells.

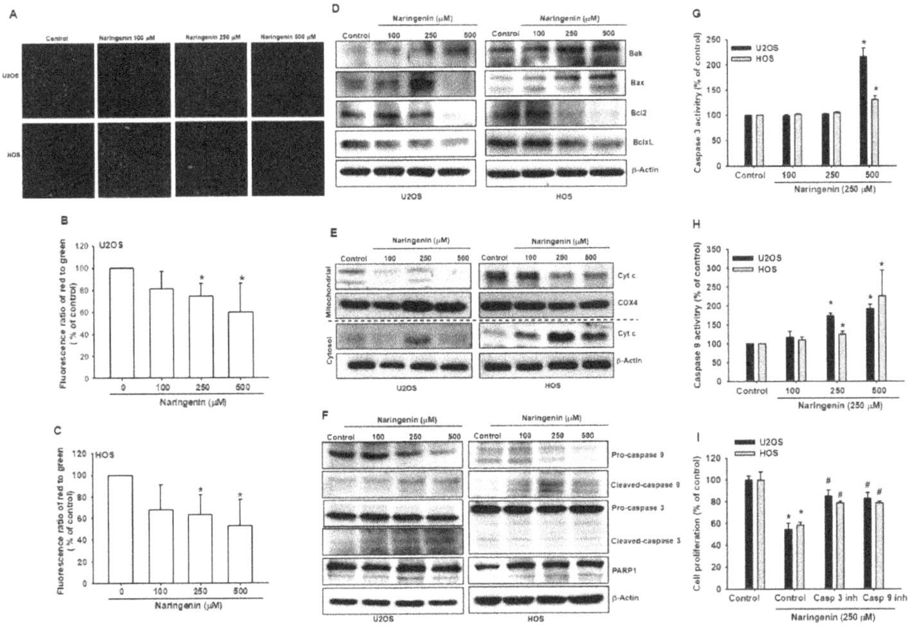

Figure 5. Naringenin triggers endoplasmic reticulum (ER) stress-related mitochondrial dysfunction in human osteosarcoma cells. (**A–C**) Cells were treated with naringenin at various concentrations (100, 250, and 500 μM) for 24 h and stained with JC-1 to determine the reduction in mitochondrial membrane potential. The percentage of cells was analyzed by using flow cytometry, and images were obtained using a fluorescence microscope. Magnification, ×200. (**D– F**) Cells were treated as described in A–B. The expression of mitochondrial dysfunction-related proteins, mitochondrial and cytosolic cytochrome c, Bax, Bak, Bcl-2, Bcl-xl, caspase-3, caspase-9, and PARP was determined by western blotting. (**G,H**) Cells were treated as described in A–B. The upregulation of caspase-3 and caspase-9 activities was determined and analyzed using the caspase activity assay. (**I**) Cells were pretreated with caspase inhibitors separately, followed by naringenin treatment (250 μM), and cell proliferation was determined using the CCK-8 assay. Results are expressed as the mean ± SD of four independent experiments. * $p < 0.05$ compared with the control group. # $p < 0.05$ compared with the naringenin-treated group.

3.4. Triggering of Autophagy in Human Osteosarcoma Cells by Naringenin

Because cell autophagy regulates cell death, we examined whether naringenin can induce autophagy. The autophagy phenomenon was observed in HOS and U2OS cells treated with naringenin. Using a transmission electronic microscope, it was observed that naringenin induced the formation of numerous intracytoplasmic vacuoles in HOS cells (Figure 6A). In addition, to observe naringenin-induced autophagy, HOS and U2OS cells were stained with DAPgreen, and the green fluorescence intensity within cells was detected. The intensity increased with the concentration of naringenin (Figure 6B). Moreover, AO staining revealed the accumulation of acidic vesicles in HOS and U2OS cells (Figure 6C). Furthermore, we examined the expression of several proteins that serve as markers of autophagy. Naringenin increased the level of microtubule-associated protein light chain 3 (LC3)-II and the expression of ATG5, Beclin 1, and p62 in a dose-dependent manner (Figure 6D). To determine the role of autophagy in naringenin-induced cell death, siRNAs of Beclin 1 and ATG5 were transfected in HOS and U2OS cells. The results of the CCK-8 assay revealed that Beclin 1 and ATG5 siRNAs suppressed the naringenin-induced loss of cell viability (Figure 6E), indicating that the autophagic mechanism of ATG5 and Beclin 1 involved in naringenin-induced cell death promotes either survival or death and is associated with signaling transduction in programmed cell apoptosis. Furthermore, we examined naringenin-induced cell autophagy by using ROS inhibitors. We performed DAPgreen staining to observe the green fluorescence intensity (Figure 6F). Cell images were obtained using a fluorescence microscope, and it was observed that after adding the inhibitors, the autophagosome (white arrow) significantly decreased in naringenin-treated cells compared with control cells. The results indicate that naringenin-induced intracellular ROS production triggered autophagic signaling in human osteosarcoma cells.

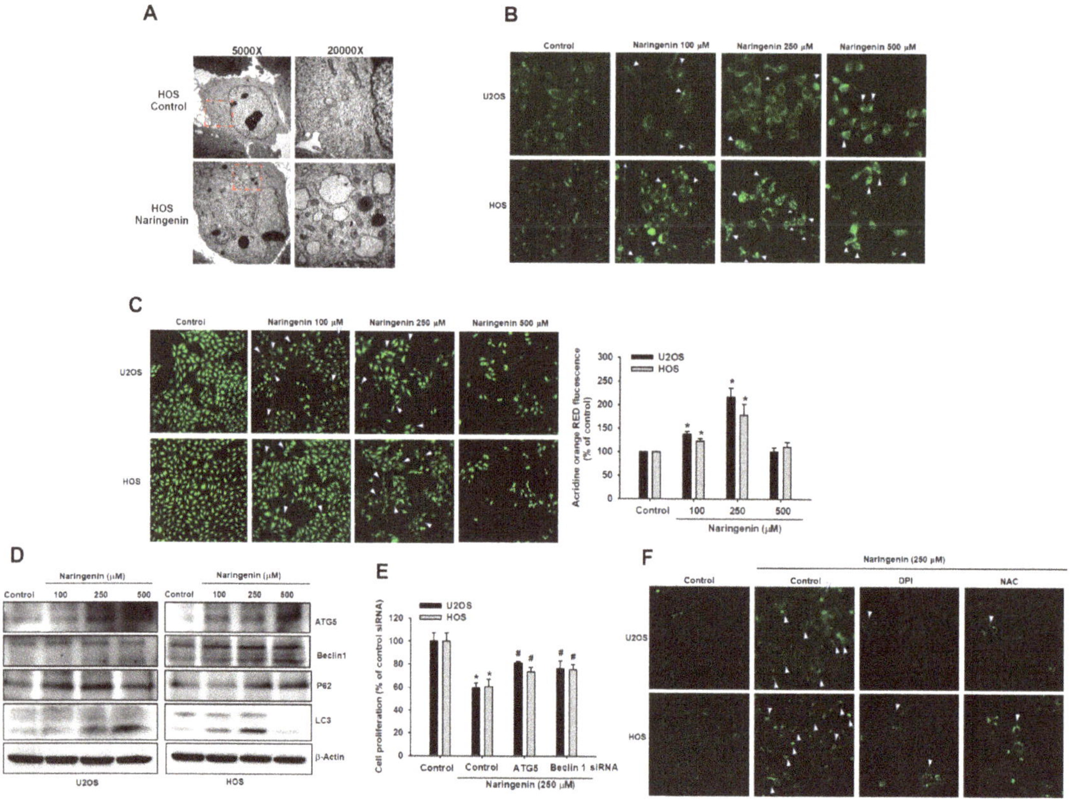

Figure 6. Naringenin initiates autophagy signaling responses in human osteosarcoma cells. (**A**) Electron microscope images of HOS cells were obtained with and without naringenin treatment for 8 h. (**B**,**C**) Cells were treated with naringenin for 8 h and stained with DAPgreen and acridine orange to determine the induction of autophagosome. (**D**) The expression of ATG5, Beclin 1, p62, and LC3-II proteins was determined by western blotting after naringenin treatment for 8 h. (**E**) siRNAs of ATG5 and Beclin 1 were transfected in cells for 24 h, followed by treatment with naringenin for 24 h. Cell

proliferation was determined using the CCK-8 assay. (**F**) Cells were pretreated with ROS inhibitors, followed by naringenin treatment (250 μM) for 8 h, and were then stained with DAPgreen to determine autophagosome formation. Images were obtained using a fluorescence microscope. Magnification, ×200. Results are expressed as the mean ± SD of four independent experiments. * $p < 0.05$ compared with the control group. # $p < 0.05$ compared with the naringenin-treated group.

4. Discussion

Grapefruit is important not only as a fruit but also in traditional medicine [27]. It is a rich source of bioactive compounds that may serve as chemopreventive agents in cancer therapy [28]. Naringenin is the active component in grapefruit extract [29]. Naringenin exhibits various bioactivities including antioxidative, anti-inflammatory, and anticancer effects. Naringenin exerts in vitro effects on various cancer cell lines at high concentration; eg., on breast cancer (SKBR3 and MDA-MD-231; 250 μM) [16–18,30,31] and liver cancer cell lines (HepG2; 100–200 μM) [31]. In this study, we found that the IC_{50} values for HOS and U2OS osteosarcoma cells after 24 h treatment with naringenin were 276 and 389 μM, respectively. Although a high naringenin concentration was used during the study, naringenin selectively inhibited the growth of osteosarcoma cells and exhibited less cytotoxicity in normal human bone cells. This is the first study to report the cytotoxic and antiproliferative effects of naringenin on human osteosarcoma HOS and U2OS cells. In our study, the results of CCK-8 and colony formation assays and DAPI staining revealed that naringenin significantly inhibited the proliferation of HOS and U2OS cells. Moreover, naringenin induced caspase-dependent apoptosis in HOS and U2OS cells, as determined through flow cytometry, TUNEL staining, and the expression of apoptosis-related proteins. In this study, we observed that naringenin exhibited potent antitumor activity against osteosarcoma in vitro. Naringenin-induced ROS overproduction resulted in ER stress activation and autophagy, which caused cell apoptosis. The findings indicated that naringenin plays multiple roles in signaling pathways in human osteosarcoma cells.

Cell apoptosis and autophagy, which are involved in the regulation of cancer cell death, have been widely studied. Autophagy plays a critical role in maintaining homeostasis and the pathogenesis of many human diseases, including cancer [32]. Autophagy, a dynamic and conserved catabolic process, plays dual roles in cell survival and death [20]. In some circumstances, autophagy can lead to cell death either in collaboration with apoptosis or as a backup mechanism when apoptosis is defective [33]. Autophagy can be induced through acute exposure to resveratrol while prolonging the activation of the caspase-mediated cell death pathway [34]. Because Bcl-2 family proteins control the release of cytochrome c during mitochondrial dysfunction, Beclin 1 can be upregulated to activate the autophagy pathway with autophagy-related genes (ATG) and their protein products [35]. Cleavage of LC3 is an essential step in autophagosome formation [36]. Thus, in this study, we examined whether naringenin induces autophagy in osteosarcoma cells. We evaluated the expression levels of the autophagy markers LC3-I/LC3-II, Beclin 1, and p62. Transmission electron microscopy was mainly used for the detection of autophagosomes and the autophagy level. Our results indicated that naringenin induced autophagy in osteosarcoma cells.

Programmed cell apoptosis can be triggered by certain endogenous or exogenous signals through various ER stress-induced signals. Condensation of chromatin, fragmentation of DNA, and shedding of small fragments from cells are the innate cellular responses of apoptosis [37]. We examined mitochondrial dysfunction-induced apoptosis and observed that caspases play critical roles in the apoptotic cascade [38]. Our results revealed that naringenin increased the protein expression of caspase-3 and caspase-9. Moreover, Bcl-2 family proteins are mitochondrial apoptotic regulators that can be either proapoptotic proteins, such as Bax, or antiapoptotic proteins, such as Bcl-2. Naringenin upregulated Bax and Bak protein expression and downregulated Bcl-2 protein expression. Our results indicated that naringenin induced apoptosis through caspase-dependent apoptotic pathways by activating caspase-3/-9 and altering Bax/Bcl-2 in HOS and U2OS cells.

ROS acts as a second messenger in cell signaling by regulating various biological processes in normal and cancer cells [39]. The ROS production is shared by most chemotherapeutics due to its role in triggering cell death [40,41]. Excessive ROS generation can result in oxidative stress, DNA damage, and cell death through apoptosis [42]. For example, Zhang et al. [43] reported that libertellenone H treatment significantly increased intracellular ROS generation and induced apoptosis in human pancreatic cancer cells in a ROS-dependent manner. Similarly, curcumin derivatives markedly increased ROS production in colon cancer cells [44]. Furthermore, naringenin-inhibited cell growth was effectively increased through intercellular ROS production [45]. ROS is crucial in cellular apoptosis and autophagy pathways [46]. For example, Zhang et al. [47] demonstrated that the induction of autophagy increased ROS-mediated apoptosis in human bladder cancer cells. They reported that apoptosis and autophagy were dependent on ROS production. The obtained results suggest that ROS plays a vital role in cancer cell apoptosis and autophagy.

Accumulating evidence has indicated that apoptosis is regulated by ER stress [48,49]. The ER is a major organelle in eukaryotic cells that is involved in protein processing, intracellular calcium storage, and lipid synthesis. In the present study, naringenin exerted potent cytotoxic and apoptotic effects on human osteosarcoma cells. We found that the intracellular ROS level was increased after naringenin treatment and that the antioxidant NAC rescued the apoptosis level. In addition, naringenin stimulated the expression of ER stress-related proteins including GRP78, GRP94, calpain I, and calpain II. Naringenin-induced apoptosis was suppressed by the ER stress inhibitor BAPTA-AM in HOS and U2OS cells. These results suggest that apoptosis is regulated by ER stress signaling. Moreover, ROS inhibitors (DPI and NAC) suppressed calcium release and autophagosome formation. Therefore, naringenin induced ER stress and autophagy by promoting ROS generation, thus resulting in cellular apoptosis.

In conclusion, we demonstrated that naringenin induced apoptosis through the ROS-mediated ER stress signaling pathway and autophagy. Although a high concentration of naringenin was required in vitro to cause cancer cell death, naringenin is a potential agent against human osteosarcoma cells. Future studies examining novel naringenin-based derivatives are warranted.

Author Contributions: C.-W.L., Y.-C.C. and J.-F.L. conceived and designed the experiments, which were performed by C.-W.L., C.C.-Y.H., K.-T.P. and M.-C.C., K.-H.L., M.-L.F. and K.-H.L. analyzed the data. K.-H.L., Y.-C.C. and J.-F.L. contributed reagents/materials/analysis tools. C.-W.L., Y.-C.C., K.-H.L. and J.-F.L. wrote the paper. All authors have read and agreed to the published version of the manuscript.

Funding: This work was supported by grants from The Ministry of Science and Technology (MOST-106-2314-B-038-099-MY3, MOST-109-2320-B-255-004-MY3), Taipei Medical University (TMU108-AE1-B47). Chang Gung University of Science and Technology Chang Gung University of Science Foundation (grants ZRRPF6L0011, ZRRPF3L0091, ZORPF6L0011) and Chang Gung Medical Research Program Foundation (grants CMRPF6K0041).

Institutional Review Board Statement: Not applicable.

Informed Consent Statement: Informed consent was obtained from all subjects involved in the study.

Data Availability Statement: All data and materials are available for verification as needed.

Conflicts of Interest: The authors declare no conflict of interest.

Sample Availability: Samples of the compounds are not available from the authors.

References

1. Li, T.; Jiang, S.; Yang, Y. Database Selection and Heterogeneity-More Details, More Credibility. *JAMA Oncol.* **2018**, *4*, 1295. [CrossRef]
2. Sung, H.; Ferlay, J.; Siegel, R.L.; Laversanne, M.; Soerjomataram, I.; Jemal, A.; Bray, F. Global Cancer Statistics 2020: GLOBOCAN Estimates of Incidence and Mortality Worldwide for 36 Cancers in 185 Countries. *CA Cancer J. Clin.* **2021**, *71*, 209–249. [CrossRef]
3. Mirabello, L.; Troisi, R.J.; Savage, S.A. Osteosarcoma incidence and survival rates from 1973 to 2004: Data from the Surveillance, Epidemiology, and End Results Program. *Cancer* **2009**, *115*, 1531–1543. [CrossRef]
4. Whelan, J.S.; Jinks, R.C.; McTiernan, A.; Sydes, M.R.; Hook, J.M.; Trani, L.; Uscinska, B.; Bramwell, V.; Lewis, I.J.; Nooij, M.A.; et al. Survival from high-grade localised extremity osteosarcoma: Combined results and prognostic factors from three European Osteosarcoma Intergroup randomised controlled trials. *Ann. Oncol.* **2012**, *23*, 1607–1616. [CrossRef] [PubMed]
5. Bielack, S.S.; Kempf-Bielack, B.; Delling, G.; Exner, G.U.; Flege, S.; Helmke, K.; Kotz, R.; Salzer-Kuntschik, M.; Werner, M.; Winkelmann, W.; et al. Prognostic factors in high-grade osteosarcoma of the extremities or trunk: An analysis of 1702 patients treated on neoadjuvant cooperative osteosarcoma study group protocols. *J. Clin. Oncol.* **2002**, *20*, 776–790. [CrossRef]
6. Jawad, M.U.; Cheung, M.C.; Clarke, J.; Koniaris, L.G.; Scully, S.P. Osteosarcoma: Improvement in survival limited to high-grade patients only. *J. Cancer Res. Clin. Oncol.* **2011**, *137*, 597–607. [CrossRef] [PubMed]
7. Jiang, F.; Shi, Y.; Li, G.J.; Zhou, F. A meta-analysis of limb-salvage versus amputation in the treatment of patients with Enneking-double daggerU pathologic fracture osteosarcoma. *Indian J. Cancer* **2015**, *51* (Suppl. 2), e21–e24. [CrossRef]
8. Gill, J.; Gorlick, R. Advancing therapy for osteosarcoma. *Nat. Rev. Clin. Oncol.* **2021**, *18*, 609–624. [CrossRef]
9. Naveen Kumar, D.R.; Cijo George, V.; Suresh, P.K.; Ashok Kumar, R. Cytotoxicity, apoptosis induction and anti-metastatic potential of Oroxylum indicum in human breast cancer cells. *Asian Pac. J. Cancer Prev.* **2012**, *13*, 2729–2734. [CrossRef] [PubMed]
10. Cragg, G.M.; Newman, D.J. Plants as a source of anti-cancer agents. *J. Ethnopharmacol.* **2005**, *100*, 72–79. [CrossRef] [PubMed]
11. Yanez, J.A.; Andrews, P.K.; Davies, N.M. Methods of analysis and separation of chiral flavonoids. *J. Chromatogr. B Anal. Technol. Biomed. Life Sci.* **2007**, *848*, 159–181. [CrossRef]
12. Mandalari, G.; Bennett, R.N.; Bisignano, G.; Trombetta, D.; Saija, A.; Faulds, C.B.; Gasson, M.J.; Narbad, A. Antimicrobial activity of flavonoids extracted from bergamot (*Citrus bergamia* Risso) peel, a byproduct of the essential oil industry. *J. Appl. Microbiol.* **2007**, *103*, 2056–2064. [CrossRef] [PubMed]
13. Ferreira, R.J.; Baptista, R.; Moreno, A.; Madeira, P.G.; Khonkarn, R.; Baubichon-Cortay, H.; Dos Santos, D.J.; Falson, P.; Ferreira, M.U. Optimizing the flavanone core toward new selective nitrogen-containing modulators of ABC transporters. *Future Med. Chem.* **2018**, *10*, 725–741. [CrossRef] [PubMed]
14. Kumar, S.; Tiku, A.B. Biochemical and Molecular Mechanisms of Radioprotective Effects of Naringenin, a Phytochemical from Citrus Fruits. *J. Agric. Food Chem.* **2016**, *64*, 1676–1685. [CrossRef]
15. Kanno, S.; Tomizawa, A.; Hiura, T.; Osanai, Y.; Shouji, A.; Ujibe, M.; Ohtake, T.; Kimura, K.; Ishikawa, M. Inhibitory effects of naringenin on tumor growth in human cancer cell lines and sarcoma S-180-implanted mice. *Biol. Pharm. Bull* **2005**, *28*, 527–530. [CrossRef] [PubMed]
16. Zhao, Z.; Jin, G.; Ge, Y.; Guo, Z. Naringenin inhibits migration of breast cancer cells via inflammatory and apoptosis cell signaling pathways. *Inflammopharmacology* **2019**, *27*, 1021–1036. [CrossRef] [PubMed]
17. Song, H.M.; Park, G.H.; Eo, H.J.; Jeong, J.B. Naringenin-Mediated ATF3 Expression Contributes to Apoptosis in Human Colon Cancer. *Biomol. Ther.* **2016**, *24*, 140–146. [CrossRef]
18. Jin, C.Y.; Park, C.; Hwang, H.J.; Kim, G.Y.; Choi, B.T.; Kim, W.J.; Choi, Y.H. Naringenin up-regulates the expression of death receptor 5 and enhances TRAIL-induced apoptosis in human lung cancer A549 cells. *Mol. Nutr. Food Res.* **2011**, *55*, 300–309. [CrossRef]
19. Camuzard, O.; Santucci-Darmanin, S.; Carle, G.F.; Pierrefite-Carle, V. Role of autophagy in osteosarcoma. *J. Bone Oncol.* **2019**, *16*, 100235. [CrossRef] [PubMed]
20. Niu, J.; Yan, T.; Guo, W.; Wang, W.; Zhao, Z. Insight into the Role of Autophagy in Osteosarcoma and Its Therapeutic Implication. *Front. Oncol.* **2019**, *9*, 1232. [CrossRef]
21. Yoon, S.Y.; Kim, D.H. Alzheimer's disease genes and autophagy. *Brain Res.* **2016**, *1649*, 201–209. [CrossRef] [PubMed]
22. Liu, W.; Meng, Y.; Zong, C.; Zhang, S.; Wei, L. Autophagy and Tumorigenesis. *Adv. Exp. Med. Biol.* **2020**, *1207*, 275–299. [CrossRef]
23. Liao, Y.X.; Yu, H.Y.; Lv, J.Y.; Cai, Y.R.; Liu, F.; He, Z.M.; He, S.S. Targeting autophagy is a promising therapeutic strategy to overcome chemoresistance and reduce metastasis in osteosarcoma. *Int. J. Oncol.* **2019**, *55*, 1213–1222. [CrossRef] [PubMed]
24. Gao, L.; Loveless, J.; Shay, C.; Teng, Y. Targeting ROS-Mediated Crosstalk Between Autophagy and Apoptosis in Cancer. *Adv. Exp. Med. Biol.* **2020**, *1260*, 1–12. [CrossRef]
25. Bhandary, B.; Marahatta, A.; Kim, H.R.; Chae, H.J. An involvement of oxidative stress in endoplasmic reticulum stress and its associated diseases. *Int. J. Mol. Sci.* **2012**, *14*, 434–456. [CrossRef] [PubMed]
26. Tian, J.; Lu, Z.; Wang, Y.; Zhang, M.; Wang, X.; Tang, X.; Peng, X.; Zeng, H. Nerol triggers mitochondrial dysfunction and disruption via elevation of Ca^{2+} and ROS in Candida albicans. *Int. J. Biochem. Cell Biol.* **2017**, *85*, 114–122. [CrossRef]
27. Lu, Y.; Zhang, C.; Bucheli, P.; Wei, D. Citrus flavonoids in fruit and traditional Chinese medicinal food ingredients in China. *Plant Foods Hum. Nutr.* **2006**, *61*, 57–65. [CrossRef]
28. Miyata, M.; Takano, H.; Takahashi, K.; Sasaki, Y.F.; Yamazoe, Y. Suppression of 2-amino-1-methyl-6-phenylimidazo [4,5-*b*]pyridine-induced DNA damage in rat colon after grapefruit juice intake. *Cancer Lett.* **2002**, *183*, 17–22. [CrossRef]

29. Zobeiri, M.; Belwal, T.; Parvizi, F.; Naseri, R.; Farzaei, M.H.; Nabavi, S.F.; Sureda, A.; Nabavi, S.M. Naringenin and its Nano-formulations for Fatty Liver: Cellular Modes of Action and Clinical Perspective. *Curr. Pharm. Biotechnol.* **2018**, *19*, 196–205. [CrossRef]
30. Chandrika, B.B.; Steephan, M.; Kumar, T.R.S.; Sabu, A.; Haridas, M. Hesperetin and Naringenin sensitize HER2 positive cancer cells to death by serving as HER2 Tyrosine Kinase inhibitors. *Life Sci.* **2016**, *160*, 47–56. [CrossRef]
31. Arul, D.; Subramanian, P. Naringenin (citrus flavonone) induces growth inhibition, cell cycle arrest and apoptosis in human hepatocellular carcinoma cells. *Pathol. Oncol. Res.* **2013**, *19*, 763–770. [CrossRef]
32. Guo, S.; Pridham, K.J.; Virbasius, C.M.; He, B.; Zhang, L.; Varmark, H.; Green, M.R.; Sheng, Z. A large-scale RNA interference screen identifies genes that regulate autophagy at different stages. *Sci. Rep.* **2018**, *8*, 2822. [CrossRef] [PubMed]
33. Huang, W.P.; Klionsky, D.J. Autophagy in yeast: A review of the molecular machinery. *Cell Struct. Funct.* **2002**, *27*, 409–420. [CrossRef]
34. Hasima, N.; Ozpolat, B. Regulation of autophagy by polyphenolic compounds as a potential therapeutic strategy for cancer. *Cell Death Dis.* **2014**, *5*, e1509. [CrossRef] [PubMed]
35. Xiong, X.; Wu, M.; Zhang, H.; Li, J.; Lu, B.; Guo, Y.; Zhou, T.; Guo, H.; Peng, R.; Li, X.; et al. Atg5 siRNA inhibits autophagy and enhances norcantharidin-induced apoptosis in hepatocellular carcinoma. *Int. J. Oncol.* **2015**, *47*, 1321–1328. [CrossRef]
36. Graf, M.R.; Jia, W.; Johnson, R.S.; Dent, P.; Mitchell, C.; Loria, R.M. Autophagy and the functional roles of Atg_5 and beclin-1 in the anti-tumor effects of 3beta androstene 17alpha diol neuro-steroid on malignant glioma cells. *J. Steroid Biochem. Mol. Biol.* **2009**, *115*, 137–145. [CrossRef] [PubMed]
37. Dai, G.; Zheng, D.; Guo, W.; Yang, J.; Cheng, A.Y. Cinobufagin Induces Apoptosis in Osteosarcoma Cells Via the Mitochondria-Mediated Apoptotic Pathway. *Cell Physiol. Biochem.* **2018**, *46*, 1134–1147. [CrossRef]
38. Yu, X.; Zhou, X.; Fu, C.; Wang, Q.; Nie, T.; Zou, F.; Guo, R.; Liu, H.; Zhang, B.; Dai, M. Celastrol induces apoptosis of human osteosarcoma cells via the mitochondrial apoptotic pathway. *Oncol. Rep.* **2015**, *34*, 1129–1136. [CrossRef]
39. Sauer, H.; Wartenberg, M.; Hescheler, J. Reactive oxygen species as intracellular messengers during cell growth and differentiation. *Cell Physiol. Biochem.* **2001**, *11*, 173–186. [CrossRef]
40. Bossis, G.; Sarry, J.E.; Kifagi, C.; Ristic, M.; Saland, E.; Vergez, F.; Salem, T.; Boutzen, H.; Baik, H.; Brockly, F.; et al. The ROS/SUMO axis contributes to the response of acute myeloid leukemia cells to chemotherapeutic drugs. *Cell Rep.* **2014**, *7*, 1815–1823. [CrossRef]
41. Ivanova, D.; Zhelev, Z.; Aoki, I.; Bakalova, R.; Higashi, T. Overproduction of reactive oxygen species—Obligatory or not for induction of apoptosis by anticancer drugs. *Chin. J. Cancer Res.* **2016**, *28*, 383–396. [CrossRef]
42. Magnano, S.; Hannon Barroeta, P.; Duffy, R.; O'Sullivan, J.; Zisterer, D.M. Cisplatin induces autophagy-associated apoptosis in human oral squamous cell carcinoma (OSCC) mediated in part through reactive oxygen species. *Toxicol. Appl. Pharmacol.* **2021**, *427*, 115646. [CrossRef] [PubMed]
43. Zhang, W.; Zhu, Y.; Yu, H.; Liu, X.; Jiao, B.; Lu, X. Libertellenone H, a Natural Pimarane Diterpenoid, Inhibits Thioredoxin System and Induces ROS-Mediated Apoptosis in Human Pancreatic Cancer Cells. *Molecules* **2021**, *26*, 315. [CrossRef] [PubMed]
44. Wang, H.; Xu, Y.; Sun, J.; Sui, Z. The Novel Curcumin Derivative 1g Induces Mitochondrial and ER-Stress-Dependent Apoptosis in Colon Cancer Cells by Induction of ROS Production. *Front. Oncol.* **2021**, *11*, 644197. [CrossRef]
45. Park, H.J.; Choi, Y.J.; Lee, J.H.; Nam, M.J. Naringenin causes ASK1-induced apoptosis via reactive oxygen species in human pancreatic cancer cells. *Food Chem. Toxicol.* **2017**, *99*, 1–8. [CrossRef] [PubMed]
46. Datta, K.; Babbar, P.; Srivastava, T.; Sinha, S.; Chattopadhyay, P. p53 dependent apoptosis in glioma cell lines in response to hydrogen peroxide induced oxidative stress. *Int. J. Biochem. Cell Biol.* **2002**, *34*, 148–157. [CrossRef]
47. Zhang, Y.; He, N.; Zhou, X.; Wang, F.; Cai, H.; Huang, S.H.; Chen, X.; Hu, Z.; Jin, X. Betulinic acid induces autophagy-dependent apoptosis via Bmi-1/ROS/AMPK-mTOR-ULK1 axis in human bladder cancer cells. *Aging* **2021**, *13*, 21251–21267. [CrossRef]
48. Xipell, E.; Gonzalez-Huarriz, M.; Martinez de Irujo, J.J.; Garcia-Garzon, A.; Lang, F.F.; Jiang, H.; Fueyo, J.; Gomez-Manzano, C.; Alonso, M.M. Salinomycin induced ROS results in abortive autophagy and leads to regulated necrosis in glioblastoma. *Oncotarget* **2016**, *7*, 30626–30641. [CrossRef]
49. Yoon, M.J.; Kang, Y.J.; Kim, I.Y.; Kim, E.H.; Lee, J.A.; Lim, J.H.; Kwon, T.K.; Choi, K.S. Monensin, a polyether ionophore antibiotic, overcomes TRAIL resistance in glioma cells via endoplasmic reticulum stress, DR5 upregulation and c-FLIP downregulation. *Carcinogenesis* **2013**, *34*, 1918–1928. [CrossRef]

molecules

MDPI

Article

In Vitro Effects of Papaverine on Cell Proliferation, Reactive Oxygen Species, and Cell Cycle Progression in Cancer Cells

Daniella A. Gomes, Anna M. Joubert and Michelle H. Visagie *

Department of Physiology, Faculty of Health Sciences, School of Medicine, University of Pretoria, Private Bag X323, Gezina, Pretoria 0031, South Africa; daniella.a.d.gomes@gmail.com (D.A.G.); annie.joubert@up.ac.za (A.M.J.)

* Correspondence: michelle.visagie@up.ac.za; Tel.: +27-12-319-2245

Abstract: Papaverine (PPV) is an alkaloid isolated from the *Papaver somniferum*. Research has shown that PPV inhibits proliferation. However, several questions remain regarding the effects of PPV in tumorigenic cells. In this study, the influence of PPV was investigated on the proliferation (spectrophotometry), morphology (light microscopy), oxidative stress (fluorescent microscopy), and cell cycle progression (flow cytometry) in MDA-MB-231, A549, and DU145 cell lines. Exposure to 150 µM PPV resulted in time- and dose-dependent antiproliferative activity with reduced cell growth to 56%, 53%, and 64% in the MDA-MB-231, A549, and DU145 cell lines, respectively. Light microscopy revealed that PPV exposure increased cellular protrusions in MDA-MB-231 and A549 cells to 34% and 23%. Hydrogen peroxide production increased to 1.04-, 1.02-, and 1.44-fold in PPV-treated MDA-MB-231, A549, and DU145 cells, respectively, compared to cells propagated in growth medium. Furthermore, exposure to PPV resulted in an increase of cells in the sub-G_1 phase by 46% and endoreduplication by 10% compared to cells propagated in growth medium that presented with 2.8% cells in the sub-G_1 phase and less than 1% in endoreduplication. The results of this study contribute to understanding of effects of PPV on cancer cell lines.

Keywords: papaverine; cancer; morphology; proliferation; cell cycle

Citation: Gomes, D.A.; Joubert, A.M.; Visagie, M.H. In Vitro Effects of Papaverine on Cell Proliferation, Reactive Oxygen Species, and Cell Cycle Progression in Cancer Cells. *Molecules* **2021**, *26*, 6388. https://doi.org/10.3390/molecules26216388

Academic Editors: Višnja Stepanić and Marta Kučerová-Chlupáčová

Received: 20 August 2021
Accepted: 18 October 2021
Published: 22 October 2021

1. Introduction

Cancer is one of the leading causes of death globally, with mortality rates increasing from 9.6 million in 2018 to 10 million in 2020 [1,2]. Simultaneously, the prevalence of cancer increased from 18.1 million to 19.3 million new cases, while death rates rose to approximately 19% in females and 43% in males [1,2]. It is projected that 28.4 million new cases will occur by 2040 should the rate trajectory remain consistent with the 2020 estimates [1]. Lung and breast cancer are the most common cancers globally, with lung cancer being the most common cause of cancer-related deaths, and breast cancer the fifth most common cause of cancer-related deaths. In 2020, approximately 2.2 million individuals were diagnosed with lung cancer, resulting in 1.8 million deaths [1]. In addition, 2.3 million individuals were diagnosed with breast cancer in 2020, resulting in 684,996 deaths [1]. Furthermore, approximately 1.4 million individuals were diagnosed with prostate cancer in 2020, resulting in 375,304 deaths [1].

The use of traditional plant-based medicine has been well documented and dates as far back as 2800 BC. Moreover, in today's modern age, the use of plant-derived compounds has grown and is creating a separate industry focused on phytomedicine rather than synthetic compounds [3]. Phytomedicines account for approximately 60% of the total of anticancer agents currently in use [4]. Naturally occurring plant derived treatments have therefore become a large avenue of research to develop novel cancer treatment options with a higher therapeutic index [5–8].

Papaverine (PPV) is a naturally occurring non-narcotic alkaloid obtained from *Papaver somniferum*, commonly known as the opium poppy seed (poppies) [9]. Poppies have been used as an herbal medicine in Chinese and Indian medicine for their analgesic effects [4,9,10]. Despite being extracted from the poppy seed along with other opioids and alkaloids, the pharmacological activity of PPV does not possess any narcotic characteristics and is unrelated to the morphine classification of opioids, and it does not exert any analgesic effects [11,12]. PPV is approved by the Food and Drug Administration (FDA) of the United States of America as a vasodilator for the treatment of cerebral vasospasms and several coronary procedures, including subendocardial ischemia and erectile dysfunction [13–17]. The bioavailability of PPV is approximately 30% when taken orally [16,18–20]. Furthermore, these effects of PPV appear to be dose-dependent [19,21–23]. In addition, research studies have indicated that a 24-h and 48-h exposure to PPV at doses ranging from 0.01 to 1000 μM exhibited a dose-dependent cytotoxic effect in breast ductal-carcinoma (T47D), a triple negative breast carcinoma cell line, M.D. Anderson-Metastatic breast cancer (MDA-MB-231), an estrogen receptor positive breast carcinoma cell line, Michigan Cancer Foundation cell line 7 (MCF-7), colorectal carcinoma (HT 29), prostate carcinoma (PC-3), and fibrosarcoma (HT1080) cells. In addition, no cytotoxic effects were exhibited in non-tumorigenic human fibroblast (NHF) and mouse non-tumorigenic embryonic fibroblasts (NIH 3T3) T-cells at doses ranging from 0.01 to 1000 μM [4,9,24]. Cytotoxicity assays conducted on NIH 3T3 cells showed that exposure to high doses of PPV (100–1000 μM) reduced the percentage of cell growth to 90%. However, exposure to 0.01–1000 μM PPV in the tumorigenic cell lines, T47D, HT 29, and HT1080 resulted in a more prominent decrease in the percentage of cell growth to 20%, 30%, and 10%, respectively [9]. Additionally, research comparing the effects of PPV on PC-3 and NHF cells indicated that at a concentration of 200 μM PPV, cell viability was reduced to 10% and 90%, respectively [24]. This indicates that PPV may have a selective cytotoxicity towards tumorigenic cells while leaving non-tumorigenic cells either unaffected or less prominently affected [4,9,24].

Previous research has reported that PPV functions as a phosphodiesterase 10A (PDE10A) inhibitor which accounts for the anti-spasmodic effects observed in blood vessels when exposed to PPV [20,25]. Inhibition of PDE10A results in the increase in 3′,5′-cyclic adenosine monophosphate (cAMP) which has several downstream effects, including the alteration of the mitochondrial complex 1 [20,25–27]. Consequently, these effects may alter the production of reactive oxygen species (ROS) as the mitochondrial complex I is one of the main sources of ROS through phosphorylation by nicotinamide adenine dinucleotide hydrogen (NADH) [28]. It is therefore possible that the effects exerted by PPV may alter ROS production. Currently, there is limited research avalaible regarding the effects that PPV exerts on ROS production.

Previous studies have indicated that the naturally occurring compound, PPV, currently in clinical use for vasodilation purposes, might inhibit cell growth and potentially induces cell death in cancer cell lines. However, specific effects on biochemical pathways remain unclear. Therefore, this study investigated the effects of PPV on cell proliferation, morphology, oxidative stress, cell cycle progression, and cell death induction in a triple negative breast cancer cell line (MDA-MB-231), adenocarcinoma alveolar cancer cell line (A549), and a prostate cancer cell line (DU145). Although previous research has extensively explored the cytotoxic effects of PPV, little research has been conducted on the effects that PPV exerts in tumorigenic cell lines on morphology and oxidative stress whilst research of the effects of PPV on cell cycle progression has yielded contradicting results [4,24]. Therefore, in the present study, we conducted a cytotoxicity assay to determine the optimal dose range in the selected tumorigenic cell lines which were then implemented in further experimentation to establish the effects that PPV exerts on morphology, oxidative stress, and cell cycle progression, which can aid future anticancer studies.

The data obtained in this study aided in the understanding of PPV's antiproliferative influence on cancer cell lines. Furthermore, contributing to the existing knowledge regarding the influence of a naturally occurring compound in cancer cell lines will improve cancer researchers' understanding of phytomedicinal compounds. The present study possibly provides new insight into the repurposing of non-addictive alkaloid, since it indicates that

PPV exerts anti-proliferative activity, and induces oxidative stress and cell cycle abnormalities. The present study explores the drug repurposing of a natural vasodilator in cancer research, determining antiproliferative and anticancer effects to provide insights into the novel application of PPV, a non-addictive, non-narcotic alkaloid, which may have reduced side effects compared to current therapeutic cancer treatments. Understanding the phytomedicinal compounds and determining the benefits of these compounds in comparison to current synthetic treatments may help develop novel treatment options with reduced side effects and potentially improved survival.

2. Results

2.1. *Cell Proliferation*

Cell Number Determination Using Crystal Violet Staining (Spectrophotometry)

The crystal violet assay (spectrophotometry) results indicated that PPV exerted differential time- and concentration-dependent effects on cell growth in all three cell lines (Figure 1, Supplementary Materials 1). Exposure to PPV at 50, 100, 150, and 300 µM for 48 h in MDA-MB-231 cells resulted in a change in cell growth of 89%, 56%, 55% and 29%, respectively. In comparison, exposure to PPV at 50, 100, 150, and 300 µM for 48 h in A549 cells resulted in a change in cell growth of 76%, 61%, 53% and 32%, respectively, and exposure to PPV at 50, 100, 150, and 300 µM for 48 h in DU145 cells resulted in a change in cell growth of 80%, 80%, 64% and 31%, respectively (Figure 1).

Exposure to PPV at 50, 100, 150, and 300 µM for 72 h in MDA-MB-231 cells resulted in a change in cell growth of 69%, 56%, 48%, and 18%, respectively. In comparison, exposure to PPV at 50, 100, 150, and 300, µM for 72 h in A549 cells resulted in a change in cell growth of 97.4%, 67.0%, 36.3%, and 16.3%, respectively, and exposure to PPV at 50, 100, 150, and 300 µM for 72 h in DU145 cells resulted in a change in cell growth of 72%, 55%, 42%, and 16%, respectively (Figure 1).

These results indicate that the PPV exerts differential time- and dose-dependent effects on cell proliferation in all three cell lines. Furthermore, the data demonstrates that PPV exerts optimal antiproliferative effects that are more prominently observed in the MDA-MB-231 and A549 cell lines after 48-h and 72-h exposure compared to the DU145 cell line. Thus, for all subsequent experiments, cell lines were exposed to PPV (10 µM, 50 µM, 100 µM, and 150 µM) for 48 h and 72 h to determine the effect of PPV on the morphology, H_2O_2 production, and cell cycle and cell death induction.

2.2. *Cell Morphology*

Morphology Observation Using Light Microscopy

The effects of PPV on cell morphology were investigated using light microscopy on MDA-MB-231, A549, and DU145 cell lines at 48 h and 72 h. This study was the first to demonstrate the effects of PPV in MDA-MB-231 and A549 cell lines. Light microscopy revealed that PPV decreased cell density and increased cell debris and abnormal morphological changes in a dose- and time-dependent manner in all three cell lines (Figures 2–4, Supplementary Materials 2). Aberrant morphological observations after exposure to PPV included shrunken cells demonstrating rounded morphology, cells demonstrating rounded morphology, cells demonstrating membrane blebbing, cells demonstrating lamellipodia-like protrusions, and cells demonstrating enlarged rounded morphology and demonstrating enlarged morphology. Lamellipodia-like protrusions referring to cellular protrusions or extensions which act with the extracellular environment typically during cellular migration [29]. Cells demonstrating enlarged and some cells exhibiting enlarged and rounded morphology were only observed after 72 h in MDA-MB-231 cells. Furthermore, after exposure to 100 and 150 µM of PPV for 48 h and 72 h in A549 cells, enlarged rounded morphology and enlarged cells were observed.

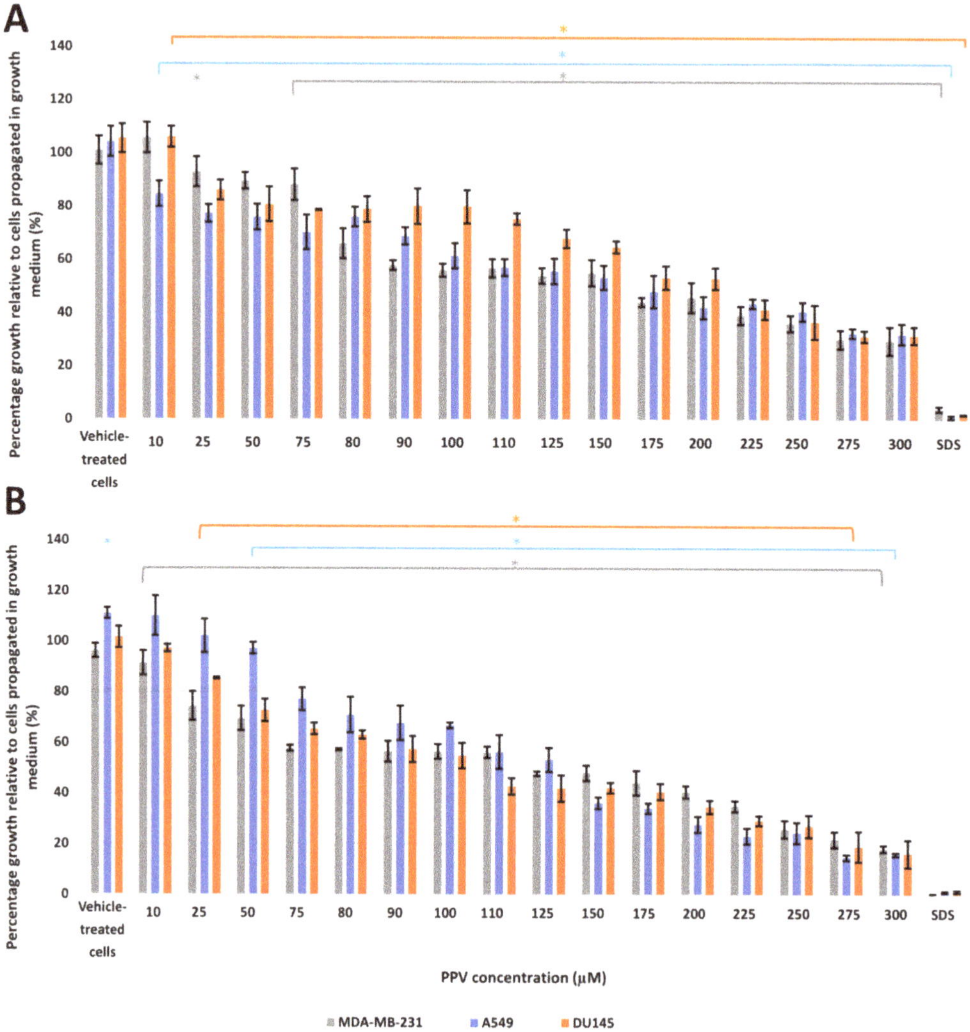

Figure 1. Spectrophotometry results of crystal violet staining demonstrating the effects of PPV (10–300 μM) on proliferation on MDA-MB-231, A549, and DU145 cell lines at 48 h (**A**) and 72 h (**B**). The average of three independent experiments is represented by the graph with error bars indicating standard deviation. The statistical significance is represented by an * when using the Student t-test with a p value of 0.05 compared to cells propagated in growth medium.

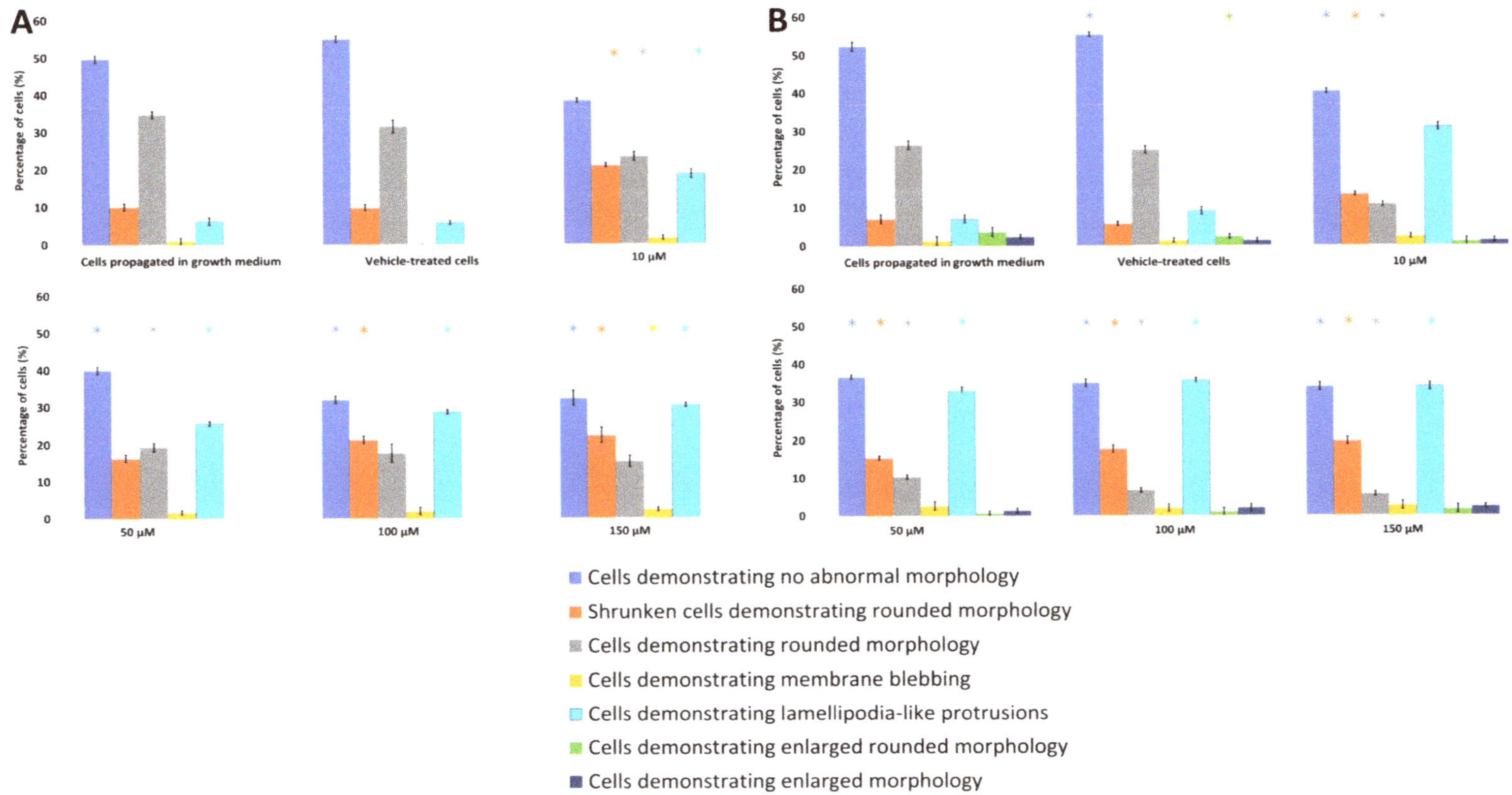

Figure 2. Light microscopy results demonstrating the effects of PPV (10–150 µM) on cell morphology on MDA-MB-231 cells at 48 h (**A**) and 72 h (**B**). Statistical significance is represented by an * when using the Student t-test with a p value of 0.05 compared to cells propagated in growth medium. Cells demonstrating enlarged and some cells exhibiting enlarged and rounded morphology were not observed after 48 h.

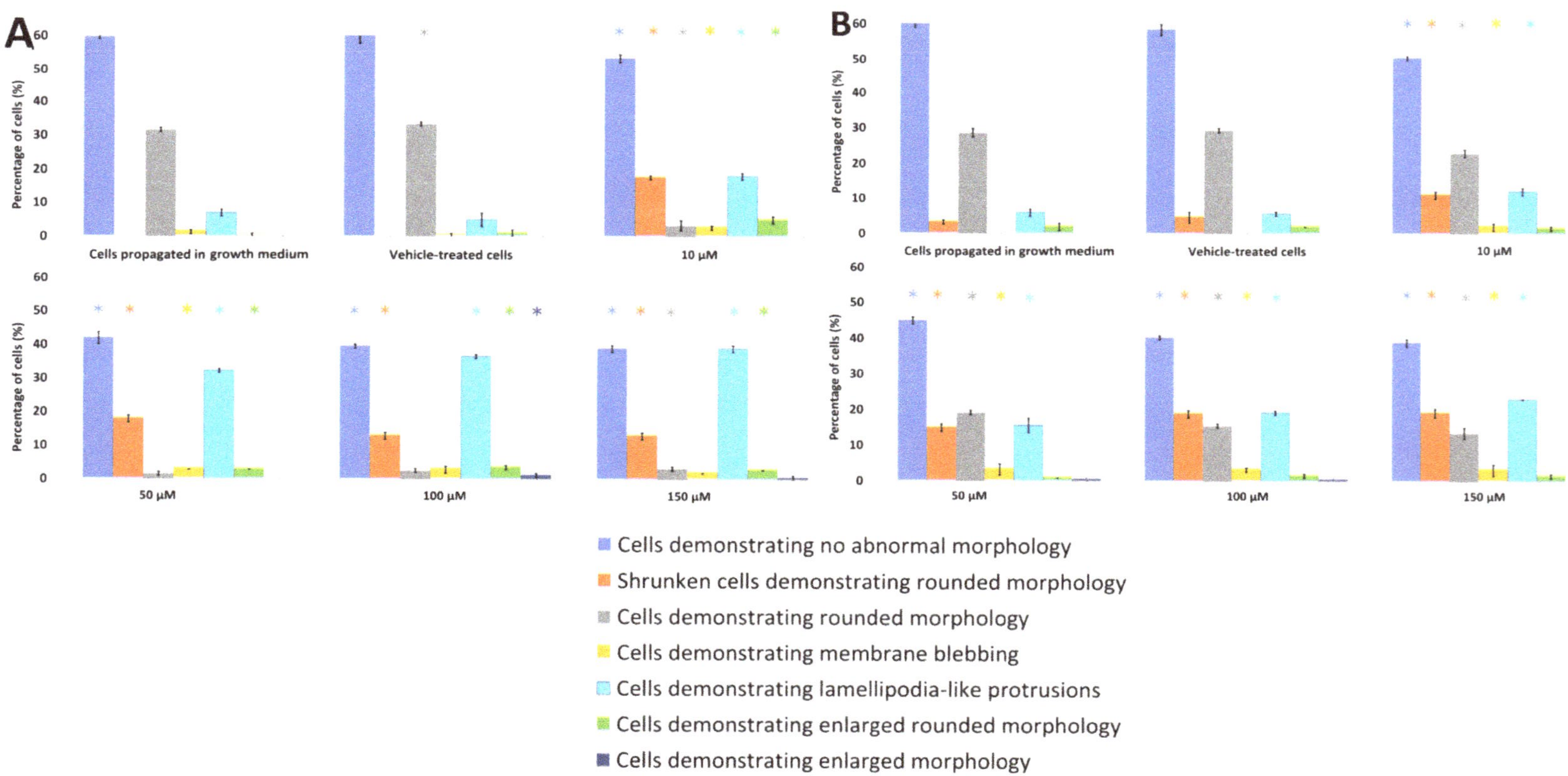

Figure 3. Light microscopy results demonstrating the effects of PPV (10–150 µM) on cell morphology on A549 cells at 48 h (**A**) and 72 h (**B**). Statistical significance is represented by an * when using the Student t-test with a p value of 0.05 compared to cells propagated in growth medium.

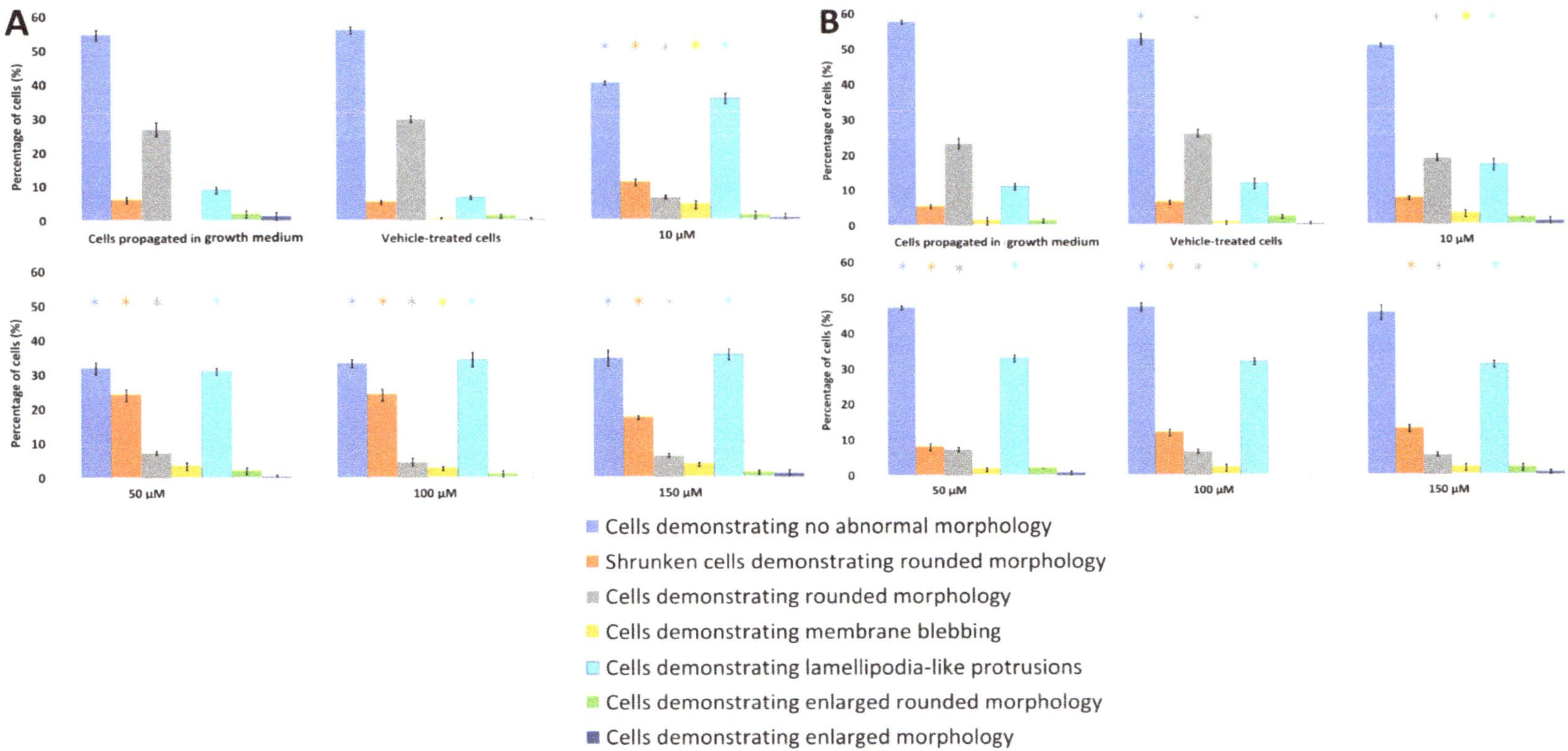

Figure 4. Light microscopy results demonstrating the effects of PPV (10–150 µM) on cell morphology on DU145 cells at 48 h (**A**) and 72 h (**B**). Statistical significance is represented by an * when using the Student t-test with a p value of 0.05 compared to cells propagated in growth medium.

Exposure to PPV at 10, 50, 100, and 150 μM for 48 h in MDA-MB-231 cells resulted in significant aberrant morphological observations (Figure 2). Cells demonstrating lamellipodia-like protrusion abnormalities increased to 18%, 25%, 28%, and 30% in 10, 50, 100, and 150 μM, respectively. Exposure to PPV at 10, 50, 100, and 150 μM for 48 h in A549 cells resulted in significant aberrant morphological observations (Figure 3). Cells demonstrating lamellipodia-like protrusion abnormalities increased to 18%, 32%, 36%, and 39% in 10, 50, 100, and 150 μM, respectively, and exposure to PPV at 10, 50, 100, and 150 μM for 48 h in DU145 cells resulted in significant aberrant morphological observations (Figure 4). Cells demonstrating lamellipodia-like protrusion abnormalities increased to 35%, 31%, 34%, and 35% in 10, 50, 100, and 150 μM, respectively (Figure 5).

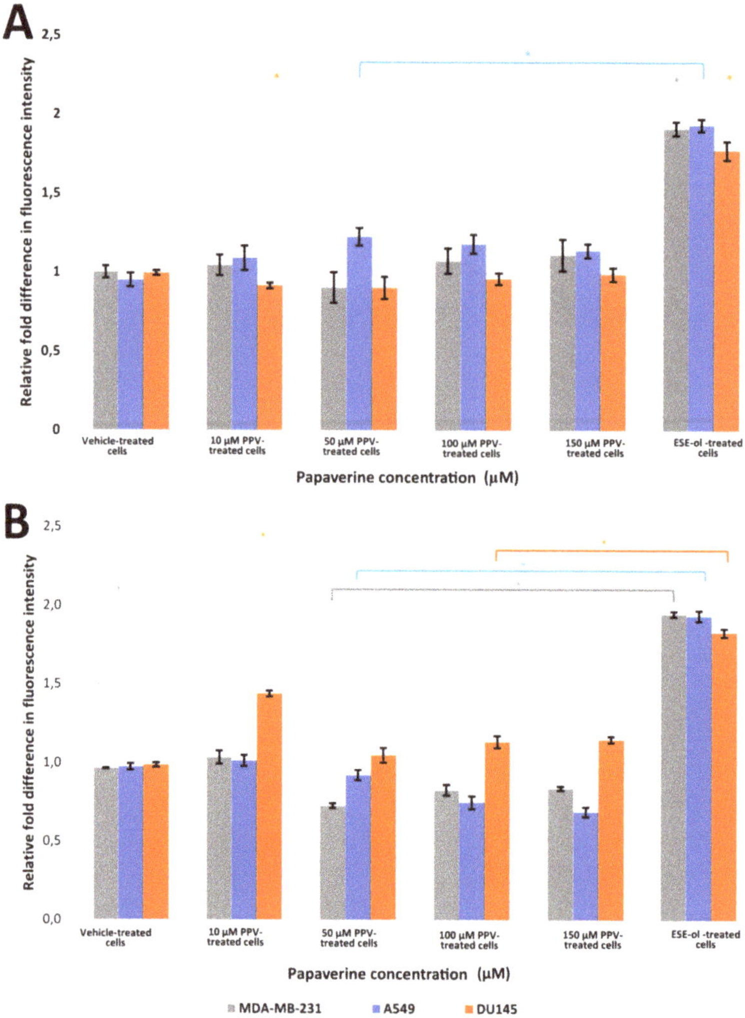

Figure 5. Fluorescence microscopy results of DCFDA staining demonstrating the effects of PPV (10–150 μM) on H_2O_2 production on MDA-MB-231, A549, and DU145 cell lines at 48 h (**A**) and 72 h (**B**). The average of three independent experiments is represented by the graph with error bars indicating standard deviation. Statistical significance is represented by an * when using the Student t-test with a *p* value of 0.05 compared to cells propagated in growth medium.

Exposure to PPV at 10, 50, 100, and 150 μM for 72 h in MDA-MB-231 cells resulted in significant aberrant morphological observations (Figure 2). Cells demonstrating lamellipodia-like protrusion abnormalities increased to 31%, 33%, 35%, and 34% in 10, 50, 100, and 150 μM, respectively. In comparison, exposure to PPV at 10, 50, 100, and 150 μM for 72 h in A549 cells resulted in significant aberrant morphological observations (Figure 3). Cells demonstrating lamellipodia-like protrusion abnormalities increased to 12%, 15%, 19%, and 23% in 10, 50, 100, and 150 μM, respectively, and exposure to PPV at 10, 50, 100, and 150 μM for 72 h in DU145 cells resulted in significant aberrant morphological observations (Figure 5). Cells demonstrating lamellipodia-like protrusion abnormalities increased to 17%, 33%, 32%, and 31% in 10, 50, 100, and 150 μM, respectively (Figure 4).

These data suggest that an increase in the concentration of PPV correlates with an increase in cells presenting with aberrant morphological manifestations such as lamellipodia-like protrusions. Furthermore, light microscopy confirmed the spectrophotometry results that PPV reduced cell growth in a time- and dose-dependent manner in MDA-MB-231, A549, and DU145 cell lines. Furthermore, studies have indicated that cells undergoing endoreduplication exhibit altered morphology including protrusions and enlarged cells [30]. It is therefore possible that some of the morphological abnormalities observed may be due to endoreduplication. However, further experimentation must be conducted to confirm these findings.

2.3. *Oxidative Stress*

Hydrogen Peroxide Production Using 2,7-Dichlorofluoresceindiacetate (DCFDA) (Fluorescent Microscopy)

The effects of PPV on hydrogen peroxide (H_2O_2) production was used as an indicator of oxidative stress. Exposure to PPV for 48 h resulted in a statistically significant increase in the fluorescent intensity in A549 cells when compared to cells propagated in growth medium (Figure 5, Supplementary Materials 3). A549 cells exposed to 10, 50, 100, and 150 μM of PPV for 48 h exhibited a fold increase to 1.09, 1.23, 1.18, and 1.14, respectively, relative to cells propagated in growth medium. However, MDA-MB-231 and DU145 cells exposed to PPV for 48 h exhibited no significant change in fluorescent intensity when compared to cells propagated in growth medium (Figure 5). Exposure to PPV for 72 h resulted in statistically significant decrease in fluorescent intensity in MDA-MB-231 and A549 cells when compared to cells propagated in growth medium (Figure 5). A549 cells exposed to 50, 100, and 150 μM of PPV for 72 h exhibited a fold decrease to 0.92, 0.75, and 0.69, respectively, relative to cells propagated in growth medium. MDA-MB-231 cells exposed to 50, 100, and 150 μM of PPV for 72 h exhibited a fold decrease to 0.73, 0.83, and 0.84 respectively relative to cells propagated in growth medium (Figure 5). A statistically significant fold increase to 1.44, 1.14, and 1.15 was observed in DU145 cells exposed to PPV for 72 h at a concentration of 10, 100, and 150 μM, respectively, relative to cells propagated in growth medium (Figure 5).

2.4. *Cell Cycle Progression and Cell Death Induction*

Cell Cycle Analysis Using Propidium Iodide Staining (Flow Cytometry)

Flow cytometry using propidium iodide staining and ethanol fixation allowed for the quantification of cell cycle distributions and cell death after exposure to PPV. The data obtained is the first to show the effects of PPV in A549 and DU145 cells on cell cycle progression. MDA-MB-231 cells exposed to 10, 50, 100, and 150 μM of PPV for 48 h exhibited a statistically significant increase of 4%, 8%, 10%, and 8% of cells occupying the sub-G_1 phase, respectively, when compared to cells propagated in growth medium (Table 1). Similarly, MDA-MB-231 cells exposed to 10, 50, 100, and 150 μM of PPV for 72 h exhibited a statistically significant increase of 9%, 5%, 9%, and 46% of cells in the sub-G_1 phase, respectively, compared to cells propagated in growth medium (Table 1, Supplementary Materials 4).

Table 1. The effects of PPV on cell cycle and cell death induction as a percentage of cells in each phase of the cell cycle on MDA-MB-231 cells at 48 h // 72 h. Statistical significance is represented by an * when using the Student *t*-test with a *p* value of 0.05 compared to cells propagated in growth medium.

	48 h // 72 h				
	sub-G$_1$	**G$_1$**	**S**	**G$_2$/M**	**Endoreduplication**
Cells propagated in growth medium	2.09 ± 0.61 // 1.65 ± 0.24	61.80 ± 2.4 2// 76.00 ± 0.80	6.90 ± 0.68 // 5.63 ± 0.48	15.77 ± 1.06 // 15.60 ± 0.70	0.41 ± 0.26 // 0.85 ± 0.13
Vehicle-treated cells	2.35 ± 0.49 // 2.19 ± 0.22	65.43 ± 2.22 // 70.77 ± 0.65 *	10.70 ± 0.60 * // 7.12 ± 1.87	19.40 ± 1.28 // 18.80 ± 0.46 *	0.48 ± 0.22 // 0.61 ± 0.22
10 µM PPV-treated cells	7.32 ± 0.72 *// 10.32 ± 0.57 *	59.10 ± 1.37 // 60.93 ± 1.01 *	7.32 ± 0.25 // 6.18 ± 0.65	17.53 ± 1.27 // 13.47 ± 0.99	1.50 ± 0.22 * // 4.28 ± 0.46 *
50 µM PPV-treated cells	9.00 ± 0.71 *// 7.12 ± 0.78 *	65.77 ± 1.53 *// 53.43 ± 1.32 *	10.16 ± 0.47 // 6.09 ± 0.68	13.60 ± 0.70 // 18.37 ± 1.00 *	1.47 ± 0.54 * // 10.83 ± 0.95 *
100 µM PPV-treated cells	7.36 ± 1.23 *// 11.25 ± 1.30 *	56.50 ± 0.85 // 54.07 ± 1.20 *	10.16 ± 0.47 * // 6.09 ± 0.68 *	21.80 ± 0.20 * // 16.37 ± 0.61	2.55 ± 1.27 * // 10.83 ± 0.95 *
150 µM PPV-treated cells	14.07 ± 0.65 *// 47.73 ± 1.46 *	60.33 ± 0.78 // 34.80 ± 1.57 *	6.71 ± 0.49 // 3.88 ± 0.32 *	16.03 ± 0.86 // 10.77 ± 0.21 *	1.47 ± 0.85 // 5.06 ± 0.53 *
ESE-ol-treated cells	28.90 ± 0.87 *// 25.87 ± 0.67 *	30.83 ± 1.07 * // 40.03 ± 0.72 *	10.47 ± 0.72 * // 8.39 ± 0.54 *	29.97 ± 0.81 * // 22.60 ± 1.39 *	0.18 ± 0.18 // 0.41 ± 0.08

Furthermore, a statistically significant increase of 1.09%, 1.06%, and 2.14% of cells in endoreduplication was observed when exposed to 10, 50, and 100 µM of PPV for 48 h, respectively, when compared to cells propagated in growth medium. Whilst a statistically significant increase of 3.4%, 10%, 10%, and 4.2% of cells in endoreduplication was observed when exposed to 10, 50, and 100 µM of PPV for 72 h, respectively, when compared to cells propagated in growth medium.

A549 cells exposed to 10, 50, 100, and 150 µM of PPV for 48 h exhibited a statistically significant increase of 4%, 8%, 10%, and 8% of cells in the sub-G$_1$ phase, respectively, when compared to cells propagated in growth medium. A549 cells exposed to 50, 100, and 150 µM of PPV for 72 h exhibited a statistically significant increase of 6%, 2%, and 6% of cells in the sub-G$_1$ phase, respectively, when compared to cells propagated in growth medium (Table 2). Additionally, a statistically significant increase of 2.3% of cells in endoreduplication was observed when exposed to PPV for 48 h when compared to cells propagated in growth medium. Furthermore, a statistically significant increase of 3%, 8%, 2%, and 3% of cells in endoreduplication was observed when exposed to 10, 50, 100, and 150 µM of PPV for 72 h, respectively, when compared to cells propagated in growth medium.

DU145 cells exposed to 10, 50, 100, and 150 µM of PPV for 48 h exhibited a statistically significant increase of 9%, 7%, 4%, and 5% of cells in the sub-G$_1$ phase, respectively, when compared to cells propagated in growth medium. DU145 cells exposed to 10, 50, 100, and 150 µM of PPV for 72 h exhibited a statistically significant increase of 13%, 10%, 14%, and 23% of cells in the sub-G$_1$ phase, respectively, when compared to cells propagated in growth medium (Table 3). Furthermore, a statistically significant increase of 1.2%, 1.6%, and 0.8% of cells in endoreduplication was observed when exposed to 10, 50, and 100 µM of PPV for 48 h, respectively, when compared to cells propagated in growth medium, whilst a statistically significant increase of 3%, 9%, 4%, and 4% of cells in endoreduplication was observed when exposed to 10, 50, 100, and 150 µM of PPV for 72 h, respectively, when compared to cells propagated in growth medium. These results confirm the findings observed in the morphology studies, as it was suggested that the abnormal morphology observed in this study is indicative of endoreduplication.

Table 2. The effects of PPV on cell cycle and cell death induction as a percentage of cells in each phase of the cell cycle on A549 cells at 48 h // 72 h. Statistical significance is represented by an * when using the Student *t*-test with a *p* value of 0.05 compared to cells propagated in growth medium.

	48 h // 72 h				
	sub-G_1	G_1	S	G_2/M	Endoreduplication
Cells propagated in growth medium	2.93 ± 0.07 // 2.87 ± 0.74	71.63 ± 0.70 // 70.87 ± 1.01	6.78 ± 0.69 // 11.27 ± 1.26	18.30 ± 0.36 // 10.65 ± 2.09	0.89 ± 0.11 // 0.08 ± 0.01
Vehicle-treated cells	3.42 ± 0.59 // 2.40 ± 0.02	70.73 ± 1.37 // 74.30 ± 1.68 *	9.04 ± 0.90 // 9.18 ± 0.72 *	17.87 ± 0.76 // 14.77 ± 1.29 *	0.59 ± 0.09 // 0.08 ± 0.01
10 µM PPV-treated cells	7.53 ± 0.10 * // 4.24 ± 2.29	57.90 ± 0.82 * // 63.00 ± 1.39 *	5.97 ± 0.77 // 6.38 ± 0.37 *	25.17 ± 0.38 * // 16.63 ± 1.39 *	3.22 ± 0.31 * // 3.03 ± 0.90 *
50 µM PPV-treated cells	11.23 ± 0.15 * // 8.45 ± 1.23 *	55.40 ± 0.66 *// 62.27 ± 1.62 *	7.78 ± 0.46 // 5.05 ± 0.61 *	23.30 ± 1.35 * // 12.03 ± 1.96	3.09 ± 0.66 * // 8.37 ± 2.07 *
100 µM PPV-treated cells	13.33 ± 0.90 * // 5.12 ± 0.35 *	60.00 ± 0.53 * // 72.30 ± 2.50	7.78 ± 0.46 // 5.05 ± 0.61 *	16.43 ± 0.64 // 13.10 ± 1.65	2.58 ± 0.28 * // 1.73 ± 0.40 *
150 µM PPV-treated cells	11.77 ± 0.32 * // 9.60 ± 0.53 *	68.77 ± 1.21 * // 66.23 ± 1.91 *	4.33 ± 0.42 * // 5.29 ± 0.78 *	14.23 ± 1.37 *// 11.80 ± 2.26 *	1.66 ± 0.74 // 3.21 ± 0.97 *
ESE-ol-treated cells	35.97 ± 1.80 * // 24.50 ± 0.75 *	22.43 ± 0.21 * // 56.20 ± 1.31 *	7.72 ± 0.54 // 8.08 ± 2.33	29.50 ± 1.35 * // 13.87 ± 1.17	0.42 ± 0.42 // 0.03 ± 0.02

Table 3. The effects of PPV on cell cycle and cell death induction as a percentage of cells in each phase of the cell cycle on DU145 cells at 48 h // 72 h. Statistical significance is represented by an * when using the Student t-test with a *p* value of 0.05 compared to cells propagated in growth medium.

	48 h // 72 h				
	sub-G_1	G_1	S	G_2/M	Endoreduplication
Cells propagated in growth medium	3.22 ± 0.91 // 1.41 ± 0.19	64.03 ± 0.96 // 76.70 ± 0.30	10.56 ± 0.82 // 5.47 ± 0.36	19.67 ± 0.78 // 16.67 ± 0.49	0.95 ± 0.03 // 0.52 ± 0.28
Vehicle-treated cells	3.16 ± 0.94 // 1.44 ± 0.07	65.90 ± 1.97 // 66.60 ± 2.10 *	11.30 ± 0.85 * // 9.80 ± 0.54 *	17.60 ± 0.89 // 18.90 ± 1.74	0.52 ± 0.36 // 0.75 ± 0.14
10 µM PPV-treated cells	12.83 ± 1.70 * // 14.30 ± 0.53 *	58.10 ± 0.26 * // 60.77 ± 2.31 *	7.47 ± 0.57 * // 9.34 ± 0.36 *	17.90 ± 0.17 * // 16.37 ± 0.93	2.15 ± 0.60 * // 3.63 ± 0.04 *
50 µM PPV-treated cells	10.91 ± 0.94 * // 10.90 ± 0.85 *	54.47 ± 0.81 * // 50.73 ± 1.26 *	7.15 ± 0.84 * // 7.35 ± 0.45 *	22.13 ± 1.43 // 20.27 ± 0.65 *	2.58 ± 0.35 * // 9.34 ± 0.51 *
100 µM PPV-treated cells	8.16 ± 0.27 * // 15.13 ± 0.32 *	63.10 ± 3.05 // 57.83 ± 0.70 *	7.15 ± 0.84 * // 7.35 ± 0.45 *	18.23 ± 1.31 * // 17.27 ± 0.78 *	1.76 ± 0.22 * // 4.57 ± 1.18 *
150 µM PPV-treated cells	8.87 ± 0.56 * // 23.97 ± 0.38 *	56.57 ± 0.55 * // 51.50 ± 0.87 *	9.62 ± 0.64 * // 5.21 ± 0.63	21.23 ± 0.91 // 15.67 ± 0.58	2.30 ± 0.82 // 4.62 ± 1.25 *
ESE-ol-treated cells	32.70 ± 1.51 * // 36.23 ± 0.81 *	27.67 ± 1.19 * // 38.03 ± 1.40 *	8.75 ± 0.62 * // 7.66 ± 0.33 *	28.53 ± 3.54 // 17.37 ± 0.76	0.52 ± 0.64 // 0.08 ± 0.05

3. Discussion

Several studies have indicated that PPV exerts antiproliferative effects in tumorigenic cell lines while leaving non-tumorigenic cell lines less affected [4,9,24]. However, there is limited literature demonstrating the influence of PPV on other cellular phenomena and signal transduction in tumorigenic cell lines. Therefore, the effects of the benzylisoquinoline alkaloid, PPV, were investigated on cell proliferation, morphology, H_2O_2 production, and cell cycle progression in MDA-MB-231, A549, and DU145 cells. The proliferation study using crystal violet staining (10–300 µM) at 24 h, 48 h, 72 h, and 96 h was implemented as a time and dose study and revealed that PPV causes time- and dose-dependent cytotoxic effects in MDA-MB-231, A459, and DU145 cells. Previous studies have indicated that exposure to PPV for 48 h reduces cell viability with a half-maximal inhibitory concentration (IC50) of more than 10 μM in MDA-MB-231, MCF7, and PC-3 cells [4,24]. Furthermore, cytotoxicity assays using the 3-(4,5-dimethylthiazol-2-yl)-2,5-diphenyltetrazolium bromide (MTT) assay revealed that cell growth was reduced to 38%, 35%, 20%, and 15% in human hepatoma (HepG-2), HT 29, T47D, and HT 1080 cells after 48 h of exposure to PPV [9]. PPV exerted cytotoxic effects in tumorigenic cell lines, indicating that these effects of PPV were cell line-dependent [9,31]. These results are supported by the current study which indicated that PPV reduced cell viability in MDA-MB-231 and A549 cells more at 48 h and 72 h when compared to DU145 cells.

Light microscopy revealed that at exposure times of 48 h and 72 h, all three cell lines were affected morphologically by PPV (10–150 μM) and indicated a reduction in cell density. An increase in aberrant morphological changes was observed that correlated with an increase in PPV concentration in all three cell lines, further supporting the suggestion that the effects exerted by PPV are dose-dependent. Additionally, a more prominent increase in abnormal morphology was present after exposure to PPV for 72 h, with more statistically significant morphological alterations seen in MDA-MB-231 cells compared to the A549 and DU145 cells. Previous research indicated that PPV exhibited no significant changes on the morphology of DU145 cells after 48 h [32]. Whilst these findings are supported by the present study, DU145 cells did exhibit more notable morphological alterations after 72 h. Furthermore, studies have indicated that cells undergoing endoreduplication exhibit aberrant morphological changes including enlarged morphology and in some cases, membrane projections; these alterations are similar to the morphological abnormalities observed in the present study [30].

A statistically significant increase in H_2O_2 production as an indicator of oxidative stress in A549 cells in comparison to MDA-MB-231 and DU145 cells was observed after 48 h exposure to PPV. Findings indicated that an increase in PPV concentration resulted in a decrease in fluorescent intensity in A549 cells; however, PPV does result in an increase in fluorescent intensity in comparison to cells propagated in growth medium only. Furthermore, exposure to PPV for 72 h indicated that an increase in PPV concentration resulted in a decrease in fluorescent intensity in all three cell lines. However, the fluorescent intensity in DU145 cells was higher than cells propagated in growth medium whilst the fluorescent intensity in MDA-MB-231 and A549 cells was lower than cells propagated in growth medium.

Previous research has suggested that PPV inhibits PDE10A; consequently, the levels of available cAMP decrease [20,25–27]. Alterations to cAMP have been shown to affect the mitochondrial complex 1 which is the start point for the electron transport chain [28,33]. Therefore, the inhibitory effect PPV exerts on PDE10A may subsequently influence H_2O_2 production via the cAMP and mitochondrial complex 1 signalling cascade [20,25–28,33]. Prior research reported that mitochondrial respiration and the mitochondrial complex 1 has been affected by PPV, implicating the inhibition of PDE10A as a potential cause [17,20]. As indicated by the DCFDA staining, ROS production is affected by PPV. It is possible that these measurable affects are a result of the inhibition of PDE10A. However, further investigation must be conducted to confirm if these effects are connected.

Cell cycle progression revealed that PPV induced a marked increase of cells in the sub-G_1 peak, variable changes in the percentage of cells in the S and G_2M phase, and a change in the percentage of cells in the endoreduplication phase when compared to cells propagated in growth medium at 48 h and 72 h. Endoreduplication has been described as a process by which cells that have undergone DNA damage continue to enter the cell cycle without dividing, resulting in polyploid cells [30]. This results in cells that can avoid programmed cell death. It has been suggested that when cells undergo endoreduplication, an initial period of inhibited cell proliferation occurs. However, subsequent to the initial cell growth inhibition, cells are still able to progress through the cell cycle and enter the S and G_2M phase before undergoing endoreduplication [30]. Cells therefore have an increase in DNA and are ultimately larger in size. This leads to a peak in the cell cycle beyond the G_2M peak [30]. Cell cycle progression after exposure to PPV showed an increase in cells undergoing endoreduplication when compared to cells propagated in growth medium and the vehicle-treated cells. The present study indicated that the effects exerted by PPV on cell cycle in all three cell lines are time- and dose-dependent. These results indicated that the effects are cell line specific, supporting previous studies [4,24].

The present study therefore contributes to the application of an existing natural vasodilator to cancer research and treatment by establishing its effects on proliferation, H_2O_2 production, and cell death induction. The results indicate time, dose, and cell line specific effects of PPV on cell proliferation, morphology, oxidative stress, and cell cycle progression. Developing an understanding of this natural herbal compound used in traditional and conventional medicine may aid in the development of novel phytomedicinal treatments which can potentially reduce the side effects observed in current treatments in cancer care and cancer research [4]. Understanding the compound's cell line specificity may aid in its use as an antiproliferative agent; however, this must be further investigated.

4. Materials and Methods

4.1. Materials

4.1.1. Cell Lines

Triple negative breast cancer (TNBC) is a subtype of breast cancer that is highly invasive and characterised by the lack of estrogen receptors (ER), progesterone receptors (PR) and does not overproduce human epidermal growth factor receptor 2 (HER2) [34,35]. M.D. Anderson-Metastasis breast cancer-231 (MDA-MB-231) is a TNBC cell line that is highly invasive and tumorigenic with limited therapeutic targets. Previous research indicated cytotoxic effects of PPV on MDA-MB-231 cells with little insight into the effects on morphology, H_2O_2 production, and limited research on cell cycle progression [4]. Type II alveolar epithelium cells are found within the lungs, despite covering a small surface area of the alveolus; there are more type II alveolar epithelium cells than type I alveolar epithelium cells, as a result, type II alveolar epithelium adenocarcinomas are typically more common [36]. The A549 cell line is an alveolar adenocarcinoma cell line that exhibits type II alveolar cell characteristics, including larger pores to allow for increased diffusion [36]. Currently, there is limited research on the effects of PPV on A549 cells with studies focusing more on cytotoxicity and mitochondrial effects than morphology and cell cycle progression [17]. Human prostate adenocarcinoma (DU145) is a metastatic prostate adenocarcinoma cell line isolated from brain lesions in a 69-year-old male in 1975 [37]. Initial cultures of this cell line did not indicate any sensitivity to hormones as cells propagated in foetal calf serum (FCS) grew at the same rate as cells propagated in bull serum [37]. This cell line is an androgen receptor (AR) negative cell line that does not express prostate specific antigen (PSA) [37,38]. Currently, most research on the effects of PPV on prostate cancers has focused on PC-3 cells with few studies exploring the effects of PPV in DU145 cells [24,32].

MDA-MB-231, A549, and DU145 cells were obtained from the American Type Culture Collection (Manassas, Virginia, United States of America). Cells were maintained in Dulbecco's Modified Eagle Growth medium (DMEM) containing 5 mM L-glutamine, 4 mM sodium pyruvate, 3 g/L glucose, 10% heat-inactivated FCS (56 °C, 30 min), 100 U/mL penicillin G, 100 mg/mL streptomycin and fungizone (250 mg/l) at 37 °C and 5% CO_2 in a humidified atmosphere in 75-cm^2 tissue flasks.

4.1.2. Chemicals and Materials

All reagents and chemicals were purchased from Sigma Chemical Co. (St. Louis, MO, USA) and all plasticware were purchased from Lasec® SA (Pty) Ltd. (Johannesburg, Gauteng) and supplied by Cellstar®, (Greiner, Germany) unless otherwise specified. PPV was purchased from Merck (Darmstadt, Germany) and was dissolved in dimethyl sulfoxide (DMSO) to a concentration of 50 mM. Appropriate controls were used including a negative control where cells were propagated in complete growth media and a vehicle-treated control (DMSO) where cells were exposed to equal amounts of the vehicle solvent solution as in PPV-treated cells, where the *v/v*% of DMSO did not exceed 0.35%.

4.2. Methods

4.2.1. Cell Proliferation

Cell Number Determination Using Crystal Violet Staining (Spectrophotometry)

The crystal violet staining technique involves a powdered triphenylmethane cation dye which binds to the deoxyribonucleic acid (DNA) of proliferating cells allowing for the rapid quantification of proliferating cells in a monolayer [39]. The intensity of the colour of the dye correlates with cell numbers which will be quantified as absorbance by means of a spectrophotometer at a wavelength of 570 nm [40]. Therefore, the effects of PPV on cell viability were determined by crystal violet staining on MDA-MB-231, A549, and DU145 cell lines.

Cells were seeded in a sterile 96-well culture plate at a cell density of 5000 cells per well prior to incubation at 37 °C and 5% CO_2 in a humified atmosphere for 24 h to allow for cell attachment. Subsequently, cells were exposed to PPV (10–300 μm) for 24 h, 48 h, 72 h, or 96 h since previous studies have indicated optimal antiproliferative activity within this concentration range after exposure for similar periods of time [9,17,18,24]. Negative controls for this experiment included cells propagated in complete growth medium and vehicle-treated cells. Positive controls included cells exposed to 50% sodium lauryl sulphate (SDS) for 48 h since previous studies indicated that SDS induces a significant decrease in cell numbers and cell proliferation [41]. Subsequently, growth medium and PPV was discarded, and cells were fixed with 1% glutaraldehyde (100 μL) purchased from Merk (Darmstadt, Germany) before incubation for 15 min at room temperature. Glutaraldehyde was removed, and cells were stained using 0.1% crystal violet solution (100 μL) purchased from Merk (Darmstadt, Germany) and incubated at room temperature for 30 min. Afterwards, the crystal violet solution was discarded, and the 96-well plate was submersed under running water for 15 min [42]. The plate was then left to dry for 24 h and 0.2% Triton X-100 (200 μL) was added to solubilise the crystal violet stain at room temperature for 30 min [42]. The absorbance was then read at 570 nm using an EPOCH Microplate Reader (Biotek Instruments, Inc. (Winooski, Vermont, United States of America)) [42]. The data obtained were analysed using Microsoft Excel 2016 (Microsoft corporation, Washington, United States of America).

4.2.2. Cell Morphology

Morphology Observation Using Light Microscopy

Light microscopy was used to evaluate and visualise the effects of PPV on MDA-MB-231, A549, and DU145 cells which were seeded into 24-well culture plates, at a cell density of 20,000 cells per well. The cells were incubated at 37 °C and 5% CO_2 in a humified atmosphere for 24 h to allow for attachment. Subsequently, cells were exposed to PPV (10–100 μM) for 48 h or 72 h since previous research showed optimal activity in cancer cell lines [9,17,18,24]. The morphology of at least 100 cells was examined per condition in each experiment to quantify morphology. Aberrant morphological observations after exposure to PPV included shrunken cells, rounded cells, membrane blebbing, cells with lamellipodia-like protrusions, and cells revealing enlarged rounded morphology. An Axiovert 40 CFL microscope (Zeiss, Oberkochen, Germany) was used to capture images. Negative controls for this experiment included cells propagated in complete growth medium and vehicle-treated cells. Positive controls included cells exposed to 0.4 μM 2-Ethyl-17-hydroxy-13-methyl-7,8,9,11,12,13,14,15,16,17-decahydro-6-cyclopenta[a]phenanthren-3-yl sulphamate (ESE-ol) for 48 h since previous studies indicated that ESE-ol induces significant changes in cell morphology [43,44].

4.2.3. Oxidative Stress

Hydrogen Peroxide Production Using 2,7 Dichlorofluoresceindiacetate (DCFDA) (Fluorescent Microscopy)

The effects of PPV on hydrogen peroxide (H_2O_2) production was used as an indicator of oxidative stress. A non-fluorescent probe, 2,7 dichlorofluoresceindiacetate (DCFDA), is oxidised by reactive oxygen species (ROS) to a fluorescent derivative, 2,7-dichlorofluorescein (DCF). Thus, DCFDA was used in this study as an indicator of oxidative stress and the effect of PPV on hydrogen peroxide production through detection of DCF using fluorescent microscopy with a maximum excitation and emission spectra of 495 nm and 529 nm, respectively [45].

MDA-MB-231, A549, and DU145 cells were seeded into 24-well culture plates at a density of 20,000 cells per well. The cells were incubated at 37 °C and 5% CO_2 in a humified atmosphere for 24 h to allow for cell attachment. Subsequently, cells were exposed to PPV (10–150 μM) for 48 h or 72 h since previous research showed optimal activity in cancer cell lines [9,17,18,24]. Negative controls for this experiment included cells propagated in complete growth medium and vehicle-treated cells. Positive controls included cells exposed to 0.4 μM ESE-ol since previous studies have shown a significant increase in hydrogen peroxide production after exposure to ESE-ol [44]. Subsequently, cells were washed with phosphate buffer solution (PBS) before incubation with DCFDA (20 μM) for 25 min at 37 °C and 5% CO_2 in a humified atmosphere. The wells were washed with PBS (0.5 mL) and PBS (500 μL) was added to each well. A Zeiss Axiovert CFL40 microscope, Zeiss Axiovert MRm monochrome camera (Zeiss, Oberkochen, Germany) and Zeiss filter 9 was operated to capture images of DCFDA-stained (green) cells. Fluorescence images were analysed using ImageJ software developed by the National Institutes of Health (Bethesda, Maryland, United States of America). The fluorescent intensity of at least 100 cells was evaluated per condition in each experiment using ImageJ software.

4.3. *Cell Cycle Progression and Cell Death Induction*

Cell Cycle Analysis Using Propidium Iodide Staining (Flow Cytometry)

The effects of PPV on cell cycle progression was evaluated using flow cytometry. Propidium iodide (PI) is used to stain DNA in order to quantify DNA correlated to each phase of the cell cycle (sub-G_1, G_1, S, G_2M and endoreduplication) [46].

MDA-MB-231, A549, and DU145 cells were seeded into T25 cm^2 culture flask at a density of 1,000,000 cells per flask. Thereafter, the flasks were incubated at 37 °C and 5% CO_2 in a humified atmosphere for 24 h to allow for attachment. Subsequently, cells were exposed to PPV (10–150 μM) for 48 h or 72 h since previous research showed optimal activity in cancer cell lines [9,17,18,24]. Negative controls for this experiment included cells propagated in complete growth medium and vehicle-treated cells. Positive controls included cells exposed to 0.4 μM ESE-ol for 48 h since previous studies indicated that ESE-ol induces significant cell death as indicated by a sub-G_1 peak [44]. Cells were then trypsinised and resuspended in 1 mL of complete growth medium [47]. Thereafter, samples were centrifugated for 5 min at 300× *g*, the supernatant was removed and the pellet of each sample was resuspended in 1 mL of ice-cold PBS containing 0.1% FCS [47]. Ice-cold ethanol (70%, 4 mL) was then added in a dropwise manner after which samples were stored at 4 °C for at least 24 h [47,48]. Samples were centrifuged for 5 min at 300× *g*; the supernatant was discarded and the pellet then resuspended in 1 mL PBS containing 40 μg/mL of PI, 100 μg/mL RNAse A and 0.1% triton X-100 [47]. Subsequently, samples were incubated at 37 °C and 5% CO_2 in a humified atmosphere for 45 min. Propidium iodide fluorescence was measured with the cytoFLEX flow cytometer (Beckman Coulter, Inc. (Brea, California, United States of America)) available from the Institute for Cellular and Molecular Medicine (ICMM), University of Pretoria, South Africa. At least 10,000 events in each sample and data was analysed. Data from cell debris and aggregated cells was excluded from analyses [47]. Cell cycle distributions were calculated using FlowJo™ Software Version 10 (Becton, Dickinson and Company, 2019 (Ashland, Oregon, United

States of America)) by assigning relative DNA content per cell to sub-G_1, G_1, S, and G_2M phases [47]. As propidium iodide emits light at 617 nm, the data collected from the log forward detector number 3 were represented on the histograms derived on the *x*-axis [47].

4.4. Statistical Analysis

Three independent experiments were conducted for all techniques performed, where the mean and the standard deviation were calculated. Means are illustrated by using bar charts and standard deviations are shown with errors bars. A p-value < 0.05 calculated by means of the Student t-test was used for statistical significance and is indicated by an asterisk (*) using the Jamovi statistical software version 1.6 (The Jamovi project (2021) (Sydney, Australia). The fluorescent intensity of at least 100 cells was evaluated per condition using Image J software developed by the National Institutes of Health (Bethesda, Maryland, United States of America). Flow cytometry analysis involved at least 10,000 events in each sample and the data were analysed using FlowJo™ Software Version 10 (Becton, Dickinson, and Company, 2019 (Ashland, Oregon, United States of America)).

5. Conclusions

This study demonstrated that PPV exerts antiproliferative effects in a time- and dose-dependent manner in MDA-MB-231, A549, and DU145 cells. An increase in aberrant morphological changes including lamellipodia-like protrusions was observed in all three cell lines. H_2O_2 production increased in A549 cells at 48 h and in MDA-MB-231 and A549 at 72 h. Cell cycle analysis revealed that PPV exerted a cell line specific and time- and dose-dependent effect that increased the percentage of cells in the sub-G_1 and endoreduplication peaks. Understanding these cell line specific effects will aid in the development of this compound and potential derivatives of this compound as an antiproliferative agent in cancer research. Future studies will involve further investigation into the molecular mechanism of PPV to clarify how PPV exerts these effects.

Supplementary Materials: The following are available online at www.mdpi.com/1420-3049/26/21/6388, Supplementary 1; Figure S1: Spectrophotometry results of crystal violet staining demonstrating the effects of PPV (10–300 μM) on proliferation on MDA-MB-231 cells compared to A549- and DU145 cell lines at 24 h, Figure S2: Spectrophotometry results of crystal violet staining demonstrating the effects of PPV (10–300 μM) on proliferation on MDA-MB-231 cells compared to A549- and DU145 cell lines at 96 h. Supplementary 2; Figure S1: Light microscopy images of cell morphology demonstrating the effects of PPV ((10–150 μM) on cell morphology on MDA-MB-231 cells at 48 h at a magnification of ×10. Table S1: table displaying the effects of papaverine on morphology as percentage change when compared to cells propagated in growth medium on MDA-MB-231 at 48 h. Figure S2: Light microscopy images of cell morphology demonstrating the effects of PPV ((10–150 μM) on cell morphology on A549 cells at 48 h at a magnification of ×10. Table S2: table displaying the effects of papaverine on morphology as percentage change when compared to cells propagated in growth medium on A549 at 48 h. Figure S3: Light microscopy images of cell morphology demonstrating the effects of PPV ((10–150 μM) on cell morphology on DU145 cells at 48 h at a magnification of ×10 Table S3: table displaying the effects of papaverine on morphology as percentage change when compared to cells propagated in growth medium on DU145 at 48 h. Figure S4: Light microscopy images of cell morphology demonstrating the effects of PPV ((10–150 μM) on cell morphology on MDA-MB-231 cells at 72 h at a magnification of x10 Table S4: table displaying the effects of papaverine on morphology as percentage change when compared to cells propagated in growth medium on MDA-MB-231 at 72 h. Figure S5: Light microscopy images of cell morphology demonstrating the effects of PPV ((10–150 μM) on cell morphology on A549 cells at 72 h at a magnification of ×10 Table S5: table displaying the effects of papaverine on morphology as percentage change when compared to cells propagated in growth medium on A549 at 72 h. Figure S6: Light microscopy images of cell morphology demonstrating the effects of PPV ((10–150 μM) on cell morphology on DU145 cells at 72 h at a magnification of ×10 Table S6: table displaying the effects of papaverine on morphology as percentage change when compared to cells propagated in growth medium on DU145 at 72 h. Figure S7. Light microscopy results demonstrating the effects of ESE-ol used as a positive control on cell morphology. Supplementary 3; Table S1. table displaying the effects of papaverine on oxidative stress as a change of fluorescence intensity relative to the fluorescence intensity of cells propagated in growth medium on MDA-MB-231-, A549- and DU145 cell lines at 48 h. Figure S1. Fluorescence staining showing H_2O_2 production in MDA-MB-231 cells after 48 h. Figure S2. Fluorescence staining showing H_2O_2 production in A549 cells after 48 h. Figure S3. Fluorescence staining showing H_2O_2 production in MDA-MB-231 cells after 48 h. Table S2. table displaying the effects of papaverine on oxidative stress as a change of fluorescence intensity relative to the fluorescence intensity of cells propagated in growth medium on MDA-MB-231-, A549- and DU145 cell lines at 72 h. Figure S4. Fluorescence staining showing H_2O_2 production in MDA-MB-231 cells after 72 h. Figure S5. Fluorescence staining showing H_2O_2 production in A549 cells after 72 h. Figure S6. Fluorescence staining showing H_2O_2 production in MDA-MB-231 cells after 72 h. Supplementary 4; Figure S1. Flow cytometry results demonstrating the effects of PPV (10–150 μM) on the cell cycle on MDA-MB-231-, A549- and DU145 cells at 48 h. Figure S2. Cell cycle progression of MDA-MB-231 cells treated with PPV (10–150 μM) at 48 h. Figure S3. Cell cycle progression of A549 cells treated with PPV (10–150 μM) at 48 h. Figure S4. Cell cycle progression of DU145 cells treated with PPV (10–150 μM) at 48 h. Figure S5. Flow cytometry results demonstrating the effects of PPV (10–150 μM) on the cell cycle on MDA-MB-231-, A549- and DU145 cells at 72 h. Figure S6. Cell cycle progression of MDA-MB-231 cells treated with PPV (10–150 μM) at 72 h. Figure S7. Cell cycle progression of A549 cells treated with PPV (10–150 μM) at 72 h. Figure S8. Cell cycle progression of DU145 cells treated with PPV (10–150 μM) at 72 h.

Author Contributions: Conceptualization, D.A.G. and M.H.V.; Methodology, D.A.G. and M.H.V.; Software, D.A.G.; Validation, D.A.G., M.H.V. and A.M.J.; Formal Analysis, D.A.G. and M.H.V.; Investigation, D.A.G., M.H.V. and A.M.J.; Resources, M.H.V. and A.M.J.; Data Curation, D.A.G. and M.H.V.; Writing—Original Draft Preparation, D.A.G.; Writing—Review & Editing, D.A.G., M.H.V. and A.M.J.; Visualization, D.A.G. and M.H.V.; Supervision, M.H.V. and A.M.J.; Project Administration, M.H.V. and A.M.J.; Funding Acquisition, M.H.V. and A.M.J. All authors have read and agreed to the published version of the manuscript.

Funding: This research was funded by grants from the Cancer Association of South Africa and Medical Research Council awarded to A.M. Joubert from the Department of Physiology. This research was also funded by grants received from the Struwig Germeshuysen Trust, School of Medicine Research Committee of the University of Pretoria and the South African National Research Foundation provided by A.M. Joubert and M.H. Visagie from the Department of Physiology.

Institutional Review Board Statement: The study was conducted according to the guidelines of the Declaration of Helsinki, and approved by the the Research Ethics Committee of the Faculty of Health Sciences (University of Pretoria) (398/2020) (21/06/2021).

Informed Consent Statement: Not applicable.

Data Availability Statement: Data is contained within the article and Supplementary Material.

Acknowledgments: Statistical considerations and support were provided by P.J. Becker from the Research Office of the Health Sciences Faculty of the University of Pretoria, Pretoria, South Africa. Access to the CytoFLEX flow cytometer (Beckman Coulter, Inc. (Brea, California, United States of America)) was provided by Institute for Cellular & Molecular Medicine (ICMM), University of Pretoria, Pretoria, South Africa. In addition, we are thankful for the English editing services provided to us for this manuscript by Abe Kasonga (Department of physiology, Faculty of Health Sciences, University of Pretoria, Pretoria, South Africa).

Conflicts of Interest: The authors declare no conflict of interest.

Sample Availability: Samples of the compounds are not available from the authors.

References

1. Sung, H.; Ferlay, J.; Siegel, R.L.; Laversanne, M.; Soerjomataram, I.; Jemal, A.; Bray, F. Global cancer statistics 2020: GLOBOCAN estimates of incidence and mortality worldwide for 36 cancers in 185 countries. *CA Cancer J. Clin.* **2021**, *71*, 209–249.
2. Bray, F.; Ferlay, J.; Soerjomataram, I.; Siegel, R.L.; Torre, L.A.; Jemal, A. Global cancer statistics 2018: GLOBOCAN estimates of incidence and mortality worldwide for 36 cancers in 185 countries. *CA Cancer J. Clin.* **2018**, *68*, 394–424. [CrossRef] [PubMed]
3. Mukeshwar, P.; Debnath, M.; Gupta, S.; Chikara, S.K. Phytomedicine: An ancient approach turning into future potential source of therapeutics. *J. Pharmacognosy Phytother.* **2011**, *3*, 27–37.

4. Sajadian, S.; Vatankhah, M.; Majdzadeh, M.; Kouhsari, S.M.; Ghahremani, M.H. Cell cycle arrest and apoptogenic properties of opium alkaloids noscapine and papaverine on breast cancer stem cells. *Toxicol. Mech. Methods* **2015**, *25*, 388–395. [CrossRef]
5. Kingham, T.P.; Alatise, O.I.; Vanderpuye, V.; Casper, C.; Abantanga, F.A.; Kamara, T.B.; Olopade, O.I.; Habeebu, M.; Abdulkareem, F.B.; Denny, L. Treatment of cancer in sub-Saharan Africa. *Lancet Oncol.* **2013**, *14*, e158–e167. [CrossRef]
6. Balunas, M.J.; Kinghorn, A.D. Drug discovery from medicinal plants. *Life Sci.* **2005**, *78*, 431–441. [CrossRef]
7. Singh, R. Medicinal plants: A review. *J. Plant. Sci.* **2015**, *3*, 50–55.
8. Rehman, J.u.; Zahra; Ahmad, N.; Khalid, M.; Noor ul Huda Khan Asghar, H.; Gilani, Z.A.; Ullah, I.; Nasar, G.; Akhtar, M.M.; Usmani, M.N. Intensity modulated radiation therapy: A review of current practice and future outlooks. *J. Radiat. Res. Appl. Sci.* **2018**, *11*, 361–367. [CrossRef]
9. Afzali, M.; Ghaeli, P.; Khanavi, M.; Parsa, M.; Montazeri, H.; Ghahremani, M.H.; Ostad, S.N. Non-addictive opium alkaloids selectively induce apoptosis in cancer cells compared to normal cells. *DARU J. Pharm Sci.* **2015**, *23*, 16. [CrossRef]
10. Gümüşçü, A.; Arslan, N.; Sarıhan, E.O. Evaluation of selected poppy (Papaver somniferum L.) lines by their morphine and other alkaloids contents. *Eur. Food Res. Technol.* **2008**, *226*, 1213–1220. [CrossRef]
11. Lawrence, P.F. Chapter 78-Pharmacologic Adjuncts to Endovascular Procedures. In *Endovascular Surgery*, 4th ed.; Moore, W.S., Ahn, S.S., Eds.; W.B. Saunders: Philadelphia, PA, USA, 2011; pp. 807–813.
12. Vardanyan, R.S.; Hruby, V.J. 19-Antianginal Drugs. In *Synthesis of Essential Drugs*; Vardanyan, R.S., Hruby, V.J., Eds.; Elsevier: Amsterdam, The Netherlands, 2006; pp. 257–267.
13. Kassell, N.F.; Helm, G.; Simmons, N.; Phillips, C.D.; Cail, W.S. Treatment of cerebral vasospasm with intra-arterial papaverine. *J. Neurosurg.* **1992**, *77*, 848–852. [CrossRef]
14. Wilson, R.F.; White, C.W. Intracoronary papaverine: An ideal coronary vasodilator for studies of the coronary circulation in conscious humans. *Circulation* **1986**, *73*, 444–451. [CrossRef]
15. Virag, R.; Frydman, D.; Legman, M.; Virag, H. Intracavernous Injection of Papaverine as a Diagnostic and Therapeutic Method in Erectile Failure. *Angiology* **1984**, *35*, 79–87. [CrossRef]
16. Clouston, J.E.; Numaguchi, Y.; Zoarski, G.H.; Aldrich, E.F.; Simard, J.M.; Zitnay, K.M. Intraarterial papaverine infusion for cerebral vasospasm after subarachnoid hemorrhage. *Am. J. Neuroradiol.* **1995**, *16*, 27–38.
17. Benej, M.; Hong, X.; Vibhute, S.; Scott, S.; Wu, J.; Graves, E.; Le, Q.-T.; Koong, A.C.; Giaccia, A.J.; Yu, B. Papaverine and its derivatives radiosensitize solid tumors by inhibiting mitochondrial metabolism. *Proc. Natl. Acad. Sci. USA* **2018**, *115*, 10756–10761. [CrossRef]
18. Inada, M.; Shindo, M.; Kobayashi, K.; Sato, A.; Yamamoto, Y.; Akasaki, Y.; Ichimura, K.; Tanuma, S.-I. Anticancer effects of a non-narcotic opium alkaloid medicine, papaverine, in human glioblastoma cells. *PLoS ONE* **2019**, *14*, e0216358. [CrossRef]
19. Berg, G.; Jonsson, K.A.; Hammar, M.; Norlander, B. Variable bioavailability of papaverine. *Pharmacol. Toxicol.* **1988**, *62*, 308–310. [CrossRef]
20. Pöch, G.; Kukovetz, W. Papaverine-induced inhibition of phosphodiesterase activity in various mammalian tissues. *Life Sci.* **1971**, *10*, 133–144. [CrossRef]
21. Hodgson, E. Chapter Fourteen-Toxins and Venoms. In *Progress in Molecular Biology and Translational Science*; Hodgson, E., Ed.; Academic Press: Cambridge, MA, USA, 2012; Volume 112, pp. 373–415.
22. Vodušek, D.B.; Aminoff, M.J. Chapter 30-Sexual Dysfunction in Patients with Neurologic Disorders. In *Aminoff's Neurology and General Medicine*, 5th ed.; Aminoff, M.J., Josephson, S.A., Eds.; Academic Press: Boston, MA, USA, 2014; pp. 633–656.
23. Meyer, M.C.; Gollamudi, R.; Straughn, A.B. The influence of dosage form on papaverine bioavailability. *J.Clin. Pharm.* **1979**, *19*, 435–444. [CrossRef]
24. Huang, H.; Li, L.-J.; Zhang, H.-B.; Wei, A.-Y. Papaverine selectively inhibits human prostate cancer cell (PC-3) growth by inducing mitochondrial mediated apoptosis, cell cycle arrest and downregulation of NF-κB/PI3K/Akt signalling pathway. *J. BUON* **2017**, *22*, 112–118.
25. Triner, L.; Vulliemoz, Y.; Schwartz, I.; Nahas, G.G. Cyclic phosphodiesterase activity and the action of papaverine. *Biochem. Biophys. Res. Comm.* **1970**, *40*, 64–69. [CrossRef]
26. Hebb, A.L.O.; Robertson, H.A.; Denovan-Wright, E.M. Phosphodiesterase 10A inhibition is associated with locomotor and cognitive deficits and increased anxiety in mice. *Eur. Neuropsychopharm.* **2008**, *18*, 339–363. [CrossRef] [PubMed]
27. Fujishige, K.; Kotera, J.; Michibata, H.; Yuasa, K.; Takebayashi, S.-i.; Okumura, K.; Omori, K. Cloning and characterization of a novel human phosphodiesterase that hydrolyzes both cAMP and cGMP (PDE10A). *J. Biol. Chem.* **1999**, *274*, 18438–18445. [CrossRef] [PubMed]
28. Lenaz, G.; Fato, R.; Genova, M.L.; Bergamini, C.; Bianchi, C.; Biondi, A. Mitochondrial Complex I: Structural and functional aspects. *Biochimica Biophysica Acta (BBA)-Bioenerg.* **2006**, *1757*, 1406–1420. [CrossRef] [PubMed]
29. Olson, H.M.; Nechiporuk, A.V. Lamellipodia-like protrusions and focal adhesions contribute to collective cell migration in zebrafish. *Dev. Biol.* **2021**, *469*, 125–134. [CrossRef]
30. Puig, P.E.; Guilly, M.N.; Bouchot, A.; Droin, N.; Cathelin, D.; Bouyer, F.; Favier, L.; Ghiringhelli, F.; Kroemer, G.; Solary, E. Tumor cells can escape DNA-damaging cisplatin through DNA endoreduplication and reversible polyploidy. *Cell Biol. Int.* **2008**, *32*, 1031–1043. [CrossRef]
31. Noureini, S.; Wink, M. Antiproliferative effect of the isoquinoline alkaloid papaverine in hepatocarcinoma HepG-2 cells—Inhibition of telomerase and induction of senescence. *Molecules* **2014**, *19*, 11846–11859. [CrossRef]

32. Goto, T.; Matsushima, H.; Kasuya, Y.; Hosaka, Y.; Kitamura, T.; Kawabe, K.; Hida, A.; Ohta, Y.; Simizu, T.; Takeda, K. The effect of papaverine on morphologic differentiation, proliferation and invasive potential of human prostatic cancer LNCaP cells. *Int. J. Urol.* **1999**, *6*, 314–319. [CrossRef]
33. Valsecchi, F.; Ramos-Espiritu, L.S.; Buck, J.; Levin, L.R.; Manfredi, G. cAMP and Mitochondria. *Physiology* **2013**, *28*, 199–209. [CrossRef]
34. Synnott, N.C.; Murray, A.; McGowan, P.M.; Kiely, M.; Kiely, P.A.; O'Donovan, N.; O'Connor, D.P.; Gallagher, W.M.; Crown, J.; Duffy, M.J. Mutant p53: A novel target for the treatment of patients with triple-negative breast cancer? *Int. J. Cancer* **2017**, *140*, 234–246. [CrossRef]
35. Tsai, C.-H.; Yang, C.-W.; Wang, J.-Y.; Tsai, Y.-F.; Tseng, L.-M.; King, K.-L.; Chen, W.-S.; Chiu, J.-H.; Shyr, Y.-M. Timosaponin AIII Suppresses Hepatocyte Growth Factor-Induced Invasive Activity through Sustained ERK Activation in Breast Cancer MDA-MB-231 Cells. *Evid. Based Complement. Alternat. Med.* **2013**, *2013*, 10. [CrossRef]
36. Foster, K.A.; Oster, C.G.; Mayer, M.M.; Avery, M.L.; Audus, K.L. Characterization of the A549 Cell Line as a Type II Pulmonary Epithelial Cell Model for Drug Metabolism. *Exp. Cell Res.* **1998**, *243*, 359–366. [CrossRef]
37. Stone, K.R.; Mickey, D.D.; Wunderli, H.; Mickey, G.H.; Paulson, D.F. Isolation of a human prostate carcinoma cell line (DU 145). *Int. J. Cancer* **1978**, *21*, 274–281. [CrossRef]
38. Alimirah, F.; Chen, J.; Basrawala, Z.; Xin, H.; Choubey, D. DU-145 and PC-3 human prostate cancer cell lines express androgen receptor: Implications for the androgen receptor functions and regulation. *FEBS Lett.* **2006**, *580*, 2294–2300. [CrossRef]
39. Al-Kadhemy, M. Absorption spectrum of Crystal Violet in Chloroform solution and doped PMMA thin films. *Atti della Fondazione Giorgio Ronchi* **2012**, *3*, 359.
40. Feoktistova Maria, M.; Geserick, P.; Leverkus, M. Crystal Violet Assay for Determining Viability of Cultured Cells. *Cold Spring Harb Protoc.* **2016**, *2016*, 343–346. [CrossRef]
41. Organisation for Economic Co-operation and Development. Guidance document on using cytotoxicity tests to estimate starting doses for acute oral systemic toxicity tests. *OECD Ser. Test. Assess.* **2010**, *20*, 1–54.
42. Gillies, R.J.; Didier, N.; Denton, M. Determination of cell number in monolayer-cultures. *Anal. Biochem.* **1986**, *159*, 109–113. [CrossRef]
43. Visagie, M.H.; van den Bout, I.; Joubert, A.M. A bis-sulphamoylated estradiol derivative induces ROS-dependent cell cycle abnormalities and subsequent apoptosis. *PLoS ONE* **2017**, *12*, e0176006. [CrossRef]
44. Visagie, M.; Theron, A.; Mqoco, T.; Vieira, W.; Prudent, R.; Martinez, A.; Lafanechère, L.; Joubert, A. Sulphamoylated 2-Methoxyestradiol Analogues Induce Apoptosis in Adenocarcinoma Cell Lines. *PLoS ONE* **2013**, *8*, e71935. [CrossRef]
45. Costa, A.; Scholer-Dahirel, A.; Mechta-Grigoriou, F. The role of reactive oxygen species and metabolism on cancer cells and their microenvironment. *Sem. Cancer Biol.* **2014**, *25*, 23–32. [CrossRef] [PubMed]
46. Rieseberg, M.; Kasper, C.; Reardon, K.F.; Scheper, T. Flow cytometry in biotechnology. *Appl. Microbiol. Biotechnol.* **2001**, *56*, 350–360. [CrossRef] [PubMed]
47. Boyd, L.S.; Gozuacik, D.; Joubert, A.M. The in vitro effects of a novel estradiol analog on cell proliferation and morphology in human epithelial cervical carcinoma. *Cell. Mol. Biol. Lett.* **2018**, *23*, 10. [CrossRef] [PubMed]
48. Visagie, M.H.; Birkholtz, L.-M.; Joubert, A.M. A 2-methoxyestradiol bis-sulphamoylated derivative induces apoptosis in breast cell lines. *Cell Biosci.* **2015**, *5*, 19. [CrossRef] [PubMed]

molecules

MDPI

Communication

Antiproliferative Effect of Colonic Fermented Phenolic Compounds from Jaboticaba (*Myrciaria trunciflora*) Fruit Peel in a 3D Cell Model of Colorectal Cancer

Paula Rossini Augusti [1,*], Andréia Quatrin [2], Renius Mello [2], Vivian Caetano Bochi [3], Eliseu Rodrigues [1], Inês D. Prazeres [4,5], Ana Catarina Macedo [4,5], Sheila Cristina Oliveira-Alves [4,5], Tatiana Emanuelli [2], Maria Rosário Bronze [4,5,6] and Ana Teresa Serra [4,5]

1 Departamento de Ciências dos Alimentos, Instituto de Ciência e Tecnologia de Alimentos (ICTA), Universidade Federal do Rio Grande do Sul (UFRGS), Porto Alegre 91501-970, Brazil; eliseu.rodrigues@ufrgs.br

2 Núcleo Integrado de Desenvolvimento em Análises Laboratoriais (NIDAL), Centro de Ciências Rurais, Departamento de Tecnologia e Ciência dos Alimentos, Universidade Federal de Santa Maria, Santa Maria 97105-900, Brazil; deiaquatrin@hotmail.com (A.Q.); reniusmello@gmail.com (R.M.); tatiana.emanuelli@ufsm.br (T.E.)

3 Departamento de Nutrição, Universidade Federal de Ciências da Saúde de Porto Alegre (UFCSPA), Porto Alegre 91501-970, Brazil; vivian_bochi@yahoo.com.br

4 iBET, Instituto de Biologia Experimental e Tecnológica, Apartado 12, 2780-901 Oeiras, Portugal; ines.duarte.96@gmail.com (I.D.P.); catarina.macedo@ibet.pt (A.C.M.); sheilacris.oliveira.alves@gmail.com (S.C.O.-A.); mbronze@ibet.pt (M.R.B.); tserra@ibet.pt (A.T.S.)

5 Instituto de Tecnologia Química e Biológica António Xavier, Universidade Nova de Lisboa (ITQB NOVA), 2780-157 Oeiras, Portugal

6 iMed, Faculdade de Farmácia da Universidade de Lisboa, Av das Forças Armadas, 1649-019 Lisboa, Portugal

* Correspondence: paularaugusti@gmail.com; Tel.: +55-51-3308-6676

Citation: Augusti, P.R.; Quatrin, A.; Mello, R.; Bochi, V.C.; Rodrigues, E.; Prazeres, I.D.; Macedo, A.C.; Oliveira-Alves, S.C.; Emanuelli, T.; Bronze, M.R.; et al. Antiproliferative Effect of Colonic Fermented Phenolic Compounds from Jaboticaba (*Myrciaria trunciflora*) Fruit Peel in a 3D Cell Model of Colorectal Cancer. *Molecules* **2021**, *26*, 4469. https://doi.org/10.3390/molecules26154469

Academic Editors: Višnja Stepanić and Marta Kučerová-Chlupáčová

Received: 14 June 2021
Accepted: 21 July 2021
Published: 24 July 2021

Abstract: Jaboticaba is a Brazilian native berry described as a rich source of phenolic compounds (PC) with health promoting effects. PC from jaboticaba peel powder (JPP) have low intestinal bio-accessibility and are catabolized by gut microbiota. However, the biological implication of PC-derived metabolites produced during JPP digestion remains unclear. This study aimed to evaluate the antiproliferative effects of colonic fermented JPP (FJPP) in a 3D model of colorectal cancer (CRC) composed by HT29 spheroids. JPP samples fermented with human feces during 0, 2, 8, 24 or 48 h were incubated (10,000 $\mu g\ mL^{-1}$) with spheroids, and cell viability was assessed after 72 h. Chemometric analyses (cluster and principal component analyses) were used to identify the main compounds responsible for the bioactive effect. The antiproliferative effect of FJPP in the CRC 3D model was increased between 8 h and 24 h of incubation, and this effect was associated with HHDP-digalloylglucose isomer and dihydroxyphenyl-γ-valerolactone. At 48 h of fermentation, the antiproliferative effect of FJPP was negligible, indicating that the presence of urolithins did not improve the bioactivity of JPP. These findings provide relevant knowledge on the role of colonic microbiota fermentation to generate active phenolic metabolites from JPP with positive impact on CRC.

Keywords: spheroids; HT29 cells; cluster analysis; principal component analysis; hydrolysable tannins; HHDP-digalloylglucose isomer; dihydroxyphenyl-γ-valerolactone

1. Introduction

Colorectal cancer (CRC) is the third most common type of cancer and the worldwide incidence in the year 2018 was nearly two million people [1]. Almost 75% of all sporadic cases of CRC are directly affected by dietary intake, and the consumption of plant phytochemicals has been reported to be beneficial in CRC management [2]. Phytochemicals, such as phenolic compounds (PC), have chemopreventive properties mostly due to their capacity to inactivate reactive oxygen species, which plays a vital role in the initiation and progression of CRC [2]. Besides, PC may also modulate cell gene expression, apoptosis or differentiation [3]. Studies regarding CRC are usually carried out using monolayer cells. In vitro 3D cell culture models using human and patient-derived CRC cell lines have been described as to be promising tools to evaluate the role of many antitumoral compounds, since these models provide cell–cell and cell–matrix interactions and, in this way, a cellular context similar to the cancer microenvironment. In particular, spheroid cultures of human CRC cells produced in stirred culture systems have been shown to be enriched in cancer stem cells and better recapitulate the in vivo tumor behavior [4,5]. Of note, these models have been used to ensure an accurate evaluation of the anticancer potential of phytochemicals derived from brassicas, nuts and citrus fruits [4,6–8].

Jaboticaba fruit (*Myrciaria trunciflora*) is a Brazilian berry with great potential for the food industry, being consumed naturally or as juices, jams, wines and liqueurs [9]. This fruit is rich in PC, in particular anthocyanins, flavonols, and ellagitannins, which are mostly concentrated in the peel [10]. Following jaboticaba processing for juice production, the peel is discarded and becomes a large food-waste. Thus, studies regarding the bioactivity of jaboticaba peel compounds will contribute to the sustainability of this agri-food system, supporting waste reuse as a nutraceutical or pharma chemical source. Accordingly, some PC extracted from jaboticaba exhibited antiproliferative effects against HT29 and HCT116 colon cell lines [11]. However, PC from jaboticaba peel powder (JPP) have low intestinal bio-accessibility and undergo extensive catabolism by gut microbiota [12].

The biological implication of the gut-derived PC metabolites of JPP remains unknown as previous studies on JPP fermentation were limited to the description of PC transformation during simulated digestion [12]. Although the antiproliferative effect of JPP extract in HT29 monolayer cells has been recently reported [13], no studies were found regarding the antiproliferative effects in CRC 3D cell models or the effects of colonic fermented JPP (FJPP).

This study aimed to evaluate the antiproliferative effect of FJPP in a 3D model of CRC (HT29 spheroids). Moreover, PC composition data of FJPP were processed by multivariate analyses, such as cluster analysis (CA) and principal component analysis, to determine the relationship between PC generated during fermentation and the antiproliferative effect of FJPP.

2. Results and Discussion

The inhibition of colon cancer cell line growth by PC has gained attention as having potential for a candidate compound for cancer therapeutics [3]. However, some caution in data interpretation is needed, since PC have low bioavailability. Given that a large portion of PC are eliminated in feces, the transformation of PC by gut microbiota may influence the therapeutic potential of these compounds [14]. Besides being substrates for colonic microbiota, PC fermentation may also generate products that benefit the intestinal environment [15].

PC from JPP are poorly absorbed in the small intestine, and most of these compounds reach the colon and suffer fermentation by gut microbiota [12]. After simulated salivary, gastric and intestinal digestion, the fraction of JPP that was not bio-accessible (JPP-IN) was used for the colonic fermentation assay. In vitro colonic fermentation is a simple model to simulate the catabolism of compounds by colonic microbiota, and several reports have demonstrated the catabolism of PC during colonic fermentation using this model [16,17].

In vitro gut fermentation assays are relatively simple and fast procedures that present an unmatched opportunity for performing studies frequently challenged in humans and animals owing to ethical concerns. Fresh feces are the usual source of gut microbiota, but the large inter-individual variability of the gut microbiota poses a great challenge for biological replications [18]. This issue can be partially overcome by pooling fecal samples from different donors, as conducted in the present study.

2.1. Antiproliferative Effect of FJPP in Monolayer Cultures and Spheroids of HT29 Cell Line

All antiproliferative assays were carried out using a feces control group, i.e., a fecal suspension alone (no JPP-IN) that was fermented for the same time as JPP, to eliminate the interference of PC already present in the feces. Before antiproliferative assays, the cytotoxicity of FJPP was tested in confluent Caco-2 cells at concentrations ranging from 125 to 10,000 μg mL^{-1}. None of the evaluated concentrations caused toxicity to these cells after exposure for 72 h (Figure S1, supplementary material), indicating their safety for intestinal epithelial cells. The antiproliferative effect of FJPP (125 to 2000 μg mL^{-1}) was firstly screened in monolayer cultures of HT29 cells, and all samples inhibited cell proliferation with EC_{50} values ranging from 769 to >2000 μg mL^{-1} (Figure 1). However, at the end of fermentation, (24 h and 48 h) the antiproliferative effect decreased when compared with time 0 (non-fermented samples) (Figure 1, $p < 0.05$).

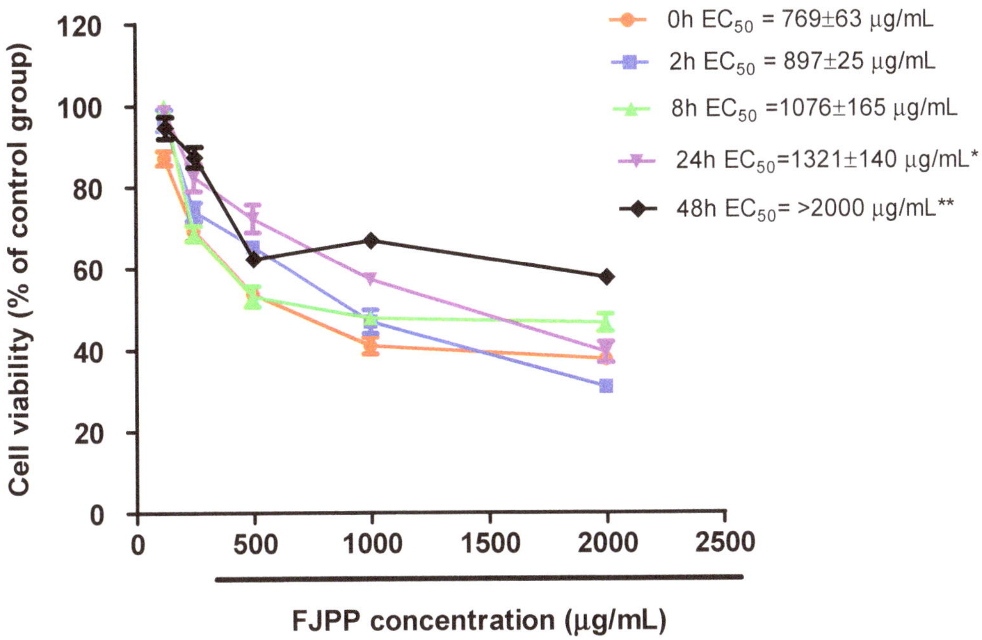

Figure 1. Antiproliferative effect of FJPP (0 h to 48 h of fermentation) in HT29 monolayer cells. Cell growth was evaluated after exposure to FJPP at 125 to 2000 μg mL^{-1} for 72 h. Results are means of at least 6 independent experiments performed in triplicate ± SEM. Statistical analyses were performed using GraphPad Prism 5 software (GraphPad Software, Inc., La Jolla, CA, USA). * Different from time 0. ** Different from all groups.

After confirming the antiproliferative potential of FJPP in HT29 monolayer cultures, we investigated whether FJPP samples were able to impair cell proliferation in a more complex model. The HT29 spheroids cultured for seven days in a stirred culture system were selected for the study because they display characteristics observed in in vivo carcinomas, such as the hypoxic regions, the apoptotic/necrotic core, less differentiated cells in the surrounding area and a higher percentage of cancer stem cells, which has been associated with chemotherapeutic resistance [19].

The antiproliferative effect of FJPP was evaluated in HT29 spheroids by analyzing cell viability after 72 h of incubation with 10,000 μg mL^{-1}. Time-course colonic fermentation of JPP resulted in a bell-shaped antiproliferative profile that was best described by a quadratic equation (Figure 2, Panel A, $p < 0.05$). The maximum point obtained from the first derivative showed that peak cell growth inhibition (69.6%) would occur for the FJPP obtained after 22.4 h of colonic fermentation (Figure 2, Panel A). This result is in agreement with a previous report [20], where colonic metabolites were detected in urine and plasma of subjects after 24 h of pomegranate juice intake, suggesting that this time would be enough for colon bacteria to metabolize the juice PC. After 48 h of fermentation, no significant antiproliferative effect of JPP was observed when compared with the control fermentation medium (Figure 2, Panel A, $p > 0.05$). Although the antiproliferative effect of dried peel powder and freeze-dried extract of *Myrciaria jaboticaba* fruits in HT29 cells has been recently reported [13], our study is the first to show the antiproliferative effect of JPP in a highly complex CRC model. Moreover, our study evaluated digested and colonic-fermented samples, which resembles the in vivo transformations of JPP and its effects in the colon.

Higher PC concentration was required to inhibit spheroids proliferation compared to monolayer cells (Figure 1 vs. Figure 2, Panel A). In agreement, the antiproliferative effect of baru nuts (*Dipteryx alata* Vog) extracts was lower in HT29 spheroids when compared with the antiproliferative effect in HT29 monolayer cells [4]. This confirms a higher resistance of HT29 spheroids when compared with usual HT29 monolayer cells, most likely due to the difficulty of PC in diffusing through the cell spheroids and/or the chemo-resistant phenotype of this model, as described previously for other PC [6,8]. Interestingly, the antiproliferative profile of JPP during time-course colonic fermentation was different in HT29 monolayer cells (Figure 1) and HT29 spheroids (Figure 2, Panel A). While the highest antiproliferative effect against monolayer cells were observed for non-fermented JPP and up to 8 h fermentation, spheroids were mostly inhibited by JPP fermented during 8 h and 24 h. The high structural complexity of PC before JPP fermentation (see Section 2.2) likely posed a greater difficulty for their diffusion into HT29 spheroids, and limited their antiproliferative properties compared to the monolayer model. This way, 3D spheroids could represent a better model to evaluate the antiproliferative effect of PC from JPP when compared with HT29 monolayer cells.

2.2. *Chemometric Analyses to Identify Bioactive PC Metabolites*

The detailed catabolism of PC during this colonic fermentation assay of digested JPP was reported in our previous study [12] and is summarized in Figure 2, Panel B. The present study brings novel information to understand how fermentation affects the antiproliferative properties of JPP and the putative PC and metabolites associated to the antiproliferative properties.

The relationship among PC, their metabolites and the antiproliferative activity during the colonic fermentation of JPP was investigated using multivariate analysis.

2.2.1. Treatment Grouping

Cluster analysis (CA) of JPP submitted to different times of colonic fermentation revealed that samples were divided into four groups (0, 2, 8/24 and 48 h of fermentation) based on their PC and metabolite composition along with the antiproliferative effect. This grouping explained 87% of data variation (Figure 3, Panel A, red dashed line). JPP fermented for 8 h and 24 h were within the same group, indicating that they have analog characteristics concerning PC composition and antiproliferative activity (Figure 3, Panel A). Principal component analysis was also used to investigate the relation among JPP from different fermentation times according to their PC composition and antiproliferative activity. The model used was constituted by three principal components that explained 91.3% of data variance (Figure 3, Panel B) and confirmed the findings of CA, as it revealed a close relationship between JPP from 8 h and 24 h of fermentation. These fermentation times were located close to each other and within the negative-central region of axis 2 and the negative regions of axes 1 and 3 (Figure 3, Panel B).

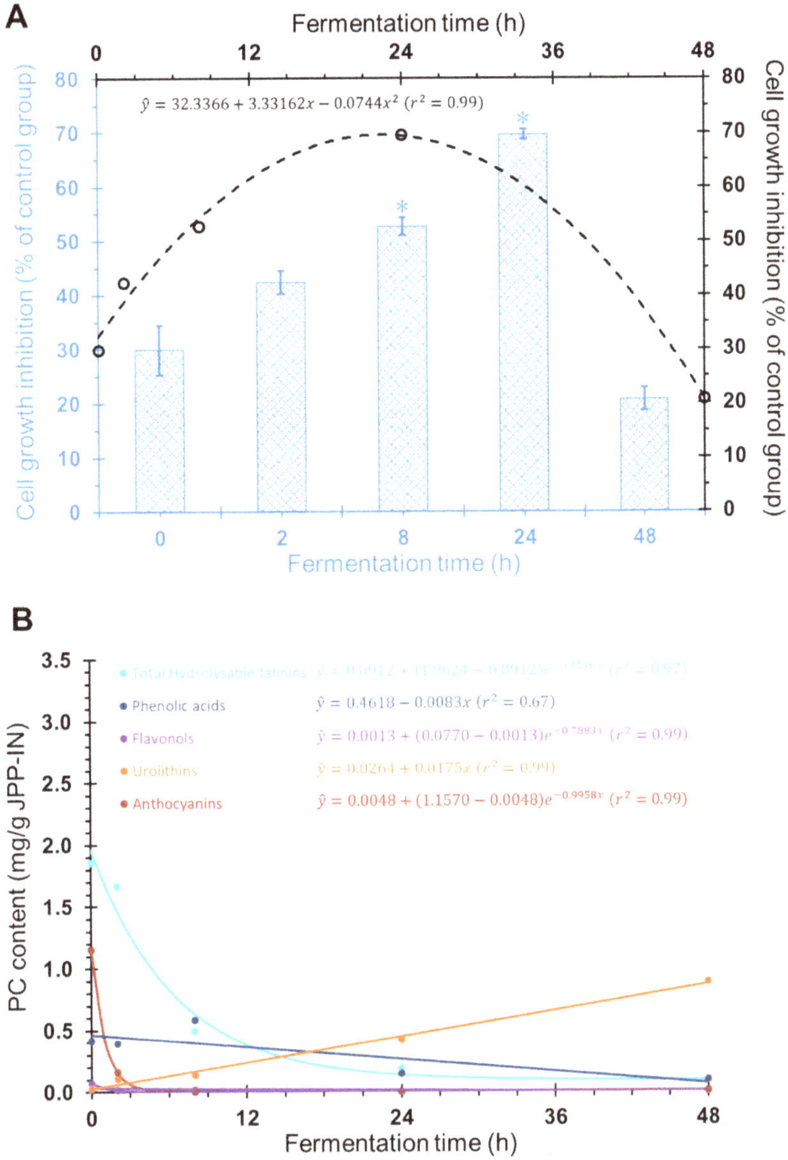

Figure 2. Changes in the antiproliferative effect (**A**) and in PC and metabolites content (**B**) of digested JPP (JPP-IN) during in vitro fermentation with human feces. Cell growth was evaluated after exposure of HT29 spheroids to FJPP at 10,000 μg mL^{-1} for 72 h. In panel A, data regarding the equation are displayed in black at the superior and right axes, whereas data regarding bars are displayed in blue at the bottom and left axes. Data displayed in panel B were obtained from Quatrin et al. (2020) [12]. Results are means of at least 6 independent experiments performed in triplicate ± SEM. Statistical analyses were performed using GraphPad Prism 5 software (GraphPad Software, Inc., La Jolla, CA, USA) and software SAS® OnDemand for Academics (SAS Institute Inc., Cary, NC, USA). * Different from medium control group.

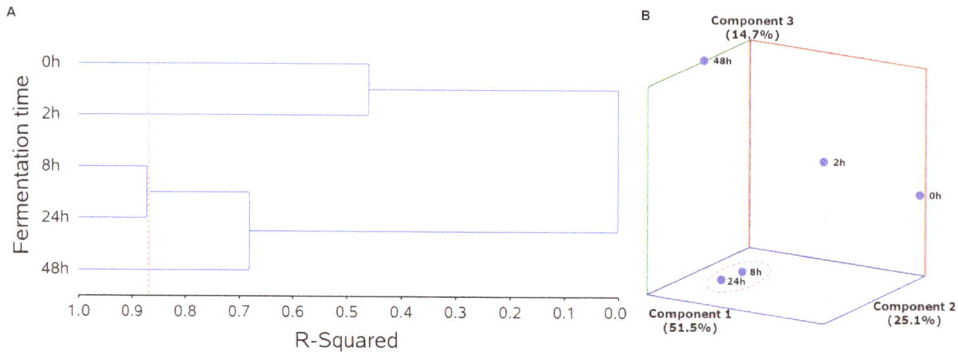

Figure 3. Dendrogram from CA of fermentation time (h, ordinate axis) in relation to the coefficient of determination (r2, abcissa axis) using Euclidean distance as a measure of dissimilarity and Ward's agglomerative hierarchical algorithm as a grouping method (**A**); three-dimensional graphic dispersion of the fermentation time in relation to the main components of principal component analysis (**B**). In panel A, the red dashed line indicates the percent of data variation explained by CA, and shows the significant groups formed. Statistical analyses were performed using software SAS® OnDemand for Academics (SAS Institute Inc., Cary, NC, USA).

2.2.2. Variable Grouping

CA of the dependent variables (PC compounds, PC metabolites and antiproliferative activity in CRC 3D cell model) allowed PC compounds to be clustered into nine groups according to their content during the different times of colonic fermentation. This clustering explained 68.5% of data variation (Figure 4, red dashed line): Groups 1 to 9 (top to bottom) contained 1, 9, 4, 2, 3, 5, 5, 5 and 7 compounds, respectively (Figure 4). Principal component analysis allowed the visualization of PC from JPP during colonic fermentation, in an n-dimensional space, by identifying the directions in which most of the information was retained. In this analysis, the biplot can show inter-unit distances among the units as well as display variances and correlations of the variables. As far as we know, this is the first study using chemometric analyses to investigate bioactive PC and metabolites in a CRC 3D model.

The antiproliferative activity (cell growth inhibition) induced by 10,000 $\mu g\ mL^{-1}$ in HT29 cell spheroids was clustered along with hexahydroxydiphenic (HHDP)-digalloylglucose isomer + dihydroxyphenyl-γ-valerolactone (Figure 4). A biplot (fermentation time vs. PC and metabolites content, Figure 5) presented encyclical behavior counterclockwise and confirmed the findings of CA. Furthermore, this biplot increased the proportion of variance explained to 91.3% using the first three principal components (Figure 5). The loadings of principal components are displayed in Table S1 (Supplementary material).

At the beginning of fermentation (0 h), the predominant PC compounds were high-molecular-weight compounds naturally found in JPP [9,11], namely hydrolysable tannins, such as penta, tetra, tri and digalloylglucose isomers and HHDP derivatives, besides anthocyanins (Figure 2, Panel B). These complex compounds were degraded over time, increasing the concentration of simpler hydrolysable tannins, such as HHDP-digalloylglucose isomer, and proanthocyanidin metabolites such as dihydroxyphenyl-γ-valerolactone, which were more closely related to the antiproliferative effect of FJPP as shown by their proximity with this bioactivity marker both in CA (Figure 4) and in principal component analysis (Figure 5).

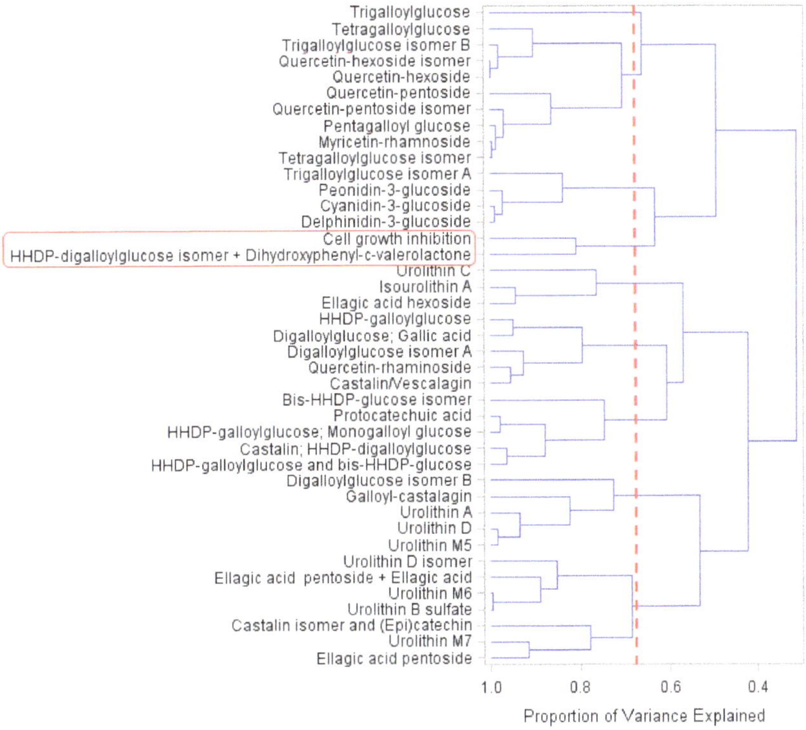

Figure 4. Dendrogram of PC content (mg compound/g equivalent to undigested JPP) and antiproliferative activity (%, ordinate axis) in relation to the coefficient of determination (r2, abcissa axis) using the correlation matrix as a measure of similarity and the main component as a grouping method. The red dashed line indicates the percent of data variation explained by CA and shows the significant groups formed. Antiproliferative activity in the CRC 3D cell model (cell growth inhibition) was highlighted using the red rectangle. Statistical analyses were performed using software SAS® OnDemand for Academics (SAS Institute Inc., Cary, NC, USA).

The biplot showed that the antiproliferative effect was closely related to 8 h and 24 h of fermentation and to HHDP-digalloylglucose isomer and dihydroxyphenyl-γ-valerolactone (Figure 5). The concentration of these compounds was transiently increased during the first stages of colonic fermentation, being thereafter decreased and not detected after 48 h of fermentation [12]. Because of the complex structure of hydrolysable tannins, they are gradually depolymerized during colonic fermentation, resulting in a transient increase in the concentration of smaller tannin polymers, such as HHDP-digalloylglucose isomer, which was already present in the undigested JPP [10]. Ellagitannins have HHDP group(s), which release ellagic acid upon hydrolysis. However, HHDP, a dimer of gallic acid from pomegranate juice, exhibited a higher antiproliferative effect by inducing necrosis in HT29 cells when compared with gallic and ellagic acids [21]. Our results suggest that hydrolysable tannins released in the colon upon consumption of JPP could potentially curtail the risk of CRC development, as previously reported for pomegranate juice [21]. Ellagitannins from *Myrciaria jaboticaba* seeds have been recently shown to exhibit chemopreventive properties against CRC by reducing inflammation and increasing proapoptotic pathways in rats [22].

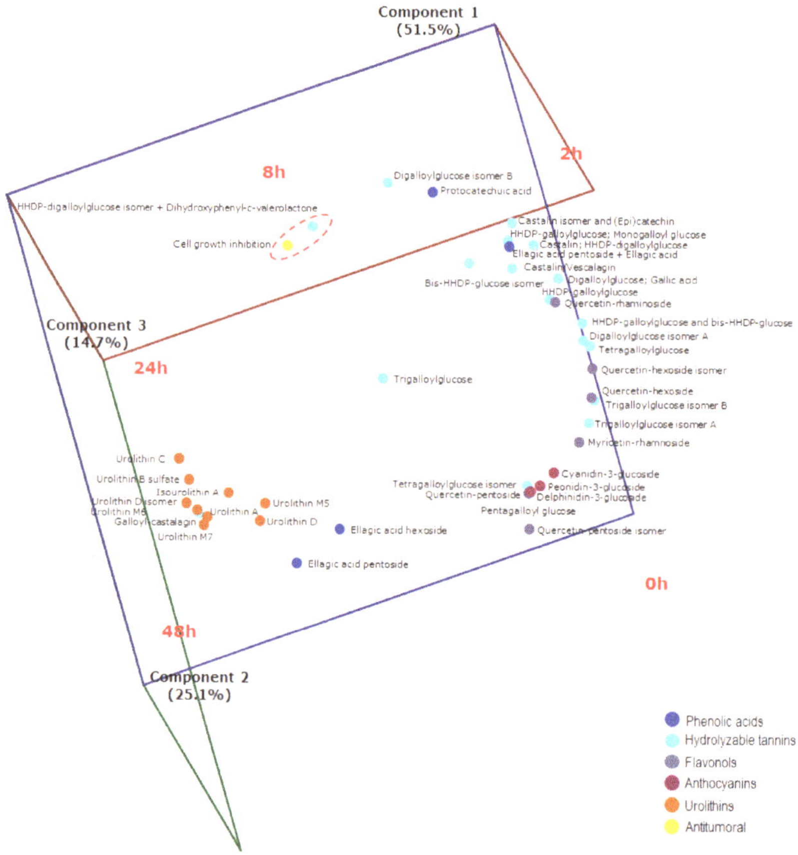

Figure 5. Three-dimensional biplot of fermentation time (scores) vs. PC, metabolites and antiproliferative activity (loadings) in relation to the main components of principal component analysis. A red dashed line indicates the proximity between HHDP-digalloylglucose isomer and dihydroxyphenyl-γ-valerolactone with the antiproliferative effect of FJPP. Statistical analyses were performed using software SAS® OnDemand for Academics (SAS Institute Inc., Cary, NC, USA).

Dihydroxyphenyl-γ-valerolactone, which was likely formed by microbial degradation of catechin and epicatechin from JPP [10], was the other compound associated to the antiproliferative activity of FJPP (Figures 4 and 5). This metabolite has been shown to contribute to the urinary antioxidant activity in rats treated with (−)-epicatechin [23]. Moreover, it has been recently shown that dihydroxyphenyl-γ-valerolactone is able to reach the brain, supporting the neuroprotective effects of PC-rich foods [24]. In fact, the plasma concentrations of dihydroxyphenyl-γ-valerolactone were positively correlated with memory improvement in mice with Alzheimer disease supplemented with polyphenolic extract from blueberries and grapes for four months [25]. As far as we know, this is the first report about the bioactivity of dihydroxyphenyl-γ-valerolactone in a cancer model.

Although HHDP-digalloylglucose isomer and dihydroxyphenyl-γ-valerolactone seem to be the major compounds associated with the increased antiproliferative activity of JPP during the intermediate stages of colonic fermentation, other PC and metabolites likely contributed to the antiproliferative effect that was already found before starting colonic fermentation and after 48 h of colonic fermentation (30% and 20% cell growth inhibition; Figure 2, Panel A).

Gut microbiota and PC have shown a synergistic effect regarding their chemopreventive properties in monolayer cell models [26], which was likely related to microorganism enzymes that convert PC into more bioavailable or bioactive forms than their parent compounds. Moreover, PC can inhibit pathogenic bacteria and favor the development of beneficial microbiota. In fact, a reduction in pathogenic bacteria count was observed concomitant with the JPP PC catabolism by human fecal microbiota [12]. Although microbiome composition has not been evaluated in the present study, the intake of a yogurt added with lyophilized seed extract of jaboticaba has been recently shown to increase the abundance of *Bacteroidetes* and decrease the number of *Firmicutes* in a rat model of chemical-induced colon cancer [27]. Lastly, as oxidative stress may be involved in the death of probiotics, PC may delay this process due to their antioxidant properties, increasing the viability of probiotics [26]. In agreement, the addition of Bifidobacterium in an extract rich in water-soluble PC from jaboticaba enhanced antioxidant activity and antiproliferative effects in monolayer cancer colon cells when compared with the group without the probiotic [26].

At the end of fermentation, a significant reduction in antiproliferative activity was observed along with the highest concentration of final metabolites (urolithins). Urolithins, which were linearly increased during fermentation (Figure 2, Panel B), are a class of compounds produced by the gut microbiota metabolism of ellagitannins that have been suggested as biomarkers of the intake of PC from berries, nuts and wines. Moreover, urolithin A was able to decrease colony formation of monolayer human colon cancer cells [28]. However, in the present study, the increase in the antiproliferative effect of FJPP was not related to the production of urolithins, since the antiproliferative activity of FJPP decreased by 48 h of fermentation (in monolayer and 3D cells) while the urolithins content reached their maximum content at this time. Since FJPP presented a mix of different urolithins, we cannot rule out that the interactions among these metabolites may have masked a possible antiproliferative effect.

3. Materials and Methods

3.1. FJPP Samples

All of the samples used in this work were prepared and characterized in our previous study [12]. Briefly, the digestion of JPP was performed with *Myrciaria trunciflora* fruits collected in São Vicente do Sul city, at Rio Grande do Sul State, Brazil (SISGEN ABD 4602). The peels were separated from the pulp, freeze-dried and triturated to produce JPP. JPP (5 g sample) was subjected to a sequential static in vitro simulation of oral, gastric and duodenal digestion as previously described [10]. The fraction containing PC that were not available for absorption (JPP-IN), was separated, freeze-dried and used for the in vitro colonic fermentation assay. Each 5 g of JPP yielded 4.2 g of JPP-IN.

In vitro colonic fermentation was carried out using fresh feces from human donors (eight men and nine women aged between 20 and 53 years-old) to provide gut microbiota. The assay was the same as that already described [10]. The JPP-IN fraction was incubated in glass bottles containing 50 mL of fecal suspension. Fecal suspension without the JPP-IN was also incubated and used to correct the results by the respective controls. After incubation, glass bottles were centrifuged at 1400× *g* for 10 min, the supernatant was collected and stored at −80 °C for chromatographic analysis or freeze-dried and stored frozen before cell culture assays. For cell culture experiments, the freeze-dried fermented supernatant (FJPP) samples were reconstituted with Milli-Q water and then centrifuged at 2000× *g* for 15 min. FJPP samples were filtered using 0.22 µm filter twice to yield the sterile FJPP samples.

3.2. Cell-Based Assays

3.2.1. Cell Lines and Culture

Human colon cancer cell lines, HT29 and Caco-2, were obtained from American Type Culture Collection (ATCC, Manassas, VA, USA) and Deutsche Sammlung von Microorganismen und Zellkulturen (Barunshweig, Germany), respectively. Both cell lines were grown in RPMI 1640 medium (Gibco, Carlsbad, CA, USA) supplemented with 10% (*v/v*) of heat-inactivated sterile filtered Fetal Bovine Serum (FBS; Biowest, Riverside, CA, USA). For Caco-2 cells, additional supplementation was made with 1% (*v/v*) of PenStrep (Gibco, Carlsbad, CA, USA). Stock cells were maintained as monolayers in 175 cm^2 culture flasks and incubated at 37 °C with 5% CO_2 in a humidified atmosphere.

3.2.2. 3D Cell Culture Using a Stirred-Tank Culture System

CRC spheroids (3D cells model) were generated as previously described [19] with some modifications. Briefly, HT29 single cells were inoculated into a 100 mL spinner flask (Corning, Tewksbury, MA, USA) in a culture medium with 10% (*v/v*) FBS, accounting for a cell density of 2.5×10^5 cells/mL. The spinner vessel was placed on a magnetic stirrer under 40 rpm, and cell culture was carried out in a humidified atmosphere with 5% CO_2 at 37 °C, with an increasing stirring speed to 50 rpm and 60 rpm at the time-point of 8 h and 28 h post-inoculation, respectively. After the 4th day post-inoculation, half of the spinner flask volume was renewed daily. Experiments were performed using spheroids collected at day eight of culture with an average diameter of 500 μm.

3.2.3. Cytotoxicity Assay in Caco-2 Cells

The cytotoxicity of FJPP was assessed using confluent and undifferentiated Caco-2 cells as a model of the human intestinal epithelium, as previously described [29]. Briefly, Caco-2 cells were seeded into 96-well plates at a density of 2×10^4 cells/well and allowed to grow for seven days, with medium renewal every 48 h. At day seven, cells were incubated with FJPP that was diluted in a culture medium at concentrations ranging from 125 μg to 10,000 μg of JPP-IN equivalents mL^{-1} Cells incubated with culture medium and cells incubated with control fermentation medium (fecal suspension without JPP-IN) were used as controls. Another batch of control cells were incubated with Milli-Q water (FJPP vehicle) to account for culture medium dilution effects during FJPP addition. After 72 h of incubation, cells were washed with Phosphate Buffered Saline (PBS, Sigma-Aldrich, St. Louis, MO, USA) and cell viability was assessed using the 3-(4,5-dimethylthiazol-2-yl)-2,5-diphenyltetrazolium bromide (MTT) assay [4]. Cell viability was calculated relative to the respective feces control (no JPP-IN) and relative to medium control. Six independent experiments were performed in triplicate.

3.2.4. Antiproliferative Assay in HT29 Cell Monolayers

HT29 cells were seeded in 96-well plates at a density of 1×10^5 cells per well. After incubation for 24 h, cells were treated with several concentrations (0-control, 125, 250, 500, 1000 and 2000 μg of JPP-IN equivalents mL^{-1}) of JPP previously fermented for 0, 2, 8, 24 and 48 h (FJPP). Cells incubated with culture medium and cells incubated with control fermentation medium (fecal suspension without JPP-IN) were used as controls. After 72 h of incubation, cells were washed with PBS and cell viability was assessed using the MTT assay, as described by [4]. Cell viability was calculated relative to the respective feces control (no JPP-IN) and relative to the medium control. Six independent experiments were performed in triplicate.

3.2.5. Antiproliferative Assay in HT29 Cell Spheroids (3D Cells)

HT 29 spheroids were seeded at a density of approximately five spheroids/well, in 96-well plates and incubated with PrestoBlue Viability Reagent (Molecular Probes, Invitrogen, CA, USA) to determine the basal viability. Then, spheroids were treated with FJPP at 10,000 μg of JPP-IN equivalents.mL^{-1} (five times the highest concentration used in the 2D cells antiproliferative assay, but still non-toxic). After 72 h of incubation, the spheroids were washed with PBS and cell viability was assessed using the MTT assay, as described above. Cell viability was calculated relative to the respective feces control (no JPP-IN) as previously described [8]. Six independent experiments were performed in triplicate.

3.3. HPLC-Q-TOF-MS/MS Analysis of PC during Colonic Fermentation

The extraction, identification and quantification of individual PC and metabolites from the FJPP of this assay were described in our previous study [12]. Briefly, anthocyanins were separated in a C-18 Core-Shell Kinetex column (2.6 µm particle size, 100 mm, 4.6 mm; Phenomenex, Torrance, CA, USA) at 38 °C using a gradient of 3% formic acid in water and 100% acetonitrile at a flow rate of 0.9 mL·min^{-1}. Non-anthocyanin PC and metabolites were separated in a C-18 Hypersil Gold column (5 µm particle size, 150 mm, 4.6 mm; Thermo Fisher Scientific, Waltham, MA, USA) at 38 °C using a gradient of 5% methanol in acidified water (0.1%, *v/v*, of formic acid) and 0.1% acetonitrile at a flow rate of 1.0 mL·min^{-1}. The identification of PC and metabolites was performed in a HPLC system connected to a diode array detector (DAD) and a Q-TOF mass spectrometer analyzer and electrospray ionization (ESI) source (micrOTOF-QIII, Bruker Daltonics, Bremen, Germany). Compounds were identified based on their elution order and the comparison of their UV to visible spectra and mass spectrometry fragmentation patterns with authentic standards and literature data. The quantification of PC and metabolites was conducted using DAD peak area data using the method previously validated [10]. Hydroxybenzoates were quantified at 280 nm as equivalents of gallic acid or protocatechuic acid, tannins were quantified at 280 nm as equivalents of gallic acid, anthocyanins were quantified at 520 nm as equivalents of cyanidin 3-glucoside, and flavonols and urolithins were quantified at 360 nm as equivalents of quercetin or myricetin. The limits of detection (LOD) and quantification (LOQ) for gallic acid, protocatechuic acid, cyanidin 3-glucoside, quercetin and myricetin were 0.012 and 0.037, 0.027 and 0.083, 0.020 and 0.068, 0.562 and 1.363, 0.166 and 0.503 mg·L^{-1}.

3.4. Statistical Analysis

Antiproliferative activity data were expressed as mean ± SEM. Statistical analyses were performed using GraphPad Prism 5 software (GraphPad Software, Inc., La Jolla, CA, USA). Data were submitted to one-way analysis of variance (ANOVA) and the means were compared by Tukey's test at a 5% significance level. Additionally, antiproliferative activity data was also submitted to regression analysis. For antiproliferative assay in monolayer HT29 cells, the EC_{50} values were calculated using the former software.

The bioactivity of PC metabolites formed during colonic fermentation of JPP was investigated using chemometric analyses. Principal Component Analysis and Cluster Analysis (CA) were used to investigate the association between HPLC-MS-fingerprinting assessment of PC (parent compounds and metabolites) and the antiproliferative activity during the colonic fermentation of JPP. Data were processed using the software SAS® OnDemand for Academics (SAS Institute Inc., Cary, NC, USA).

4. Conclusions

This study demonstrated the antiproliferative effect of JPP fermented by colonic microbiota against CRC using a complex 3D cell model. The potential effects of JPP against CRC were increased in the intermediate times of fermentation, and were associated to HHDP-digalloylglucose isomer and dihydroxyphenyl-γ-valerolactone rather than to other colonic PC metabolites or to the PC found at highest concentrations in the undigested fruit. Studies regarding the antiproliferative effect of these isolated compounds in CRC 3D models should be carried out in the near future.

Supplementary Materials: The following are available online, Figure S1: Effect of FJPP (10,000 µg mL^{-1}) on viability of Caco-2 cells. Cell viability was evaluated after exposure of Caco-2 cells to FJPP at 10,000 µg mL^{-1} for 72 h. Results are means of at least six independent experiments performed in triplicate ± SEM., Table S1: The loadings of the first three principal components.

Author Contributions: Conceptualization, P.R.A., T.E. and A.T.S.; formal analysis, P.R.A., A.Q., V.C.B., E.R., I.D.P., A.C.M., S.C.O.-A.; resources, T.E., A.T.S. and M.R.B.; data curation, P.R.A., A.T.S., T.E. and R.M.; writing—original draft preparation, P.R.A., T.E. and A.T.S.; funding acquisition, T.E., A.T.S. and M.R.B. All authors have read and agreed to the published version of the manuscript.

Funding: This research was funded by the Brazilian National Council for Scientific and Technological Development (CNPq) [grant number 303654/2017-1 and postdoctoral scholarship 205295/2018-5] and Coordination for the Improvement of Higher Education Personnel (CAPES) [finance code 001]. This research was also funded by the Fundação para a Ciência e Tecnologia/Ministério da Ciência, Tecnologia e Ensino Superior [Grant CEECIND/04801/2017] and iNOVA4Health–UIDB/04462/2020 and UIDP/04462/2020, a program financially supported by Fundação para a Ciência e Tecnologia/Ministério da Ciência, Tecnologia e Ensino Superior.

Institutional Review Board Statement: This study was conducted according to the guidelines of the Declaration of Helsinki and approved by the Institutional Ethics Committee of Federal University of Santa Maria (protocol of study CAAE 50151015.6.0000.5346). Informed consent was obtained from all subjects involved in the study.

Informed Consent Statement: Informed consent was obtained from all subjects involved in the study.

Data Availability Statement: The data supporting the findings of this study are available on request from the corresponding author. Supporting information is provided in the supplementary material.

Conflicts of Interest: The authors declare no conflict of interest. The funders had no role in the design of the study; in the collection, analyses, or interpretation of data; in the writing of the manuscript, or in the decision to publish the results.

Sample Availability: Samples used in this study are not available from the authors.

References

1. International Agency for Research on Cancer, Global Cancer Observatory. Cancer Today. 2018. Available online: https.//gco.iarc.fr (accessed on 12 June 2021).
2. Jaganathan, S.K.; Vellayappan, M.V.; Narasimhan, G.; Supriyanto, E.; Dewi, D.E.O.; Narayanan, A.L.T.; Balaji, A.; Subramanian, A.P.; Yusof, M. Chemopreventive effect of apple and berry fruits against colon cancer. *World J. Gastroenterol.* **2014**, *20*, 17029–17036. [CrossRef]
3. Costea, T.; Hudită, A.; Ciolac, O.-A.; Gălăteanu, B.; Ginghină, O.; Costache, M.; Ganea, C.; Mocanu, M.M. Chemoprevention of colorectal cancer by dietary compounds. *Int. J. Mol. Sci.* **2018**, *19*, 3787. [CrossRef]
4. Oliveira-Alves, S.C.; Pereira, R.S.; Pereira, A.B.; Ferreira, A.; Mecha, E.; Silva, A.B.; Serra, A.T.; Bronze, M.R. Identification of functional compounds in baru (*Dipteryx alata* Vog.) nuts. Nutritional value, volatile and phenolic composition, antioxidant activity and antiproliferative effect. *Food Res. Int.* **2020**, *131*, 109026. [CrossRef] [PubMed]
5. Serra, A.T.; Serra, M.; Silva, A.C.; Brckalo, T.; Seshire, A.; Brito, C.; Wolf, M.; Alves, P.M. Scalable culture strategies for the expansion of patient-derived cancer stem cell lines. *Stem Cells Int.* **2019**, *2019*, 8347595. [CrossRef] [PubMed]
6. Silva, I.; Estrada, M.F.; Pereira, V.C.; da Silva, A.B.; Bronze, M.R.; Alves, P.M.; Duarte, C.M.M.; Brito, C.; Serra, A.T. polymethoxylated flavones from orange peels inhibit cell proliferation in a 3D cell model of human colorectal Ccncer. *Nutr. Cancer* **2018**, *70*, 257–266. [CrossRef] [PubMed]
7. Pereira, C.; Duarte, M.; Silva, P.; Bento da Silva, A.; Duarte, C.; Cifuentes, A.; García-Cañas, V.; Bronze, M.R.; Albuquerque, C.; Serra, A.T. Polymethoxylated flavones target cancer stemness and improve the antiproliferative effect of 5-fluorouracil in a 3D cell model of colorectal cancer. *Nutrients* **2019**, *11*, 326. [CrossRef]
8. Pereira, L.; Silva, P.; Duarte, M.; Rodrigues, L.; Duarte, C.; Albuquerque, C.; Serra, A. Targeting colorectal cancer proliferation, stemness and metastatic potential using Brassicaceae extracts enriched in isothiocyanates. A 3D cell model-based study. *Nutrients* **2017**, *9*, 368. [CrossRef] [PubMed]
9. Mannino, G.; Perrone, A.; Campobenedetto, C.; Schittone, A.; Bertea, C.M.; Gentile, C. Phytochemical profile and antioxidative properties of *Plinia trunciflora* fruits. A new source of nutraceuticals. *Food Chem.* **2020**, *307*, 125515. [CrossRef] [PubMed]
10. Quatrin, A.; Pauletto, R.; Maurer, L.H.; Minuzzi, N.; Nichelle, S.M.; Carvalho, J.F.C.; Maróstica Junior, M.R.; Rodrigues, E.; Bochi, V.C.; Emanuelli, T. Characterization and quantification of tannins, flavonols, anthocyanins and matrix-bound polyphenols from jaboticaba fruit peel. A comparison between *Myrciaria trunciflora* and *M. jaboticaba*. *J. Food Compos. Anal.* **2019**, *78*, 59–74. [CrossRef]
11. Reynertson, K.A.; Wallace, A.M.; Adachi, S.; Gil, R.R.; Yang, H.; Basile, M.J.; D'Armiento, J.; Weinstein, I.B.; Kennelly, E.J. Bioactive depsides and anthocyanins from jaboticaba (*Myrciaria cauliflora*). *J. Nat. Prod.* **2006**, *69*, 1228–1230. [CrossRef] [PubMed]
12. Quatrin, A.; Rampelotto, C.; Pauletto, R.; Maurer, L.H.; Nichelle, S.M.; Klein, B.; Rodrigues, R.F.; Junior, M.R.M.; Fonseca, B.d.S.; Ragagnin de Menezes, C.; et al. Bioaccessibility and catabolism of phenolic compounds from jaboticaba (*Myrciaria trunciflora*) fruit peel during in vitro gastrointestinal digestion and colonic fermentation. *J. Funct. Foods* **2020**, *65*, 103714. [CrossRef]
13. Frauches, N.S.; Montenegro, J.; Amaral, T.; Abreu, J.P.; Laiber, G.; Junior, J.; Borguini, R.; Santiago, M.; Pacheco, S.; Nakajima, V.M.; et al. Antiproliferative Activity on Human Colon Adenocarcinoma Cells and In Vitro Antioxidant Effect of Anthocyanin-Rich Extracts from Peels of Species of the *Myrtaceae* Family. *Molecules* **2021**, *26*, 564. [CrossRef]

14. Augusti, P.R.; Conterato, G.M.M.; Denardin, C.C.; Prazeres, I.D.; Serra, A.T.; Bronze, M.R.; Emanuelli, T. Bioactivity, Bioavailability, and Gut Microbiota Transformations of Dietary Phenolic Compounds. Implications for Covid-19. *J. Nutr. Biochem.* **2021**, 108787. [CrossRef]
15. Wojtunik-Kulesza, K.; Oniszczuk, A.; Oniszczuk, T.; Combrzyński, M.; Nowakowska, D.; Matwijczuk, A. Influence of in vitro digestion on composition, bioaccessibility and antioxidant activity of food polyphenols—a non-systematic review. *Nutrients* **2020**, *12*, 1401. [CrossRef] [PubMed]
16. Attri, S.; Goel, G. Influence of polyphenol rich seabuckthorn berries juice on release of polyphenols and colonic microbiota on exposure to simulated human digestion model. *Food Res. Int.* **2021**, *111*, 314–323. [CrossRef] [PubMed]
17. Cárdenas-Castro, A.P.; Zamora-Gasga, V.M.; Alvarez-Parrilla, E.; Ruíz-Valdiviezo, V.M.; Venema, K.; Sáyago-Ayerdi, S.G. In vitro gastrointestinal digestion and colonic fermentation of tomato (*Solanum lycopersicum* L.) and husk tomato (Physalis ixocarpa Brot.). Phenolic compounds released and bioconverted by gut microbiota. *Food Chem.* **2021**, *360*, 130051. [CrossRef] [PubMed]
18. Pham, V.T.; Mohajeri, M.H. The application of in vitro human intestinal models on the screening and development of pre- and probiotics. *Benef. Microbes.* **2018**, *9*, 725–742. [CrossRef]
19. Santo, V.E.; Estrada, M.F.; Rebelo, S.P.; Abreu, S.; Silva, I.; Pinto, C.; Veloso, S.C.; Serra, A.T.; Boghaert, E.; Alves, P.M.; et al. Adaptable stirred-tank culture strategies for large scale production of multicellular spheroid-based tumor cell models. *J. Biotechnol.* **2016**, *221*, 118–129. [CrossRef]
20. Cerdá, B.; Espín, J.C.; Parra, S.; Martínez, P.; Tomás-Barberán, F.A. The potent in vitro antioxidant ellagitannins from pomegranate juice are metabolised into bioavailable but poor antioxidant hydroxy-6H-dibenzopyran-6-one derivatives by the colonic microflora of healthy humans. *Eur. J. Nutr.* **2004**, *43*, 205–220. [CrossRef]
21. Kasimsetty, S.G.; Bialonska, D.; Reddy, M.K.; Ma, G.; Khan, S.I.; Ferreira, D. Colon cancer chemopreventive activities of pomegranate ellagitannins and Urolithins. *J. Agric. Food Chem.* **2010**, *58*, 2180–2187. [CrossRef]
22. Vieira do Carmo, M.A.; Fidelis, M.; Francielli de Oliveira, P.; Feitoza, L.Q.; Marques, M.J.; Ferreira, E.B.; Oh, W.Y.; Shahidi, F.; Hellström, J.; Almeida, L.A.; et al. Ellagitannins from jabuticaba (*Myrciaria jaboticaba*) seeds attenuated inflammation, oxidative stress, aberrant crypt foci, and modulated gut microbiota in rats with 1, 2 dimethyl hydrazine-induced colon carcinogenesis. *Food Chem. Toxicol.* **2021**, *154*, 112287. [CrossRef] [PubMed]
23. Unno, T.; Tamemoto, K.; Yayabe, F.; Kakuda, T. Urinary excretion of 5-(3′, 4′-Dihydroxyphenyl)-γ -valerolactone, a ring-fission metabolite of (-)-epicatechin, in rats and its in vitro antioxidant activity. *J. Agric. Food Chem.* **2003**, *51*, 6893–6898. [CrossRef] [PubMed]
24. Angelino, D.; Carregosa, D.; Domenech-Coca, C.; Savi, M.; Figueira, I.; Brindani, N.; Jang, S.; Lakshman, S.; Molokin, A.; Urban, J.F., Jr.; et al. 5-(Hydroxyphenyl)-γ-valerolactone-sulfate, a key microbial metabolite of flavan-3-ols, is able to reach the brain. Evidence from different *in silico, in vitro* and in vivo experimental models. *Nutrients* **2019**, *11*, 2678. [CrossRef] [PubMed]
25. Dal-Pan, A.; Dudonné, S.; Bourassa, P.; Bourdoulous, M.; Tremblay, C.; Desjardins, Y.; Calon, F. Neurophenols consortium, Cognitive-enhancing effects of a polyphenols-rich extract from fruits without changes in neuropathology in an animal model of Alzheimer's disease. *J. Alzheimer's Dis.* **2016**, *55*, 115–135. [CrossRef]
26. Holkem, A.T.; Robichaud, V.; Favaro-Trindade, C.S.; Lacroix, M. Chemopreventive properties of extracts obtained from blueberry (*Vaccinium myrtillus* L.) and jabuticaba (*Myrciaria cauliflora* Berg.) in combination with probiotics. *Nutr. Cancer* **2021**, *73*, 671–685. [CrossRef] [PubMed]
27. Fidelis, M.; Santos, J.S.; Escher, G.B.; Rocha, R.S.; Cruz, A.G.; Cruz, T.M.; Marques, M.B.; Nunes, J.B.; do Carmo, M.A.V.; de Almeida, L.A.; et al. Polyphenols of jabuticaba [*Myrciaria jaboticaba* (Vell.) O.Berg] seeds incorporated in a yogurt model exert antioxidant activity and modulate gut microbiota of 1, 2-dimethylhydrazine-induced colon cancer in rats. *Food Chem.* **2021**, *334*, 127565. [CrossRef] [PubMed]
28. Giménez-Bastida, J.A.; Ávila-Gálvez, M.Á.; Espín, J.C.; González-Sarrías, A. The gut microbiota metabolite urolithin A, but not other relevant urolithins, induces p53-dependent cellular senescence in human colon cancer cells. *Food Chem. Toxicol.* **2020**, *139*, 111260. [CrossRef]
29. Rodrigues, L.; Silva, I.; Poejo, J.; Serra, A.T.; Matias, A.A.; Simplício, A.L.; Bronze, M.R.; Duarte, C.M.M. Recovery of antioxidant and antiproliferative compounds from watercress using pressurized fluid extraction. *RSC Adv.* **2016**, *6*, 30905–30918. [CrossRef]

molecules

MDPI

Article

Activity Guided Isolation of Phenolic Compositions from *Anneslea fragrans* Wall. and Their Cytoprotective Effect against Hydrogen Peroxide Induced Oxidative Stress in HepG2 Cells

Shuyue He [1,†], Xiaoyan Cui [2,†], Afsar Khan [3], Yaping Liu [1], Yudan Wang [1,4], Qimin Cui [1], Tianrui Zhao [1], Jianxin Cao [1] and Guiguang Cheng [1,*]

1 Faculty of Agriculture and Food, Kunming University of Science and Technology, Kunming 650500, China; kmlgdx303@163.com (S.H.); liuyaping@mail.kib.ac.cn (Y.L.); sdlcwyd@163.com (Y.W.); 18908742179@163.com (Q.C.); food363@163.com (T.Z.); jxcao321@hotmail.com (J.C.)

2 Foundation Department, Hai Yuan College, Kunming Medical University, Kunming 650106, China; cuixiaoyan910@163.com

3 Department of Chemistry, Abbottabad Campus, COMSATS University Islamabad, Abbottabad 22060, Pakistan; afsarhej@gmail.com

4 National and Local Joint Engineering Research Center for Green Preparation Technology of Biobased Materials, Yunnan Minzu University, Kunming 650500, China

* Correspondence: ggcheng@kmust.edu.cn; Tel.: +86-871-65920216

† These authors contributed equally to this work.

Abstract: *Anneslea fragrans* Wall., commonly known as "Pangpo Tea", is traditionally used as a folk medicine and healthy tea for the treatment of liver and intestine diseases. The aim of this study was to purify the antioxidative and cytoprotective polyphenols from *A. fragrans* leaves. After fractionation with polar and nonpolar organic solvents, the fractions of aqueous ethanol extract were evaluated for their total phenolic (TPC) and flavonoid contents (TFC) and antioxidant activities (DPPH, ABTS, and FRAP assays). The *n*-butanol fraction (BF) showed the highest TPC and TFC with the strongest antioxidant activity. The bio-guided chromatography of BF led to the purification of six flavonoids (**1**–**6**) and one benzoquinolethanoid (**7**). The structures of these compounds were determined by NMR and MS techniques. Compound **6** had the strongest antioxidant capacity, which was followed by **5** and **2**. The protective effect of the isolated compounds on hydrogen peroxide (H_2O_2)-induced oxidative stress in HepG2 cells revealed that the compounds **5** and **6** exhibited better protective effects by inhibiting ROS productions, having no significant difference with vitamin C ($p > 0.05$), whereas **6** showed the best anti-apoptosis activity. The results suggest that *A. fragrans* could serve as a valuable antioxidant phytochemical source for developing functional food and health nutraceutical products.

Keywords: *Anneslea fragrans*; antioxidant; guided isolation; oxidative stress; flavonoid glycosides

Citation: He, S.; Cui, X.; Khan, A.; Liu, Y.; Wang, Y.; Cui, Q.; Zhao, T.; Cao, J.; Cheng, G. Activity Guided Isolation of Phenolic Compositions from *Anneslea fragrans* Wall. and Their Cytoprotective Effect against Hydrogen Peroxide Induced Oxidative Stress in HepG2 Cells. *Molecules* **2021**, *26*, 3690. https://doi.org/10.3390/molecules26123690

Academic Editors: Višnja Stepanić and Marta Kučerová-Chlupáčová

Received: 29 April 2021
Accepted: 13 June 2021
Published: 17 June 2021

1. Introduction

Reactive oxygen species (ROS) are produced as by-products though oxidative respiration in normal physiological and biochemical processes [1]. As endogenous free radicals, ROS plays an important role in cell signaling and the maintenance of body constancy in a normal range [2]. If the ROS could not be effectively scavenged, excessive ROS could lead to the occurrence of oxidative stress, which may influence cell proliferation and apoptosis [3]. Oxidative stress is closely related to a variety of diseases including diabetes, hyperlipemia, obesity, cancer, and cardiovascular and neurodegenerative diseases [4]. Recently, many studies have evidenced that a diet enriched with antioxidants possesses a series of beneficial effects owing to their scavenging ability on excessive ROS [5,6].

Natural antioxidants contain a variety of molecules such as polyphenols [7], carotenoids [8], vitamins [9], nitrogen-containing compounds [10], and coumarins [9]. Polyphenols are distributed widely throughout the plant kingdom and promote health benefits owing to their antioxidant properties [11]. Many of them, such as resveratrol and some derivatives [12], anthocyanidins [13], isoflavones [14], catechin [15], and quercetin [16] are well known for their protective effect by scavenging ROS [1]. For the past few years, polyphenols have been attracting attention in the prevention of cancer [17], cardiovascular dysfunction [18], neurodegenerative diseases [19], and aging [20]. Therefore, searching for effective antioxidants is an urgent need to promote human health.

Anneslea fragrans Wall. is an evergreen plant, which is mainly distributed in the southern of China [21]. In addition to its ornamental purpose, the leaves of *A. fragrans* are also used as a folk medicine to treat fever, liver protection, invigorating stomach and intestines in China and Cambodia [22], which had been recorded in "Yunnan Simao Chinese Herbal Medicine". In addition, the leaves have also been processed as a tea beverage, which is known as "Pangpo Tea". In previous reports [23], the extract of *A. fragrans* has shown antioxidant and antimalarial activities. However, to date, its antioxidant phenolic compounds from *A. fragrans* have not been reported.

Thus, the purpose of this research was to isolate and identify the antioxidant compounds from *A. fragrans* leaves that are responsible for its traditional use for the treatment of liver diseases. The four fractions, dichloromethane fraction (DF), ethyl acetate fraction (EAF), *n*-butanol fraction (BF), and residual water fraction (RWF) of aqueous ethanol extract were assessed for their total phenolic (TPC) and total flavonoid contents (TFC) and for their antioxidant capacity. The *n*-butanol fraction (BF) had the highest TPC and TFC with strongest antioxidant activity. The bio-guided fractionation of BF allowed the purification of compounds. Furthermore, the cytoprotective effect of the isolated compounds was performed on hydrogen peroxide (H_2O_2)-induced oxidative stress in human liver cancer HepG2 cells. The intracellular ROS production and cell apoptosis were determined using flow cytometry. Thus, this research afforded a valuable antioxidant phytochemical ingredient for the development and utilization of *A. fragrans* leaves as a functional supplement (healthy tea) in food and health industry.

2. Results and Discussion

2.1. Yield Efficiency of Fractions and Subfractions

The liquid–liquid partitioning by organic solvents and column fractionation are important techniques for enriching the bioactive compounds from crude extract [24]. Using liquid–liquid partitioning, ethanol extract (CE) from *A. fragrans* leaves was successively fractionated with dichloromethane, ethyl acetate, and *n*-butanol to give four fractions (DF, EAF, BF, and RWF), respectively. Yield percentages of fractions were found to vary from 11 to 27%. The residual water fraction (RWF, 27%) had the highest yield, and the yields of other fractions were in the order as follows: *n*-butanol fraction (BF, 24%), ethyl acetate fraction (EAF, 21%), dichloromethane fraction (DF, 11%).

The yield percentage of subfractions (BF-A to E) was found to vary from 3.82 to 48.4%. The BF-E presented the highest yield of 48.4% followed by BF-D (24.96%), which has significantly higher yields than the other fractions.

2.2. HPLC Analysis

High-performance liquid chromatography (HPLC) detection has been demonstrated to be a powerful technique for quantitative determination. HPLC analysis revealed that BF had the most compounds with the strongest antioxidant activity. Under the guidance of antioxidant assays and HPLC analysis, the antioxidative fraction was further chromatographed for the isolation of subfractions. The BF was subjected to a hydrated resin D101 column to yield five subfractions (BF-A to E). HPLC analysis revealed that BF-C to E had the most antioxidant compounds. Bio-guided fractionation of these fractions (BF-C to E) allowed the purification of seven pure compounds.

As shown in Figure 1, the BF and seven pure compounds were profiled by HPLC analysis. Based on comparison with the retention times and UV absorption curves, these seven compounds were confirmed through retention times at 7.51 min (cornoside, **7**), 7.92 min ((epi)-

catechin, **6**), 9.38 min (confusoside, **1**), 9.57 min ((S)-naringenin-7-*O*-β-D-glucopyranoside, **4**), 10.32 min (vacciniifolin, **2**), 12.64 min (2′,3,4,4′-tetrahydroxydihydrochalcone, **5**), and 13.95 min (1-[4-(β-D-glucopyranosyloxy)-2-hydroxyphenyl]-3-(4-hydroxy-3-methoxyphenyl)-1-propanone, **3**) (Figure 1A). Among them, compounds **1–6** belong to flavonoids, and compound **7** is benzoquinolethanoid.

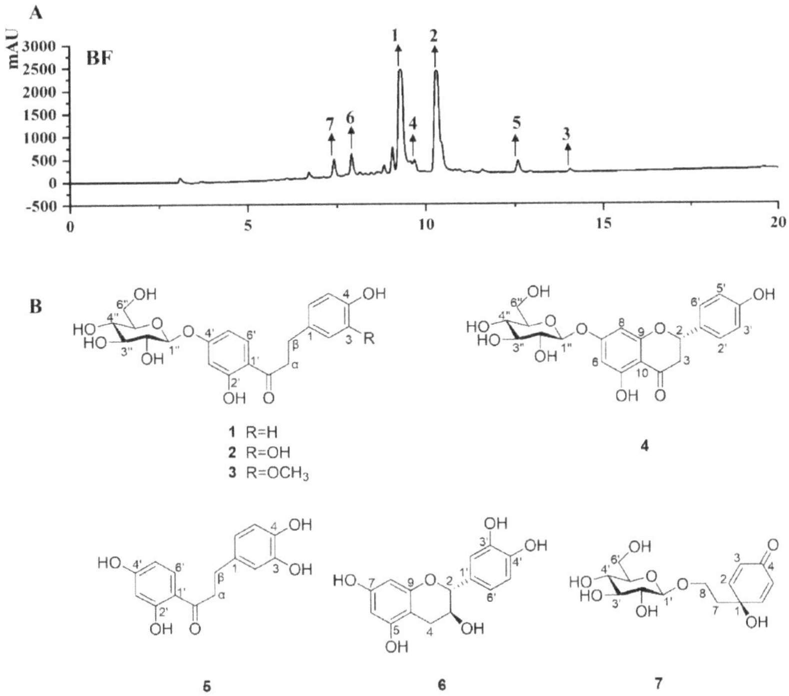

Figure 1. (**A**). HPLC chromatogram of BF from *Anneslea fragrans*. (**B**). Chemical structures of the compounds **1–7**.

2.3. Total Phenolic Contents (TPC) and Total Flavonoid Contents (TFC)

The *A. fragrans* leaves are traditionally used as processed health tea and have been proven to be good resource of phenolics and flavonoids [25]. According to spectrophotometric assays, the TPC and TFC were tested in different fractions and subfractions from *A. fragrans* leaves. As shown in Table 1, the BF had the highest TPC value with 238.12 ± 12.05 mg GAE/g extract followed by EAF. The DF showed the lowest TPC value as 66.52 ± 0.57 mg GAE/g extract. Similarly, the highest TFC concentrations were also found in BF with TFC value as 165.19 ± 5.21 mg RE/g extract. With regard to the RWF and EAF (66.92 ± 1.38 and 116.12 ± 2.89 mg GAE/g extract, respectively), they also had lower TFC concentrations than the BF (165.19 ± 5.21 mg RE/g extract), while DF had the lowest TFC concentration as 47.94 ± 2.84 mg RE/g extract. The contents of TPC and TFC in BF were approximately four times higher than those in DF.

Table 1. TPC, TFC, and antioxidant capacity values of four fractions and five subfractions from *Anneslea fragrans* Wall [1].

Samples	TPC	TFC	FRAP [2]	ABTS [3]	DPPH [4]
Vc [5]	-	-	3126.08 ± 197.63 [h]	2932.91 ± 93.63 [g]	1873.56 ± 121.68 [a]
DF	66.52 ± 0.57 [b]	47.94 ± 2.84 [b]	180.05 ± 9.46 [b]	139.23 ± 8.62 [a]	779.94 ± 84.76 [b]
EAF	123.49 ± 1.14 [d]	116.12 ± 2.89 [d]	975.24 ± 16.73 [e]	796.92 ± 19.26 [c]	684.31 ± 31.26 [b]
BF	238.12 ± 12.05 [g]	165.19 ± 5.21 [f]	1822.96 ± 29.24 [f]	1808.46 ± 96.52 [d]	951.42 ± 87.75 [b]
RWF	81.29 ± 1.48 [c]	66.92 ± 1.38 [c]	592.65 ± 11.55 [d]	135.38 ± 6.75 [a]	613.38 ± 45.35 [b]
BF-A	22.22 ± 1.63 [a]	25.46 ± 2.89 [a]	71.27 ± 5.42 [a]	130.01 ± 4.16 [a]	128.60 ± 6.31 [a]
BF-B	76.37 ± 0.94 [c]	63.63 ± 3.82 [c]	427.61 ± 13.78 [c]	246.92 ± 7.03 [b]	615.31 ± 38.21 [b]
BF-C	210.69 ± 6.31 [e]	151.01 ± 4.33 [e]	1777.94 ± 72.93 [f]	2020.87 ± 97.29 [e]	1249.87 ± 81.59 [c]
BF-D	219.84 ± 4.01 [f]	157.01 ± 2.50 [e]	1839.38± 12.87 [f]	1881.53 ± 28.57 [d]	1431.35 ± 98.32 [cd]
BF-E	262.03 ± 1.72 [h]	180.52 ± 6.30 [g]	2096.77 ± 49.13 [g]	2162.31 ± 64.49 [f]	1660.42 ± 119.72 [de]

[1] TPC: total phenolic content expressed as mg GAE/g extract, TFC: total flavonoid content expressed as mg RE/g extract. All the values are expressed as mean ± SD ($n = 3$). Data are obtained from three replicates and presented as mean ± SD; different numbers in the same column with different letters as superscript are significantly different ($p < 0.05$). DF, EAF, BF, and RWF mean dichloromethane fraction, ethyl acetate fraction, *n*-butanol fraction, and residual water fraction. BF-A to E mean that BF was subjected to D101 column chromatography to yield five subfractions. [2] FRAP: µmol Trolox equivalents (TE)/g extract. [3] ABTS: µmol Trolox equivalents (TE)/g extract. [4] DPPH: µmol Trolox equivalents (TE)/g extract. [5] Vc means the standard of vitamin C.

The data depicted the presence of highest TPC and TFC in BF-E with values as 262.03 ± 1.72 mg GAE/g extract; 180.52 ± 6.30 mg RE/g extract, respectively, followed by BF-D (219.84 ± 4.01 and 157.01 ± 2.50 mg RE/g extract, respectively), BF-C (210.69 ± 6.31 and 151.01 ± 4.33 mg RE/g extract, respectively), while BF-B and BF-A had the lowest TPC and TFC values (Table 1).

2.4. Antioxidant Activity

Most of the polyphenols, especially flavonoids and phenolic acids, are exploited into popular antioxidant foods (nutraceuticals) and present a series of human benefits [17]. In our previous study, the *A. fragrans* leaves have been proven to be a good resource of polyphenols [26]. However, the antioxidant activity of the extract from *A. fragrans* leaves and its phytochemicals have not been investigated yet. Due to different antioxidative reaction mechanisms, three assays of ABTS, DPPH, and FRAP were combined to evaluate the antioxidant activity of the fractions and subfractions from *A. fragrans* leaves.

Among the fractions, BF showed the most potent antioxidant activity in ABTS, DPPH, and FRAP radical-scavenging activities with 1808.46 ± 96.52, 951.42 ± 87.75, and 1822.96 ± 29.24 µmol TE/g extract (Table 1). DF showed the lowest antioxidative activity in ABTS and FRAP assays (139.23 ± 8.62 µmol TE/g extract and 180.05 ± 9.46 µmol TE/g extract, respectively) (Table 1). Whereas, the radical scavenging activities of four fractions (DF, EAF, BF, and RWF) in the DPPH radical-scavenging assay were found to vary from 613.38 ± 45.35 to 951.42 ± 87.75 µmol TE/g extract, respectively, having no significant difference ($p > 0.05$). These results suggested that the phytochemicals from *A. fragrans* leaves might be insensitive to DPPH. To further obtain the active metabolites, BF was selected for further fractionation. The BF was subjected to D101 macroporous adsorbing resin column chromatography eluting by a gradient of the methanol–aqueous system to yield five subfractions (BF-A to E). Hence, the ABTS radical-scavenging activity of these subfractions can be ranked as BF-E > BF-C > BF-D. The reducing activity in the FRAP assay revealed that BF-D had the strongest antioxidative activity, which was followed by BF-C and BF-E. It is noteworthy that three subfractions (BF-C to E) presented higher antioxidative activity than the mother fraction (BF). Taken together, BF-C to E were submitted to column chromatography for isolating and identifying the antioxidative phytochemicals.

2.5. Antioxidant Activity of the Isolated Compounds

All the isolated compounds (compounds **1–7**) were evaluated for antioxidant capacity by ABTS, DPPH radical scavenging activity, and FRAP antioxidant activity. All the data are described in Table 2. Compounds **6** and **2** showed the highest antioxidant activity, followed by **5**. Compound **4** displayed moderate antioxidant activity and compounds **1**, **3**, and **7** were considered inactive with ABTS and DPPH radical scavenging activities less than 200 μmol TE/g extract. Using Vc (the radical scavenging activities with 2932.91 ± 93.63 and 1873.56 ± 121.68 μmol TE/g extract, respectively) as the positive control, compound **6** had the strongest antioxidant activity in ABTS and DPPH radical scavenging assays with 1580.37 ± 89.32 and 1953.31 ± 109.93 μmol TE/g extract ($p < 0.05$), respectively. Furthermore, compound **6** had the best antioxidant capacity in the FRAP assay as 5027.43 ± 620.75 μmol TE/g extract, which was much higher than that of Vc as 3291.28 ± 241.02 μmol TE/g extract. Moreover, compounds **2** and **6** showed a significant antioxidant activity, which was equivalent to positive control (Vc).

Table 2. Antioxidant capacity values of seven compounds from *Anneslea fragrans* Wall [1].

Compound	FRAP [2]	ABTS [3]	DPPH [4]
Vc [5]	3126.08 ± 197.63 [e]	2932.91 ± 93.63 [e]	1873.56 ± 121.68 [c]
1	714.17 ± 64.63 [b]	174.81 ± 14.38 [a]	133.33 ± 4.98 [a]
2	3065.37 ± 283.87 [e]	1312.41 ± 108.63 [c]	1888.31 ± 162.94 [c]
3	562.09 ± 63.98 [a]	126.37 ± 9.32 [a]	128.33 ± 9.93 [a]
4	2806.15 ±171.92 [d]	1021.92 ± 51.82 [b]	1418.31 ± 84.62 [b]
5	2448.36 ± 150.38 [c]	1132.16 ± 93.85 [b]	1533.31 ± 114.96 [b]
6	5027.43 ± 620.75 [f]	1580.37 ± 89.32 [d]	1953.31 ± 109.93 [c]
7	613.94 ± 87.29 [a]	54.18 ± 4.38 [a]	108.33 ± 8.71 [a]

[1] TPC: total phenolic content expressed as mg GAE/g extract, TFC: total flavonoid content expressed as mg RE/g extract. Data are obtained from three replicates and presented as mean ± SD; different numbers in the same column with different letters as superscript are significantly different ($p < 0.05$). [2] FRAP: μmol Trolox equivalents (TE)/g extract. [3] ABTS: μmol Trolox equivalents (TE)/g extract. [4] DPPH: μmol Trolox equivalents (TE)/g extract. [5] Vc means the standard of vitamin C.

Additionally, the antioxidant structure–activity relationship of the flavonoids (**1–6**) is discussed by varying degrees of inhibitory effects. Compound **2** presented more antioxidant activity than compounds **1**, **3**, and **5**, which suggested that the 3,4-dihydroxy groups in dihydrochalcones in **2** may play a critical role against ABTS, DPPH radical scavenging activity, and FRAP antioxidant activity, and the glucose moiety substituted at C-4′ in **2** may play an important role on its antioxidant capacity [4]. Furthermore, compound **6** exhibited the highest antioxidant activity compared to other compounds, which suggested that flavan-3-ols (catechin) probably exhibit a better radical scavenging activity than dihydrochalcones and flavanones [27]. In summary, compounds **2**, **5**, and **6** with good antioxidant activity were selected further for cytoprotective effects against oxidative stress by H_2O_2 in the next study.

2.6. Relationship between Antioxidant Activity and TPC/TFC Contents

Concentrations of the TPC and TFC highly correlated with antioxidant activity from the values of FRAP (r = 0.982 and 0.977) and ABTS (r = 0.959 and 0.965), respectively. The correlation matrix also showed strong correlation between the ABTS and FRAP values (r = 0.992). Furthermore, a multivariate analysis (PCA), which was extracted from the data of Table 1, was carried out. As shown in Figure 2, PCA explained 94.44% of total variation, in which PC1 accounted for 84.73% of the variance and PC2 accounted for 9.71%. The FRAP, ABTS, and DPPH assays with EAF and RWF are placed at the upper right quadrants, and TPC and TFC concentrations with BF and BF-C to E are located at the lower right quadrants in the PC1 positive scores, respectively. Meanwhile, the DF and BF-A to B with low TPC and TFC concentrations are located along the axis of PC1 negative scores. These findings showed that the TPC and/or TFC concentrations are closely associated with antioxidant capacity, and the greater the TPC or TFC concentrations in the fractions and subfractions, the higher their antioxidant capacity values. These results revealed that the high phenolic and flavonoid contents in different fractions and subfractions from *A. fragrans* leaves might contribute to antioxidant activity.

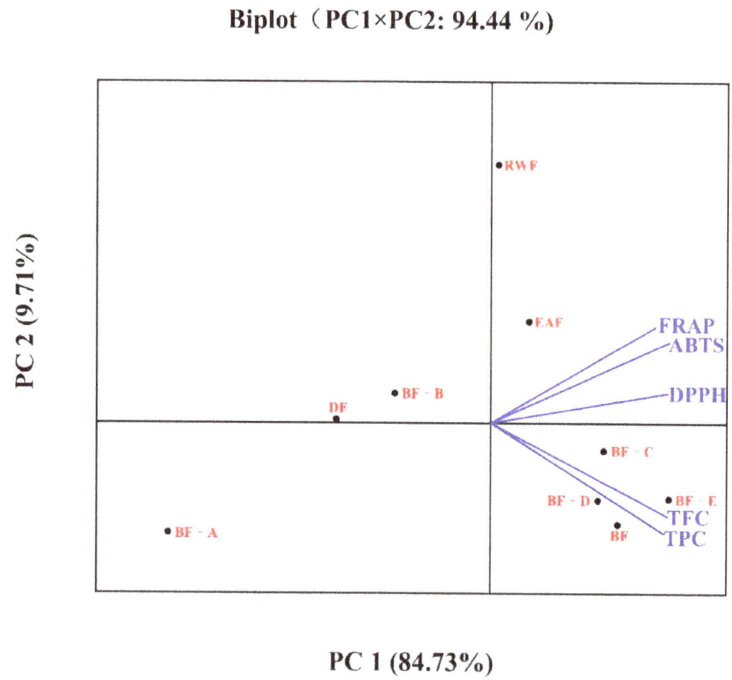

Figure 2. PCA analysis on total phenolics, flavonoids, and antioxidant activity. Means with different letters indicate significant differences ($p < 0.05$). DF, EAF, BF, RWF mean dichloromethane fraction, ethyl acetate fraction, *n*-butanol fraction, and residual water fraction. BF-A to E means BF was subjected to D101 column chromatography to yield five subfractions.

2.7. Inhibitory Effect on Intracellular ROS Generation

Oxidative stress is a kind of imbalance between oxidation and antioxidation in the body [28]. Excessive accumulation of ROS can lead to oxidative stress that may cause damage to cells and tissues such as lipids, membranes, and DNA [29]. Antioxidants including polyphenols and flavonoids can help the body reduce these damages caused by ROS. Generally, H_2O_2 is widely used to induce intracellular ROS disordered production and impair the antioxidant defense of cells [30]. In our present study, H_2O_2 was chosen to induce the abnormal accumulation of intracellular ROS and impair the antioxidant defense of cells. To evaluate the inhibitory effect on intracellular ROS generation in H_2O_2-induced HepG2 cells, the levels of intracellular ROS were tested by flow cytometry. Briefly, HepG2 cells were cultured in a 6-well plate (1×10^5 cells/well). After incubation for 24 h of 50 μg/mL of different extracts or 8 μg/mL of Vc (positive group), all the groups except for the control group were induced by H_2O_2 for 6 h. The intracellular ROS levels of each compound were tested (Figure 3). The ROS generation ratio significantly increased to $215.64 \pm 9.80\%$ in H_2O_2-treated group compared with the control group (100%). Compared with the H_2O_2-treated group, compounds **2**, **5**, and **6** remarkably suppressed intracellular ROS production ($p < 0.05$), and their inhibitory effect was equal to the Vc group (Figure 3). Many phenolics have been proven to have protective effect against intracellular ROS by H_2O_2 induction [31]. Additionally, compound **6** displayed the strongest suppressive effect on intracellular ROS production, which suggests that flavonoid compounds play a crucial role in inhibiting intracellular ROS production.

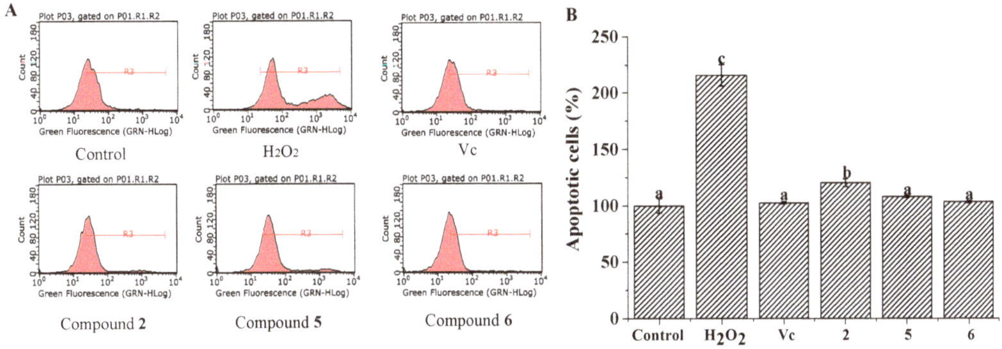

Figure 3. The inhibitory effects of compounds **2**, **5**, and **6** on intracellular ROS in H_2O_2-induced HepG2 cells. (**A**) Flow cytometry analysis; (**B**) the ROS intensity of different groups. All the values are presented as mean ± SD ($n = 3$). Means (bar values) with different letters indicate significant differences ($p < 0.05$).

2.8. Cytoprotective Effect Against H_2O_2-Induced Cell Apoptosis

Apoptosis is a basic biological phenomenon of cells, which plays an important role in the regulatory mechanism of cells' proliferation, growth, and mutation, and the stability of the internal environment. Apoptosis, different from necrosis, is a special type of cell death. The disorder of apoptotic process has bad effects on the body and causes many diseases [32]. H_2O_2, as an important signaling molecule, regulates the process of cell proliferation, growth, and apoptosis [5]. The present study measured the apoptosis of H_2O_2-induced HepG2 cells and evaluated the cytoprotective effects of compounds **2**, **5**, and **6**. After treating HepG2 cells with 1.0 mM H_2O_2, the apoptosis ratio remarkably augmented (57.20 ± 1.97%), compared with that of the control group (9.10 ± 0.62%, $p < 0.05$) (Figure 4). The ratios of apoptotic cells in the treated groups of compounds **2**, **5**, and **6** significantly decreased compared with the H_2O_2-treated group (model group, $p < 0.05$) (Figure 4). Moreover, compound **6** had significantly efficiency on protecting HepG-2 cells from H_2O_2 toxicity, and the cell apoptosis ratio of **6** (10.56 ± 1.15%) was lower than that of the Vc group (positive control), which was equal to that of the control group (Figure 4). Meanwhile, compounds **2** and **5** showed moderate cytoprotective effect with cell apoptosis ratios of **2** (26.76 ± 2.60%) and **5** (27.64 ± 0.83%). The differences of antioxidant capacity may be attributed to the number of phenolic hydroxyl moieties and the link positions.

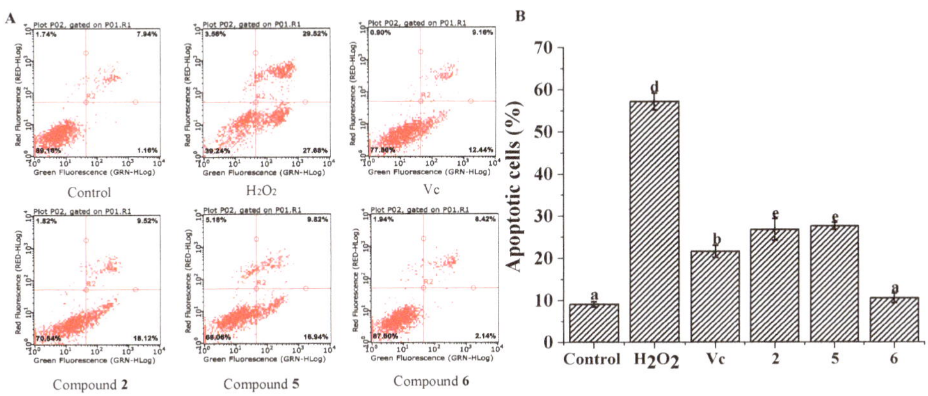

Figure 4. Cytoprotective effect of the compounds **2**, **5**, and **6** on apoptosis in H_2O_2-induced HepG-2 cells. (**A**) Flow cytometry analysis; (**B**) the apoptotic cell percentage of different groups. All the values are presented as mean ± SD ($n = 3$). Means (bar value) with different letters are significantly different ($p < 0.05$).

3. Materials and Methods

3.1. Chemicals and Reagents

Methanol, acetonitrile, and formic acid for high-performance liquid chromatography (HPLC) were of HPLC grade and purchased from Merck (Darmstadt, Germany). Solvents for sample extraction including ethanol, dichloromethane, and *n*-butanol were of analytical grade. Deionized water was purified using a Milli-Q ultrapure water system (Millipore, Bedford, Massachusetts, MA, USA) and employed in all the experiments. Phenolic standard compounds of gallic acid, rutin, Trolox, and vitamin C were purchased from Chengdu Must Bio-Technology Co., Ltd. (Chengdu, China). Methylthiazol-2-yl-2,5-diphenyl tetrazolium bromide (MTT), Folin–Ciocalteu reagent, 2,2′-azino-bis(3-ethylbenzo-thiazoline-6-sulfonic acid) (ABTS), 2,2-diphenyl-1-picrylhydrazyl radical (DPPH), 1,3,5-tri(2-pyridyl)-2,4,6-triazine (TPTZ), 2′,7′-dichlorofluorescin diacetate (DCFH-DA), and $FeSO_4{\cdot}7H_2O$ were purchased from Sigma–Aldrich (Shanghai, China). The NMR spectra were obtained using Bruker AV-400, and/or DRX-500 spectrometers. ESIMS spectra were recorded on An Agilent 1290 UPLC/6540 Q-TOF spectrometer.

3.2. Sample Preparation

Anneslea fragrans Wall. leaves were collected from the Lincang city of China in July 2020. The leaves were dried in a shade room until constant weight and then were powdered with an electric grinder. The extraction and fractionation were performed as our previously reported method with slight modification [33]. The powdered sample (100 g) was mixed with 1000 mL of 80% aqueous ethanol solvent and ultrasonicated in an ultrasonic cleaning bath at 200 W for three times (0.5 h per time). Then, the sonicated slurry was collected and centrifuged at 1500× *g* for 10 min by Eppendorf centrifuge (TGL-20B, Shanghai Anting Scientific Instrument Factory, Shanghai, China). The combined supernatant was concentrated at 50 °C by a rotary evaporator (Hei-VAP, Heidolph, Germany) and further dried by a vacuum drying lyophilizer (Alpha 1-2 LD plus, Christ, Germany). The crude ethanol extract (CE, 30 g) was re-suspended with water and sequentially partitioned with dichloromethane, ethyl acetate, and *n*-butanol solvents three times. After concentration and lyophilization, the dichloromethane fraction (DF), ethyl acetate fraction (EAF), *n*-butanol fraction (BF), and residual water fraction (RWF) weighing 3.2 g, 6.3 g, 7.2 g, and 8.2 g were obtained, respectively. According to the antioxidant activities of different fractions, the BF was chromatographed on glass columns (30 mm × 400 mm) wet-packed with 20 g (dry resin) of the selected hydrated resin D101. The bed volume (BV) of the resin was about 40 mL. After reaching the adsorptive saturation, the column was first washed by distilled water with 4 × BV and then eluted by ethanol–water (0:100, 20:80, 50:50, 80:20, 100:0, *v*/*v*, each 4 × BV), to yield five subfractions (BF-A to E). Each part of the desorption solutions was concentrated to dryness under vacuum. The CE, four fractions (DF, EAF, BF, and RWF) and five subfractions (BF-A to E) were stored in a refrigerator (−20 °C) for further experimentation.

3.3. Bio-Guided Isolation of Active Constituents

Under the guidance of antioxidant assays and HPLC analysis, the antioxidative fraction was further chromatographed for the isolation of pure compounds. In brief, the BF was subjected to a hydrated resin D101 column to yield five subfractions (BF-A to E). The BF-C (1.5 g) was subjected to a silica gel column, eluting with DCM/MeOH (15:1), and then was separated using preparative TLC (DCM/MeOH, 10:1) to obtain compounds **6** (118 mg) and **7** (10 mg). BF-D (1.1 g) was subjected to silica gel column ($CHCl_3$/MeOH 10:1, 5:1) to give compounds **1** (129 mg) and **4** (15 mg). BF-E (1.2 g) was subjected to silica gel column ($CHCl_3$/MeOH, 30:1, 10:1, 5:1) to yield compounds **1** (216 mg), **2** (135 mg), and a mixture. The later was purified by silica gel column ($CHCl_3$/MeOH, 12:1) to afford compounds **3** (19 mg) and **5** (15 mg) (Figure 5).

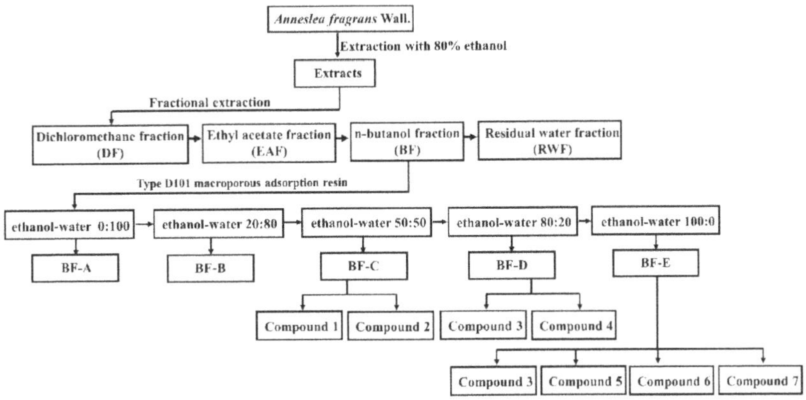

Figure 5. Extraction and isolation procedure of the compounds from 80% aqueous ethanol extract from *Anneslea fragrans* leaves.

3.4. Structure Elucidation of Compounds **1**–**7**

According to the antioxidant activities of different fractions, three subfractions, BF-C to E were submitted to column chromatography. In this way, the bioactivity-guided fractionation of BF-C to E led to the isolation of seven individual phenolic compounds. Their structures were identified as confusoside (**1**) [34], vacciniifolin (**2**) [34], 1-[4-(β-D-glucopyranosyloxy)-2-hydroxyphenyl]-3-(4-hydroxy-3-methoxyphenyl)-1-propanone (**3**) [35], (*S*)-naringenin-7-*O*-*β*-D-glucopyranoside (**4**) [22], 2′,3,4,4′-tetrahydroxydihydrochalcone (**5**) [26], (epi)-catechin (**6**) [26], and cornoside (**7**) [36] (Figure 1B) by the analysis of 1D-NMR and ESI-MS data and comparison with the previously reported compounds in the literature. Among them, compounds **3**, **4**, and **7** were isolated from this plant for the first time.

Confusoside (**1**). The molecular formula was assigned as $C_{21}H_{24}O_9$, and the separation gave 129 mg of pale-yellow needles. ^{1}H NMR (500 MHz, DMSO-d_6) and ESI-MS (m/z 421 $[M + H]^+$) data were identical to the previously reported compound in the literature [34]. The identification was further supported by ^{13}C NMR (125 MHz, DMSO-d_6): δ 204.4 (s, C=O), 163.5 (s, C-2′), 163.3 (s, C-4′), 155.5 (s, C-4), 132.6 (d, C-6′), 130.9 (s, C-1), 129.2 (d, C-2, 6), 115.0 (d, C-3/C-5), 114.4 (s, C-1′), 108.3 (d, C-5′), 103.4 (d, C-1″), 99.6 (d, C-3′), 77.1 (d, C-5″), 76.4 (d, C-3″), 73.1 (d, C-2″), 69.5 (d, C-4″), 60.5 (t, C-6″), 39.8 (t, C-α), 28.9 (t, C-β).

Vacciniifolin (**2**) was obtained as yellow amorphous powder (153 mg) and assigned a molecular formula of $C_{21}H_{24}O_{10}$. ^{1}H NMR (500 MHz, DMSO-d_6) and ESI-MS (m/z 437 $[M+H]^+$) data were the same as the previous reported data [34]. The identification was further supported by ^{13}C NMR (125 MHz, DMSO-d_6): δ 204.4 (s, C=O), 163.5 (s, C-2′), 163.3 (s, C-4′), 145.0 (s, C-3), 143.3 (s, C-4), 132.6 (d, C-6′), 131.7 (s, C-1), 118.9 (d, C-6), 115.8 (d, C-2), 115.4 (d, C-5), 114.4 (s, C-1′), 108.3 (d, C-5′), 103.4 (d, C-1″), 99.6 (d, C-3′), 77.1 (d, C-5″), 76.4 (d, C-3″), 73.1 (d, C-2″), 69.5 (d, C-4″), 60.5 (t, C-6″), 39.4 (t, C-α), 29.1 (t, C-β).

1-[4-(β-D-Glucopyranosyloxy)-2-hydroxyphenyl]-3-(4-hydroxy-3-methoxyphenyl)-1-propanone (**3**). The compound was obtained as colorless needles (19 mg), and the molecular formula was assigned as $C_{22}H_{26}O_{10}$. ^{1}H NMR (400 MHz, DMSO-d_6) and ESI-MS (m/z 451 $[M + H]^+$) data agreed with the literature [35]. The identification was further supported by ^{13}C NMR (100 MHz, DMSO-d_6): δ 204.5 (s, C=O), 163.5 (s, C-2′), 163.3 (s, C-4′), 147.4 (s, C-3), 144.6 (s, C-4), 132.6 (d, C-6′), 131.6 (s, C-1), 120.4 (d, C-6), 115.2 (d, C-5), 114.5 (s, C-1′), 112.6 (d, C-2), 108.3 (d, C-5′), 103.3 (d, C-1″), 99.5 (d, C-3′), 77.0 (d, C-5″), 76.3 (d, C-3″), 73.0 (d, C-2″), 69.5 (d, C-4″), 60.5 (t, C-6″), 55.5 (q, C-OCH_3), 39.5 (t, C-α), 29.4 (t, C-β).

(*S*)-Naringenin-7-*O*-*β*-D-glucopyranoside (**4**) had the molecular formula of $C_{21}H_{22}O_{10}$, which was obtained as 15 mg of white powder. ^{1}H NMR (500 MHz, DMSO-d_6) and ESI-MS (m/z 435 $[M + H]^+$) data were in agreement with the previous work [22]. The identification was further supported by ^{13}C NMR (125 MHz, DMSO-d_6): δ 197.2 (s, C-4), 165.3 (s, C-7),

165.2 (s, C-5), 162.9 (s, C-9), 157.9 (s, C-4′), 128.6 (s, C-1′), 128.4 (d, C-2′, 6′), 115.2 (d, C-3′/C-5′), 103.2 (s, C-10), 99.6 (d, C-1″), 96.5 (d, C-6), 95.4 (d, C-8), 78.7 (d, C-2), 77.0 (d, C-5″), 76.3 (d, C-3″), 73.0 (d, C-2″), 69.5 (d, C-4″), 60.5 (t, C-6″), 42.0 (t, C-3).

2′,3,4,4′-Tetrahydroxydihydrochalcone (**5**) possessed the molecular formula as $C_{15}H_{14}O_5$ and separated (15 mg) as white powder. ^{1}H NMR (400 MHz, DMSO-d_6) and ESI-MS (*m/z* 275 [M + H]$^+$) data were consistent with that reported in the literature [26]. The identification was further supported by ^{13}C NMR (125 MHz, DMSO-d_6): δ 203.9 (s, C=O), 164.7 (s, C-2′), 164.2 (s, C-4′), 144.9 (s, C-3), 143.3 (s, C-4), 133.0 (d, C-6′), 131.8 (s, C-1), 118.9 (d, C-6), 115.7 (d, C-2), 115.4 (d, C-5), 112.5 (s, C-1′), 108.2 (d, C-5′), 102.4 (d, C-3′), 39.5 (t, C-α), 29.2 (t, C-β).

(Epi)-catechin (**6**) was isolated as white powder (118 mg) and established the molecular formula as $C_{11}H_6O_4$. ^{1}H NMR (400 MHz, DMSO-d_6) and ESI-MS (*m/z* 291 [M + H]$^+$) data agreed well with that reported in the literature [26]. The identification was further supported by ^{13}C NMR (125 MHz, DMSO-d_6): δ 156.4 (s, C-7), 156.1 (s, C-5), 155.3 (s, C-9), 144.8 (s, C-3′), 144.8 (s, C-4′), 130.5 (s, C-1′), 118.4 (d, C-6′), 115.0 (d, C-5′), 114.4 (d, C-2′), 99.0 (s, C-10), 95.1 (d, C-6), 93.8 (d, C-8), 80.9 (d, C-2), 66.2 (d, C-3), 27.8 (t, C-4).

Cornoside (**7**) was obtained as amorphous solid (10 mg) and determined the molecular formula as $C_{14}H_{20}O_8$. ^{1}H NMR (500 MHz, DMSO-d_6) and ESI-MS (*m/z* 317 [M + H]$^+$) data corresponded with the published data [36]. The identification was further supported by ^{13}C NMR (100 MHz, DMSO-d_6): δ 185.3 (s, C-4), 153.3 (d, C-2), 153.2 (d, C-6), 126.4 (d, C-3), 126.4 (d, C-5), 102.8 (d, C-1′), 76.8 (d, C-5′), 76.6 (d, C-3′), 73.3 (d, C-2′), 70.0 (d, C-4′), 67.3 (s, C-1), 63.8 (t, C-8), 61.0 (t, C-6′), 39.7 (t, C-7).

3.5. *Determination of Total Phenolic (TPC) and Total Flavonoid Contents (TFC)*

The TPC and TFC of four fractions (DF, EAF, BF, and RWF) and five subfractions (BF-A to E) were measured according to our previously reported method [37]. For TFC, 1.0 mL of each sample (with the concentration at 1.0 mg/mL) dissolved in methanol was mixed with 0.5 mL of Folin–Ciocalteu reagent in a centrifuge tube and incubated for 1 min. Then, 20% Na_2CO_3 solution (*m/v*) (1.5 mL) and deionized water (7.0 mL) were added to the tube and kept at 70 °C in a water bath for 10 min. After being cooled to room temperature, 200 μL of the solution was transferred to a 96-well microplate and the absorbance was determined at 765 nm by a SpectraMax M5 microplate reader (Molecular Devices, Sunnyvale, CA, USA).

For TFC, 1.2 mL of sample solutions (with the concentration at 1.0 mg/mL) were mixed with 0.3 mL of $NaNO_2$ (5% *m/v*) and 3.8 mL of 70% aqueous ethanol and incubated for 8 min. Subsequently, 0.3 mL 10% aqueous $Al(NO_3)_3$, 4 mL 4% aqueous NaOH, and 0.4 mL 70% aqueous ethanol were added to the mixture and allowed to react at room temperature for 30 min. Then, 200 μL of the solution was transferred to a 96-well microplate, for which the absorbance was measured at 510 nm by using a microplate reader. The TFC and TPC was expressed as milligrams of gallic acid equivalents (mg GAE/g extract) and rutin equivalents per gram of extract (mg RE/g extract).

3.6. *Determination of Antioxidant Activity*

The antioxidant activity of four fractions (DF, EAF, BF, and RWF) and five subfractions (BF-A to E) were evaluated in a combination of DPPH and ABTS radical scavenging assays and FRAP assay based on the method described in our previous study [33]. For DPPH assay, 50 μL of the sample solution (50, 100, 200 μg/mL) was mixed with 0.2 mL DPPH solution (0.1 mmol/L) in a 96-well plate and allowed to incubate for 30 min. The absorbance was measured at 517 nm with SpectraMax M5 microplate reader (Molecular Devices, Sunnyvale, CA, USA). For ABTS, 25 μL of the sample solution (50, 100, 200 μg/mL) were added to 0.2 mL ABTS solution (7 mmol/L). The mixture was kept in the dark for 6 min, and then, the absorbance was recorded at 734 nm. For FRAP, 20 μL sample solution (50, 100, 200 μg/mL) was mixed with 0.18 mL of FRAP reagent (7 mmol/L). After incubating for 10 min in the dark at 37 °C, the absorbance was determined at 593 nm. All the tests were performed in triplicate. The results of DPPH, ABTS, and FRAP values were expressed as μmol Trolox equivalents per gram of extract (μmol TE/g extract).

3.7. HPLC Analysis

The BF and compounds **1–7** were analyzed on an Agilent 1260 HPLC system coupled with a diode array detector. Before analysis, the freshly prepared sample solution was filtered through a 0.45 μm nylon membrane. The separation was performed using a Reprosil-Pur Basic C18 column (5 μm, 4.6 × 250 mm, Germany) maintained at 35 °C. The injection volume was 5.0 μL, the flow rate was 1.0 mL/min, and the detection wavelength was set at 280 nm. The mobile phases were acidified water with 0.1% formic acid (phase A) and acetonitrile (phase B), the linear gradient elution was performed as follows: 0–3 min, 20% B; 3–10 min, 40% B; 10–15 min, 60% B; 15–20 min, 100% B.

3.8. Cell Culture and Cell Viability

Human liver cancer HepG2 cells were purchased from Kunming Cell Bank (Kunming, China). HepG2 cells were grown in DMEM supplemented with 1% penicillin–streptomycin and 10% fetal bovine serum in an atmosphere of 5% CO_2/95% air at 37 °C. When the cells were incubated to an appropriate density (approximately 80%), they were treated with positive control (Vc) and the isolated compounds for further experiments.

Cell viability was determined by MTT assay for evaluating the cytotoxicity of each sample [29]. The cells at a density of 1×10^4 cells per well were seeded in a 96-well plate and allowed to incubate for 24 h. Each compound (prepared as four doses from 50 to 200 μg/mL) was added to each well for 20 h. Then, the cells were treated using MTT solution with a final concentration for 4 h. The medium with MTT was removed, and 200 μL of the DMSO was added to dissolve the formazan. The absorbance was recorded at 570 nm by a microplate reader. The results demonstrated that each compound was nontoxic to HepG2 cells at the tested concentrations.

3.9. Inhibition of ROS Generation in H_2O_2-Induced HepG2 Cells

H_2O_2-induced HepG2 cells were employed to determine the inhibitory effect on ROS production [3]. HepG2 cells (1.0×10^5 cells per well) were seeded in a 6-well plate and co-cultured with isolated compounds with 50 μg/mL and Vc (8 μg/mL). After incubation for 20 h, the medium was removed, and 2 μL of H_2O_2 (0.5 mM) was added to each well for another 6 h. At the end of experiment, the cells were labeled with 2 μL DCFH-DA (10 mM) in the dark at 37 °C for 20 min. The absorbance was recorded by flow cytometry (Guava easyCyte 6-2L, Millipore, Billerica, Massachusetts, MA, USA).

3.10. Determination of Cell Apoptosis

The protective effect of each compound on H_2O_2-induced apoptosis of HepG2 cells was determined using a human annexin VFITC/PI apoptosis kit [38]. HepG2 cells were pre-treated with or without isolated compounds for 48 h. After incubation, 100 μL of the binding buffer was added to the cells, and the cells reacted in the dark with 10 μL annexin V-FITC for 5 min at room temperature and with 10 μL propidium iodide (PI) in an ice bath for 5 min, successively. Cell apoptosis was immediately analyzed using flow cytometry.

3.11. Statistical Analysis

All the experiments were performed in triplicate. All the values are expressed as mean ± standard deviation (SD). The differences within and between the groups were analyzed using one-way analysis of variance (ANOVA) followed by Tukey's test. Difference was considered statistically significant at $p < 0.05$. All analyses were performed using Origin 2019b software (OriginLab, Northampton, MA, USA).

4. Conclusions

In this study, different fractions from *A. fragrans* leaves were fractionated, and their TPC and TFC were analyzed. Under antioxidant activity guided isolation, compounds **1**−**7**, including four flavonoid glycosides (**1**–**4**) and two flavonoids (**5** and **6**), were isolated and identified from *A. fragrans* leaves, which suggested that this species is rich in flavonoid compounds. Compounds **2**, **5**, and **6** showed significant antioxidant activity in DPPH, ABTS radical scavenging, and FRAP assays. Furthermore, they visibly prevented the oxidative stress damage through a decrease in ROS content and cell apoptosis in H_2O_2-induced HepG2 cells. According to these results, polyphenol compounds, especially flavonoids, have considerable antioxidant capacity because of their phenolic hydroxyl groups. Furthermore, compound **2**, possessing the glycoside moiety and three phenolic hydroxyl groups, was the main antioxidant component with the highest content from *A. fragrans* leaves. Compound **6** displayed the best antioxidant activity, which may be a major contribution to the activity of *A. fragrans*. Furthermore, the extracts of *A. fragrans* could be served as a feasible natural source of antioxidants in promising health beverages. The study on the compounds from *A. fragrans* leaves suggest that these could be served as antioxidant healthy tea for treating oxidative stress-induced cell damage and could serve as nutritional supplements applied in the food and health industry.

Author Contributions: S.H.: investigation, methodology, data curation, visualization, writing—original draft; X.C.: conceptualization, methodology, data curation, visualization, writing—original draft; A.K.: writing—review and editing; Y.L.: formal analysis, writing—review and editing, funding acquisition; Y.W.: visualization; Q.C.: software, data curation, supervision; T.Z.: software, validation; J.C.: resources, supervision; G.C.: conceptualization, methodology, validation, writing—review and editing, supervision, project administration, funding acquisition; All authors have read and agreed to the published version of the manuscript.

Funding: This project was financially supported by Yunnan Major Science and Technology Project in China (2019ZF010), the National Natural Science Foundation of China (31872676 and 32060542) and China Postdoctoral Science Foundation (2020M673586XB).

Institutional Review Board Statement: Not applicable.

Informed Consent Statement: Not applicable.

Data Availability Statement: Data is contained within the article.

Acknowledgments: This project was supported in parts by Yunnan Major Science and Technology Project (Grant No. 2019ZF010) and the National Natural Science Foundation of China (Grant No. 32060542).

Conflicts of Interest: The authors declare no conflict of interest.

Sample Availability: Samples of the compounds are available from the authors.

References

1. Jiao, Y.; Ji, L.; Kuang, Y.; Yang, Q. Cytotoxic effect of oxaloacetate on HepG2-human hepatic carcinoma cells via apoptosis and ROS accumulation. *Neoplasma* **2017**, *64*, 192–198. [CrossRef]
2. Imen, B.E.; Fatma, T.; Abdennacer, B.; Mohamed-Amine, J.; Mohamed, B.; Chokri, M.; Hichem, S. Bioactive compounds from Tunisian Pelargonium graveolens (L'Hér.) essential oils and extracts: α-amylase and acethylcholinesterase inhibitory and antioxidant, antibacterial and phytotoxic activities. *Ind. Crops Prod.* **2020**, *158*, 112951.
3. Cao, J.; Liu, Y.; Jia, L.; Jiang, L.P.; Geng, C.Y.; Yao, X.F.; Kong, Y.; Jiang, B.N.; Zhong, L.F. Curcumin attenuates acrylamide-induced cytotoxicity and genotoxicity in HepG2 cells by ROS scavenging. *J. Agric. Food Chem.* **2008**, *56*, 12059–12063. [CrossRef] [PubMed]
4. Cai, Y.; Luo, Q.; Mei, S.; Corke, H. Antioxidant activity and phenolic compounds of 112 traditional Chinese medicinal plants associated with anticancer. *Life Sci.* **2004**, *74*, 2157–2184. [CrossRef]
5. Kim, H.R.; Kim, J.E.; Yun, W.B.; Lee, M.R.; Choi, J.Y.; Park, J.J.; Song, B.R.; Song, H.K.; Hwang, D.Y. Protective effect of oil extracted from *Neophocaena asiaeorientalis* against hydgrogen peroxide-induced oxidative stress in HepG2 cells. *Fish. Sci.* **2019**, *85*, 867–876. [CrossRef]
6. Martin, M.A.; Ramos, S.; Mateos, R.; Granado, S.; Izquierdo-Pulido, M.A.; Bravo, L.; Goya, L. Protection of Human HepG2 Cells against Oxidative Stress by Cocoa Phenolic Extract. *J. Agric. Food Chem.* **2008**, *56*, 7765–7772. [CrossRef]

7. Berson, D.S. Natural antioxidants. *J. Drugs Dermatal.* **2008**, *7*, 7–12.
8. Osawa, T.; Namiki, M. Natural antioxidants isolated from Eucalyptus leaf waxes. *J. Agric. Food Chem.* **1985**, *5*, 777–780. [CrossRef]
9. Anna, P. Natural antioxidants and antioxidant capacity of *Brassica* vegetables: A review. *LWT-Food Sci. Technol.* **2007**, *44*, 1–11.
10. Xu, D.P.; Zheng, J.; Zhou, Y.; Li, Y.; Li, S. Ultrasound-assisted extraction of natural antioxidants from the flower of *Limonium sinuatum*: Optimization and comparison with conventional methods. *Food Chem.* **2017**, *217*, 552–559. [CrossRef] [PubMed]
11. Benabderrahim, M.A.; Yahia, Y.; Bettaieb, I.; Elfalleh, W.; Nagaz, K. Antioxidant activity and phenolic profile of a collection of medicinal plants from Tunisian arid and Saharan regions. *Ind. Crops Prod.* **2019**, *138*, 111427. [CrossRef]
12. Biais, B.; Krisa, S.; Cluzet, S.; Costa, G.D.; Waffo-Teguo, P.; Mérillo, J.M.; Richar, T. Antioxidant and cytoprotective activities of grapevine stilbenes. *J. Agric. Food Chem.* **2017**, *65*, 4592–4960. [CrossRef]
13. Noda, Y.; Kaneyuki, T.; Akitane, M.; Packer, L. Antioxidant activities of pomegranate fruit extract and its anthocyanidins: Delphinidin, cyanidin, and pelargonidin. *J. Agric. Food Chem.* **2002**, *50*, 166–171. [CrossRef] [PubMed]
14. Wei, H.; Bowen, R.; Cai, Q.; Barnes, S.; Wang, Y. Antioxidant and Antipromotional Effects of the Soybean Isoflavone Genistein. *Proc. Soc. Exp. Biol. Med.* **1995**, *208*, 124–130. [CrossRef]
15. Tian, L.; Huang, J.A.; Tian, L.; Huang, J.A. Antioxidant effects of tea catechins on the shelf life of raw minced duck meat. *LWT-Food Sci. Technol.* **2019**, *39*, 59–65. [CrossRef]
16. Fuentes, J.; Atala, E.; Pastene, E.; Carrasco-Pozo, C.; Speisky, H. Quercetin oxidation paradoxically enhances its antioxidant and cytoprotective properties. *J. Agric. Food Chem.* **2017**, *65*, 11002–11010. [CrossRef]
17. Khan, N.; Mukhtar, H. Multitargeted therapy of cancer by green tea polyphenols. *Cancer Lett.* **2008**, *269*, 269–280. [CrossRef] [PubMed]
18. Mehmood, A.; Usman, M.; Patil, P.J.; Zhao, L.; Wang, C. A review on management of cardiovascular diseases by olive polyphenols. *Food Sci. Nutr.* **2020**, *8*, 4639–4655. [CrossRef] [PubMed]
19. Silva, R.F.; Poganik, L. Polyphenols from Food and Natural Products: Neuroprotection and Safety. *Antioxidants* **2020**, *9*, 61. [CrossRef]
20. Zhang, C.; Suen, L.C.; Yang, C.; Quek, S.Y. Antioxidant capacity and major polyphenol composition of teas as affected by geographical location, plantation elevation and leaf grade. *Food Chem.* **2018**, *244*, 109–119. [CrossRef]
21. Yan, Z.; Dingfeng, G.; Simin, L.; Feng, S. Response Surface Optimization of Antioxidant Activity and Composition Analysis of Anneslea fragrans Wall. Bark Soaked Wine. *Food Res. Dev.* **2020**, *41*, 72–78.
22. Sun, L.; Hu, X.; Cheng, X.; Luo, L.; Liu, H.; Chen, L. NMR Characterization of Flavanone Naringenin 7-*O*-Glycoside Diastereomer. *Chin. J. Magn. Reson.* **2017**, *34*, 465–473.
23. Huang, H.Y.; Ko, H.; Jin, Y.J.; Yang, S.Z.; Shih, Y.A.; Chen, I.S. Dihydrochalcone glucosides and antioxidant activity from the roots of *Anneslea fragrans* var *lanceolata*. *Phytochemistry* **2012**, *78*, 120–125. [CrossRef] [PubMed]
24. Yeilada, E.; Tsuchiya, K.; Takaishi, Y.; Kawazoe, K. Isolation and characterization of free radical scavenging flavonoid glycosides from the flowers of Spartium junceum by activity-guided fractionation. *J. Ethnopharmacol.* **2000**, *73*, 471–478. [CrossRef]
25. Zhang, H. Theoretical solution of the relationship between the structural activity of flavonoid antioxidants. *Sci. China Ser. B* **1999**, *29*, 91–96.
26. Omar, A.M.; Dibwe, D.F.; Tawila, A.M.; Sun, S.; Phrutivorapongkul, A.; Awale, S. Chemical constituents of Anneslea fragrans and their antiausterity activity against the PANC-1 human pancreatic cancer cell line. *J. Nat. Prod.* **2019**, *82*, 3133–3139. [CrossRef]
27. Sheiraz, A.B.; Nathalie, M.; Michèle, L. Chemically Synthesized Glycosides of Hydroxylated Flavylium Ions as Suitable Models of Anthocyanins: Binding to Iron Ions and Human Serum Albumin, Antioxidant Activity in Model Gastric Conditions. *Molecules* **2014**, *19*, 20709–20730.
28. Aruoma, O.I. Free radicals, oxidative stress, and antioxidants in human health and disease. *J. Am. Oil Chem. Soc.* **1998**, *2*, 199–212. [CrossRef]
29. Marian, V.; Dieter, L.; Jan, M.; Milan, M.; Joshua, T. Free radicals and antioxidants in normal physiological functions and human disease. *Int. J. Biochem. Cell Biol.* **2007**, *39*, 44–84.
30. Li, X.L.; Zhou, A.G.; Li, X.M. Inhibition of Lycium barbarum polysaccharides and Ganoderma lucidum polysaccharides against oxidative injury induced by γ-irradiation in rat liver mitochondria. *Carb. Polym.* **2007**, *69*, 172–178. [CrossRef]
31. Cui, Q.; Wang, Y.; Zhou, W.; He, S.; Yang, M.; Xue, Q.; Wang, Y.; Zhao, T.; Cao, J.; Khan, A.; et al. Phenolic composition, antioxidant and cytoprotective effects of aqueous-methanol extract from Anneslea fragrans leaves as affected by drying methods. *Int. J. Food Sci. Technol.* **2021**. [CrossRef]
32. Hengartner, M.O. The biochemistry of apoptosis. *Nature* **2000**, *407*, 770–771. [CrossRef] [PubMed]
33. Yang, M.L.; Ma, Y.L.; Wang, Z.Y.; Khan, A.; Zhou, W.B.; Zhao, T.R.; Cao, J.X.; Cheng, G.G.; Cai, S.B. Phenolic constituents, antioxidant and cytoprotective activities of crude extract and fractions from cultivated artichoke inflorescence. *Ind. Crops Prod.* **2020**, *143*, 111433. [CrossRef]
34. Ling, T. Dihydrochalcones from Symplocos vacciniifolia. *Chin. Chem. Lett.* **2004**, *10*, 1182–1184.
35. Antus, S.; Farkas, L.; Gottsegen, Á.; Nógrádi, M.; Strelisky, J.; Pfliegel, T. Dihydrochalcone type sweetening agents. II. The synthesis of some dihydrochalcones. *Acta Chim.* **1978**, *98*, 231–240.
36. Li, H.M.; Kim, J.K.; Jang, J.M.; Cui, C.B.; Lim, S.S. Analysis of the inhibitory activity of *Abeliophyllum distichum* leaf constituents against aldose reductase by using high-speed counter current chromatography. *Arch. Pharm. Res.* **2013**, *36*, 1104–1112. [CrossRef]

37. Fan, Z.; Wang, Y.; Yang, M.; Cao, J.; Khan, A.; Cheng, G. UHPLC-ESI-HRMS/MS analysis on phenolic compositions of different E Se tea extracts and their antioxidant and cytoprotective activities. *Food Chem.* **2020**, *318*, 126512. [CrossRef]
38. Zhuang, Y.L.; Ma, Q.Y.; Guo, Y.; Sun, L.P. Protective effects of rambutan (*Nephelium lappaceum*) peel phenolics on H_2O_2-induced oxidative damages in HepG2 cells and D-galactose-induced aging mice. *Food Chem. Toxicol.* **2017**, *108*, 554–562. [CrossRef]

molecules

MDPI

Article

Computational Screening for the Anticancer Potential of Seed-Derived Antioxidant Peptides: A Cheminformatic Approach

Tsun-Thai Chai [1,2,*], Jiun-An Koh [1], Clara Chia-Ci Wong [1], Mohamad Zulkeflee Sabri [3] and Fai-Chu Wong [1,2]

1 Department of Chemical Science, Faculty of Science, Universiti Tunku Abdul Rahman, Kampar 31900, Malaysia; jiunan30@1utar.my (J.-A.K.); clara2000genesis@1utar.my (C.C.-C.W.); wongfc@utar.edu.my (F.-C.W.)
2 Center for Agriculture and Food Research, Universiti Tunku Abdul Rahman, Kampar 31900, Malaysia
3 Green Chemistry and Sustainable Technology Cluster, Bioengineering Section, Universiti Kuala Lumpur Malaysian Institute of Chemical and Bioengineering Technology (UniKL MICET), Lot 1988, Bandar Vendor Taboh Naning, Alor Gajah 78000, Malaysia; mzulkeflee@unikl.edu.my
* Correspondence: chaitt@utar.edu.my; Tel.: +60-5-468-8888

Abstract: Some seed-derived antioxidant peptides are known to regulate cellular modulators of ROS production, including those proposed to be promising targets of anticancer therapy. Nevertheless, research in this direction is relatively slow owing to the inevitable time-consuming nature of wet-lab experimentations. To help expedite such explorations, we performed structure-based virtual screening on seed-derived antioxidant peptides in the literature for anticancer potential. The ability of the peptides to interact with myeloperoxidase, xanthine oxidase, Keap1, and $p47^{phox}$ was examined. We generated a virtual library of 677 peptides based on a database and literature search. Screening for anticancer potential, non-toxicity, non-allergenicity, non-hemolyticity narrowed down the collection to five candidates. Molecular docking found LYSPH as the most promising in targeting myeloperoxidase, xanthine oxidase, and Keap1, whereas PSYLNTPLL was the best candidate to bind stably to key residues in $p47^{phox}$. Stability of the four peptide-target complexes was supported by molecular dynamics simulation. LYSPH and PSYLNTPLL were predicted to have cell- and blood-brain barrier penetrating potential, although intolerant to gastrointestinal digestion. Computational alanine scanning found tyrosine residues in both peptides as crucial to stable binding to the targets. Overall, LYSPH and PSYLNTPLL are two potential anticancer peptides that deserve deeper exploration in future.

Keywords: anticancer; cheminformatics; in silico; Keap1; molecular docking; molecular dynamics; myeloperoxidase; NADPH oxidase; seed antioxidant peptide; xanthine oxidase

Citation: Chai, T.-T.; Koh, J.-A.; Wong, C.C.-C.; Sabri, M.Z.; Wong, F.-C. Computational Screening for the Anticancer Potential of Seed-Derived Antioxidant Peptides: A Cheminformatic Approach. *Molecules* **2021**, *26*, 7396. https://doi.org/10.3390/molecules26237396

Academic Editors: Višnja Stepanić and Marta Kučerová-Chlupáčová

Received: 18 November 2021
Accepted: 2 December 2021
Published: 6 December 2021

1. Introduction

The past decade has seen a surge in scientific interest towards the exploration of bioactive peptides for potential applications in health promotion and disease management. Bioactive peptides identified from plant food and other natural origins often range between 2 and 20 residues, although this is not a hard-and-fast definition as exceptions do exist [1–3]. Plant bioactive peptides, known to exhibit diverse bioactivities, such as antioxidant, antihypertensive, antimicrobial, and antitumor activities, are often purified and identified from enzymatic hydrolysates of edible plant sources and plant-based agricultural by-products. The bioactive potency of some such peptides have also been demonstrated in cellular and animal models [2,4,5]. To date, a growing body of research has shown that plant seeds are a good source of antioxidant peptides [2,5,6]. While such peptides could be developed into natural additive for food processing and nutraceuticals for health maintenance, they may also be therapeutically relevant as some could modulate cellular and/or in vivo antioxidant status [2]. Cellular redox homeostasis is connected to the initiation and/or progression of certain cancers [7,8]. Perturbation in reactive oxygen species (ROS) homeostasis resulting from unchecked ROS production is associated with carcinogenesis; scavenging of excessive ROS accumulation may prevent early neoplasia [9]. Significant reduction in the antioxidant activity of the blood serum of patients with malignant neoplasms has also been reported [10].

In the body, cellular redox status is regulated by oxidative and antioxidative enzymes, non-enzymatic antioxidants, and certain protein-protein interactions involved in regulating antioxidant gene expression. Myeloperoxidase (MPO), xanthine oxidase (XO), and nicotinamide adenine dinucleotide phosphate oxidase (NADPH oxidase) are three examples of such oxidative enzymes. MPO, an abundant heme-containing enzyme in the human neutrophils, catalyzes the reaction between hydrogen peroxide and chloride, generating hypochlorous acid, a potent oxidant. MPO-mediated oxidative burst has been linked to the initiation and progression of cancer, including tumor cell metastasis. Notably, downregulation of MPO gene expression is connected to reduction in the risk of lung, breast, and ovarian cancers [7,8]. XO is an enzyme that catalyzes the conversion of hypoxanthine to xanthine and ultimately to uric acid, producing ROS during the reaction. The importance of XO as an anticancer target is highlighted by the discovery that XO inhibitor febuxostat could repress breast cancer cell migration and the metastasis of breast cancer to the lung in animal models [11,12]. Six isoforms of NADPH oxidase are known to date. NADPH oxidase is a membrane-bound enzyme complex in phagocytes, whose primary function is the production of superoxide anion radicals. The assembly and activation of NADPH oxidase requires protein-protein interaction between the cytosolic factor $p47^{phox}$ and transmembrane component $p22^{phox}$ [13,14]. Due to the importance of $p47^{phox}$-$p22^{phox}$ interaction in NADPH oxidase activation, the interaction can be targeted in structure-based virtual screening for NADPH oxidase inhibitors [15]. Notably, enhanced NADPH oxidase expression in multiple malignant diseases supports the recognition of the NADPH oxidase family as potential targets in cancer therapies [13,16]. The Kelch-like ECH-associated protein 1 (Keap1)-nuclear factor E2-related factor 2 (Nrf2) pathway is one of the major signaling cascades involved in protecting cells against oxidative stress. The Nrf2 transcription factor can activate the transcription of cytoprotective genes implicated in protection against cancer. However, Keap1-Nrf2 protein-protein interaction could trigger Nrf2 degradation mediated by the ubiquitin–proteasome pathway. Hence, there has been strong interest among researchers to discover inhibitors of Keap1-Nrf2 protein-protein interaction. Such inhibitors may preserve or enhance the transcription-activating role of Nrf2, counteracting ROS-mediated damage in cancers [17,18].

Although a growing number of seed-derived antioxidant peptides has been documented in the literature, knowledge of their ability to modulate cellular regulators of oxidative status (i.e., MPO, XO, NADPH oxidase, and Keap1-Nrf2), which are also promising targets of anticancer therapy, is still limited. A recent report of watermelon seed-derived antioxidant peptides targeting the Keap1-Nrf2 system [19] suggests that seed-derived antioxidant peptides should be explored more intensively as potential modulators of cellular regulators of ROS balance. Thus, this in silico study was undertaken to virtually screen the numerous seed-derived antioxidant peptides in the literature for their potential as anticancer peptides that can target two oxidative enzymes (MPO and XO) and two protein-protein interactions (Keap1-Nrf2 and $p47^{phox}$-$p22^{phox}$). *In silico* or virtual screening is a less costly and less time-consuming strategy to screen for desirable bioactive peptides and other compounds when compared with wet-lab screening [20]. In bioactive peptide screening, this approach can benefit from various freely available peptide databases (e.g., PlantPepDB [21], and other online tools, such as AntiCP 2.0 [22] and MLCPP [23], which are anticancer peptide and cell-penetrating potential prediction servers designed from machine learning models). Moreover, different molecular modelling and simulation methods [24–26] may also be used to clarify the mechanisms of action between the peptides and the protein targets of interest. Although virtual screening cannot replace wet-lab experimentation, the aforementioned benefits have driven increasing popularity of in silico research on bioactive peptides [20,27]. Notably, by narrowing down a large set of candidate peptides to a small number, in silico screening can facilitate a more focused research strategy in future wet-lab experimentation; this also allows more efficient use of limited research resources [20].

The goal of this in silico study was three-fold: (a) to compile a virtual library of seed antioxidant peptides from the literature, followed by screening for non-toxic, non-allergenic, and non-hemolytic anticancer peptides; (b) to perform structure-based screening of the predicted anticancer peptides for ability to target Keap1-Nrf2, MPO, XO, and $p47^{phox}$-$p22^{phox}$, followed by molecular dynamics validation of peptide-target interactions; and (c) to further characterize the predicted anticancer peptides based on computational alanine mutagenesis and prediction of cell- and blood-brain barrier penetrating potential, as well as plasma and gastrointestinal (GI) stability.

2. Results and Discussion

A virtual library consisting of 677 seed-derived antioxidant peptides was generated (Table S1), based on peptide sequences collected from Scopus and PlantPepDB databases, as outlined in Materials and Methods. The collection encompassed antioxidant peptides of 2–57 residues in length and 192–5338 Da in molecular mass. Seed sources in the virtual library included legumes, such as faba bean and soybean; cereals, such as wheat and rye; and seeds of plantation crop species, such as oil palm and coconut. The types of antioxidant activities reported for the seed-derived peptides included in vitro free radical scavenging activities, lipid peroxidation inhibitory activity, cellular antioxidant activity, and in vivo antioxidant activity (Table S1). Based on Figure 1a, 52% of the seed-derived antioxidant peptides contain five to ten residues. By contrast, seed-derived antioxidant peptides with more than 20 residues comprised only 0.15–0.59% of the virtual antioxidant peptide library. Among the 63 Scopus-indexed publications we examined for the preparation of the virtual library, 42 (67%) reported peptides of 5–10 residues. The prevalence of such peptide length could be accounted by many seed-derived antioxidant peptides being purified and identified from protein fractions of a relatively low molecular mass range, such as <3 kDa fractions [28–30].

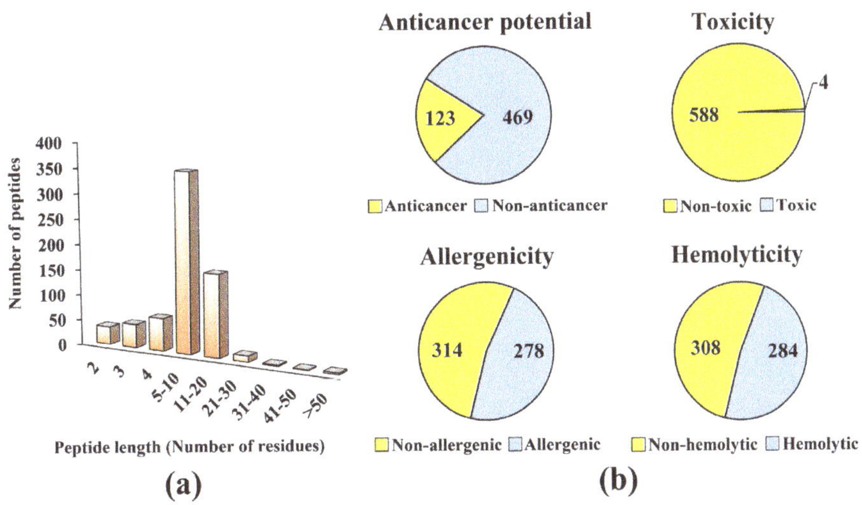

Figure 1. (**a**) The distribution of peptide length in the virtual library of seed-derived antioxidant peptides. (**b**) Outcome of the screening of 592 seed-derived antioxidant peptides of 4–50 residues for anticancer potential, toxicity, allergenicity, and hemolyticity. Numbers in the pie charts represent numbers of peptides with the predicted properties.

Next, we proceeded to screening the virtual library for potential anticancer peptides. Only 592 peptides were screened since dipeptides, tripeptides, peptides with more than 50 residues, and peptides with unnatural or modified amino acid residues could not be analyzed by the AntiCP 2.0 tool. Among the 592 peptides, only 123 (21%) were predicted as anticancer (Figure 1b). We carefully examined the publications reporting the 123 peptides and found that none of the peptides had been tested for anticancer activity experimentally. The 123 predicted anticancer peptides averaged 6 residues in length and 750 Da in mass (data not shown). Safety is an important consideration in the design or discovery of anticancer peptides. A functional anticancer peptide should not exhibit toxicity, elicit immune response, and induce the lysis of erythrocytes [31–33]. Our screening found that at least 50% of the 592 seed-derived antioxidant peptides were predicted to be safe (i.e., non-toxic, non-allergenic, and non-hemolytic) (Figure 1b). Among the 592 peptides screened, only 0.7% (4 peptides) were predicted to be toxic (Figure 1b). An in silico study also predicted that all 253 antioxidant peptides liberated from the flaxseed proteome were non-toxic [34]. This agrees with our observation of high abundance (99%) of non-toxic peptides in our antioxidant peptide virtual library (Figure 1b). In comparison with toxicity prediction, 47–48% of our antioxidant peptide virtual library comprised allergenic and hemolytic peptides. In an in silico study of 26 antimicrobial peptides of rapeseed, 54% were predicted as non-allergenic and 46% allergenic [35]. This relative distribution of allergenicity and non-allergenicity resembles that observed in our virtual screening. Among the 308 non-hemolytic peptides (Figure 1b), the greatest proportion (41%) originated from legumes, which included soybean and chickpea (data not shown). Fourteen soybean-derived multifunctional cationic peptides were shown to have no hemolytic effect on sheep red blood cells [36]. Meanwhile, two chickpea-derived antioxidant peptides also did not cause any hemolysis in bovine red blood cells [37]. These findings support our observation of legumes being a potential source of non-hemolytic peptides.

Based on our in silico screening, five seed-derived antioxidant peptides were predicted to be anticancer, non-toxic, non-allergenic, and non-hemolytic. The two-dimensional (2D) structures and molecular weight of the five peptides are shown in Figure 2. The five peptides, identified from chickpeas, cherry seeds, and tomato seeds, are 5–9 residues in length and 615–1016 Da in mass. The five peptides each contain at least one imidazole functional group or one aromatic ring among their amino acid side chains. Notably, LPHFNS and LYSPH each contain both an imidazole functional group and an aromatic ring in their structures. This is characteristic of many food-derived antioxidant peptides; imidazole groups and aromatic rings are associated with the ability of the peptides to scavenge free radicals by electron transfer/proton donation [38]. On the other hand, among the five peptides (Figure 2), FGPEMEQ has Phe (F) at the N-terminus, whereas Leu (L), His (H), and Phe (F) are present in four, three, and two of the peptides, respectively. The N-terminal preference for Phe and the abundance of Leu, His, and Phe are both characteristics of experimentally validated anticancer peptides [22].

In this in silico study, to investigate whether the five predicted anticancer peptides (Figure 2) could modulate cellular targets of cancer treatments, we docked the five peptides on Keap1, MPO, XO, and $p47^{phox}$. To the best of our knowledge, structure-based virtual screening of the five peptides on the four targets has not been reported. Molecular docking analysis found that LYSPH, a cherry seed peptide, had the strongest binding affinity to Keap1, whereas PSYLNTPLL, a tomato seed peptide, had the weakest (Table 1). LYSPH, LPHFNS, and AEHGSLH also had binding affinity values more negative than that of ETGE (−7.1 kcal/mol) (data not shown). ETGE is the key motif of the co-crystalized 16-mer Nrf2 peptide that is involved in Keap1-Nrf2 interaction [39]. Thus, LYSPH, LPHFNS, and AEHGSLH could form similarly stable or more stable binding to Keap1 when compared with Nrf2. Furthermore, all five peptides could bind to the key residues of Keap1 that are required for stable Keap1-Nrf2 complex formation, mostly accomplished via hydrogen bonds and hydrophobic interactions (Table 1). Two tripeptides (DKK and DDW) that could bind to the key residues of Keap1 have been shown experimentally to inhibit Keap1-Nrf2 interaction in vitro [40]. DKK, which possessed a stronger activity than DDW, was reported to bind to key residues Arg380 and Asn382 [40]. Similar to DKK, all five seed-derived peptides in Table 1 were predicted to bind to Arg380 and Asn382. Thus, our binding affinity and intermolecular interaction results suggest that LYSPH, LPHFNS, and AEHGSLH may serve as potential inhibitors of Keap1-Nrf2 interaction. Specifically, at the molecular level, LYSPH was predicted to bind with the same Keap1 residues as did ETGE, namely, Arg380, Arg415, Arg483, and Ser508 [39]. This observation, in addition to LYSPH having the most negative binding affinity to Keap1 among the five peptides, suggests that the peptide is the most promising for targeting Keap1-Nrf2 interaction. A graphical representation of a LYSPH-Keap1 docked model and the intermolecular interactions between LYSPH and Keap1 is shown in Figure 3.

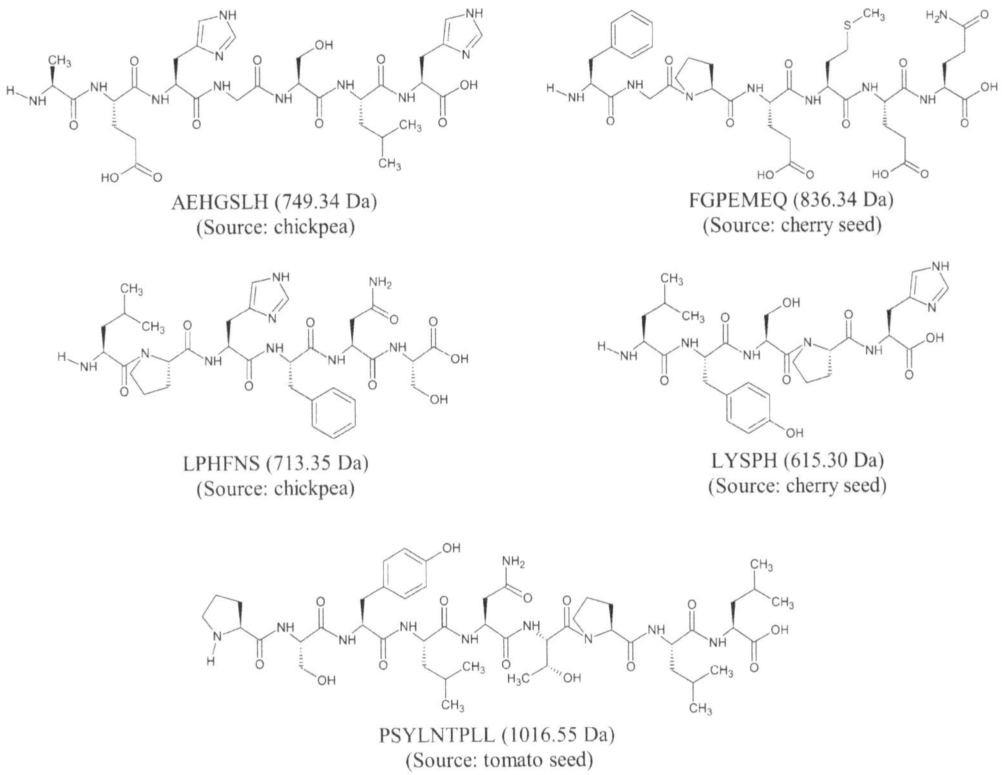

Figure 2. 2D structures of five selected seed-derived antioxidant peptides predicted to be anticancer, non-toxic, non-allergenic, and non-hemolytic.

Table 1. Intermolecular interactions between seed-derived antioxidant peptides and Keap1.

Peptide	Binding Affinity (kcal/mol)	Interaction with Keap1 [a]		
		Hydrogen Bond	Hydrophobic Interaction	Salt Bridge
LYSPH	−7.6	**Arg380, Arg415(2), Arg483, Ser508**	**Tyr334, Ser363, Arg380, Asn382, Arg415, Ser508, Tyr525, Gln530,** Ala556, **Tyr572, Phe577, Ser602**	-
LPHFNS	−7.4	**Ser363, Arg380,** Asn414, **Arg415(2), Gln530**	**Tyr334, Ser363, Arg380, Asn382,** Asn414, **Arg415, Arg483, Tyr525, Gln530, Ser555,** Ala556, **Tyr572, Phe577, Ser602**	-

Table 1. *Cont.*

Peptide	Binding Affinity (kcal/mol)	Interaction with Keap1 [a]		
		Hydrogen Bond	Hydrophobic Interaction	Salt Bridge
AEHGSLH	−7.2	**Tyr334(2)**, **Arg380**, **Asn382(2)**, **Arg415**	**Tyr334**, Arg336, **Ser363**, **Arg380**, **Asn382**, Ser383, Pro384, **Arg415**, Ile461, **Arg483**, **Ser508**, **Tyr525**, **Gln530**, Ala556, **Tyr572**, **Ser602**	-
FGPEMEQ	−7.0	**Ser363**, **Arg380(2)**, **Asn382(2)**, **Asn387**, Asn414	**Tyr334**, **Ser363**, **Arg380**, **Asn382**, **Asn387**, Asp389, **Arg415**, Gly433, Ile461, **Ser555**, Ala556, **Tyr572**, **Phe577**	**Arg380(2)**
PSYLNTPLL	−6.4	**Arg380(2)**, **Asn382**, **Arg415**, **Arg483**, **Ser555**, **Tyr572**	**Tyr334**, **Ser363**, Gly364, **Arg380**, **Asn382**, **Arg415**, **Arg483**, **Tyr525**, **Gln530**, **Ser555**, Ala556, **Tyr572**, **Phe577**	-

[a] Number in brackets indicates the number of hydrogen bonds or salt bridges formed with the same residue of Keap1. Keap1 residues that were reported to bind to ETGE (the key motif of Nrf2 peptide) [39] are marked in boldface type. Residues in the Keap1 binding pocket that were reported to contribute to stability of the Keap1:Nrf2 complex as evidenced by mutagenesis studies [39] are underlined.

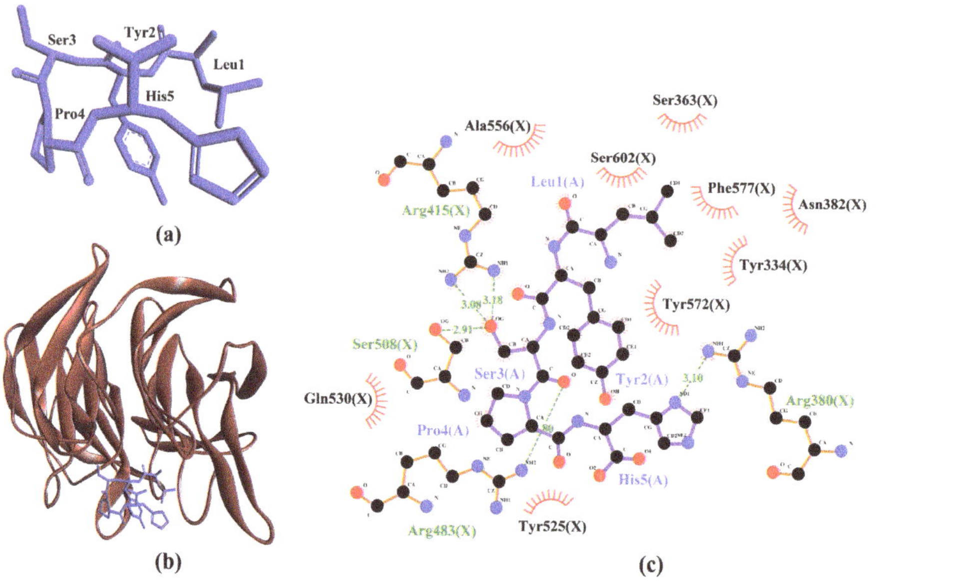

Figure 3. (**a**) Three-dimensional (3D) diagram of LYSPH; (**b**) 3D diagram of LYSPH-Keap1 docked model; (**c**) 2D LYSPH-Keap1 interaction diagram. In (**a**,**b**), LYSPH is displayed in a blue-stick style. In (**b**), Keap1 is displayed as red ribbon. In (**c**), green dashed lines and red spoked arcs represent hydrogen bonds and hydrophobic interactions, respectively. Residues of LYSPH are shown in purple bonds, whereas residues of Keap1 are shown in brown bonds and also represented by the red spoked arcs.

LYSPH showed the strongest binding to MPO, whereas PSYLNTPLL the weakest (Table 2), similar to our observations when the five peptides were docked to Keap1 (Table 1). None of the peptides showed better binding affinity to MPO than did 7-benzyl-1H-1H-[1–3]triazolo[4,5-b]pyridin-5-amine (7GD)(−7.1 kcal/mol) (data not shown), a co-crystalized inhibitor of MPO [41]. However, all five peptides could form hydrophobic interactions with one of the catalytic residues (Arg239) of MPO. Besides, all five peptides could interact with the heme group (Hec606) through hydrophobic interactions (Table 2); the heme group is a cofactor in the active site of MPO [42]. Based on the interactions with both catalytic residue Arg239 and the heme group of MPO, all the five peptides are potential MPO inhibitors. Supporting this possibility is the finding that two experimentally-validated anti-MPO peptides (TDY and FAPQY) could also bind to Arg239 and the heme group of MPO [43]. Analysis of intermolecular interactions revealed that LYSPH could form hydrophobic interactions with Phe99, Thr238, Arg239, Glu242, Phe366, Phe407, and Hec606 of MPO (Table 2), resembling to the binding pattern of 7GD [41]. Hence, LYSPH is the most promising MPO inhibitor among the five peptides as it showed the best binding affinity to MPO and could interact with MPO similarly as 7GD. A graphical representation of LYSPH-MPO docked model and the intermolecular interactions between LYSPH and MPO is depicted in Figure 4.

Table 2. Intermolecular interactions between seed-derived antioxidant peptides and MPO.

Peptide	Binding Affinity (kcal/mol)	Interaction with MPO	
		Hydrogen Bond	Hydrophobic Interaction
LYSPH	−7.1	-	**Phe99**, Glu102, Phe146, Pro220, **Thr238**, **Arg239**, **Glu242**, **Phe366**, Phe407, Leu415, Leu420, Hec606
LPHFNS	−6.9	-	**Phe99**, Glu102, Glu116, Phe147, Pro220, **Thr238**, **Arg239**, **Phe366**, Phe407, Met411, Hec606
FGPEMEQ	−6.8	Glu102	**Phe99**, Thr100, Glu102, Glu116, Pro145, Phe147, Leu216, Pro220, **Thr238**, **Arg239**, **Glu242**, **Phe366**, Phe407, Met411, Leu415, Arg424, Hec606
AEHGSLH	−6.3	-	**Phe99**, Thr100, Glu102, Glu116, Pro145, Phe146, Phe147, **Thr238**, **Arg239**, **Glu242**, **Phe366**, Phe407, Val410, Met411, Leu415, Arg424, Hec606
PSYLNTPLL	−3.0	Thr100	**Phe99**, Glu102, Glu116, Pro145, Phe147, Leu216, Pro220, **Thr238**, **Arg239**, **Phe366**, Phe407, Val410, Met411, Arg412, Leu415, Hec606

MPO residues that were observed to interact with 7GD (co-crystalized inhibitor) based on LigPlot+ analysis of the crystal (PDB ID: 6WYD) are marked in boldface type. MPO residues that were reported to be involved in catalysis [44] are underlined.

Comparison of binding affinities found LYSPH (−6.2 kcal/mol) to have the most stable binding to XO among the five peptides analyzed (Table 3). Nevertheless, all of the five peptides had less negative binding affinities to XO than quercetin (−8.2 kcal/mol) (data not shown), a co-crystalized inhibitor of XO [12]. This implies that none of the peptides could bind more stably to XO when compared with quercetin. On the other hand, analysis of intermolecular interactions showed that all five peptides could bind to at least nine of the XO residues known to bind to quercetin, mainly through hydrophobic interactions. Each of the peptides could also bind to at least one catalytic residue (Glu802 or Arg880) of XO [12] through hydrophobic interactions. FGPEMEQ and PSYLNTPLL could also hydrogen bond to Glu802 and Arg880, respectively. However, despite additional interactions with Glu802 and Arg880, FGPEMEQ-XO interaction was predicted to be slightly less favorable than LYSPH-XO interaction based on comparison of their binding affinity values. Meanwhile, PSYLNTPLL-XO interaction was likely non-favorable or non-spontaneous considering the positive value predicted for its binding affinity (Table 3). Previous studies found that experimentally-proven XO-inhibitory peptides, KGFP [45] and EEAK [46] could both bind to the catalytic residue Glu802. Thus, the aforementioned binding patterns of the five seed peptides to XO, particularly their binding to XO catalytic residues, suggest that the peptides are potential XO inhibitors. LYSPH could be the most promising XO inhibitor among the five peptides considering its strongest binding affinity and its binding to XO residues that known XO inhibitors bind to (Table 3). A graphical representation of a LYSPH-XO docked model and the intermolecular interactions between LYSPH and XO is shown in Figure 5.

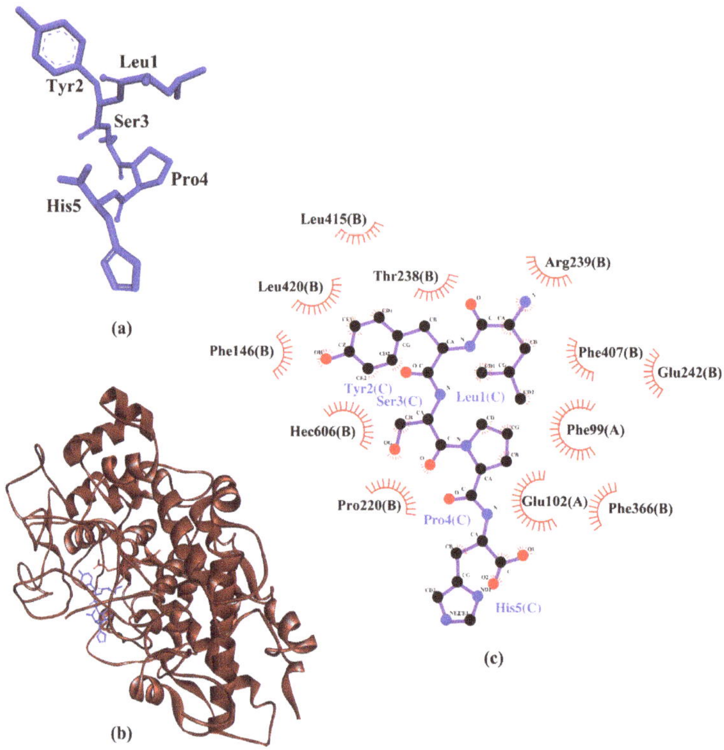

Figure 4. (**a**) 3D diagram of LYSPH; (**b**) 3D diagram of LYSPH-MPO docked model; (**c**) 2D LYSPH-MPO interaction diagram. In (**a**,**b**), LYSPH is displayed in a blue-stick style. In (**b**), MPO is displayed as red ribbon. In (**c**), red spoked arcs represent hydrophobic interactions. Residues of LYSPH are shown in purple bonds. Residues of MPO are represented by red spoked arcs.

Table 3. Intermolecular interactions between seed-derived antioxidant peptides and XO.

Peptide	Binding Affinity (kcal/mol)	Interaction with XO			
		Hydrogen Bond [a]	Hydrophobic Interaction	Salt Bridge	External Bond
LYSPH	−6.2	Ser876	**Leu648**, Phe649, **Glu802**, **Leu873**, His875, Ser876, Glu879, **Phe914**, **Phe1009**, **Thr1010**, **Val1011**, Pro1012, **Phe1013**, **Leu1014**, Ala1078, Ala1079	-	-
FGPEMEQ	−5.9	**Glu802**, Ser876(2), Ala1079	**Leu648**, Phe649, Gln767, Phe798, Gly799, **Glu802**, Thr803, **Leu873**, His875, Ser876, Glu879, **Arg880**, Ala910, Phe911, Arg912, **Phe914**, **Phe1009**, **Thr1010**, **Val1011**, Pro1012, **Leu1014**, Pro1076, Ala1078, Ala1079, Ser1080, Glu1261	His875	-

Table 3. *Cont.*

Peptide	Binding Affinity (kcal/mol)	Interaction with XO			
		Hydrogen Bond [a]	Hydrophobic Interaction	Salt Bridge	External Bond
LPHFNS	−4.7	His875, Ser876	**Leu648**, Phe649, **Glu802**, **Leu873**, His875, Ser876, Glu879, **Arg880**, **Phe914**, **Phe1009**, **Thr1010**, **Val1011**, Pro1012, **Phe1013**, **Leu1014**, Ala1078, Ala1079, Glu1261	-	-
AEHGSLH	−3.4	Glu879	**Leu648**, Phe649, Leu712, **Glu802**, **Leu873**, His875, Ser876, Glu879, **Phe914**, **Phe1009**, **Thr1010**, **Val1011**, Pro1012, **Phe1013**, **Leu1014**, Pro1076, Tyr1140, Phe1142	His875	-
PSYLNTPLL	3.0	Asn768, Asp872, Ser876(2), **Arg880**, **Thr1010(2)**	**Leu648**, Phe649, Leu712, Asn768, **Glu802**, Thr803, Arg871, Asp872, **Leu873**, Ser874, His875, Ser876, Glu879, **Arg880**, **Phe914**, Ser1008, **Phe1009**, **Thr1010**, **Val1011**, Pro1012, **Phe1013**, **Leu1014**, Pro1076, Ala1079, Tyr1140, Phe1142, Glu1261	-	Ala1079

[a] Number in brackets indicates the number of hydrogen bonds formed with the same residue of XO. XO residues that were reported to bind to quercetin (co-crystalized inhibitor in the crystal PDB ID 3NVY) [12] are marked in boldface type. XO residues that were reported to be involved in catalysis [12] are underlined.

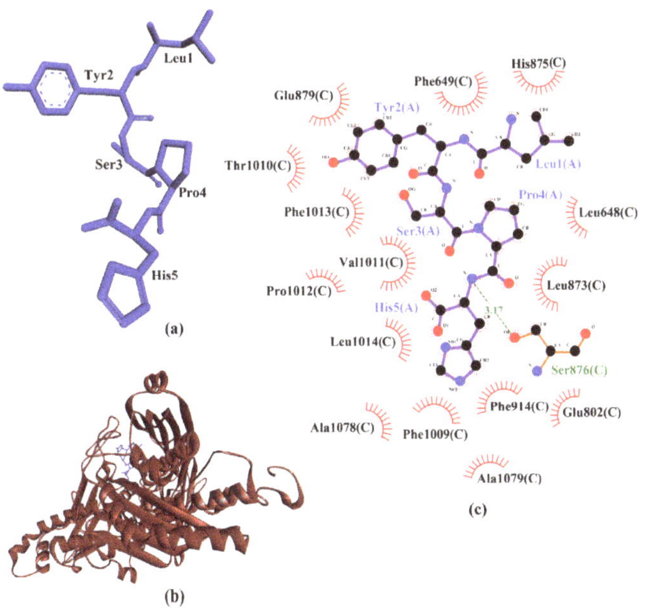

Figure 5. (**a**) 3D diagram of LYSPH; (**b**) 3D diagram of LYSPH-XO docked model; (**c**) 2D LYSPH-XO interaction diagram. In (**a**,**b**), LYSPH is displayed in a blue-stick style. In (**b**), XO is displayed as red ribbon. In (**c**), green dashed line and red spoked arcs represent hydrogen bond and hydrophobic interactions, respectively. Residues of LYSPH are shown in purple bonds, whereas residues of XO are shown in brown bonds and also represented by the red spoked arcs.

In the molecular docking to p47^{phox}, tomato seed-derived PSYLNTPLL had the best docking score, whereas the cherry seed-derived FGPEMEQ had the worst (Table 4). All peptides had docking score less negative than that of proline-rich peptide derived from p22^{phox}

(−309.862) (data not shown). Thus, none of the five peptides could bind more stably to p47phox than the p22phox-derived peptide. Nevertheless, the potential of the five peptides as inhibitors of p47phox-p22phox interaction could not be completely ruled out solely based on this. Supporting this proposition is the observation that four peptides (RRSSIRNAHSIHQRSRKRLS, ISNSESGPRGVHFIFNKENF, RSRKRLSQDAYRRNSVRFLQQR, and AGGPPGGPQVNPIPVTDEVV) that were experimentally demonstrated to inhibit p47phox-p22phox interaction [47,48] were also predicted to bind less strongly to p47phox than was p22phox (Table S4). In short, a peptide predicted to bind less strongly to p47phox than p22phox may still inhibit p47phox-p22phox interaction.

Table 4. Intermolecular interactions between seed-derived antioxidant peptides and p47phox.

Peptide	Docking Score	Interaction with p47phox	
		Hydrogen Bond	Hydrophobic Interaction
PSYLNTPLL	−216.493	**Trp193**, **Trp263**	Thr170, Ser171, Ser173, Glu174, Gly192, **Trp193**, **Trp204**, **Pro206**, Ser208, **<u>Phe209</u>**, Lys235, **Tyr237**, Asp261, Gly262, **Trp263**, **Met278**, **Tyr279**
LPHFNS	−195.377	**Tyr279**	**Tyr167**, Thr170, Ser191, **Trp193**, **Pro206**, **Tyr237**, **Asp243**, **Glu244**, Asp261, Gly262, **Trp263**, **Tyr274**, **Pro276**, **Tyr279**
LYSPH	−185.715	Thr170, Ser208	**Tyr167**, Thr170, Ser171, Ser173, Glu174, **Trp193**, **Trp204**, **Pro206**, Ser208, **<u>Phe209</u>**, **Trp263**, **Met278**
AEHGSLH	−181.729	**Trp263**	**Tyr167**, Thr170, Ser171, Glu174, **Trp193**, Glu241, **Asp243**, **Trp263**, **Tyr274**, **Pro276**, **Met278**
FGPEMEQ	−175.680	Thr170, **Trp193**	Thr170, Ser173, Glu174, Gly192, **Trp193**, **Trp204**, **Pro206**, **<u>Phe209</u>**, Asp261, **Met278**, **Tyr279**

p47phox residues that were reported to bind to the ligand p22phox-derived proline-rich peptide in the crystal (PDB ID: 1WLP) [14] are marked in boldface type. Key residues that were reported for high-affinity binding between p47phox and p22phox as evidenced by mutagenesis studies [14] are underlined.

As shown in Table 4, each of the peptides could bind to at least six of the 17 p47phox residues known to bind to the p22phox-derived peptide. However, only PSYLNTPLL, LYSPH, and FGPEMEQ could interact with Phe209, a key residue of p47phox which accounts for high-affinity binding between p47phox and p22phox (Table 4). Furthermore, PSYLNTPLL could bind to p47phox in a similar manner as the co-crystalized p22phox-derived peptide, by binding to Trp193, Trp204, Pro206, Phe209, Tyr237, Trp263, Met278, and Tyr279. Hence, PSYLNTPLL is the most promising among the five peptides to target p47phox-p22phox interaction considering its docking score and pattern of binding to p47phox. A graphical representation of PSYLNTPLL-p47phox docked model and the intermolecular interactions between PSYLNTPLL and p47phox is shown in Figure 6.

Based on binding affinities and similarity of binding patterns to those of co-crystalized inhibitors/ligands and reported peptide-based inhibitors, our analyses found LYSPH and PSYLNTPLL to have the greatest potential as modulators of the four targets of cancer treatments that we investigated. Specifically, LYSPH may be a multi-target peptide which could bind to, thus inhibiting the activity of MPO and XO, as well as interrupting Keap1-Nrf2 complex formation. On the other hand, PSYLNTPLL is the most promising peptide that could bind to p47phox, thus precluding p47phox-p22phox interaction and the subsequent activation of NADPH oxidase. Inhibition of the four targets could potentially dampen ROS overproduction which is associated with the initiation and/or progression of certain cancers [11,49–52]. To our knowledge, the Keap1-, MPO-, XO-, and p47phox-binding activity of the two peptides have not been previously reported. Considering the in silico evidence presented here, future investigations of the effectiveness of LYSPH and PSYLNTPLL in modulating the four targets, thus repressing ROS production and even cancer initiation and/or progression are warranted.

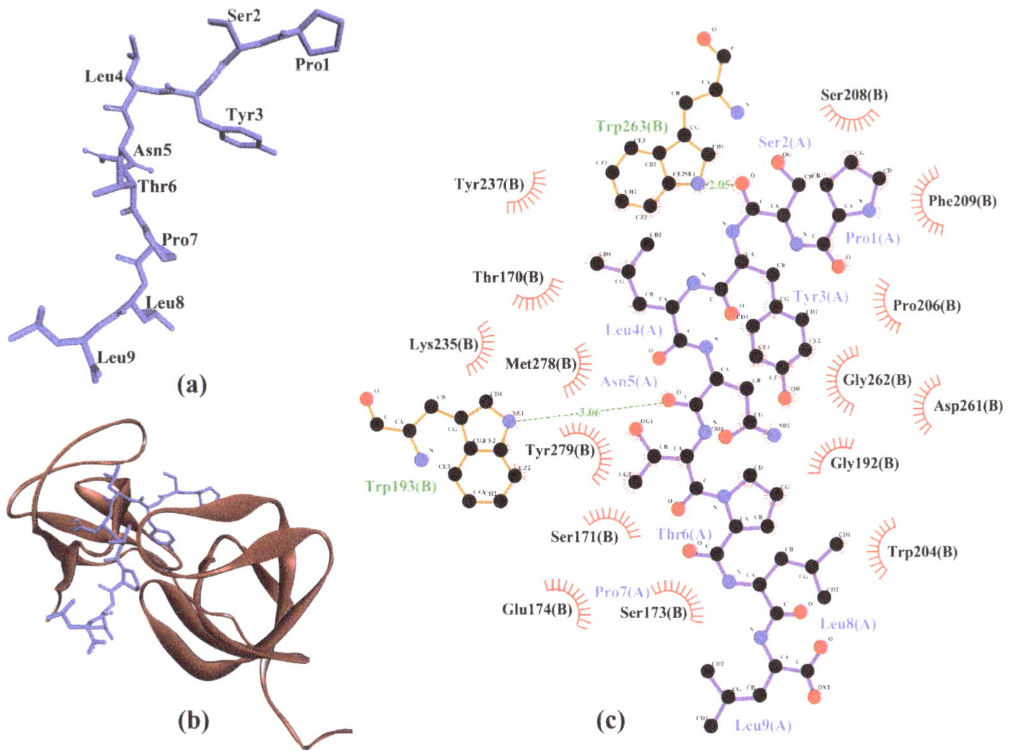

Figure 6. (**a**) 3D diagram of PSYLNTPLL; (**b**) 3D diagram of PSYLNTPLL-p47^{phox} docked model; (**c**) 2D PSYLNTPLL-p47^{phox} interaction diagram. In (**a**,**b**), PSYLNTPLL is displayed in a blue-stick style. In (**b**), p47^{phox} is displayed as red ribbon. In (**c**), green dashed lines and red spoked arcs represent hydrogen bonds and hydrophobic interactions, respectively. Residues of PSYLNTPLL are shown in purple bonds, whereas residues of p47^{phox} are shown in brown bonds and also represented by the red spoked arcs.

The five anticancer peptides predicted from our virtual library were also screened for cell-penetrating potential, blood-brain barrier penetrating potential, plasma half-life, and tolerance to in silico GI digestion, which can shed light on their potential as anticancer agents. Among the five peptides, LYSPH and PSYLNTPLL had the top two best plasma half-life (Table 5). The two peptides predictably had cell- and blood-brain barrier penetrating potential, although both were susceptible to GI digestion. The predicted cell-penetrating potential of the two peptides supports their potential in entering body cells and binding to the four intracellular targets: MPO, XO, Keap1, and p47^{phox}, modulating the functions of the four proteins. The predicted ability of LYSPH and PSYLNTPLL to cross the blood-brain barrier may also facilitate their development as brain-tumor targeting peptides. Plasma half-life and tolerance to in silico GI digestion are related to the bioavailability of a peptide. Our results suggests that the two peptides were similar in their level of susceptibility to plasma peptidases, thus not differing much in their stability during systemic circulation. When compared with other natural anticancer peptides, such as KENPVLSLVNGMF identified from the giant barrel sponge *Xestospongia testudinaria* (half-life of 3.2 h in human serum in vitro) [53], the half-life of LYSPH and PSYLNTPLL was relatively short (about 14 min). Meanwhile, LYSPH and PSYLNTPLL were similarly susceptible to degradation by GI proteases. So, poor stability in blood and susceptibility to GI digestion is a key potential weakness of the two peptides, despite their ability to target MPO, XO, Keap1, and p47^{phox}, as well as predictably having cell- and blood-brain barrier penetrating potential.

The stability issue may limit the potential effectiveness of LYSPH and PSYLNTPLL as anticancer agents in the body, whether introduced into the body through oral or non-oral routes. To enhance the in vivo bioavailability of LYSPH and PSYLNTPLL, structural modifications that could improve their resistance to plasma and GI peptidases, such as cyclization of peptides [54] could be considered in future research. Moreover, the application of innovative technology such as mucoadhesive nanoparticles [55] may also be explored for oral delivery of the peptides with reduced risk of GI degradation and enhanced bioavailability.

Table 5. The predictions of the cell-penetrating potential, blood-brain barrier penetrating potential, plasma half-life, and tolerance to in silico GI digestion of the five selected seed-derived antioxidant peptides.

Peptide	Cell-Penetrating Potential	Blood-Brain Barrier Penetrating Potential	Plasma Half-Life (Seconds)	Tolerance to In Silico GI Digestion
AEHGSLH	No	No	828.91	No
FGPEMEQ	No	Yes	796.21	No
LPHFNS	Yes	Yes	823.51	No
LYSPH	Yes	Yes	832.41	No
PSYLNTPLL	Yes	Yes	833.41	No

Based on computational alanine scanning, Tyr played the most significant role in the binding and stabilizing of peptide-protein complexes for Keap1, MPO, XO, and $p47^{phox}$. This can be observed from the drastically elevated ΔΔG values after the substitution of Tyr to Ala in both LYSPH and PSYLNTPLL (Table 6). This suggests that the hydrophobic interactions between Tyr and the residues of Keap1 (Tyr572), of XO (Glu879, Thr1010, Phe1013), of MPO (Phe99, Glu102, Phe146, Leu415, Leu420) and of $p47^{phox}$ (Gly192, Asp261, Gly262, Met278) (Figures 3–6) are critical to the formation of stable peptide-protein complexes. In line with our findings, Wu and co-workers [56] found that the only Tyr-containing peptide in their study had the highest XO inhibitory activity; Tyr in the peptide also interacted hydrophobically with Phe1013 of XO. On the other hand, Ala substitution of His in LYSPH also led to the second largest increase in ΔΔG by 14.2670 kJ/mol (Table 6) when the LYSPH-XO complex was analyzed. By contrast, Ala substitution of His in LYSPH led to only a minor increase in ΔΔG of the LYSPH-Keap1 and LYSPH-MPO complexes. A possible explanation is that the His residue of LYSPH could bind to more key residues in XO (Glu802, Phe914, Phe1009, and Leu1014) (Figure 5c). By contrast, the His residue of LYSPH interacted with only one key residue (Arg380) in Keap1 (Figure 3c) and with none in MPO (Figure 4c). Our analysis suggests that future research that considers re-designing LYSPH and PSYLNTPLL for enhanced interactions with Keap1, MPO, and $p47^{phox}$ should avoid replacing or removing the Tyr residue. For stable binding to XO, the Tyr and His residues of LYSPH both should not be replaced or removed.

Molecular dynamics (MD) is a simulation technique which applied to derive the statements about the structural, dynamical, and thermodynamic properties of a molecular system [57]. The approach is able to observe minor conformational changes corresponds to the residue side chains which affect the binding site of a protein and ligand complementarity [57]. In the current study MD was applied to observe the dynamic level stability of each peptide ligand against the targeted proteins, as the peptides can functions either as the receptor inhibitors [58,59] or as the mediator such as the peptide mediated interactions in cell signaling [60].

Table 6. Changes in the binding free energies (ΔΔG) of the LYSPH- and PSYLNTPLL-protein complexes as revealed by the computational alanine scanning of the peptide residues.

Peptide	Residue	ΔΔG (kJ/mol)			
		Keap1	**MPO**	**XO**	**p47^{phox}**
LYSPH	Leu	8.4596	7.7764	6.3704	-
	Tyr	14.3232	23.0458	16.8951	-
	Ser	2.5540	−0.1909	−0.1396	-
	Pro	5.0370	5.2764	6.3965	-
	His	0.6399	3.1014	14.2670	-
PSYLNTPLL	Pro	-	-	-	0.0856
	Ser	-	-	-	0.5437
	Tyr	-	-	-	27.1615
	Leu	-	-	-	−0.4367
	Asn	-	-	-	0.7800
	Thr	-	-	-	1.1903
	Pro	-	-	-	2.9660
	Leu	-	-	-	2.7811
	Leu	-	-	-	0.4836

The MD simulations results in Figure 7a–d determines the protein-ligand complexes stability during the 50 ns duration. In Figure 7a, the all-atom averaged root mean square deviation (RMSD) value for protein target Keap1, XO and MPO (chain A and B) in the complex were shown to be low at 2.14 ± 0.11 Å, 3.15 ± 0.33 Å, 2.59 ± 0.18 Å and 3.17 ± 0.36Å, respectively. In comparison, Figure 7b shows the all-atom averaged RMSD value of ligand LYSPH docked on each Keap1, XO and MPO were 1.73 ± 0.26 Å, 3.07 ± 0.33 Å and 3.34 ± 0.47 Å, respectively. This shows that RMSD values of both the docked proteins and the ligands are below the allowed limit [61], confirming that the protein-ligand complexes are stable over time. The plotted RMSD graphs also shows that receptor p47^{phox} took longer time to reach complex stability compared to the other docked proteins with the averaged RMSD value of 5.14 ± 0.13 Å, while all-atom averaged RMSD for its ligand, PSYLNTPLL was 4.36 ± 0.45 Å. The high p47^{phox} RMSD value was contributed by the flexibility of the N- and also C-terminal residues of the protein which reached up to 6.00 Å due to the loop structure of both terminals, visible by the root mean square fluctuations RMSF plot (Figure S1). The ligand interacted residues, however, were not affected and gave relatively low fluctuations during the 50 ns duration. In addition, the RMSD of PSYLNTPLL was also similarly low with LYSPH docked on other protein target (Figure 7b) during the 50 ns duration.

The dynamic intermolecular hydrogen bonds formed between the docked peptide and receptor protein were summarized in Figure 7c. The figure shows that highest number of intermolecular hydrogen bonds formed was in between LYSPH-XO (ave: 7), followed by PSYLNTPLL-p47^{phox} (ave: 5) and LYSPH-Keap1 (ave: 3). MPO protein was consists of chain A and chain B domain, where chain A formed only one intermolecular hydrogen bond with the ligand in average while chain B has the average of three intermolecular hydrogen bonds formed within the 50 ns duration. The polar group of the XO hot-spot region, and LYSPH peptide both contributed to the higher number of hydrogen bonds formed making the complex more stable [62]. Higher surface of interactions between PSYLNTPLL-p47^{phox} due to the longer sequence of the peptide had stabilized its docking on the active site of p47^{phox} [63]. The intermolecular hydrogen bonds formed in the complex had also correlated with the distance formed between each ligand and protein, as summarized in Figure 7d. The result shows that all complexes were tightly packed with the average protein-ligand distance of 1.43 Å–1.82 Å, except for LYSPH and the chain A of MPO which varied from 1.43 Å up to 5.31 Å. This was contributed by the binding site of the ligand which located closer to the chain B of MPO. Overall, the duration 50 ns were shown to be sufficient to evaluate the stability of protein-ligand complex formation where most of the ligand tends to reach conformational stability after 3 ns. The RMSF and radius of gyration (Rg) plots of each complex are available in Figures S1 and S2.

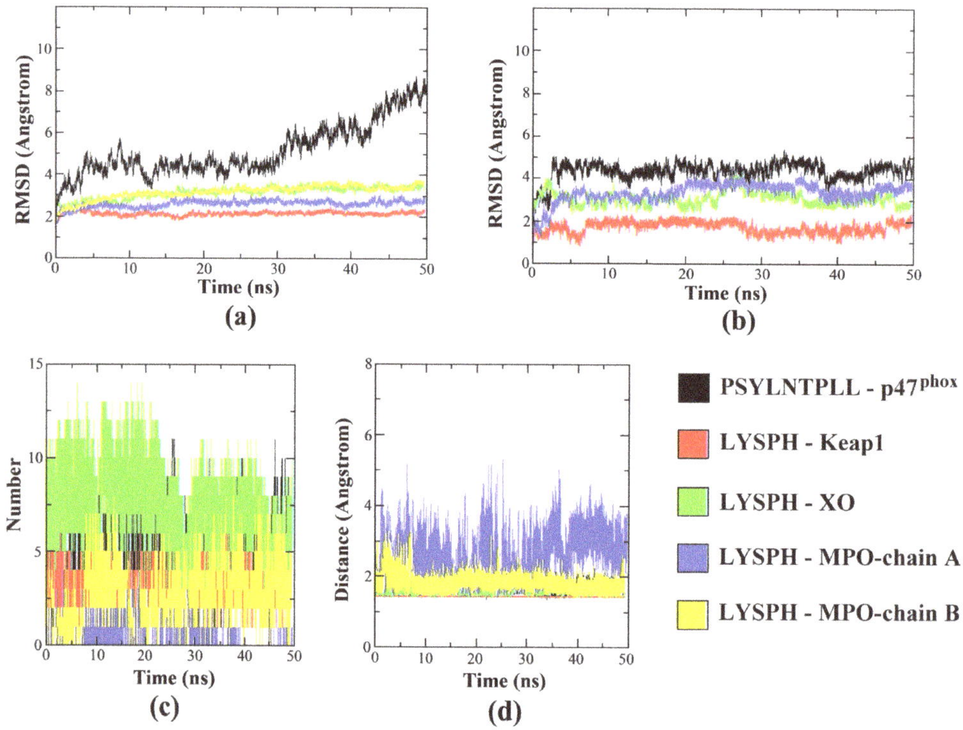

Figure 7. (**a**) all-atom RMSD value for proteins; (**b**) all-atom RMSD value for ligands; (**c**) number of intermolecular hydrogen bonds; and (**d**) minimum distance between the proteins and each peptide ligand for the 50 ns duration.

3. Materials and Methods

3.1. Compilation of a Virtual Library of Seed-Derived Antioxidant Peptides

Seed-derived antioxidant peptides were compiled from the publications in the Scopus database by using the search words listed in Table S2 (Accessed: 5–7 October 2021). A total of 63 publications were carefully examined to find antioxidant peptides identified from different seed sources. In addition, peptides were compiled from the PlantPepDB database (http://14.139.61.8/PlantPepDB/index.php) [21] by using "Simple Search", searching "Antioxidant" and selecting "Peptide Activity" as search field (Accessed: 5–6 October 2021). Following the exclusion of redundant sequences, the resulting collection of seed-derived antioxidant peptide sequences was used in subsequent screening and molecular modelling analyses, as depicted in Figure 8.

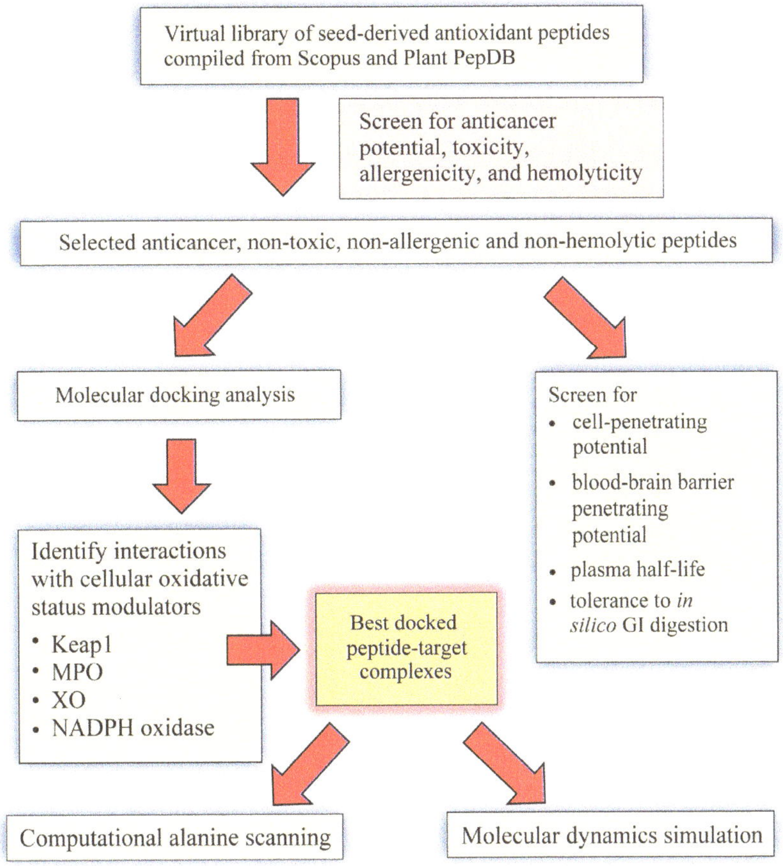

Figure 8. An overview of the computational approach used in this study.

3.2. *Virtual Screening for Anticancer Potential, Toxicity, Allergenicity and Hemolyticity*

Anticancer potential was predicted by using AntiCP 2.0 (https://webs.iiitd.edu.in/raghava/anticp2/index.html) [22] with the default SVM threshold of 0.45. Seed-derived antioxidant peptide sequences that were predicted as anticancer peptides by both Model 1 and Model 2 in the AntiCP 2.0 tool were noted. Toxicity was predicted by using ToxinPred (https://webs.iiitd.edu.in/raghava/toxinpred/index.html) [33] with the SVM threshold of 0.0, by using two methods: (i) SVM (Swiss-Prot) + Motif, and (ii) SVM (TrEMBL) + Motif. Only peptide sequences that were predicted as non-toxic by both of the aforementioned methods are regarded as non-toxic. Allergenicity was predicted by using AllerTOP v. 2.0 (https://www.ddg-pharmfac.net/AllerTOP/index.html) [32]. Hemolyticity was predicted by using HemoPI (https://webs.iiitd.edu.in/raghava/hemopi/index.php) [31] with the SVM + Motif (HemoPI-1) method. The aforementioned tools were accessed on 8–9 October 2021. Only seed-derived antioxidant peptides of 4–50 residues were screened as the peptides outside this range cannot be analyzed by AntiCP 2.0. The 2D structures of selected anticancer peptides were drawn by using the ACD/ChemSketch freeware (ACD/ChemSketch, version 2019.2.1, Advanced Chemistry Development, Inc., Toronto, ON, Canada, www.acdlabs.com, 2019). The molecular masses of peptides were calculated by using PepDraw (https://pepdraw.com/) (Accessed: 9 October 2021).

3.3. Molecular Docking Analysis

The 3D structures of peptides predicted to be anticancer, non-toxic, non-allergenic, and non-hemolytic were constructed by using PEP-FOLD 3 (https://bioserv.rpbs.univ-paris-diderot.fr/services/PEP-FOLD3/) [64–66] (Accessed: 9 October 2021). Two hundred simulations were run and the resulting models were sorted by the sOPEP method. The best output model of each peptide was downloaded and used in molecular docking.

The crystal structures of human Keap1 complexed with 16-mer peptide of Nrf2 (PDB ID: 2FLU) [39], human MPO complexed with 7GD (PDB ID: 6WYD) [41], bovine XO complexed with quercetin (PDB ID: 3NVY) [12], and $p47^{phox}$ complexed with $p22^{phox}$-derived proline-rich peptide (PDB ID: 1WLP) [14] were downloaded from the RCSB Protein Data Bank (https://www.rcsb.org/) [67] (Accessed: 9 October 2021). The separation of proteins and ligands from the crystals were performed by using BIOVIA Discovery Studio Visualizer (BIOVIA, Dassault Systèmes, BIOVIA Discovery Studio Visualizer, Version 20.1.0.192, San Diego: Dassault Systèmes, CA, USA, 2020).

For Keap1, MPO, and XO, the proteins were prepared as receptors in the PDBQT format after deleting water, adding polar hydrogen, and adding Kollman charges by using the AutoDock Tools 1.5.6 [25]. The prepared receptors were used for molecular docking on Webina 1.0.3 (https://durrantlab.pitt.edu/webina/) [24]. The co-crystalized ligands and the selected seed-derived peptides to be docked to the proteins were prepared as ligands and saved in the PDBQT format by using AutoDock Tools 1.5.6. Redocking of the co-crystalized ligands to Keap1, MPO and XO was performed using Webina 1.0.3 and RMSD was predicted by using LigRMSD v1.0 (https://ligrmsd.appsbio.utalca.cl/) [68]. For Keap1, redocking was performed by using tetrapeptide ETGE, the key motif of the 16-mer peptide of Nrf2, as recommended previously [40]. The coordinates of box center and box size used in molecular docking on Webina 1.0.3 were tabulated in Table S3. Molecular docking between $p47^{phox}$ and the peptide ligands was performed by using the HPEPDOCK Server (http://huanglab.phys.hust.edu.cn/hpepdock/) [69–73]. Redocking of the co-crystalized $p22^{phox}$-derived proline-rich peptide (GPLGSKQPPSNPPPRPPAEARKKPS) to $p47^{phox}$ was also performed on HPEPDOCK and RMSD was predicted by using LigRMSD v1.0. Webina 1.0.3 and HPEPDOCK were accessed between 9 and 12 October 2021. Intermolecular interactions between proteins and peptides in selected docked models were analyzed and 2D interaction diagrams were generated by using LigPlot+ v.2.2.4 [74,75].

3.4. Prediction of Cell-Penetrating Potential, Blood-Brain Barrier Penetrating Potential, Plasma Half-Life, and Tolerance to In Silico GI Digestion

Cell-penetrating potential was predicted by using MLCPP (http://www.thegleelab.org/MLCPP/MLCPP.html) [23]. Blood-brain barrier penetrating potential was predicted by using B3Pred (https://webs.iiitd.edu.in/raghava/b3pred/index.html) [76] with Random-Forest (RF)-based prediction model and RF probability threshold of 0.1. Plasma half-life was predicted by using PlifePred (https://webs.iiitd.edu.in/raghava/plifepred/index.php) [77]. Tolerance to in silico GI digestion was predicted using the "enzyme(s) action" tool on BIOPEP-UWM (http://www.uwm.edu.pl/biochemia/index.php/en/biopep) [78] as previously reported [29]. The aforementioned tools were accessed on 18 October 2021.

3.5. Computational Alanine Scanning Mutagenesis

To assess the energetic contribution of individual residues in the selected seed-derived antioxidant peptides in conferring stability of binding to Keap1, MPO, XO, and $p47^{phox}$, computational alanine scanning was performed by using BUDE Alanine Scan (https://pragmaticproteindesign.bio.ed.ac.uk/balas/) [79,80] (Accessed: 18–19 October 2021) as previously reported [27].

3.6. Molecular Dynamics Simulation

For a comprehensive analysis of the biomolecular dynamics, molecular dynamics (MD) simulation has evolved as the most powerful technique [26]. The detailed MD simulations of the complexes were conducted in GROMACS 2020 using the GROMOS96 54a7 force field [81]. The 54a7 force field was shown to improve the stability of α-helical structures in proteins and widely used in peptide simulations [82]. Molecular dynamics simulation was performed on each peptide ligand and protein complex of LYSPH-Keap1, LYSPH-MPO, LYSPH-XO, and PSYLNTPLL-p47^{phox} for 50 ns duration. In the MD, each complex was solvated in a cubic box with the distance of 1.2 nm between the complex and each side of the solvated box [83]. Sodium and chloride ions were added to neutralize the total charge of the system. The complex then was energy-minimized using the steepest descent algorithm [84]. The simulation condition was set at the room temperature (300 K) and the atmospheric pressure (1 bar) to closely mimic the general experiment conditions. The NVT thermal equilibration was carried out with a constrained structure and a velocity rescale thermostat specific to GROMACS, followed by NPT pressure equilibration was applied with the same velocity-rescale temperature coupling in addition to the Parrinello–Rahman pressure coupling [85]. The fully temperature and pressure equilibrated system was then used as the initial configuration for the MD production dynamic analysis. All simulations were conducted using a 2 fs time step [86]. The results were then analyzed using GROMACS functions such as RMSD and RMSF, while the formation of hydrogen bonds between each peptide and target proteins were analyzed using GROMACS "gmx_hbond" functions. Additionally, the distance between each protein and its ligand peptide was measured using the "gmx_pairdist" function.

4. Conclusions

Our computational study narrowed down the 677 peptides in the virtual library to five candidates predicted to have anticancer potential, in addition to non-toxicity, non-allergenicity and non-hemolyticity. Structure-based virtual screening found that LYSPH was the most promising peptide in targeting MPO, XO, and Keap1. On the other hand, PSYLNTPLL was the candidate that interacted most stably with p47^{phox}. LYSPH and PSYLNTPLL were predicted to have cell- and blood-brain barrier penetrating potential. Taken together, LYSPH and PSYLNTPLL are two potential candidates of anticancer peptides that deserve more in-depth explorations, particularly wet-lab experimental validations, in future.

Supplementary Materials: Table S1: Seed-derived antioxidant peptides compiled from Scopus and PlantPepDB databases; Table S2: Search words used in Scopus to compile seed-derived antioxidant peptides; Table S3: Coordinates of box center and box size for different targets in molecular docking, and RMSD values; Table S4: Docking scores for peptides that were experimentally demonstrated to inhibit p47phox-p22^{phox} interaction and NADPH oxidase, in comparison with p22^{phox}; Figure S1: RMSF plots for (**a**) PSYLNTPLL-p47phox, (**b**) LYSPH-Keap1, (**c**) LYSPH-XO, and (**d**) LYSPH-MPO; Figure S2: Gyration (Rg) plots of each complex.

Author Contributions: Conceptualization, T.-T.C. and F.-C.W.; methodology, T.-T.C. and F.-C.W.; software, T.-T.C. and M.Z.S.; validation, T.-T.C., F.-C.W., and M.Z.S.; formal analysis, C.C.-C.W., M.Z.S., and J.-A.K.; investigation, T.-T.C., M.Z.S., C.C.-C.W., and J.-A.K.; resources, T.-T.C., F.-C.W.; data curation, C.C.-C.W., J.-A.K., T.-T.C., and F.-C.W.; writing—original draft preparation, T.-T.C., J.-A.K., and M.Z.S.; writing—review and editing, T.-T.C., J.-A.K., F.-C.W., C.C.-C.W., and M.Z.S.; visualization, C.C.-C.W., J.-A.K., F.-C.W., and M.Z.S.; supervision, T.-T.C. and F.-C.W.; project administration, T.-T.C. All authors have read and agreed to the published version of the manuscript.

Funding: This research received no external funding.

Institutional Review Board Statement: Not applicable.

Informed Consent Statement: Not applicable.

Data Availability Statement: The data presented in this study are available on request from the corresponding author.

Conflicts of Interest: The authors declare no conflict of interest.

References

1. Jakubczyk, A.; Karaś, M.; Rybczyńska-Tkaczyk, K.; Zielińska, E.; Zieliński, D. Current trends of bioactive peptides—New sources and therapeutic effect. *Foods* **2020**, *9*, 846. [CrossRef]
2. Wong, F.-C.; Xiao, J.; Wang, S.; Ee, K.-Y.; Chai, T.-T. Advances on the antioxidant peptides from edible plant sources. *Trends Food Sci. Technol.* **2020**, *99*, 44–57. [CrossRef]
3. Apostolopoulos, V.; Bojarska, J.; Chai, T.-T.; Elnagdy, S.; Kaczmarek, K.; Matsoukas, J.; New, R.; Parang, K.; Lopez, O.P.; Parhiz, H.; et al. A global review on short peptides: Frontiers and perspectives. *Molecules* **2021**, *26*, 430. [CrossRef]
4. Chai, T.-T.; Tan, Y.-N.; Ee, K.-Y.; Xiao, J.; Wong, F.-C. Seeds, fermented foods, and agricultural by-products as sources of plant-derived antibacterial peptides. *Crit. Rev. Food Sci. Nutr.* **2019**, *59*, S162–S177. [CrossRef]
5. Díaz-Gómez, J.L.; Castorena-Torres, F.; Preciado-Ortiz, R.E.; García-Lara, S. Anti-cancer activity of maize bioactive peptides. *Front. Chem.* **2017**, *5*, 44. [CrossRef]
6. Ramkisson, S.; Dwarka, D.; Venter, S.; Mellem, J.J. In vitro anticancer and antioxidant potential of Amaranthus cruentus protein and its hydrolysates. *Food Sci. Technol.* **2020**, *40*, 634–639. [CrossRef]
7. Mika, D.; Guruvayoorappan, C. Myeloperoxidase: The yin and yang in tumour progression. *J. Exp. Ther. Oncol.* **2011**, *9*, 93–100.
8. Kargapolova, Y.; Geißen, S.; Zheng, R.; Baldus, S.; Winkels, H.; Adam, M. The enzymatic and non-enzymatic function of myeloperoxidase (MPO) in inflammatory communication. *Antioxidants* **2021**, *10*, 562. [CrossRef]
9. Wang, Y.; Qi, H.; Liu, Y.; Duan, C.; Liu, X.; Xia, T.; Chen, D.; Piao, H.-L.; Liu, H.-X. The double-edged roles of ROS in cancer prevention and therapy. *Theranostics* **2021**, *11*, 4839–4857. [CrossRef]
10. Kazakov, Y.; Tarasov, A.; Alyoshina, L.; Brainina, K. Interplay between antioxidant activity, health and disease. *Biointerface Res. Appl. Chem.* **2020**, *10*, 4893–4901. [CrossRef]
11. Oh, S.-H.; Choi, S.-Y.; Choi, H.-J.; Ryu, H.-M.; Kim, Y.-J.; Jung, H.-Y.; Cho, J.-H.; Kim, C.-D.; Park, S.-H.; Kwon, T.-H.; et al. The emerging role of xanthine oxidase inhibition for suppression of breast cancer cell migration and metastasis associated with hypercholesterolemia. *FASEB J.* **2019**, *33*, 7301–7314. [CrossRef]
12. Cao, H.; Pauff, J.M.; Hille, R. X-ray crystal structure of a xanthine oxidase complex with the flavonoid inhibitor quercetin. *J. Nat. Prod.* **2014**, *77*, 1693–1699. [CrossRef]
13. Vermot, A.; Petit-Härtlein, I.; Smith, S.M.E.; Fieschi, F. NADPH oxidases (NOX): An overview from discovery, molecular mechanisms to physiology and pathology. *Antioxidants* **2021**, *10*, 890. [CrossRef] [PubMed]
14. Ogura, K.; Nobuhisa, I.; Yuzawa, S.; Takeya, R.; Torikai, S.; Saikawa, K.; Sumimoto, H.; Inagaki, F. NMR solution structure of the tandem Src homology 3 domains of $p47^{phox}$ complexed with a $p22^{phox}$-derived proline-rich peptide. *J. Biol. Chem.* **2006**, *281*, 3660–3668. [CrossRef]
15. Macías Pérez, M.E.; Hernández Rodríguez, M.; Cabrera Pérez, L.C.; Fragoso-Vázquez, M.J.; Correa-Basurto, J.; Padilla-Martínez, I.I.; Méndez Luna, D.; Mera Jiménez, E.; Flores Sandoval, C.; Tamay Cach, F.; et al. Aromatic regions govern the recognition of NADPH oxidase inhibitors as diapocynin and its analogues. *Arch. Der Pharm.* **2017**, *350*, 1700041. [CrossRef]
16. Konaté, M.M.; Antony, S.; Doroshow, J.H. Inhibiting the activity of NADPH oxidase in cancer. *Antioxid Redox Signal.* **2020**, *33*, 435–454. [CrossRef] [PubMed]
17. Deshmukh, P.; Unni, S.; Krishnappa, G.; Padmanabhan, B. The Keap1–Nrf2 pathway: Promising therapeutic target to counteract ROS-mediated damage in cancers and neurodegenerative diseases. *Biophys. Rev.* **2017**, *9*, 41–56. [CrossRef]
18. Tascioglu Aliyev, A.; Panieri, E.; Stepanić, V.; Gurer-Orhan, H.; Saso, L. Involvement of NRF2 in breast cancer and possible therapeutical role of polyphenols and melatonin. *Molecules* **2021**, *26*, 1853. [CrossRef] [PubMed]
19. Wen, C.; Zhang, J.; Zhang, H.; Duan, Y.; Ma, H. Study on the structure–activity relationship of watermelon seed antioxidant peptides by using molecular simulations. *Food Chem.* **2021**, *364*, 130432. [CrossRef]
20. Koh, J.-A.; Ong, J.-H.; Abd Manan, F.; Ee, K.-Y.; Wong, F.-C.; Chai, T.-T. Discovery of bifunctional anti-DPP-IV and anti-ACE peptides from housefly larval proteins after in silico gastrointestinal digestion. *Biointerface Res. Appl. Chem.* **2022**, *12*, 4929–4944. [CrossRef]
21. Das, D.; Jaiswal, M.; Khan, F.N.; Ahamad, S.; Kumar, S. PlantPepDB: A manually curated plant peptide database. *Sci. Rep.* **2020**, *10*, 2194. [CrossRef]
22. Agrawal, P.; Bhagat, D.; Mahalwal, M.; Sharma, N.; Raghava, G.P.S. AntiCP 2.0: An updated model for predicting anticancer peptides. *Brief. Bioinform.* **2020**, *22*, bbaa153. [CrossRef] [PubMed]
23. Manavalan, B.; Subramaniyam, S.; Shin, T.H.; Kim, M.O.; Lee, G. Machine-learning-based prediction of cell-penetrating peptides and their uptake efficiency with improved accuracy. *J. Proteome Res.* **2018**, *17*, 2715–2726. [CrossRef]
24. Kochnev, Y.; Hellemann, E.; Cassidy, K.C.; Durrant, J.D. Webina: An open-source library and web app that runs AutoDock Vina entirely in the web browser. *Bioinformatics* **2020**, *36*, 4513–4515. [CrossRef] [PubMed]
25. Morris, G.M.; Ruth, H.; Lindstrom, W.; Sanner, M.F.; Belew, R.K.; Goodsell, D.S.; Olson, A.J. Software news and updates AutoDock4 and AutoDockTools4: Automated docking with selective receptor flexibility. *J. Comput. Chem.* **2009**, *30*, 2785–2791. [CrossRef]
26. Shukla, R.; Tripathi, T. Molecular dynamics simulation of protein and protein-ligand complexes. In *Computer-Aided Drug Design*; Springer: Singapore, 2020; Chapter 7; pp. 133–161.
27. Ong, J.-H.; Liang, C.-E.; Wong, W.-L.; Wong, F.-C.; Chai, T.-T. Multi-target anti-SARS-CoV-2 peptides from mealworm proteins: An in silico study. *Malays. J. Biochem. Mol. Biol.* **2021**, *24*, 83–91.

28. Mohana, D.S.; Chai, T.-T.; Wong, F.-C. Antioxidant and protein protection potentials of fennel seed-derived protein hydrolysates and peptides. *Mod. Food Sci. Technol.* **2019**, *35*, 22–29. [CrossRef]
29. Chai, T.-T.; Soo, Z.-Y.; Hsu, K.-C.; Li, J.-C.; Abd Manan, F.; Wong, F.-C. Antioxidant activity of semen cassiae protein hydrolysate: Thermal and gastrointestinal stability, peptide identification, and in silico analysis. *Mod. Food Sci. Technol.* **2019**, *35*, 38–48.
30. Chai, T.-T.; Xiao, J.; Mohana Dass, S.; Teoh, J.-Y.; Ee, K.-Y.; Ng, W.-J.; Wong, F.-C. Identification of antioxidant peptides derived from tropical jackfruit seed and investigation of the stability profiles. *Food Chem.* **2021**, *340*, 127876. [CrossRef] [PubMed]
31. Chaudhary, K.; Kumar, R.; Singh, S.; Tuknait, A.; Gautam, A.; Mathur, D.; Anand, P.; Varshney, G.C.; Raghava, G.P.S. A web server and mobile app for computing hemolytic potency of peptides. *Sci. Rep.* **2016**, *6*, 22843. [CrossRef]
32. Dimitrov, I.; Bangov, I.; Flower, D.R.; Doytchinova, I. AllerTOP v.2—A server for in silico prediction of allergens. *J. Mol. Model.* **2014**, *20*, 2278. [CrossRef]
33. Gupta, S.; Kapoor, P.; Chaudhary, K.; Gautam, A.; Kumar, R.; Open Source Drug Discovery, C.; Raghava, G.P.S. In silico approach for predicting toxicity of peptides and proteins. *PLoS ONE* **2013**, *8*, e73957. [CrossRef]
34. Ji, D.; Udenigwe, C.C.; Agyei, D. Antioxidant peptides encrypted in flaxseed proteome: An in silico assessment. *Food Sci. Hum. Wellness* **2019**, *8*, 306–314. [CrossRef]
35. Duan, X.; Zhang, M.; Chen, F. Prediction and analysis of antimicrobial peptides from rapeseed protein using in silico approach. *J. Food Biochem.* **2021**, *45*, e13598. [CrossRef]
36. Taniguchi, M.; Aida, R.; Saito, K.; Kikura, T.; Ochiai, A.; Saitoh, E.; Tanaka, T. Identification and characterization of multifunctional cationic peptides from enzymatic hydrolysates of soybean proteins. *J. Biosci. Bioeng.* **2020**, *129*, 59–66. [CrossRef]
37. Ghribi, A.M.; Sila, A.; Przybylski, R.; Nedjar-Arroume, N.; Makhlouf, I.; Blecker, C.; Attia, H.; Dhulster, P.; Bougatef, A.; Besbes, S. Purification and identification of novel antioxidant peptides from enzymatic hydrolysate of chickpea (*Cicer arietinum* L.) protein concentrate. *J. Funct. Foods* **2015**, *12*, 516–525. [CrossRef]
38. Chai, T.-T.; Law, Y.-C.; Wong, F.-C.; Kim, S.-K. Enzyme-assisted discovery of antioxidant peptides from edible marine invertebrates: A review. *Mar. Drugs* **2017**, *15*, 42. [CrossRef] [PubMed]
39. Lo, S.-C.; Li, X.; Henzl, M.T.; Beamer, L.J.; Hannink, M. Structure of the Keap1:Nrf2 interface provides mechanistic insight into Nrf2 signaling. *EMBO J.* **2006**, *25*, 3605–3617. [CrossRef]
40. Li, L.; Liu, J.; Nie, S.; Ding, L.; Wang, L.; Liu, J.; Liu, W.; Zhang, T. Direct inhibition of Keap1–Nrf2 interaction by egg-derived peptides DKK and DDW revealed by molecular docking and fluorescence polarization. *RSC Adv.* **2017**, *7*, 34963–34971. [CrossRef]
41. Shaw, S.A.; Vokits, B.P.; Dilger, A.K.; Viet, A.; Clark, C.G.; Abell, L.M.; Locke, G.A.; Duke, G.; Kopcho, L.M.; Dongre, A.; et al. Discovery and structure activity relationships of 7-benzyl triazolopyridines as stable, selective, and reversible inhibitors of myeloperoxidase. *Bioorganic Med. Chem.* **2020**, *28*, 115723. [CrossRef]
42. Maiocchi, S.L.; Ku, J.; Thai, T.; Chan, E.; Rees, M.D.; Thomas, S.R. Myeloperoxidase: A versatile mediator of endothelial dysfunction and therapeutic target during cardiovascular disease. *Pharmacol. Ther.* **2021**, *221*, 107711. [CrossRef]
43. Kang, N.; Kim, E.-A.; Kim, J.; Lee, S.-H.; Heo, S.-J. Identifying potential antioxidant properties from the viscera of sea snails (*Turbo cornutus*). *Mar. Drugs* **2021**, *19*, 567. [CrossRef] [PubMed]
44. Davey, C.A.; Fenna, R.E. 2.3 Å resolution x-ray crystal structure of the bisubstrate analogue inhibitor salicylhydroxamic acid bound to human myeloperoxidase: A model for a prereaction complex with hydrogen peroxide. *Biochemistry* **1996**, *35*, 10967–10973. [CrossRef] [PubMed]
45. Hu, X.; Zhou, Y.; Zhou, S.; Chen, S.; Wu, Y.; Li, L.; Yang, X. Purification and identification of novel xanthine oxidase inhibitory peptides derived from round scad (*Decapterus maruadsi*) protein hydrolysates. *Mar. Drugs* **2021**, *19*, 538. [CrossRef]
46. Yu, Z.; Kan, R.; Wu, S.; Guo, H.; Zhao, W.; Ding, L.; Zheng, F.; Liu, J. Xanthine oxidase inhibitory peptides derived from tuna protein: Virtual screening, inhibitory activity, and molecular mechanisms. *J. Sci. Food Agric.* **2021**, *101*, 1349–1354. [CrossRef]
47. Nakanishi, A.; Imajoh-Ohmi, S.; Fujinawa, T.; Kikuchi, H.; Kanegasaki, S. Direct evidence for interaction between COOH-terminal regions of cytochrome b_{558} subunits and cytosolic 47-kDa protein during activation of an $O(_2{}^-)$-generating system in neutrophils. *J. Biol. Chem.* **1992**, *267*, 19072–19074. [CrossRef]
48. Huang, J.; Kleinberg, M.E. Activation of the phagocyte NADPH oxidase protein p47^{phox}: Phosphorylation controls SH3 domain-dependent binding to p22^{phox} *. *J. Biol. Chem.* **1999**, *274*, 19731–19737. [CrossRef]
49. Arora, R.; Sawney, S.; Saini, V.; Steffi, C.; Tiwari, M.; Saluja, D. Esculetin induces antiproliferative and apoptotic response in pancreatic cancer cells by directly binding to KEAP1. *Mol. Cancer* **2016**, *15*, 64. [CrossRef]
50. Zhang, H.; Jing, X.; Shi, Y.; Xu, H.; Du, J.; Guan, T.; Weihrauch, D.; Jones, D.W.; Wang, W.; Gourlay, D.; et al. N-acetyl lysyltyrosylcysteine amide inhibits myeloperoxidase, a novel tripeptide inhibitor[1][S]. *J. Lipid Res.* **2013**, *54*, 3016–3029. [CrossRef]
51. Wang, Z.; Tang, T.; Wang, S.; Cai, T.; Tao, H.; Zhang, Q.; Qi, S.; Qi, Z. Aloin inhibits the proliferation and migration of gastric cancer cells by regulating NOX2-ROS-mediated pro-survival signal pathways. *Drug Des. Devel.* **2020**, *14*, 145–155. [CrossRef]
52. Rymaszewski, A.L.; Tate, E.; Yimbesalu, J.P.; Gelman, A.E.; Jarzembowski, J.A.; Zhang, H.; Pritchard, K.A., Jr.; Vikis, H.G. The role of neutrophil myeloperoxidase in models of lung tumor development. *Cancers* **2014**, *6*, 1111–1127. [CrossRef] [PubMed]
53. Quah, Y.; Mohd Ismail, N.I.; Ooi, J.L.S.; Affendi, Y.A.; Abd Manan, F.; Wong, F.-C.; Chai, T.-T. Identification of novel cytotoxic peptide KENPVLSLVNGMF from marine sponge *Xestospongia testudinaria*, with characterization of stability in human serum. *Int. J. Pept. Res. Ther.* **2018**, *24*, 189–199. [CrossRef]
54. Amigo, L.; Hernández-Ledesma, B. Current evidence on the bioavailability of food bioactive peptides. *Molecules* **2020**, *25*, 4479. [CrossRef]

55. Sharifi, S.; Samani, A.A.; Ahmadian, E.; Eftekhari, A.; Derakhshankhah, H.; Jafari, S.; Mokhtarpour, M.; Vahed, S.Z.; Salatin, S.; Dizaj, S.M. Oral delivery of proteins and peptides by mucoadhesive nanoparticles. *Biointerface Res. Appl. Chem.* **2019**, *9*, 3849–3852. [CrossRef]
56. Wu, Y.; He, H.; Hou, T. Purification, identification, and computational analysis of xanthine oxidase inhibitory peptides from kidney bean. *J. Food Sci.* **2021**, *86*, 1081–1088. [CrossRef]
57. Salo-Ahen, O.M.H.; Alanko, I.; Bhadane, R.; Bonvin, A.M.J.J.; Honorato, R.V.; Hossain, S.; Juffer, A.H.; Kabedev, A.; Lahtela-Kakkonen, M.; Larsen, A.S.; et al. Molecular dynamics simulations in drug discovery and pharmaceutical development. *Processes* **2021**, *9*, 71. [CrossRef]
58. Mangold, M.; Gütschow, M.; Stirnberg, M. A short peptide inhibitor as an activity-based probe for matriptase-2. *Pharmaceuticals* **2018**, *11*, 49. [CrossRef]
59. Baig, M.S.; Alagumuthu, M.; Rajpoot, S.; Saqib, U. Identification of a potential peptide inhibitor of SARS-CoV-2 targeting its entry into the host cells. *Drugs R D* **2020**, *20*, 161–169. [CrossRef]
60. Zhou, P.; Miao, Q.; Yan, F.; Li, Z.; Jiang, Q.; Wen, L.; Meng, Y. Is protein context responsible for peptide-mediated interactions? *Mol. Omics* **2019**, *15*, 280–295. [CrossRef] [PubMed]
61. Azizian, H.; Nabati, F.; Sharifi, A.; Siavoshi, F.; Mahdavi, M.; Amanlou, M. Large-scale virtual screening for the identification of new *Helicobacter pylori* urease inhibitor scaffolds. *J. Mol. Model.* **2012**, *18*, 2917–2927. [CrossRef] [PubMed]
62. Keskin, O.; Ma, B.; Nussinov, R. Hot regions in protein–protein interactions: The organization and contribution of structurally conserved hot spot residues. *J. Mol. Biol.* **2005**, *345*, 1281–1294. [CrossRef]
63. London, N.; Movshovitz-Attias, D.; Schueler-Furman, O. The structural basis of peptide-protein binding strategies. *Structure* **2010**, *18*, 188–199. [CrossRef]
64. Thévenet, P.; Shen, Y.; Maupetit, J.; Guyon, F.; Derreumaux, P.; Tufféry, P. PEP-FOLD: An updated de novo structure prediction server for both linear and disulfide bonded cyclic peptides. *Nucleic Acids Res.* **2012**, *40*, W288–W293. [CrossRef]
65. Lamiable, A.; Thévenet, P.; Rey, J.; Vavrusa, M.; Derreumaux, P.; Tufféry, P. PEP-FOLD3: Faster de novo structure prediction for linear peptides in solution and in complex. *Nucleic Acids Res.* **2016**, *44*, W449–W454. [CrossRef]
66. Shen, Y.; Maupetit, J.; Derreumaux, P.; Tufféry, P. Improved PEP-FOLD approach for peptide and miniprotein structure prediction. *J. Chem. Theory Comput.* **2014**, *10*, 4745–4758. [CrossRef]
67. Burley, S.K.; Berman, H.M.; Bhikadiya, C.; Bi, C.; Chen, L.; Di Costanzo, L.; Christie, C.; Dalenberg, K.; Duarte, J.M.; Dutta, S.; et al. RCSB Protein Data Bank: Biological macromolecular structures enabling research and education in fundamental biology, biomedicine, biotechnology and energy. *Nucleic Acids Res.* **2018**, *47*, D464–D474. [CrossRef]
68. Velázquez-Libera, J.L.; Durán-Verdugo, F.; Valdés-Jiménez, A.; Valdés-Jiménez, A.; Núñez-Vivanco, G.; Caballero, J. LigRMSD: A web server for automatic structure matching and RMSD calculations among identical and similar compounds in protein-ligand docking. *Bioinformatics* **2020**, *36*, 2912–2914. [CrossRef]
69. Zhou, P.; Jin, B.; Li, H.; Huang, S.Y. HPEPDOCK: A web server for blind peptide-protein docking based on a hierarchical algorithm. *Nucleic Acids Res.* **2018**, *46*, W443–W450. [CrossRef]
70. Huang, S.-Y.; Zou, X. Ensemble docking of multiple protein structures: Considering protein structural variations in molecular docking. *Proteins Struct. Funct. Bioinform.* **2007**, *66*, 399–421. [CrossRef]
71. Huang, S.-Y.; Zou, X. An iterative knowledge-based scoring function for protein–protein recognition. *Proteins: Struct. Funct. Bioinform.* **2008**, *72*, 557–579. [CrossRef]
72. Yan, Y.; Zhang, D.; Huang, S.-Y. Efficient conformational ensemble generation of protein-bound peptides. *J. Cheminform.* **2017**, *9*, 59. [CrossRef]
73. Zhou, P.; Li, B.; Yan, Y.; Jin, B.; Wang, L.; Huang, S.Y. Hierarchical flexible peptide docking by conformer generation and ensemble docking of peptides. *J. Chem. Inf. Model.* **2018**, *58*, 1292–1302. [CrossRef]
74. Laskowski, R.A.; Swindells, M.B. LigPlot+: Multiple ligand-protein interaction diagrams for drug discovery. *J. Chem. Inf. Model.* **2011**, *51*, 2778–2786. [CrossRef]
75. Wallace, A.C.; Laskowski, R.A.; Thornton, J.M. LIGPLOT: A program to generate schematic diagrams of protein-ligand interactions. *Protein Eng. Des. Sel.* **1995**, *8*, 127–134. [CrossRef] [PubMed]
76. Kumar, V.; Patiyal, S.; Dhall, A.; Sharma, N.; Raghava, G.P.S. B3pred: A random-forest-based method for predicting and designing blood–brain barrier penetrating peptides. *Pharmaceutics* **2021**, *13*, 1237. [CrossRef]
77. Mathur, D.; Singh, S.; Mehta, A.; Agrawal, P.; Raghava, G.P.S. In silico approaches for predicting the half-life of natural and modified peptides in blood. *PLoS ONE* **2018**, *13*, e0196829. [CrossRef]
78. Minkiewicz, P.; Iwaniak, A.; Darewicz, M. BIOPEP-UWM database of bioactive peptides: Current opportunities. *Int. J. Mol. Sci.* **2019**, *20*, 5978. [CrossRef]
79. Ibarra, A.A.; Bartlett, G.J.; Hegedüs, Z.; Dutt, S.; Hobor, F.; Horner, K.A.; Hetherington, K.; Spence, K.; Nelson, A.; Edwards, T.A.; et al. Predicting and experimentally validating hot-spot residues at protein–protein interfaces. *ACS Chem. Biol.* **2019**, *14*, 2252–2263. [CrossRef] [PubMed]
80. Wood, C.W.; Ibarra, A.A.; Bartlett, G.J.; Wilson, A.J.; Woolfson, D.N.; Sessions, R.B. BAlaS: Fast, interactive and accessible computational alanine-scanning using BudeAlaScan. *Bioinformatics* **2020**, *36*, 2917–2919. [CrossRef]
81. Schmid, N.; Eichenberger, A.P.; Choutko, A.; Riniker, S.; Winger, M.; Mark, A.E.; Van Gunsteren, W.F. Definition and testing of the GROMOS force-field versions 54A7 and 54B7. *Eur. Biophys. J.* **2011**, *40*, 843–856. [CrossRef]

82. Huang, W.; Lin, Z.; van Gunsteren, W.F. Validation of the GROMOS 54A7 force field with respect to β-peptide folding. *J. Chem. Theory Comput.* **2011**, *7*, 1237–1243. [CrossRef]
83. Leherte, L.; Vercauteren, D.P. Reduced point charge models of proteins: Effect of protein–water interactions in molecular dynamics simulations of ubiquitin systems. *J. Phys. Chem. B* **2017**, *121*, 9771–9784. [CrossRef] [PubMed]
84. Nunes-Alves, A.; Ormersbach, F.; Wade, R.C. Prediction of the drug–target binding kinetics for flexible proteins by comparative binding energy analysis. *J. Chem. Inf. Model.* **2021**, *61*, 3708–3721. [CrossRef]
85. Sabri, M.Z.; Hamid, A.A.A.; Hitam, S.M.S.; Rahim, M.Z.A. The assessment of three dimensional modelling design for single strand DNA aptamers for computational chemistry application. *Biophys. Chem.* **2020**, *267*, 106492. [CrossRef] [PubMed]
86. Sabri, M.Z.; Abdul Hamid, A.A.; Sayed Hitam, S.M.; Abdul Rahim, M.Z. *In silico* screening of aptamers configuration against hepatitis b surface antigen. *Adv. Bioinform.* **2019**, *2019*, 6912914. [CrossRef]

molecules

Article

Nomad Jellyfish *Rhopilema nomadica* Venom Induces Apoptotic Cell Death and Cell Cycle Arrest in Human Hepatocellular Carcinoma HepG2 Cells

Mohamed M. Tawfik [1,*], Nourhan Eissa [1], Fayez Althobaiti [2], Eman Fayad [2,*] and Ali H. Abu Almaaty [1]

1 Department of Zoology, Faculty of Science, Port Said University, Port Said 42526, Egypt; n.essa@sci.psu.edu.eg (N.E.); ali_hussein@sci.psu.edu.eg (A.H.A.A.)
2 Department of Biotechnology, Faculty of Sciences, Taif University, P.O. Box 11099, Taif 21944, Saudi Arabia; faiz@tu.edu.sa
* Correspondence: tawfik@sci.psu.edu.eg (M.M.T.); e.esmail@tu.edu.sa (E.F.)

Abstract: Jellyfish venom is a rich source of bioactive proteins and peptides with various biological activities including antioxidant, antimicrobial and antitumor effects. However, the anti-proliferative activity of the crude extract of *Rhopilema nomadica* jellyfish venom has not been examined yet. The present study aimed at the investigation of the in vitro effect of *R. nomadica* venom on liver cancer cells (HepG2), breast cancer cells (MDA-MB231), human normal fibroblast (HFB4), and human normal lung cells (WI-38) proliferation by using MTT assay. The apoptotic cell death in HepG2 cells was investigated using Annexin V-FITC/PI double staining-based flow cytometry analysis, western blot analysis, and DNA fragmentation assays. *R. nomadica* venom displayed significant dose-dependent cytotoxicity on HepG2 cells after 48 h of treatment with IC_{50} value of 50 μg/mL and higher toxicity (3:5-fold change) against MDA-MB231, HFB4, and WI-38 cells. *R. nomadica* venom showed a prominent increase of apoptosis as revealed by cell cycle arrest at G2/M phase, upregulation of p53, BAX, and caspase-3 proteins, and the down-regulation of anti-apoptotic Bcl-2 protein and DNA fragmentation. These findings suggest that *R. nomadica* venom induces apoptosis in hepatocellular carcinoma cells. To the best of the authors' knowledge, this is the first scientific evidence demonstrating the induction of apoptosis and cell cycle arrest of *R. nomadica* jellyfish venom.

Keywords: *Rhopilema nomadica*; apoptosis; cell cycle arrest; HepG2

Citation: Tawfik, M.M.; Eissa, N.; Althobaiti, F.; Fayad, E.; Abu Almaaty, A.H. Nomad Jellyfish *Rhopilema nomadica* Venom Induces Apoptotic Cell Death and Cell Cycle Arrest in Human Hepatocellular Carcinoma HepG2 Cells. *Molecules* **2021**, *26*, 5185. https://doi.org/10.3390/molecules26175185

Academic Editors: Višnja Stepanić and Marta Kučerová-Chlupáčová

Received: 30 June 2021
Accepted: 17 August 2021
Published: 26 August 2021

1. Introduction

Hepatocellular carcinoma (HCC) is the fourth most common cause of cancer-related mortality globally [1]. In Egypt, HCC is one of the prevalent malignancy cancer accounting for 70.48% of all liver tumors in both sexes, which may be attributed to the rising incidence rates of hepatitis C virus-related cirrhosis [2,3]. Chemotherapy and radiotherapy were determined to be the main rational therapeutic regimens for HCC [4,5]. However, the use of these routine treatments is assigned with potent toxic adverse effects in addition to the development of resistance of HCC cells to anticancer drugs [6,7]. In this perspective, there is an unmet need for novel anticancer agents that may have different mechanisms of action from current therapies, leading to higher selectivity for HCC cells.

Recently, jellyfish populations are rapidly increasing in various marine ecosystems worldwide, which may be associated with global warming, eutrophication, alien species invasions, and changes in salinity [8–10]. Jellyfish venom is primarily confined in specialized venom-containing capsules known as nematocysts that are found mainly in the tentacles. Nematocysts venom contains a mixture of bioactive proteins and peptides, which exhibits hemolytic, cardiotoxic, neurotoxic, musculotoxic, antioxidant, and cytolytic effects [11–15]. Jellyfish venom proteins are highly potent against a panel of human cancer cell lines which have attracted significant interests in tumor research [16,17].

Stimulating apoptotic pathways is the most significant non-surgical cancer treatment and HCC treatment in particular. Such strategies to trigger cancer cell death include upregulation of pro-apoptotic proteins and reduction of the expression of anti-apoptotic proteins [18–20]. Apoptosis is mainly associated with DNA fragmentation, chromatin compaction, nuclear fragmentation, and cell cycle arrest [21–23]. Interestingly, various jellyfish venoms have been reported to induce apoptosis and cell cycle arrest against diverse cancer cells [24,25]. Several studies have clearly indicated the link between the ability of jellyfish venom to produce oxidative stress and its induction of apoptosis in cancer cells [16,26,27]. ROS generation, lipid peroxidation induction, and mitochondria damage contribute to the cytotoxicity of the particular jellyfish venom, e.g., that of *Pelagia noctiluca* towards colon cancer cells and that of *Cassiopea andromeda* towards breast cancer cells, respectively [28,29].

Rhopilema nomadica jellyfish (nomad jellyfish) blooms off the eastern Mediterranean coasts of Italy, Turkey, Greece, Tunisia, and Egypt [30–32]. Nomad jellyfish was introduced in the Egyptian Mediterranean via Suez Canal in the latter half of the last century, which causes severe damage effects on the fishery industry and tourism [33,34]. Very few studies have explored the biological activities of nomad jellyfish. The present study is designed to assess the cytotoxic activity of *R. nomadica* venom on liver cancer cells (HepG2), breast cancer cells (MDA-MB231), human normal fibroblast (HFB4), and human normal lung cells (WI-38). The study also evaluates the extent of apoptosis in HepG2 cell death caused by *R. nomadica* venom.

2. Results

2.1. Electrophoretic Separation of the Protein Mixture of *R. nomadica* Venom by SDS-PAGE

Quantitative analysis of proteins from *R. nomadica* venom was performed using the SDS-PAGE gel. The results revealed the presence of proteins ranging in mass from 16 to ~250 kDa. Major six protein bands with molecular masses of ~16, ~18, ~23, ~28, ~31, and ~48 kDa have dominated the profile (Figure 1).

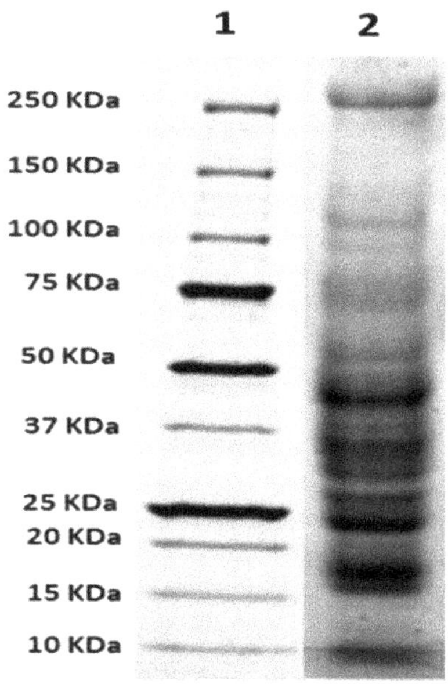

Figure 1. SDS-PAGE protein profile of *R. nomadica* venom. Lane 1: standard marker, Lane 2: protein extracts from the crude venom.

2.2. *R. nomadica* Venom Suppresses the Growth of HepG2, MDA-MB231, HFB4 and WI-38 Cells

R. nomadica venom was evaluated for cytotoxic activity on the viability of the human hepatocellular carcinoma cell line (HepG2) using MTT assay. The venom has shown a significant cytotoxic effect on HepG2 cells in a concentration-dependent manner after 48 h of cell treatment. The IC_{50} value was approximately determined to be 50 μg/mL.

The toxicity testing of the venom was extended on extra cancer cell line MDA-MB231 (breast cancer) and two normal cell lines HFB4 (human fibroblast) and WI-38 (human lung cells). Results showed that MDA-MB231, WI-38, and HFB4 cells had an IC_{50} for the venom of 216, 250, and 168 μg/mL, respectively, 3:5-fold higher than HepG2 cells. Therefore, the crude extract of *R. nomadica* venom exhibited some preference against HepG2 cells rather than other cancer or normal cell lines (Figure 2).

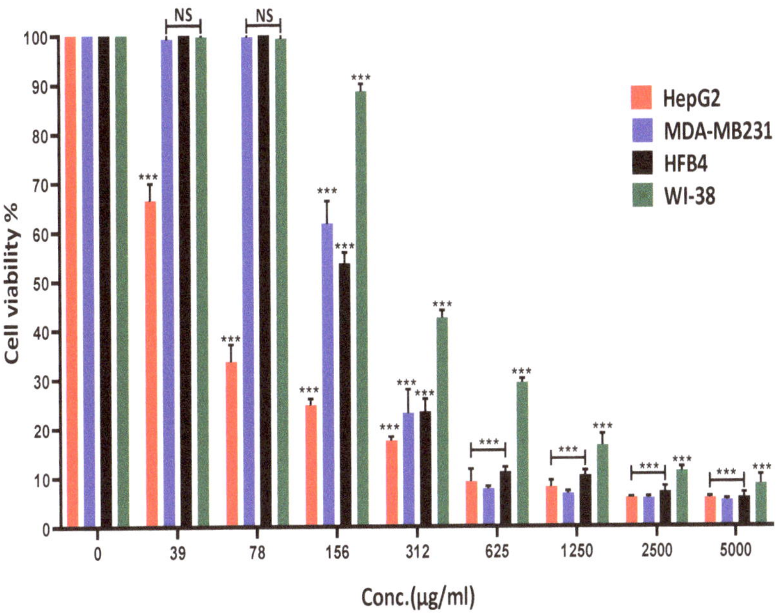

Figure 2. Effects of *R. nomadica* venom on cell proliferation of HepG2, MDA-MB231, HFB4, and WI-38 cell lines at different concentrations. The percentage of cell viability was measured using MTT assay after 48 h treatment. Non-treated cells were used as a control. Data are analyzed with one-way ANOVA followed by Tukey's test. All values are represented as mean ± SEM. *** denotes significance difference ($p < 0.001$) vs. control, otherwise noted NS: non-significant ($p > 0.05$).

2.3. *R. nomadica* Venom Induces G2/M Arrest on HepG2 Cells

Flow cytometric analysis of the cell cycle was performed to determine the cell cycle distribution of HepG2 cells treated with *R. nomadica* venom. The incubation of HepG2 cells with IC_{50} of *R. nomadica* venom for 48 h caused cell cycle arrest at the G2/M phase. The population of cells in G2/M and sub-G1 phases significantly increased from 12.28% and 1.49% in untreated cells to 28.51% and 18.17% in the treated cells with *R. nomadica* venom, respectively ($p < 0.001$). On the contrary, the treatment with *R. nomadica* venom resulted in a statistically significant decrease in the population of cells in G0–G1 and S phases ($p < 0.001$) (Figure 3).

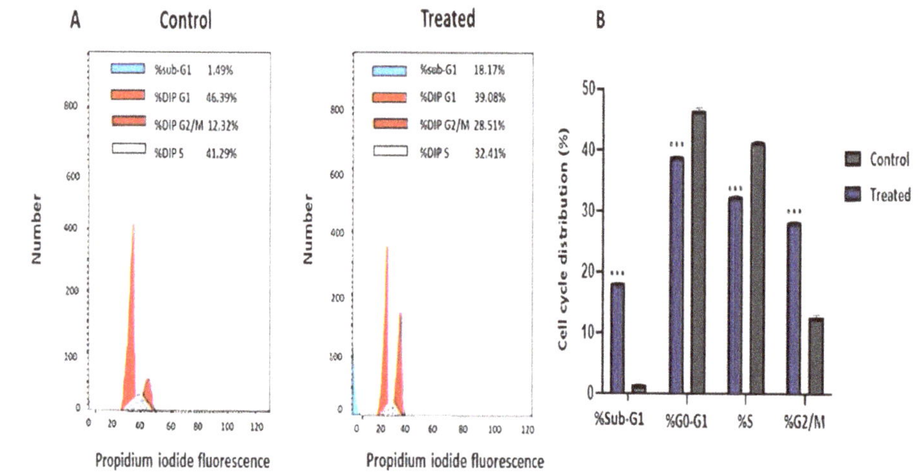

Figure 3. Cell cycle analysis of HepG2 cells treated with IC_{50} value of *R. nomadica* venom for 48 h by flow cytometry. (**A**) Representative profiles of cell cycle distribution in HepG2 cells after treatments. (**B**) The percentages of cell populations in sub-G1, G0–G1, S and G2/M phases. Data are analyzed with paired Student's t-test. Values are represented as mean ± SEM. *** denotes significance difference ($p < 0.001$) vs. control.

2.4. R. nomadica Venom Induces Apoptosis in HepG2 Cells

Annexin V-FITC/PI double staining-dependent on flow cytometry analysis showed that *R. nomadica* venom-induced apoptotic cell death in HepG2 cells (Figure 4). Populations of early and late apoptotic cells were significantly increased after 48 h treatment of IC_{50} of *R. nomadica* venom compared to untreated HepG2 cells ($p < 0.001$). Accordingly, the percentage of total apoptotic cells was significantly increased from 1.49% to 18.17% of treated cells.

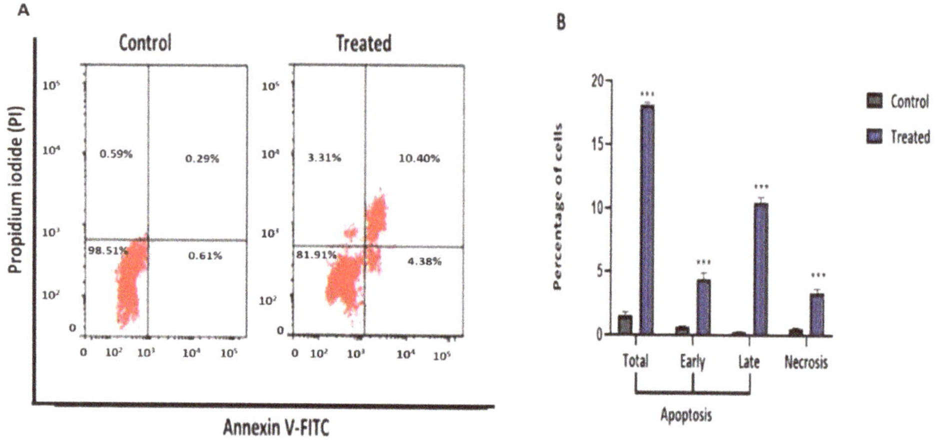

Figure 4. Apoptosis detection assay was performed using the Annexin V/PI double staining assay in HepG2 cells. HepG2 cells were treated with IC_{50} value of *R. nomadica* venom for 48 h, stained with Annexin V and PI, and analyzed on BD FACSCalibur flow cytometer. (**A**) Representative scatter plots of PI (*y*-axis) vs. Annexin V (*x*-axis). Lower left quadrants show viable cells (An −, PI −), whereas lower right quadrants represent the early apoptotic cells (An +, PI −). The upper left quadrants contain the necrotic cells (An −, PI +), while the upper right quadrants demonstrate the late apoptotic cells (An +, PI +). (**B**) Quantification graph of Annexin V/PI double staining assay obtained from BD FACSCalibur flow cytometer. Data are analyzed with paired Student's t-test. Values are represented as mean ± SEM. *** denotes significance difference ($p < 0.001$) vs. control.

2.5. *R. nomadica* Venom Induces Apoptosis in HepG2 Cells through the Regulation of the Expression of Apoptosis-Related Proteins

To understand the molecular mechanisms underlying *R. nomadica* venom-induced apoptosis, western blotting was performed to investigate the expression levels of apoptosis-related proteins. The expression of anti-apoptotic Bcl-2 protein was remarkably decreased in *R. nomadica* venom-treated HepG2 cells after 48 h treatment with IC_{50} concentration. Whereas apoptosis-inducing proteins BAX, caspase-3, and p53 were increased in the treated cells. Upon treatment of HepG2 cells with *R. nomadica* venom, a significant increase in Bax/Bcl-2 ratio compared to untreated cells ($p < 0.01$) (Figure 5).

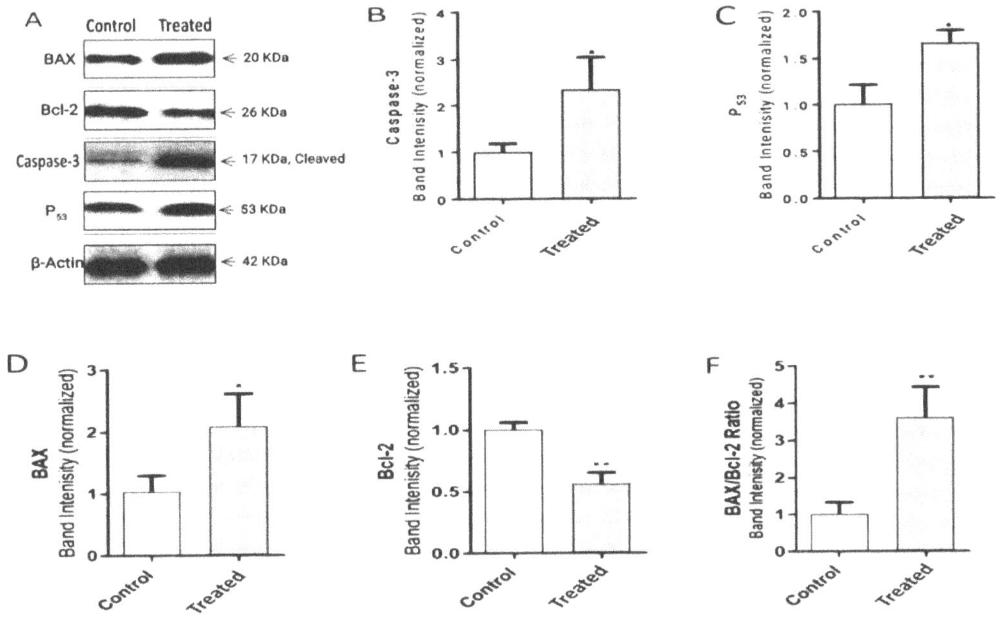

Figure 5. Western blot analysis. (**A**) shows western blot results about the expression level of apoptosis-related proteins BAX, Bcl-2, caspase-3, and p53 in HepG2 cells after being treated for 48 h with IC_{50} concentration of *R. nomadica* venom. From (**B–F**) represent the statistical graphs of the density ratios of the proteins calculated by ImageLab. Protein levels were normalized to β-Actin. Data are analyzed with paired Student's t-test Values are represented as mean ± SEM. * denotes significance difference ($p < 0.05$) vs. control. ** denotes significance ($p < 0.01$) vs. control.

2.6. *R. nomadica* Venom Induces DNA Damage

The DPA assay measured the relative quantity of DNA fragments in the treated HepG2 cells with *R. nomadica* venom (Figure 6). Compared to the untreated cells (5.27% ± 0.29%), *R. nomadica* venom at concentrations of 25 and 50 μg/mL caused a marked significant elevation in DNA fragmentation percentage (30.47% ± 1.66%) and (42.85% ± 2.33%), respectively.

Agarose gel electrophoresis of DNA from HepG2 cells treated with *R. nomadica* venom showed a ladder-like pattern of DNA fragments in a concentration-dependent manner relative to untreated cells (Figure 7). The gel pattern of the DNA samples isolated from untreated control of HepG2 cells showed clear bands of intact DNA, while *R. nomadica* venom-induced an increase in DNA smearing level with increasing dose from 25 to 50 μg/mL.

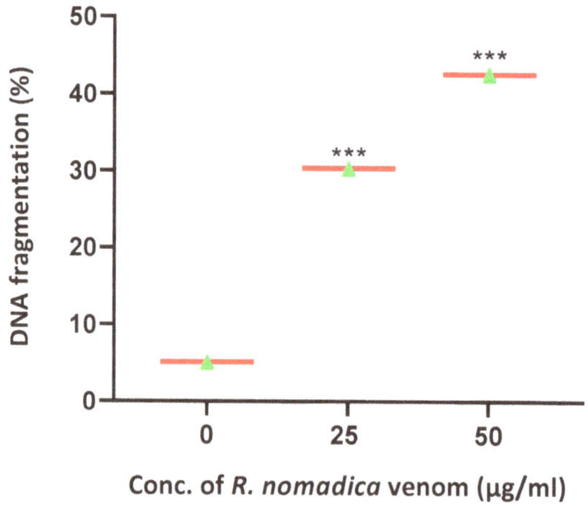

Figure 6. Quantitative estimation of DNA fragmentation by diphenylamine (DPA) assay in treated HepG2 cells with two different concentrations of *R. nomadica* venom (25–50 µg/mL) and untreated cells. Data are analyzed with one-way ANOVA followed by Tukey's test. Values are represented as mean ± SEM. *** denotes significance difference ($p < 0.001$) vs. control.

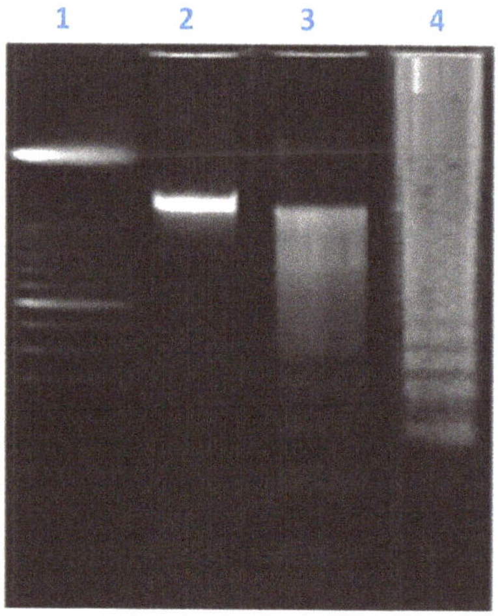

Figure 7. Electrophoretic pattern of DNA fragments isolated from HepG2 cells on 8% agarose gel electrophoresis. Lane 1: DNA marker, 100-bp. Lane 2: the intact DNA of the untreated HepG2 cells. Lane 3 and 4: the fragmented DNA of HepG2 cells treated with 25 and 50 µg/mL of *R. nomadica* venom, respectively.

2.7. *R. nomadica Venom Induces Less Hemolytic Activity against Human Erythrocytes*

The hemolysis assay is carried out to investigate whether the cytotoxic activity is related to direct damage to the cell membrane. *R. nomadica* venom exhibited weak hemolytic activity under 10% at tested concentrations from 40 to 640 µg/mL (Figure 8).

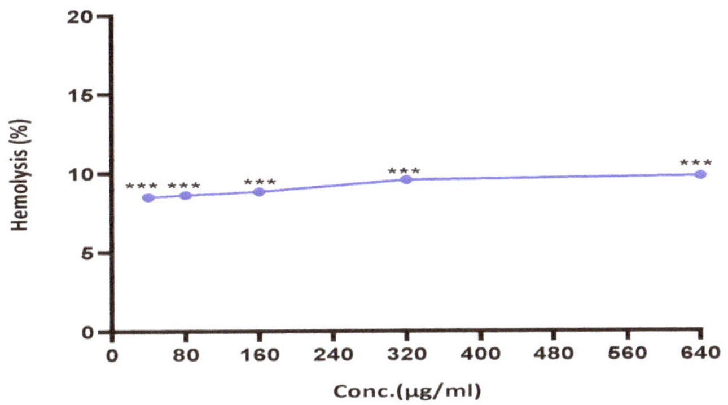

Figure 8. The hemolytic activity of *R. nomadica* venom against human RBC_s incubated with serial concentrations of the crude venom for 1 h at room temperature. PBS and 10% triton 100× were used as negative and positive controls, respectively. The absorbance of the supernatant was measured at 570 nm. Data are analyzed with one-way ANOVA followed by Tukey's test. Values are represented as mean ± SEM. *** denotes significance difference ($p < 0.001$) vs. control.

3. Discussion

Jellyfish venoms have long attracted the interest of researchers to find and develop novel anticancer agents. Jellyfish venoms either or isolated peptides have anti-proliferative effects on various cancer cell lines such as brain, colorectal, breast, lung, and liver cancer cells [26,28,35–37]. Although there are many swarms of *R. nomadica* jellyfish in the summertime of the Egyptian Mediterranean coasts, much less attention has been paid to their biological activities. The present study showed that crude venom of *R. nomadica* jellyfish inhibited the growth of HepG2 cells in a dose-dependent manner (IC_{50} value of 50 μg/mL). Less or no hemolytic activity has been detected for *R. nomadica* venom against human erythrocytes up to 10-fold greater than the IC_{50} concentration. To our knowledge, the anticancer activity of this venom is being reported herein for the first time.

Similar inhibitory effects against HepG2 cells have been reported for other jellyfish species such as *Nemopilema nomurai*, *Cyanea lamarckii*, and *Acromitus flagellates* venoms [35,37–40]. Interestingly, both *Acromitus flagellates* venom and *Cyanea lamarckii* extract share three to six protein bands with *R. nomadica* venom in their electrophoretic patterns [35,39]. A pore-forming toxin, CcTX-1 (31.17 KDa), is an isolated proteinaceous cytotoxin from *Cyanea capillata* jellyfish venom that exhibited potent inhibitory activity against HepG2 cells [41,42]. The other two cytotoxic proteins, known as CfTX-A (~40 kDa) and CfTX-B (~42 kDa), were partially purified from *Chironex fleckeri* venom [43]. Similar protein bands with a similar molecular weight of CcTX-1, CfTX-A, and CfTX-B have been observed in the protein profile of *R. nomadica* venom. These latter findings may interpret the cytotoxic properties of *R. nomadica* venom against HepG2 Cells.

In the current study, the DNA fragmentation assays revealed that treated HepG2 cells with IC_{50} value of *R. nomadica* venom showed a highly significant DNA fragmentation in comparison to non-treated cells. Similar DNA damages have been observed for *Nemopilema nomurai*, *Chiropsalmus quadrumanus* crude venoms and *Chrysaora Quinquecirrha* venom peptide against a variety of cancer cells particularly HCC cells [37,44,45].

Besides the DNA damage, cell cycle arrest is also a key event in apoptosis. Our results showed the induction of cell cycle arrest at the G2/M phase after the treatment of HepG2 cells for 48 h with IC_{50} value of *R. nomadica* venom and elevation in the sub-G1 population. These data are in accordance with the previous study that exhibited the increase of sub-G1 population in HepG2 cells treated with *Nemopilema nomurai* venom [37]. DNA fragmentation and cell cycle arrest results were denoting the strong apoptotic effect of *R. nomadica* venom.

Annexin V and PI double staining assay elucidated the efficacy of *R. nomadica* venom on triggering apoptosis in HepG2 cells treated with IC_{50} by raising the percentage of early and late apoptotic cells compared to untreated cells. Similar reports presented the venom potency of different jellyfish species such as *Stomolophus nomurai*, *Chrysaora helvola*, *Chiropsalmus quadrigatus*, *Chrysaora Quinquecirrha*, and *Cassiopea andromeda* in inducing the apoptotic cell death in various cancer cells via the formation of apoptotic bodies [36,44–47].

Apoptosis regulation is an impressive target for HCC treatment. Apoptosis is a significant mode of programmed cell death characterized by distinct hallmarks and controlled by essential extrinsic and intrinsic regulatory proteins [48,49]. The intrinsic pathway is mediated by pro-apoptotic proteins such as BAX/BAK proteins and anti-apoptotic proteins involved Bcl-2 and Bcl-xL proteins [50,51]. Terminally, apoptosis is executed by caspase-activated cascade involved the effector caspases such as caspase-3, 6, 7, and 10, which induce the activation of cytoplasmic endonuclease (CAD), which in turn causes chromatin condensation, DNA fragmentation, the formation of cytoplasmic blebs, and apoptotic bodies [52–54]. Furthermore, p53 is a tumor suppressor protein. Under Extra and intracellular stress signals, it plays a pivotal role in suppressing anti-apoptotic Bcl-2 family proteins. Moreover, p53 can directly interact with BAX and promotes the release of cytochrome c via mitochondrial outer membrane permeabilization (MOMP), which upregulates the tumor cells apoptosis [51,55].

Likewise, *R. nomadica* venom induces a potent intrinsically apoptotic effect on HepG2 cells by increasing p53 expression level, which in turn upregulates BAX (pro-apoptotic protein) and downregulates Bcl-2 (anti-apoptotic protein). We also demonstrate the intrinsic apoptotic pathway induced by *R. nomadica* venom in HepG2 cells through the elevation of BAX/Bcl-2 ratio leading to triggering of caspase-3 signaling protein, which causes cells destruction and ends up to apoptosis. Following the present results, previous studies have demonstrated the activity of other jellyfish venoms on modulating the apoptotic cell death in cancer cells via executioner and regulatory proteins [36,40,44–46]. Similarly, some apoptosis-inducing activities of other jellyfish venoms were attributed to their ability to produce high levels of ROS in different cancer cells, which may point out the potential ROS mediated cytotoxicity of *R. nomadica* venom [27–29].

Most of the jellyfish venoms which exhibited potent anticancer activities have strong hemolytic activities such as *Palythoa caribaeorum*, *Nemopilema nomurai*, and *Cassiopea xamachana* [26,56]. Even *Nemopilema nomurai* (schyphozoa) showed similar selective anticancer activities of *R. nomadica* venom against HepG2 cells rather than other cancer or normal cell lines [37]. However, erythrocytes were more susceptible to *N. nomurai* as >50% of mammalian RBCs underwent hemolysis at 100–200 µg/mL. In comparison with other jellyfish venoms and scyphozoans, *R. nomadica* showed less (weaker) hemolytic activities against RBCs. Furthermore, severe cytotoxicity on cardiac and skeletal cell lines has been reported for *N. nomurai* [12].

4. Materials and Methods

4.1. *Jellyfish Collection*

R. nomadica jellyfish specimens were captured from the eastern Egyptian Mediterranean of Port Said coast during the blooming summer months in 2019. Only tentacles were collected in tanks full of fresh seawater and transported immediately to a laboratory for further preparations.

4.2. *Nematocysts Venom Extraction and Preparation*

Nematocysts were isolated from the excised tentacles as described with slight modification [57]. Briefly, dissected tentacles were submerged in cold seawater at the mass: volume ratio of 1:3 to allow autolysis of the tissues at 4 °C with gentle swirling for 30 min every two hours. This process was repeated for 3 days. The nematocysts suspension was filtered through a plankton net to remove tissue debris and centrifuged at 10,000 rpm for 30 min at 4 °C. Then, the resultant supernatant was collected, lyophilized, and stored at −20 °C.

The crude venom was extracted from freeze-dried nematocysts with a minor modification. In which, 100 mg nematocyst powder was resuspended in 1 mL of phosphate-buffered saline (PBS, pH 7.4, 4 °C) and centrifuged at 15,000 rpm for 15 min at 4 °C [58]. The supernatant was separated and used as the extracted *Rhopilema nomadica* venom stock solution for the present study. The stock solution was aliquoted (to avoid freezing and rethawing) and stored at −20 °C. The protein concentration of the venom was estimated by the Bradford method, and the venom was used based on its protein concentration [59]. In terms of quality control, SDS-PAGE (Sodium dodecyl sulfate-polyacrylamide gel electrophoresis) (Bio-Rad Laboratories, Inc., Hercules, CA, USA) was used regularly to assess the banding pattern of venom proteins. Moreover, IC_{50} (concentration of the venom required for 50% inhibition of cell growth) against cell lines has been monitored to check the effectiveness and stability of the venom.

4.3. SDS-Polyacrylamide Gel Electrophoresis

R. nomadica crude venom proteins were separated based on their molecular weights by using SDS-PAGE. This technique was carried out using 12% polyacrylamide gel performed by TGX Stain-Free™ FastCast™ Acrylamide Kit (Bio-Rad Laboratories, Inc., Hercules, CA, USA). Briefly, the sample was boiled at 95 °C for 5 min with the sample loading buffer (4% SDS, 10% 2-mercaptoethanol, 20% glycerol, 0.004% bromophenol blue and 0.125 M Tris HCl, PH 6.8). The protein sample was electrophoresed for 60 min at 150 V, using Tris-glycine running buffer. Protein bands were visualized using stain-free technology and Chemi Doc imager (Bio-Rad Laboratories, Inc., Hercules, CA, USA) [60].

4.4. Cell Culture

HepG2 (human hepatocellular carcinoma), MDA-MB-231 (human breast adenocarcinoma), HFB4 (human normal fibroblast), and WI-38 (human normal lung) cell lines used in the present study were purchased from Holding company for biological products and vaccines (VACSERA, Giza, Egypt). Cells were maintained as monolayer in culture in complete medium (RPMI-1640 medium supplemented with 2 mM L-glutamine, 10% heat-inactivated FCS, 100 mg/mL streptomycin, and 100 U/mL penicillin) and were incubated at 37 °C with 5% CO2 in a humidified atmosphere.

4.5. Cell Viability by MTT Assay

To evaluate the potential cytotoxic effect of the *R. nomadica* venom on cell viability, an MTT (3-(4, 5-methylthiazol-2-yl)-2, 5-diphenyl-tetrazolium bromide) reduction assay was performed [61,62]. HepG2, MDA-MB-231, HFB4, and WI-38 cells (VACSERA, Giza, Egypt) were seeded in a 96-well plate at a concentration of 1.0×10^4 cells/well and incubated at 37 °C for 24 h to settle down. The cells were treated for another 48 h with different concentrations ranging from 0 to 5000 µg/mL of *R. nomadica* venom. 20 µL of MTT solution was added to each well and incubated for an additional 3 h at 37 °C. After the supernatant was removed, dimethyl sulfoxide (200 µL DMSO) was added to each well to resuspend the formazan crystals. Cell viability was determined by measuring optical density at 540 nm using Bio-Tek ELISA multi-well plate reader (Bio-Tek Instruments Inc., Burlington, VT, USA). The cytotoxic effect of the crude venom was determined by comparing the optical density of the treated cells and the untreated cells. Phosphate buffered saline (PBS) was used as a negative control. The IC_{50} value was calculated from the dose-response curve.

4.6. Cell Cycle Analysis

Flow cytometry assessed the percentage of cellular DNA content was assessed by flow cytometry [37]. HepG2 cells were seeded in a 6-well plate at a density of 1.0×10^4 cells/well and incubated at 37 °C for one day before the experiment. The cells were treated with IC_{50} concentration of *R. nomadica* venom for 48 h, harvested, and fixed with 70% cold ethanol at

−20 °C overnight. Fresh PBS washed the fixed cells, then stained with PI (50 μg/mL) in the presence of RNase A (10 μg/mL) and incubated for 30 min in the dark. The cellular DNA content was analyzed using a flow cytometer (FACSCalibur, Becton Dickinson, Franklin Lakes, NJ, USA).

4.7. *Annexin V-FITC/PI Double Staining Assay*

HepG2 cells were treated with IC_{50} concentration of *R. nomadica* venom for 48 h. Both adherent and suspended cells were collected, centrifuged (1000 rpm for 5 min), washed with PBS and resuspended with binding buffer (1×). After that, the cells were stained with 4 μL Annexin V-FITC and 2 μL propidium iodide (PI) (Annexin V-FITC Apoptosis Detection Kit) (Abcam, Cambridge, UK) and incubated for 15 min at 37 °C in the dark. The total apoptotic and necrotic cells were measured by BD FACSCalibur flow cytometer [46].

4.8. *Western Blot Analysis*

HepG2 cells were treated with IC_{50} concentration of *R. nomadica* venom for 48 h, and the cell lysate was prepared in cold lysis buffer [100 mM NaCl, 10 mM Tris, 25 mM EDTA, 25 mM EGTA, 1% Triton X-100, 1% NP-40 (pH 7.4)], with 1:300 protease inhibitor cocktail (Sigma-Aldrich, St. Louis, MO, USA) and Phosphatase inhibitor cocktail Tablet (Roche Diagnostics GmbH, Mannheim, Germany). Total protein concentration was determined using the Bradford method before proceeding to the western blotting. Equal amounts (20 μg) of protein samples were loaded into 12% SDS-polyacrylamide gel and separated by Cleaver electrophoresis unit (Cleaver Scientific Ltd., Rugby, UK). They were transferred onto polyvinylidene fluoride (PVDF) membranes (Bio-Rad Laboratories, Inc., Hercules, CA, USA) for 30 min using a Semi-dry Electroblotter (BioRad) at 2.5 A and 25 V for 30 min. The membrane was blocked with 5% nonfat dry milk in TBS-T for two hours at 37 °C. The membrane was incubated overnight at 4 °C with primary antibodies against BAX (1:1000, Cell Signaling Technology, Inc., Danvers, MA, USA), Bcl-2 (1:1500, Cell Signaling Technology), cleaved caspase-3 (1:750, Cell Signaling Technology), p53 (1:1000, Abcam, Cambridge, UK), and β-actin (1:5000, Sigma-Aldrich, St. Louis, MO, USA). Then, the membrane was washed three times with TBS-T and followed by incubation with the corresponding horseradish peroxidase-linked secondary antibodies (1:1000, Dako Ltd., High Wycombe, UK) for another hour at room temperature. β-actin was used as an internal reference protein. Finally, specific protein bands were visualized by a chemiluminescent detecting kit (Perkin Elmer, Waltham, MA, USA). The blot image was captured using a CCD camera-based imager (Chemi Doc imager, BioRad). The bands' intensities were then measured by densitometry using ImageLab software (Bio-Rad Laboratories, Inc., Hercules, CA, USA) [63,64].

4.9. *DNA Damage Evaluation*

The apoptotic effect induced in HepG2 cells was evaluated by two different techniques that included diphenylamine (DPA) assay and DNA laddering assay.

4.9.1. DNA Fragmentation Percentage by Diphenylamine (DPA) Assay

DPA assay is a colorimetrical quantitative method used for assessing the DNA fragmentation caused by the treatment of HepG2 cells by *R. nomadica* venom with the concentration of 25, 50 μg/mL for 48 h upon utilizing diphenylamine (DPA) reagent, which binds to deoxyribose [65]. The optical density was measured at 600 nm in the S (fragmented DNA) and the P (intact DNA) fractions. The percentage of fragmented DNA was calculated using the formula:

$$\text{DNA fragmentation (\%)} = [S/(S + P)] \times 100 \quad (1)$$

4.9.2. DNA Laddering Assay

Genomic DNA was extracted from HepG2 cells after the treatment with three different concentrations of *R. nomadica* venom (25, 50 µg/mL) for 48 h. Electrophoresis was performed in 8% agarose gel with ethidium bromide staining. Following this, the DNA in gel was visualized under UV and photographed by a digital camera (Canon U.S.A., Inc., Lake Success, NY, USA) [66].

4.10. Hemolysis Assay

Hemolytic activity of *R. nomadica* venom was performed on erythrocytes of humans with some modifications [56]. A blood sample was taken from a healthy adult volunteer after informed consent was obtained. The blood sample was immediately combined with anti-clot material (EDTA) and centrifuged at 5000 rpm for 5 min. Briefly, the erythrocytes suspension in sterile phosphate buffer saline (PBS) was received serial concentrations of the crude venom. After one hour of incubation at room temperature, the cells were centrifuged, and the supernatant was used to measure the absorbance of the liberated hemoglobin at 570 nm. The positive control (100% hemolysis) and the negative control (0% hemolysis) were Triton X-100 and sterile phosphate buffer saline (PBS), respectively. The following equation measured the hemolysis percentage for each sample:

$$\text{Hemolysis (\%)} = [(AS - A)/(AP.C - AN.C] \times 100 \quad (2)$$

where AS is the mean of absorbance of sample, AN.C is the mean of absorbance of negative control, and AP.C is the mean of absorbance of positive control.

4.11. Statistical Analysis

All experiments are achieved in triplicate. The values are shown as mean value ± SE and data were analyzed using paired Student's t-test and one-way ANOVA followed by Tukey's test. Statistical significance was assessed by $p < 0.05$.

5. Conclusions

In conclusion, our findings highlighted the anti-proliferative effect of *R. nomadica* venom and clearly showed that the venom extract could induce apoptosis in vitro. Therefore, these data might provide a starting point for the research on using *R. nomadica* venom in the treatment of cancer and hepatocellular carcinoma in particular. However, future investigations may be conducted to investigate the anticancer activities of *R. nomadica* venom in vivo and to characterize and identify the bioactive peptides that exert anticancer activity. Further, future experiments will be planned to investigate cytoprotective properties of *R. nomadica* venom on H_2O_2-treated cells in vitro and in vivo models. Moreover, protein and gene expressions of the antioxidative enzymes and oxidative biomarkers for cell membrane lipid peroxidation will be analyzed upon treatment with *R. nomadica* venom. Such experimental work may reveal in detail the redox activity of *R. nomadica* venom.

Author Contributions: Conceptualization, M.M.T.; A.H.A.A. and N.E.; methodology, M.M.T.; A.H.A.A.; N.E.; E.F. and F.A.; software and validation, M.M.T.; A.H.A.A. and E.F.; formal analysis investigation, resources, data curation, M.M.T.; A.H.A.A. and N.E.; writing—original draft preparation M.M.T. and N.E.; writing—review and editing, M.M.T. and F.A.; project administration, M.M.T. and A.H.A.A.; funding acquisition, F.A.; E.F. and N.E. All authors have read and agreed to the published version of the manuscript.

Funding: Taif University Researchers Supporting Project number (TURSP—2020/222), Taif University, Taif, Saudi Arabia.

Institutional Review Board Statement: This study was performed according to the guidelines of the US National Institute of Health (NIH Publication No. 85.23, revised 1985) guides and has been reviewed and approved by the animal care and use according to Zoology Department, Faculty of Science, Port Said University (approved duration: 2020 to 2021).

Informed Consent Statement: A blood sample was taken from a healthy adult volunteer after informed consent was obtained.

Data Availability Statement: The data presented in this study are available on request from the corresponding author.

Acknowledgments: The authors would like to thank the Deanship of Scientific Research at Taif University for funding this work through Taif University Researchers Supporting Project number (TURSP—2020/222), Taif University, Taif, Saudi Arabia.

Conflicts of Interest: The authors declare no conflict of interest.

Sample Availability: Samples of the compounds are available from the authors.

References

1. Villanueva, A. Hepatocellular Carcinoma. *N. Engl. J. Med.* **2019**, *380*, 1450–1462. [CrossRef] [PubMed]
2. Ibrahim, A.S.; Khaled, H.M.; Mikhail, N.N.; Baraka, H.; Kamel, H. Cancer incidence in egypt: Results of the national population-based cancer registry program. *J. Cancer Epidemiol.* **2014**, *2014*, 437971. [CrossRef] [PubMed]
3. Ziada, D.H.; El Sadany, S.; Soliman, H.; Abd-Elsalam, S.; Salama, M.; Hawash, N.; Selim, A.; Hamisa, M.; Elsabagh, H.M. Prevalence of hepatocellular carcinoma in chronic hepatitis C patients in Mid Delta, Egypt: A single center study. *J. Egypt. Natl. Cancer Inst.* **2016**, *28*, 257–262. [CrossRef]
4. Chen, K.-W.; Ou, T.-M.; Hsu, C.-W.; Horng, C.-T.; Lee, C.-C.; Tsai, Y.-Y.; Tsai, C.-C.; Liou, Y.-S.; Yang, C.-C.; Hsueh, C.-W.; et al. Current systemic treatment of hepatocellular carcinoma: A review of the literature. *World J. Hepatol.* **2015**, *7*, 1412–1420. [CrossRef] [PubMed]
5. Yim, H.J.; Suh, S.J.; Um, S.H. Current management of hepatocellular carcinoma: An Eastern perspective. *World J. Gastroenterol.* **2015**, *21*, 3826–3842. [CrossRef]
6. Le Grazie, M.; Biagini, M.R.; Tarocchi, M.; Polvani, S.; Galli, A. Chemotherapy for hepatocellular carcinoma: The present and the future. *World J. Hepatol.* **2017**, *9*, 907–920. [CrossRef] [PubMed]
7. Yu, Y.; Feng, M. Radiotherapy for Hepatocellular Carcinoma. *Semin. Radiat. Oncol.* **2018**, *28*, 277–287. [CrossRef]
8. Goldstein, J.; Steiner, U.K. Ecological drivers of jellyfish blooms—The complex life history of a 'well-known' medusa (Aurelia aurita). *J. Anim. Ecol.* **2020**, *89*, 910–920. [CrossRef]
9. Needleman, R.K.; Neylan, I.P.; Erickson, T.B. Environmental and Ecological Effects of Climate Change on Venomous Marine and Amphibious Species in the Wilderness. *Wilderness Environ.* **2018**, *29*, 343–356. [CrossRef]
10. Thé, J.; Barroso, H.d.S.; Mammone, M.; Viana, M.; Batista Melo, C.S.; Mies, M.; Banha, T.N.S.; Morandini, A.C.; Rossi, S.; Soares, M.d.O. Aquaculture facilities promote populational stability throughout seasons and increase medusae size for the invasive jellyfish Cassiopea andromeda. *Mar. Environ. Res.* **2020**, *162*, 105161. [CrossRef]
11. Ayed, Y.; Dellai, A.; Ben Mansour, H.; Bacha, H.; Abid, S. Analgesic and antibutyrylcholinestrasic activities of the venom prepared from the Mediterranean jellyfish Pelagia noctiluca (Forsskal, 1775). *Ann. Clin. Microbiol. Antimicrob.* **2012**, *11*, 15. [CrossRef] [PubMed]
12. Kang, C.; Munawir, A.; Cha, M.; Sohn, E.-T.; Lee, H.; Kim, J.-S.; Yoon, W.D.; Lim, D.; Kim, E. Cytotoxicity and hemolytic activity of jellyfish Nemopilema nomurai (Scyphozoa: Rhizostomeae) venom. *Comp. Biochem. Physiol. Toxicol. Pharmacol.* **2009**, *150*, 85–90. [CrossRef] [PubMed]
13. Ramasamy, S.; Isbister, G.K.; Seymour, J.E.; Hodgson, W.C. Pharmacologically distinct cardiovascular effects of box jellyfish (Chironex fleckeri) venom and a tentacle-only extract in rats. *Toxicol. Lett* **2005**, *155*, 219–226. [CrossRef] [PubMed]
14. Weston, A.J.; Chung, R.; Dunlap, W.C.; Morandini, A.C.; Marques, A.C.; Moura-da-Silva, A.M.; Ward, M.; Padilla, G.; da Silva, L.F.; Andreakis, N.; et al. Proteomic characterisation of toxins isolated from nematocysts of the South Atlantic jellyfish Olindias sambaquiensis. *Toxicon Off. J. Int. Soc. Toxinology* **2013**, *71*, 11–17. [CrossRef] [PubMed]
15. Yu, H.; Liu, X.; Xing, R.; Liu, S.; Li, C.; Li, P. Radical scavenging activity of protein from tentacles of jellyfish Rhopilema esculentum. *Bioorg. Med. Chem. Lett.* **2005**, *15*, 2659–2664. [CrossRef] [PubMed]
16. Leone, A.; Lecci, R.M.; Durante, M.; Piraino, S. Extract from the Zooxanthellate Jellyfish Cotylorhiza tuberculata Modulates Gap Junction Intercellular Communication in Human Cell Cultures. *Mar. Drugs* **2013**, *11*, 1728–1762. [CrossRef] [PubMed]
17. Mariottini, G.L. The Role of Cnidaria in Drug Discovery. In *The Cnidaria, Past, Present and Future: The World of Medusa and Her Sisters*; Goffredo, S., Dubinsky, Z., Eds.; Springer International Publishing: Cham, Switzerland, 2016; pp. 653–668. [CrossRef]
18. Fulda, S. Targeting apoptosis for anticancer therapy. Intracellular Signaling and Response to Anti-Cancer Therapy. *Semin. Cancer Biol.* **2015**, *31*, 84–88. [CrossRef]
19. Pfeffer, C.M.; Singh, A.T.K. Apoptosis: A Target for Anticancer Therapy. *Int. J. Mol. Sci.* **2018**, *19*, 448. [CrossRef]
20. Verena, J. The Intrinsic Apoptosis Pathways as a Target in Anticancer Therapy. *Curr. Pharm. Biotechnol.* **2012**, *13*, 1426–1438.
21. Doonan, F.; Cotter, T.G. Morphological assessment of apoptosis. *Methods* **2008**, *44*, 200–204. [CrossRef]
22. Häcker, G. The morphology of apoptosis. *Cell Tissue Res.* **2000**, *301*, 5–17. [CrossRef] [PubMed]
23. Saraste, A.; Pulkki, K. Morphologic and biochemical hallmarks of apoptosis. *Cardiovasc. Res.* **2000**, *45*, 528–537. [CrossRef]

24. Cuiping, L.; Pengcheng, L.; Jinhua, F.; Rongfeng, L.; Huahua, Y. Cytotoxicity of the venom from the nematocysts of jellyfish Cyanea nozakii Kishinouye. *Toxicol. Ind. Health* **2012**, *28*, 186–192. [CrossRef] [PubMed]
25. Kawabata, T.; Lindsay, D.J.; Kitamura, M.; Konishi, S.; Nishikawa, J.; Nishida, S.; Kamio, M.; Nagai, H. Evaluation of the bioactivities of water-soluble extracts from twelve deep-sea jellyfish species. *Fish Sci.* **2013**, *79*, 487–494. [CrossRef]
26. Ayed, Y.; Sghaier, R.M.; Laouini, D.; Bacha, H. Evaluation of anti-proliferative and anti-inflammatory activities of Pelagia noctiluca venom in Lipopolysaccharide/Interferon-γ stimulated RAW264.7 macrophages. *Biomed. Pharmacother.* **2016**, *84*, 1986–1991. [CrossRef] [PubMed]
27. Morabito, R.; Condello, S.; Currò, M.; Marino, A.; Ientile, R.; La Spada, G. Oxidative stress induced by crude venom from the jellyfish Pelagia noctiluca in neuronal-like differentiated SH-SY5Y cells. *Toxicol. In Vitro* **2012**, *26*, 694–699. [CrossRef]
28. Mirshamsi, M.R.; Omranipour, R.; Vazirizadeh, A.; Fakhri, A.; Zangeneh, F.; Mohebbi, G.H.; Seyedian, R.; Pourahmad, J. Persian Gulf Jellyfish (Cassiopea andromeda) Venom Fractions Induce Selective Injury and Cytochrome C Release in Mitochondria Obtained from Breast Adenocarcinoma Patients. *Asian Pac. J. Cancer Prev.* **2017**, *18*, 277–286. [CrossRef]
29. Ayed, Y.; Boussabbeh, M.; Zakhama, W.; Bouaziz, C.; Abid, S.; Bacha, H. Induction of cytotoxicity of Pelagia noctiluca venom causes reactive oxygen species generation, lipid peroxydation induction and DNA damage in human colon cancer cells. *Lipids Health Dis.* **2011**, *10*, 232. [CrossRef]
30. Edelist, D.; Guy-Haim, T.; Kuplik, Z.; Zuckerman, N.; Nemoy, P.; Angel, D.L. Phenological shift in swarming patterns of *Rhopilema nomadica* in the Eastern Mediterranean Sea. *J. Plankton Res.* **2020**, *42*, 211–219. [CrossRef]
31. Jafari, H.; Honari, H.; Zargan, J.; Jahromi, S.T. Identification and hemolytic activity of jellyfish (Rhopilema sp., Scyphozoa: Rhizostomeae) venom from the Persian Gulf and Oman Sea. *Biodiversitas J. Biol. Divers.* **2019**, *20*, 1228–1232. [CrossRef]
32. Yahia, M.N.D.; Yahia, O.K.-D.; Gueroun, S.K.M.; Aissi, M.; Deidun, A.; Fuentes, V.; Piraino, S. The invasive tropical scyphozoan *Rhopilema nomadica* Galil, 2013, 1990 reaches the Tunisian coast of the Mediterranean Sea. *BioInvasions Rec.* **2013**, *2*, 319–323. [CrossRef]
33. Balistreri, P.; Spiga, A.; Deidun, A.; Gueroun, S.K.; Yahia, M.N.D. Further spread of the venomous jellyfish *Rhopilema nomadica* Galil, Spannier & Ferguson, 1990 (*Rhizostomeae, Rhizostomatidae*) in the western Mediterranean. *BioInvasions Rec.* **2017**, *6*, 19–24. [CrossRef]
34. El-Regal, M.A.A.; Temraz, T.A. Blooming of the Nomad Jelly Fish *Rhopilema Nomadica* along the Egyptian Mediterranean Coasts 1. In Proceedings of the International Jellyfish Bloom Symposium, Barcelona, Spain, 30 May–3 June 2016.
35. Lazcano-Pérez, F.; Zavala-Moreno, A.; Rufino-González, Y.; Ponce-Macotela, M.; García-Arredondo, A.; Cuevas-Cruz, M.; Gómez-Manzo, S.; Marcial-Quino, J.; Arreguín-Lozano, B.; Arreguín-Espinosa, R. Hemolytic, anticancer and antigiardial activity of Palythoa caribaeorum venom. *J. Venom. Anim. Toxins Trop. Dis.* **2018**, *24*, 1–7. [CrossRef] [PubMed]
36. Maduraiveeran, H.; Raja, K.; Chinnasamy, A. Antiproliferative and antioxidant properties of nematocysts crude venom from jellyfish Acromitus flagellatus against human cancer cell lines. *Saudi J. Biol. Sci.* **2021**, *28*, 1954–1961. [CrossRef]
37. Lee, H.; Bae, S.K.; Kim, M.; Pyo, M.J.; Kim, M.Y.S.; Won, C.-K.; Yoon, W.D.; Han, C.H.; Kang, C.; Kim, E. Anticancer Effect of Nemopilema nomurai Jellyfish Venom on HepG2 Cells and a Tumor Xenograft Animal Model. *Evid. Based Complement. Altern. Med.* **2017**, *2017*, 2752716. [CrossRef]
38. Choudhary, I.; Lee, H.; Pyo, M.J.; Heo, Y.; Chae, J.; Yum, S.S.; Kang, C.; Kim, E. Proteomic Investigation to Identify Anticancer Targets of Nemopilema nomurai Jellyfish Venom in Human Hepatocarcinoma HepG2 Cells. *Toxins* **2018**, *10*, 194. [CrossRef]
39. Helmholz, H.; Ruhnau, C.; Schütt, C.; Prange, A. Comparative study on the cell toxicity and enzymatic activity of two northern scyphozoan species *Cyanea capillata* (L.) and Cyanea lamarckii (Péron & Léslieur). *Toxicon* **2007**, *50*, 53–64. [CrossRef] [PubMed]
40. Lee, H.; Pyo, M.J.; Bae, S.K.; Heo, Y.; Choudhary, I.; Hwang, D.; Yang, H.; Kim, J.; Chae, J.; Han, C.H.; et al. Nemopilema nomurai jellyfish venom exerts an anti-metastatic effect by inhibiting Smad- and NF-κB-mediated epithelial–mesenchymal transition in HepG2 cells. *Sci. Rep.* **2018**, *8*, 2808. [CrossRef]
41. Lassen, S.; Helmholz, H.; Ruhnau, C.; Prange, A. A novel proteinaceous cytotoxin from the northern Scyphozoa Cyanea capillata (L.) with structural homology to cubozoan haemolysins. *Toxicon* **2011**, *57*, 721–729. [CrossRef]
42. Remigante, A.; Costa, R.; Morabito, R.; La Spada, G.; Marino, A.; Dossena, S. Impact of Scyphozoan Venoms on Human Health and Current First Aid Options for Stings. *Toxins* **2018**, *10*, 133. [CrossRef]
43. Brinkman, D.L.; Konstantakopoulos, N.; McInerney, B.V.; Mulvenna, J.; Seymour, J.E.; Isbister, G.K.; Hodgson, W.C. Chironex fleckeri (Box Jellyfish) Venom Proteins: Expansion of a Cnidarian Toxin Family that Elicits Variable Cytolytic and Cardiovascular Effects. *J. Biol. Chem.* **2014**, *289*, 4798–4812. [CrossRef]
44. Balamurugan, E.; Ramesh Kumar, D.; Menon, V. Proapoptotic Effect of Chrysaora Quinquecirrha (Sea Nettle) Nematocyst Venom Peptide in HEp 2 and HeLa Cells. *Eur. J. Sci. Res.* **2009**, *35*, 355–367.
45. Sun, L.-K.; Yoshii, Y.; Hyodo, A.; Tsurushima, H.; Saito, A.; Harakuni, T.; Li, Y.-P.; Nozaki, M.; Morine, N. Apoptosis induced by box jellyfish (Chiropsalmus Quadrigatus) toxin in glioma and vascular endothelial cell lines. *Toxicon* **2002**, *40*, 441–446. [CrossRef]
46. Ha, S.-H.; Jin, F.; Kwak, C.-H.; Abekura, F.; Park, J.-Y.; Park, N.G.; Chang, Y.-C.; Lee, Y.-C.; Chung, T.-W.; Ha, K.-T.; et al. Jellyfish extract induces apoptotic cell death through the p38 pathway and cell cycle arrest in chronic myelogenous leukemia K562 cells. *PeerJ* **2017**, *5*, e2895. [CrossRef] [PubMed]
47. Qu, X.; Xia, X.; Lai, Z.; Zhong, T.; Li, G.; Fan, L.; Shu, W. Apoptosis-like cell death induced by nematocyst venom from Chrysaora helvola Brandt jellyfish and an in vitro evaluation of commonly used antidotes. *Comp. Biochem. Physiol. Part C Toxicol. Pharmacol.* **2016**, *180*, 31–39. [CrossRef] [PubMed]

48. Goldar, S.; Khaniani, M.S.; Derakhshan, S.M.; Baradaran, B. Molecular mechanisms of apoptosis and roles in cancer development and treatment. *Asian Pac. J. Cancer Prev.* **2015**, *16*, 2129–2144. [CrossRef] [PubMed]
49. Nikoletopoulou, V.; Markaki, M.; Palikaras, K.; Tavernarakis, N. Crosstalk between apoptosis, necrosis and autophagy. *Biochim. Et Biophys. Acta BBA Mol. Cell Res.* **2013**, *1833*, 3448–3459. [CrossRef]
50. Ghobrial, I.M.; Witzig, T.E.; Adjei, A.A. Targeting apoptosis pathways in cancer therapy. *CA Cancer J. Clin.* **2005**, *55*, 178–194. [CrossRef]
51. Kim, R. Recent advances in understanding the cell death pathways activated by anticancer therapy. *Cancer* **2005**, *103*, 1551–1560. [CrossRef] [PubMed]
52. Elmore, S. Apoptosis: A Review of Programmed Cell Death. *Toxicol. Pathol.* **2007**, *35*, 495–516. [CrossRef]
53. Hengartner, M.O. The biochemistry of apoptosis. *Nature* **2000**, *407*, 770–776. [CrossRef] [PubMed]
54. Sankari, S.L.; Masthan, K.M.K.; Babu, N.A.; Bhattacharjee, T.; Elumalai, M. Apoptosis in cancer—An update. *Asian Pac. J. Cancer Prev.* **2012**, *13*, 4873–4878. [CrossRef]
55. Chi, S.-W. Structural insights into the transcription-independent apoptotic pathway of p53. *BMB Rep.* **2014**, *47*, 167–172. [CrossRef] [PubMed]
56. Torres, M.; Aguilar, M.B.; Falcón, A.; Sánchez, L.; Radwan, F.F.Y.; Burnett, J.W.; Heimer-de la Cotera, E.P.; Arellano, R.O. Electrophysiological and hemolytic activity elicited by the venom of the jellyfish Cassiopea xamachana. *Toxicon* **2001**, *39*, 1297–1307. [CrossRef]
57. Bloom, D.; Burnett, J.; Alderslade, P. Partial purification of box jellyfish (Chironex fleckeri) nematocyst venom isolated at the beachside. *Toxicon Off. J. Int. Soc. Toxinology* **1998**, *36*, 1075–1085. [CrossRef]
58. Marino, A.; Morabito, R.; Pizzata, T.; La Spada, G. Effect of various factors on Pelagia noctiluca (Cnidaria, Scyphozoa) crude venom-induced haemolysis. *Comp. Biochem. Physiol. A Mol. Integr. Physiol.* **2008**, *151*, 144–149. [CrossRef] [PubMed]
59. Bradford, M.M. A rapid and sensitive method for the quantitation of microgram quantities of protein utilizing the principle of protein-dye binding. *Anal. Biochem.* **1976**, *72*, 248–254. [CrossRef]
60. Laemmli, U.K. Cleavage of structural proteins during the assembly of the head of bacteriophage T4. *Nature* **1970**, *227*, 680–685. [CrossRef]
61. Sebeka, A.A.H.; Osman, A.M.A.; Sayed, I.; Bahanasawy, M.; Tantawy, M. Synthesis and Antiproliferative Activity of Novel Neocryptolepine-Hydrazides Hybrids. *J. Appl. Pharm. Sci.* **2017**, *7*, 9–15. [CrossRef]
62. Tantawy, M.A.; Sroor, F.M.; Mohamed, M.F.; El-Naggar, M.E.; Saleh, F.M.; Hassaneen, H.M.; Abdelhamid, I.A. Molecular Docking Study, Cytotoxicity, Cell Cycle Arrest and Apoptotic Induction of Novel Chalcones Incorporating Thiadiazolyl Isoquinoline in Cervical Cancer. *Anticancer Agents Med. Chem.* **2020**, *20*, 70–83. [CrossRef]
63. Sambrook, J.; Fritsch, E.F.; Maniatis, T. *Molecular Cloning: A Laboratory Manual*, 2nd ed.; Cold Spring Harbor Laboratory Press: New York, NY, USA, 1989; ISBN 0879693096.
64. Burnette, W.N. "Western blotting": Electrophoretic transfer of proteins from sodium dodecyl sulfate–polyacrylamide gels to unmodified nitrocellulose and radiographic detection with antibody and radioiodinated protein A. *Anal. Biochem.* **1981**, *112*, 195–203. [CrossRef]
65. Zhao, Y.; Xiang, S.; Dai, X.; Yang, K. A simplified diphenylamine colorimetric method for growth quantification. *Appl. Microbiol. Biotechnol.* **2013**, *97*, 5069–5077. [CrossRef]
66. Gong, J.P.; Traganos, F.; Darzynkiewicz, Z. A Selective Procedure for DNA Extraction from Apoptotic Cells Applicable for Gel Electrophoresis and Flow Cytometry. *Anal. Biochem.* **1994**, *218*, 314–319. [CrossRef]

molecules

MDPI

Article

Cell Culture Characterization of Prooxidative Chain-Transfer Agents as Novel Cytostatic Drugs

Victoria Heymans [1], Sascha Kunath [1], Parvana Hajieva [2] and Bernd Moosmann [1,*]

[1] Evolutionary Biochemistry and Redox Medicine, Institute for Pathobiochemistry, University Medical Center of the Johannes Gutenberg University, 55128 Mainz, Germany; vichey2405@gmail.com (V.H.); sakunath@uni-mainz.de (S.K.)

[2] Institute for Translational Medicine, MSH Medical School Hamburg, 20457 Hamburg, Germany; parvana.hajieva@medicalschool-hamburg.de

* Correspondence: moosmann@uni-mainz.de; Tel.: +49-6131-39-26707

Abstract: Prooxidative therapy is a well-established concept in infectiology and parasitology, in which prooxidative drugs like artemisinin and metronidazole play a pivotal clinical role. Theoretical considerations and earlier studies have indicated that prooxidative therapy might also represent a promising strategy in oncology. Here, we have investigated a novel class of prooxidative drugs, namely chain-transfer agents, as cytostatic agents in a series of human tumor cell lines in vitro. We have found that different chain-transfer agents of the lipophilic thiol class (like dodecane-1-thiol) elicited half-maximal effective concentrations in the low micromolar range in SY5Y cells (human neuroblastoma), Hela cells (human cervical carcinoma), HEK293 cells (immortalized human kidney), MCF7 cells (human breast carcinoma), and C2C12 cells (mouse myoblast). In contrast, HepG2 cells (human hepatocellular carcinoma) were resistant to toxicity, presumably through their high detoxification capacity for thiol groups. Cytotoxicity was undiminished by hypoxic culture conditions, but substantially lowered after cellular differentiation. Compared to four disparate, clinically used reference compounds in vitro (doxorubicin, actinomycin D, 5-fluorouracil, and hydroxyurea), chain-transfer agents emerged as comparably potent on a molar basis and on a maximum-effect basis. Our results indicate that chain-transfer agents possess a promising baseline profile as cytostatic drugs and should be explored further for anti-tumor chemotherapy.

Keywords: chain-transfer agent; chemotherapy; free radical chain reaction; lipid peroxidation; lipophilic thiol; oxidative cell death; prooxidative drug; radical propagation; rate-limiting step

Citation: Heymans, V.; Kunath, S.; Hajieva, P.; Moosmann, B. Cell Culture Characterization of Prooxidative Chain-Transfer Agents as Novel Cytostatic Drugs. *Molecules* **2021**, *26*, 6743. https://doi.org/10.3390/molecules26216743

Academic Editors: Višnja Stepanić and Marta Kučerová-Chlupáčová

Received: 8 October 2021
Accepted: 4 November 2021
Published: 8 November 2021

1. Introduction

Despite tremendous successes in the last few decades, there is a continuing demand for new lead structures in oncology. One of the reasons behind this demand is the still sobering survival rate observed with many different types of cancer. For instance, 5-year-survival rates after cancer diagnosis in the US between 2008 and 2014 have been reported to be only 9% for pancreas, 18% for liver, 19% for esophagus, and 19% for lung [1]. Moreover, the increasing cost of many newer drugs has become a serious concern [2]. To meet these challenges, drug candidates would be particularly interesting that reach beyond the established therapeutic principles [3]. In general, the most difficult task in generating novel and tolerable cytostatic drugs for chemotherapy has been the identification of new biochemical aspects in which tumor cells are substantially and "drugably" different from normal, differentiated cells and normal, but regularly dividing cells such as stem cells.

In recent years, there has been an increasing awareness that redox metabolism in tumor cells is substantially altered, pointing at the presence of a generalized prooxidant state [4–7]. Specifically, certain tumor cells appear to exhibit reduced antioxidant enzyme activities [8] and increased production of reactive oxidative species (ROS) due to flavoprotein activation [9,10] or, potentially, mutation accumulation in the mitochondrial DNA [5].

In consequence, a prooxidative treatment strategy for cancer was proposed, based on the idea that an additional elevation of ROS levels in cancer cells would lift these cells above a toxic threshold, whereas the same lift in normal cells would perhaps damage, but not kill the cells [4,6,7]. The latter idea was rationalized by the recognition that established therapeutic regimes such as radiotherapy [11,12] or photodynamic therapy [13] also have a strong prooxidant functional component.

Despite an impressive number of different approaches towards prooxidant tumor therapy [4,6,14–16], none of those has seemingly involved the direct "sensing" of the elevated level of ROS or free radicals in tumor cells as criterion to distinguish between tumor cells and normal cells. Therefore, based on recent work describing the unique catalytic behavior of so-called "chain-transfer agents" in biological cells [17], we have investigated the cytostatic potential of these prooxidative agents in a series widely utilized tumor cell lines in vitro. Chain-transfer agents are generally reducing chemicals, whose prooxidative action in vivo only materializes after intracellular oxidation by endogenous free radicals. Thus, we hypothesized that these agents may indeed represent sensors of the elevated free radical tone in tumor cells.

Very different structural classes of compounds can exhibit chain-transfer activity in the test tube, among them metal complexes [18], thiols [19], trithiocarbonates [20] and nitroxides [21]. All of these compounds are widely used in polymer chemistry to control the outcome of radical polymerization processes [22]. In the present tumor biological study, we have focused on lipophilic thiols as lead compounds for three main reasons: (i) they appeared to be the most compatible with an aqueous, biological context, (ii) they have already been demonstrated to exhibit chain-transfer activity in cell culture and in vivo [17], and (iii) there is strong evidence that thiol-type chain-transfer activity might have significantly shaped biochemical evolution in the past [23,24].

2. Results

2.1. Comparative Evaluation of Chain-Transfer Agents as Anti-Proliferative Drugs in Four Human Tumor Cells Lines

A series of linear primary thiols with incremental lipophilicity, ranging from octane-1-thiol (8SH) to octadecane-1-thiol (18SH) (Table 1), was investigated in cell culture for potential cytostatic effects at nanomolar and micromolar concentrations during a 3-day incubation period.

Table 1. Selected properties of the thiols and thioethers investigated in this work.

Compound	Abbreviation	Purity	Lipophilicity (logP)
Octane-1-thiol	8SH	98.5%	3.47
Decane-1-thiol	10SH	99%	4.30
Dodecane-1-thiol	12SH	98%	5.14
Tetradecane-1-thiol	14SH	98%	5.97
Hexadecane-1-thiol	16SH	97%	6.80
Octadecane-1-thiol	18SH	98%	7.64
1-Methylsulfanyldodecane	12SMe	97%	5.39

The results in Figure 1 indicate that lipophilic thiols were efficient inhibitors of cell proliferation in diverse types of cultivated tumor cells, namely SY5Y human neuroblastoma cells, Hela human cervical carcinoma cells, HEK293 immortalized human kidney cells, and MCF7 human breast carcinoma cells. With respect to the inhibition of cell proliferation, half-maximal effective concentrations (EC_{50} values) in the single-digit micromolar range were attained in all cell lines (Figure 1; Table 2). In general, more lipophilic compounds were more effective in terms of their EC_{50} values. This relationship was not linear, however, as a strong increase in efficacy was noted between 8SH and 10SH, whereas only a modest additional increase was seen with the more highly lipophilic compounds. Hence, cytotoxicity was apparently restricted to compounds beyond a certain lipophilicity threshold (logP = 4) as noted before [17]. This observation probably relates to the fact that less lipophilic thiols, after conversion to chain-transferring thiyl radicals, might reversibly protrude from the lipid bilayer and react with glutathione, which would blunt chain-transfer catalysis [17]. More highly lipophilic thiyl radicals, however, are probably restricted to the lipid bilayer permanently and thus cannot be scavenged by aqueous glutathione.

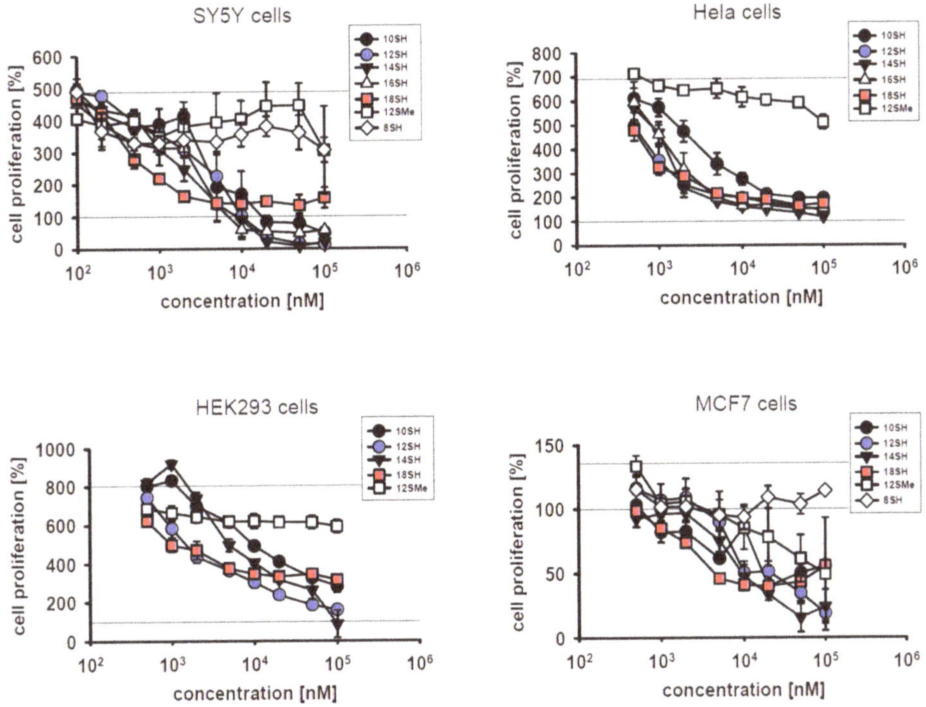

Figure 1. Inhibitory effect of different thiol-type chain-transfer agents on cellular proliferation in SY5Y cells, Hela cells, HEK293 cells and MCF7 cells. Compound abbreviations are explained in Table 1. Cellular proliferation was assessed by metabolic MTT assay as described in the Materials and Methods. The control line at 100% represents the metabolic activity of the adherent cells at the beginning of the experiment; the variable, upper control line represents the final activity of the cells after the 3-day experiment. Note that MCF7 cells exhibited a much lower cell division rate than the other cells, amounting to less than one population doubling over the course of the experiment.

Table 2. Half-maximal effective concentrations (µM) of the chain-transfer agents and the control compounds investigated in this work.

Compound	SY5Y	SY5Y, 1% O_2	Hela	HEK293	MCF7	C2C12	C2C12, Differ.	HepG2
8SH	>100	>100	-	-	>100	-	-	-
10SH	5 ± 1	4 ± 2	5 ± 3	20 ± 8	9 ± 1	5 ± 1	60 ± 40	>100
12SH	4 ± 1	4 ± 1	1 ± 0.2	4 ± 1	9 ± 1	0.7 ± 0.2	>100	>100
14SH	2 ± 0.5	2 ± 0.5	2 ± 0.5	9 ± 5	6 ± 2	3 ± 1	10 ± 2	90 ± 40
16SH	2 ± 0.5	4 ± 1	2 ± 1	-	-	-	-	-
18SH	0.8 ± 0.5	6 ± 4	1 ± 0.2	4 ± 3	2 ± 1	30 ± 8	>100	>100
12SMe	>100	>100	>100	>100	40 ± 10	>100	>100	>100
Dox	0.05 ± 0.01	-	<0.01	-	-	-	-	-
Act	2 ± 1	-	0.3 ± 0.1	-	-	-	-	-
FU	5 ± 1	-	0.8 ± 0.2	-	-	-	-	-
HU	80 ± 20	-	70 ± 20	-	-	-	-	-

On the other hand, very long-chain thiols like 18SH tended to be somewhat less effective in the killing of already present cells, potentially due to limited penetration of established cells in the culture. Still, such a cytotoxic effect (i.e., a value of less than 100% in the graphs in Figure 1) was only observed in certain cell lines like SY5Y, but not in Hela cells. Notably, the compound 12SMe, which is not a chain-transfer agent, but a chain-transfer negative control for the compound 12SH, generally did not affect cell proliferation up to the highest concentration tested (100 µM) (except in MCF7 cells; Table 2). This result verifies that the thiol group of the active agents caused their toxicity, as would be expected for chain-transfer agents [17,19]. Nonspecific alkyl group overload effects were apparently irrelevant for the noted cytostatic effects.

2.2. *Effect of Cellular Differentiation on Chain-Transfer Agent Cytotoxicity*

Cytotoxic compounds for clinical use should exhibit efficacy towards dividing cells, but should ideally spare differentiated, quiescent cells. To test the behavior of chain-transfer agents in this respect, mouse myoblast C2C12 cells were chosen because they divide very rapidly under cultivation conditions with FCS, but differentiate rapidly upon serum withdrawal at high cell densities [25]. Within 3 days of cultivation, C2C12 cells achieved approximately 4 population doublings (~1700% proliferation) (Figure 2). C2C12 cell proliferation was not inhibited by the thioether control compound 12SMe, whereas the thiol compound 12SH fully blocked cell division at a concentration of 20 µM, with half-maximal efficacy at approximately 1 µM (Figure 2). Differentiated C2C12 cells were significantly less affected by chain-transfer agent toxicity, as the obtained survival curves were shifted to the right by about one order of magnitude. This indicates an approximately 10x lower toxicity of chain-transfer agents upon cellular differentiation (Figure 2, Table 2). The highly lipophilic alkyl thiol 18SH apparently reduced the viability of the plated, differentiated cells by up to 50%, but without a clear dose-response. This finding might indicate some nonspecific toxicity of long-chain alkyl compounds in differentiated myoblasts that is unrelated to chain-transfer activity. The latter idea is supported by the fact that in differentiated cells, the formerly observed, wide gap between 12SH and 12SMe (Figure 2, left) completely collapsed, with coinciding survival curves for both compounds (Figure 2, right).

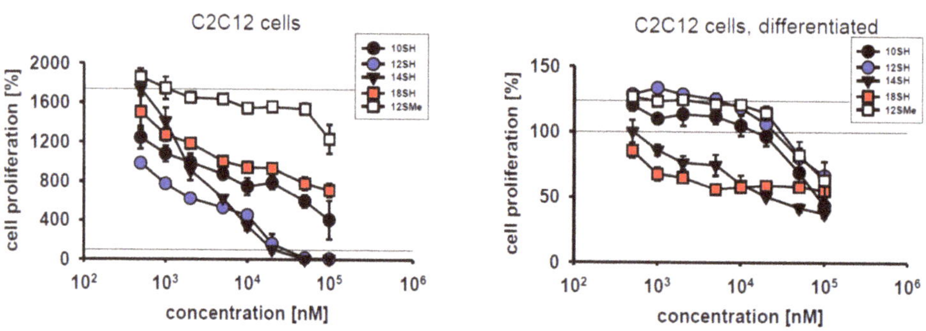

Figure 2. Cytotoxic effect of chain-transfer agents in naïve vs. differentiated C2C12 cells. Compound designations are used as in Table 1. The employed differentiation protocol involving serum withdrawal of a confluent culture led to a significant reduction of proliferation from ~1700% to ~125% as assessed by MTT assay.

2.3. *Potential Limitations of Chain-Transfer Agents as Cytostatic Drugs*

Low tumor oxygenation (hypoxia) in solid tumors is of major relevance for tumor cell behavior and treatability [26]. Specifically, tumor hypoxia may induce genomic instability of the tumor cells, exert local immunosuppressive effects, and it frequently leads to cancer cell spreading and tumor dissemination [26–28]. Importantly, hypoxia is known to limit the efficacy of radiotherapy [26]. Hence, it was investigated whether hypoxic conditions may also curtail the cytostatic potency of prooxidative chain-transfer agents. Cultivation of SY5Y cells under 1% oxygen slightly reduced their baseline proliferative capacity as expected (Figure 3). However, there were no relevant changes in the cytostatic and cytotoxic activity of the tested compounds (Figures 1 and 3); EC_{50} values were essentially identical at 1% O_2 and 20% O_2 (Table 2). This somewhat surprising result may be accounted for by the fact that even at only 1% O_2, other steps of prototypical radical chain reactions are slower (and thus rate-limiting) than steps involving the O_2 molecule itself, as detailed in the Discussion.

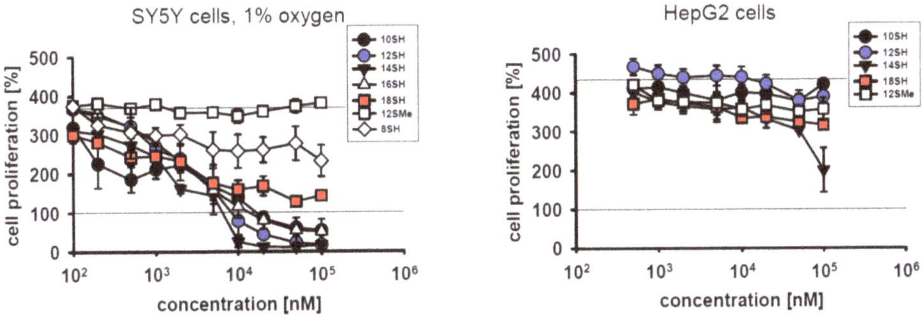

Figure 3. Cytotoxicity of chain-transfer agents in SY5Y cells under hypoxic culture conditions and in HepG2 cells. Compound-treated SY5Y cells were cultivated at 1% oxygen partial pressure under otherwise unchanged conditions for 3 days. Hypoxic culture conditions only modestly lowered baseline proliferation of the SY5Y cells from ~500% to ~400% as per MTT assay. HepG2 hepatocellular carcinoma cells were cultivated at 20% oxygen partial pressure and evaluated as in Figure 1.

Hepatocellular carcinoma is a malignant disease characterized by low 5-year survival rates of about 15% [29]. One of the origins of therapeutic futility in this cancer is cellular chemoresistance involving very effective drug expulsion and drug metabolism, among other mechanisms [29]. Human hepatocellular carcinoma cells (HepG2 cells) were thus added to the spectrum of tumor cell lines investigated in this work. The results in Figure 3 demonstrate that HepG2 cells were indeed entirely resistant to lipophilic thiol toxicity, as plausibly explained by the superior thiol detoxification capacity described for the liver [30]. Whether chain-transfer agents with other lead structures may overcome HepG2 cell resistance remains to be determined.

2.4. Comparison of 12SH and 18SH with Four Clinically Established Cytostatic Drugs

To achieve a quantitative assessment of chain-transfer agent cytostatic potential in direct comparison with established anti-tumor drugs, the compounds doxorubicin (a DNA intercalator and topoisomerase inhibitor), actinomycin D (a transcriptional inhibitor), 5-fluorouracil (a thymidylate synthase inhibitor), and hydroxyurea (a ribonucleotide reductase inhibitor) were chosen as reference standards. These compounds were investigated in SY5Y cells and Hela cells under identical conditions as the chain-transfer agents before. The results in Figure 4 demonstrate that all four clinical reference compounds acted as cytostatic drugs in both cell lines, but with vastly differing molar efficacies spanning five orders of magnitude; EC_{50} values are provided in Table 2. Notably, the chain-transfer agents 12SH and 18SH were both localized right in the middle of the efficacy spectrum, most closely resembling actinomycin D in SY5Y cells, and 5-fluorouracil in Hela cells. Certain reference compounds, namely doxorubicin and actinomycin D, were particularly effective in the killing of the initially plated cells (i.e., they achieved a value of less than 100% in the graphs in Figure 4), beyond their inhibition of cell proliferation. Hydroxyurea, however, similarly as 18SH, only inhibited proliferation, but did not kill existing cells up to the highest concentration tested (500 μM). It is unclear at present whether the killing of initially plated cells under the employed conditions should be viewed as desirable for an anti-tumor drug, because it might also predict the killing of non-tumor, preexisting cells in vivo.

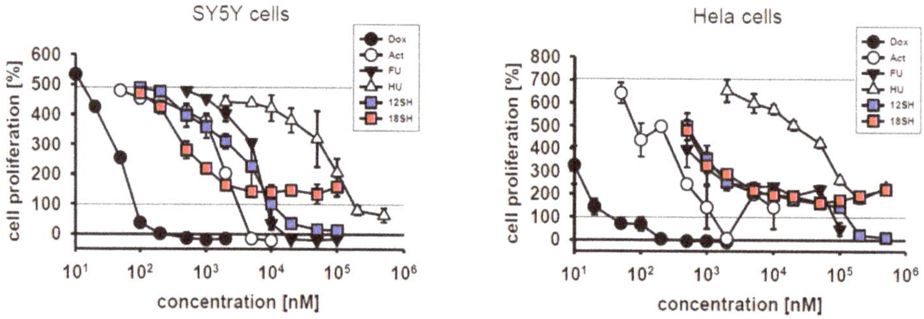

Figure 4. Cytostatic efficacy of chain-transfer agents in comparison with doxorubicin (Dox), actinomycin D (Act), 5-fluorouracil (FU) and hydroxyurea (HU). SY5Y cells and Hela cells were investigated after 3-day treatment under standard cultivation conditions as in Figure 1; the curves for 12SH and 18SH were adopted from that figure.

3. Discussion

In this work, we provide initial evidence that chain-transfer agents might become useful anti-cancer drugs of an entirely novel mechanistic class, for which we would propose the term "prooxidative amplifiers". With EC_{50} values in the low micromolar range, chain-transfer agents exerted cytostatic effects at approximately the same concentrations as traditional and clinically administered anti-cancer agents like actinomycin D and fluorouracil under identical testing conditions (Figure 4, Table 2). The cytostatic activity of the chain-transfer agents was undiminished by hypoxic culture conditions (Figure 3), which is relevant for the potential treatment of solid, hypoxic tumors. Cellular differentiation, however, led to an increase in EC_{50} in the investigated cell line (C2C12) by approximately one order of magnitude, and it was accompanied by substantially lowered maximum effects (Figure 2), which would fulfill a second, important prerequisite for anti-tumor drugs. On the other hand, the chain-transfer agents were ineffective in hepatocellular carcinoma cells (Figure 3), presumably due to rapid drug metabolism and inactivation. Therefore, chain-transfer agents are obviously not universal cytotoxins, but will require serial screening for the most promising fields of application.

In a cell biological context, chain-transfer agents of the lipophilic thiol class accelerate free radical chain reactions, which leads to a heightened toxicity of the low levels of free radicals naturally produced by the cell [17]. In particular, chain-transfer agents in normal human diploid fibroblasts expedited lipid peroxidation, as evidenced by lowered levels of phospholipid poly-unsaturated fatty acids (PUFAs), and sharply elevated the levels of 8-isoprostanes and trans-fatty acids [17]. Moreover, increased protein oxidation, especially of membrane proteins, was observed, which was accompanied by a massively induced cellular stress response. Similar findings were made in *C. elegans* in vivo [17]. Hence, chain-transfer agents in living cells evoke a well-characterized spectrum of biochemical changes and subsequent compensatory responses related to oxidative stress. Essential starting point of this prooxidative amplification is the presence of naturally formed, endogenous initiator radicals, because in contrast to many classic prooxidant pharmaceuticals like artemisinin, chain-transfer agents by themselves are reducing chemicals whose complete catalytic cycle has to be considered in order to appreciate their overall prooxidant activity, as detailed below.

As many tumor cells appear to possess higher steady-state levels of endogenous initiator radicals than normal cells [4,6,8–10], the accelerating catalysis of the chain-transfer agents might be used to achieve a specific toxic effect in tumor cells, whereas normal cells would be relatively spared [4,6,7]. Unlike other prooxidant anti-cancer strategies evaluated so far [4,6], chain-transfer agents would thus mechanistically respond to the difference in oxidant tone between tumor cells and normal cells, to proportionally potentiate this difference [17]. Thereby, they would act as "pathologically activated therapeutics" [31]. In our view, this mechanistic feature might be an important advantage compared to more traditional strategies such as antioxidant enzyme inhibition or direct prooxidation [4,6], which usually add oxidative reactivity to many cell types in a relatively non-specific fashion.

In the following, we would like to provide a brief overview of the biochemical mechanism of chain-transfer agents in vivo, to illustrate the differences between the various prooxidative strategies proposed for cancer treatment. Biological and cytotoxic damage from free radicals is foremost related to radical chain reactions, which can produce extensive damage once started (Figure 5). The arguably most important such chain reaction is lipid peroxidation [32,33]. As sketched in Figure 5, lipid peroxidation is started by the initiation step, which involves the attack of a variable initiator radical (I•) on a lipid (L), usually followed by a rapid reaction of the ensuing lipid radical (L•) with ambient molecular oxygen (O_2) to yield a lipid peroxyl radical (LOO•). During propagation, the lipid peroxyl radical (LOO•) slowly radicalizes another lipid (L′) to yield another lipid radical (L′•), which again rapidly adds oxygen to produce a lipid peroxyl radical (L′OO•). The latter product may then attack yet another lipid, resulting in a potentially endless chain reaction as long as enough substrates (lipid L′ and O_2) are present. Termination may be effectuated by a variety of mechanisms, predominantly the donation of a hydrogen radical by a low-molecular weight antioxidant (HX) to a lipid peroxyl radical (L′OO•). This step results in two relatively stable products to be disposed of or recycled, namely a lipid hydroperoxide (LOOH) and an antioxidant radical (X•).

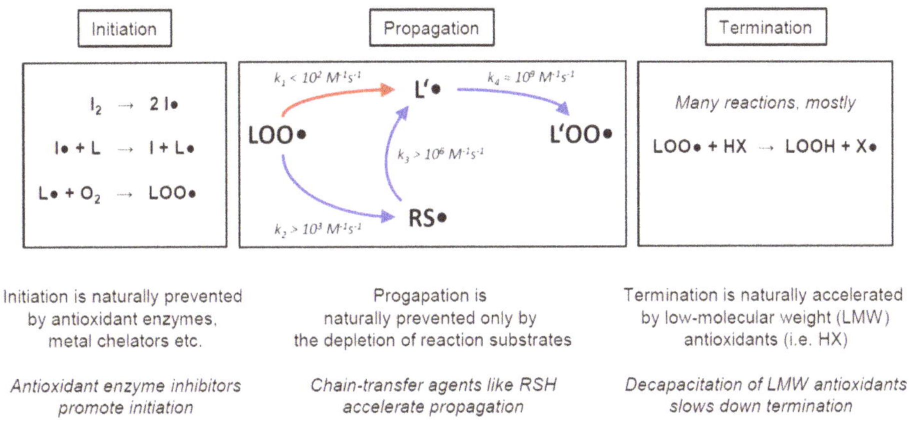

Figure 5. Prooxidative mechanism of chain-transfer agents in living cells, exemplified by the lipid peroxidation reaction. Chemical reactions involving free radicals in living cells frequently present as radical chain reactions (RCRs). RCR possess three kinetically independent elementary steps, namely initiation, propagation, and termination. Antioxidant or prooxidant chemicals and enzymes are generally characterized by their specific interference with only one of these elementary steps. For example, hydrogen peroxide typically accelerates initiation, whereas vitamin E accelerates termination; both do not affect propagation. In contrast, chain transfer agents specifically accelerate radical propagation. More details are provided in the Discussion. The abbreviations denote: initiator (I_2); initiator radical (I•); lipid (L); lipid radical (L•); molecular oxygen (O_2); lipid peroxyl radical (LOO•); a second lipid (L′, omitted for clarity); a second lipid radical (L′•); a second lipid peroxyl radical (L′OO•); lipophilic thiol (RSH, omitted for clarity); lipophilic thiol radical (RS•); low-molecular weight antioxidant (HX); lipid hydroperoxide (LOOH); antioxidant radical (X•); rate constant (k_X). The propagation scheme and the rate constants were adopted from [17].

The rate-limiting step of propagation is the formation of the lipid radical L′• (Figure 5, red arrow). Importantly, it has been argued that propagation may also constitute the rate-limiting step of the overall chain reaction in many biological systems, since it is certainly the most difficult of the three elementary steps (initiation, propagation, termination) to be modified by acute cellular intervention or long-term evolutionary adaptation [34,35]. Notably, it is this very step that is bypassed and thereby accelerated by chain-transfer agents such as lipophilic thiols (RSH). Moreover, adverse chain-transfer catalysis by lipid bilayer thiol groups may also explain why these groups appear to be negatively selected for during evolution [24,36]. Representative rate constants for the propagation reaction are given in Figure 5, indicating that thiol-type chain transfer agents would usually accelerate propagation by more than 10×, yet depending on the actual substrate concentrations present. Detailed quantitative considerations analyzing these factors and their biological implications have been published [17].

Importantly, propagation cannot be easily modified by adaptive enzymatic responses of the cell, as it formally depends only on the concentration of the lipid substrate L′, the concentration of oxygen, and the temperature. In a tumor biological context, hardly any of these factors may become relevant as a mechanism of tumor cell chemoresistance. First, temperature is widely constant in the human body. Second, oxygen concentration is certainly of interest and has been extensively discussed in terms of its impact on tumor behavior, progression, and treatability [5,26–28]. However, as regards its impact on lipid peroxidation and other radical chain reactions, the reaction rate of carbon-centered radicals with oxygen is so fast (Figure 5, $k_4 \approx 10^9\ M^{-1}s^{-1}$) that even a 100× lower oxygen concentration in tumors arguably would not make this reaction rate-relevant [17]. Experimentally, we have investigated SY5Y cells cultivated under 20% and 1% oxygen partial pressure, and we have not seen any notable differences in their susceptibility to chain-transfer agent toxicity (Figure 3, Table 2). Finally, the concentrations of the lipid substrates need to be considered. As judged from the reactivities of saturated vs. mono-unsaturated vs. polyunsaturated fatty acids, only the latter are of general relevance [17]. Because the degree and type of lipid unsaturation are largely preset by the biological species and the tissue that is analyzed [37,38], however, there is only a modest chance for a tumor cell to adaptively respond to and thus escape the toxic action of a chain-transfer agent. Altered PUFA usage has been described for a variety of tumor cell types already, but the effect sizes were generally smaller than 2× and thus negligible in a reaction rate context [39–41]. Therefore, an adaptive escape of tumor cells from chain-transfer agent toxicity is very unlikely, such that it appears paramount to assess and identify those tumor cell types whose baseline properties at the outset are the most promising [8–10].

To date, there is only basic information available about the pharmacodynamics and toxicology of the employed chain-transfer agents. According to the manufacturer-provided chemical safety record, the reference compound 12SH is non-genotoxic (as per Ames test, micronucleus test and sister chromatid exchange assay), non-teratogenic, and devoid of reproductive toxicity in mice [42]. After oral application in rats, the half-lethal dose (LD_{50}) was higher than 5000 mg/kg (~25 mmol/kg), apparently the highest dose tested. For comparison, 5-fluorouracil was positive in all genotoxicity assays and half-lethal in rats at 230 mg/kg (~1.8 mmol/kg) [43]. Actinomycin D, in turn, has been reported to be half-lethal in rats already at 7.2 mg/kg (~0.0057 mmol/kg) following oral administration [44].

The current study has two major limitations. First, in vivo data from an accepted animal model are not available yet. Such data would be essential for the assessment of the selectivity of the presented chain-transfer agents for tumor cells vs. normal cells in a whole-body context. Second, the chemical mechanism of chain-transfer catalysis elaborated before in fibroblasts [17] was not rechecked in the presently investigated tumor cells. Still, since the original report [17] has shown coherent effects in two rather different experimental systems (diploid human lung fibroblasts and *C. elegans* nematodes), it appears plausible that a related mechanism also accounts for the here described cytotoxicity in tumor cells.

4. Materials and Methods

4.1. Chemicals and Reagents

The investigational thiols and thioethers were obtained from the following sources: octane-1-thiol (8SH; CAS 111-88-6) was from Sigma-Aldrich, St. Louis, MO, USA (#471836, purity ≥98.5%); decane-1-thiol (10SH; CAS 143-10-2) was from Sigma-Aldrich (#705233, purity 99%); dodecane-1-thiol (12SH; CAS 112-55-0) was from Sigma-Aldrich (#471364, purity ≥98%); tetradecane-1-thiol (14SH; CAS 2079-95-0) was from Sigma-Aldrich (#87193, purity ≥98%); hexadecane-1-thiol (16SH; CAS 2917-26-2) was from Alfa Aesar, Ward Hill, MA, USA (#L15099, purity 97%); octadecane-1-thiol (18SH; CAS 2885-00-9) was from Sigma-Aldrich (#O1858, purity 98%); 1-methylsulfanyldodecane (12SMe; CAS 3698-89-3) was from Sigma-Aldrich (#641480, purity 97%). The lipophilicities of these compounds were calculated as octanol-water partition coefficients (logP) with the ChemPropPro tool that is part of the ChemBio3D 13.0 software package (PerkinElmer, Waltham, MA, USA).

Reference cytostatic drugs were purchased from the following suppliers: doxorubicin hydrochloride (Dox; CAS 25316-40-9) was from Cayman Chemicals, Ann Arbor, MI, USA (#15007; purity ≥98%); actinomycin D (Act; CAS 50-76-0) was from Cayman Chemicals (#11421; purity ≥95%); 5-fluorouracil (FU; CAS 51-21-8) was from Sigma-Aldrich (#F6627; purity ≥99%); hydroxyurea (HU; CAS 127-07-1) was from Sigma-Aldrich (#H8627; purity 98%).

All standard laboratory chemicals and solvents were from Sigma-Aldrich. Cell culture reagents including DMEM (#41965-039), sodium pyruvate (#11360-039), penicillin/streptomycin (#15240-062), and trypsin/EDTA (#15400-054) were from Gibco, Carlsbad, CA, USA, except for FCS (#S181BH from Biowest, Nuaillé, France), PBS (#D8537 from Sigma-Aldrich), and antibiotic-antimycotic solution (#A5955 from Sigma-Aldrich). Cell culture dished and flasks were from TPP, Trasadingen, Switzerland, and used without further surface treatment.

4.2. Cell Lines and Their Cultivation

SY5Y human neuroblastoma cells were from LGC Standards, Teddington, UK. Hela human cervical carcinoma cells were from the stocks of the Institute for Pathobiochemistry of the University of Mainz and were authenticated by short tandem repeat (STR) analysis as described [45]. HEK293 immortalized human kidney cells and MCF7 human breast carcinoma cells were from the American Type Culture Collection (ATCC), Manassas, VA, USA. C2C12 mouse myoblast cells were from LGC Standards. HepG2 human hepatocellular carcinoma cells were a kind gift from Dr. Alain Lescure (CNRS, Strasbourg, France).

Cell lines were cultivated at 37 °C in an incubator providing a humidified ambient air atmosphere containing 5% CO_2. Standard growth medium for all cell types was high-glucose DMEM supplemented with 1 mM pyruvate and 10% heat-inactivated FCS. MCF7 cells were further supplemented with 1× penicillin/streptomycin; SY5Y cells, Hela cells and HEK293 cells received 1× antibiotic-antimycotic solution. During routine culture, the cells were grown in 100 mm dishes and were passaged on reaching approximately 80% confluence (C2C12 cells at 60% confluence).

C2C12 cell differentiation was achieved in 96-well-plates in which the cells had grown to confluence over a course of approximately 3 days. Subsequently, the medium was removed and replaced by serum-free, but otherwise unaltered standard medium. Following 3 days of differentiation, the exhausted medium was exchanged, marking the beginning of the experiment. Hypoxia treatments were performed in a separate incubator that flushed the cultivation chamber with external nitrogen until reaching the desired O_2 and CO_2 concentrations. All cells were regularly tested to be negative for contamination with mycoplasma by PCR against the conserved 16S rRNA coding region of the mollicutes using a commercial test kit (Venor GeM Classic from Minerva Biolabs, Berlin, Germany).

4.3. Cell Proliferation and Cytotoxicity

The widely employed MTT reduction assay was adopted to a 96-well format in order to quantify cell proliferation and cell survival in response to standardized chemical treatments [46]. Cells were plated at low density in 96-well-plates and cultivated until approximately 25% confluence were reached (within 2–3 days). At this point, parallel plates for the investigational test agents were administered with a minimum of 8 concentrations of each test agent in multiplicates (3–5) for a fixed period of 3 days. All test agents were dissolved as 100× stocks in analytical grade ethanol. Reference plates were supplied with vehicle and analyzed immediately, to yield a control value representing the beginning of the experiment (100% proliferation). The test plates, in turn, were incubated for 3 days under the respective condition, before the same treatment applied to the control plates was identically executed on the test plates. For cell proliferation analysis, the cells were administered with 10 µL MTT solution (5 mg/mL 3-(4,5-dimethylthiazol-2-yl)-2,5-diphenyltetrazolium bromide in ultrapure water) per 100 µL cultivation medium and incubated at 37 °C for a preset time, dependent on the specific cell line (usually 3 h). Subsequently, the cells were lysed with 100 µL solubilization solution (40% dimethylformamide, 10% SDS, pH 4.0 with acetic acid) for 24 h in the dark, after which microscopic homogeneity of the solution was reached. The effectuated cellular MTT reduction was then quantified photometrically at 560 nm with a standard microplate reader. Blanking was done on medium-filled wells in which the cells had been omitted. Interference of the investigational compounds with the assay procedure was also tested and found to be negative at the employed concentrations.

5. Conclusions

Thiol-based chain-transfer agents function as prooxidant cytostatics in a variety of cancer cell lines in vitro. They show similar molar potency as different clinically established anti-cancer drugs, but they may be of lower systemic toxicity due to their mode of action requiring activation by endogenous free radicals. Chain-transfer agents target tumor cells independently of the classic mechanisms (rapid cell division, DNA synthesis, and tumor antigens), but rather exploit the higher levels of initiator free radicals found in many tumor cells. In modern combination therapy, they might thus add an extra level of specificity to standard triple-therapeutic regimens. They might also find their role in the adjuvant amplification of standard radiotherapy, which essentially acts by inducing initiator radicals in the first place.

6. Patents

The University Medical Center of the Johannes Gutenberg University, Mainz, Germany, has filed a patent pertaining to the use of chain-transfer agents as medicinal drugs (PCT Int. Appl. (2021), 44 pp., WO 2021/105435).

Author Contributions: Conceptualization, P.H. and B.M.; methodology, all authors; validation, V.H., S.K. and B.M.; formal analysis, V.H.; investigation, V.H.; resources, S.K.; data curation, V.H.; writing—original draft preparation, B.M.; writing—review and editing, all authors; visualization, V.H. and B.M.; supervision, B.M.; funding acquisition, P.H. and B.M. All authors have read and agreed to the published version of the manuscript.

Funding: This research was funded by the Volkswagen Foundation, grant number 95462.

Institutional Review Board Statement: Not applicable.

Informed Consent Statement: Not applicable.

Data Availability Statement: The data presented in this study are available on request from the corresponding author.

Acknowledgments: The authors would like to thank Heike Nagel for conducting the mycoplasma screening.

Conflicts of Interest: The authors declare no conflict of interest beyond the above-mentioned PCT patent application by the authors' scientific institution. The funders had no role in the design of the study; in the collection, analyses, or interpretation of data; in the writing of the manuscript, or in the decision to publish the results.

Sample Availability: All chemicals and cell lines employed in this work are available from commercial sources.

References

1. Siegel, R.L.; Miller, K.D.; Jemal, A. Cancer statistics, 2019. *CA Cancer J. Clin.* **2019**, *69*, 7–34. [CrossRef] [PubMed]
2. Dolgin, E. Bringing down the cost of cancer treatment. *Nature* **2018**, *555*, S26–S29. [CrossRef] [PubMed]
3. Sessa, C.; Gianni, L.; Garassino, M.; van Halteren, H. *ESMO Handbook of Clinical Pharmacology of Anticancer Agents*; European Society for Medical Oncology (ESMO): Lugano, Switzerland, 2012.
4. Trachootham, D.; Alexandre, J.; Huang, P. Targeting cancer cells by ROS-mediated mechanisms: A radical therapeutic approach? *Nat. Rev. Drug Discov.* **2009**, *8*, 579–591. [CrossRef] [PubMed]
5. Wallace, D.C. Mitochondria and cancer. *Nat. Rev. Cancer* **2012**, *12*, 685–698. [CrossRef]
6. Gorrini, C.; Harris, I.S.; Mak, T.W. Modulation of oxidative stress as an anticancer strategy. *Nat. Rev. Drug Discov.* **2013**, *12*, 931–947. [CrossRef]
7. Sosa, V.; Moliné, T.; Somoza, R.; Paciucci, R.; Kondoh, H.; LLeonart, M.E. Oxidative stress and cancer: An overview. *Ageing Res. Rev.* **2013**, *12*, 376–390. [CrossRef]
8. Doskey, C.M.; Buranasudja, V.; Wagner, B.A.; Wilkes, J.G.; Du, J.; Cullen, J.J.; Buettner, G.R. Tumor cells have decreased ability to metabolize H2O2: Implications for pharmacological ascorbate in cancer therapy. *Redox Biol.* **2016**, *10*, 274–284. [CrossRef]
9. Szatrowski, T.P.; Nathan, C.F. Production of large amounts of hydrogen peroxide by human tumor cells. *Cancer Res.* **1991**, *51*, 794–798.
10. Kumar, B.; Koul, S.; Khandrika, L.; Meacham, R.B.; Koul, H.K. Oxidative stress is inherent in prostate cancer cells and is required for aggressive phenotype. *Cancer Res.* **2008**, *68*, 1777–1785. [CrossRef]
11. Moss, R.W. Do antioxidants interfere with radiation therapy for cancer? *Integr. Cancer Ther.* **2007**, *6*, 281–292. [CrossRef]
12. Barker, H.E.; Paget, J.T.; Khan, A.A.; Harrington, K.J. The tumour microenvironment after radiotherapy: Mechanisms of resistance and recurrence. *Nat. Rev. Cancer* **2015**, *15*, 409–425. [CrossRef]
13. Dolmans, D.E.; Fukumura, D.; Jain, R.K. Photodynamic therapy for cancer. *Nat. Rev. Cancer* **2003**, *3*, 380–387. [CrossRef]
14. Toler, S.M.; Noe, D.; Sharma, A. Selective enhancement of cellular oxidative stress by chloroquine: Implications for the treatment of glioblastoma multiforme. *Neurosurg. Focus* **2006**, *21*, E10. [CrossRef]
15. Cui, Q.; Wen, S.; Huang, P. Targeting cancer cell mitochondria as a therapeutic approach: Recent updates. *Future Med. Chem.* **2017**, *9*, 929–949. [CrossRef]
16. Kubli, S.P.; Bassi, C.; Roux, C.; Wakeham, A.; Göbl, C.; Zhou, W.; Jafari, S.M.; Snow, B.; Jones, L.; Palomero, L.; et al. AhR controls redox homeostasis and shapes the tumor microenvironment in BRCA1-associated breast cancer. *Proc. Natl. Acad. Sci. USA* **2019**, *116*, 3604–3613. [CrossRef]
17. Kunath, S.; Schindeldecker, M.; De Giacomo, A.; Meyer, T.; Sohre, S.; Hajieva, P.; von Schacky, C.; Urban, J.; Moosmann, B. Prooxidative chain transfer activity by thiol groups in biological systems. *Redox Biol.* **2020**, *36*, 101628. [CrossRef]
18. Gridnev, A.A.; Ittel, S.D. Catalytic chain transfer in free-radical polymerizations. *Chem. Rev.* **2001**, *101*, 3611–3660. [CrossRef]
19. Dietrich, B.K.; Pryor, W.A.; Wu, S.J. Chain transfer constants of mercaptans in the emulsion polymerization of styrene. *J. Appl. Polym. Sci.* **1988**, *36*, 1129–1141. [CrossRef]
20. Moad, G.; Rizzardo, E.; Thang, S.H. Living Radical Polymerization by the RAFT Process—A Third Update. *Aust. J. Chem.* **2012**, *65*, 985–1076. [CrossRef]
21. Nicolas, J.; Guillaneuf, Y.; Lefay, C.; Bertin, D.; Gigmes, D.; Charleux, B. Nitroxide-mediated polymerization. *Prog. Polym. Sci.* **2013**, *38*, 63–235. [CrossRef]
22. Odian, G. Radical chain polymerization. In *Principles of Polymerization*; Odian, G., Ed.; John Wiley & Sons: Hoboken, NJ, USA, 2004.
23. Moosmann, B. Respiratory chain cysteine and methionine usage indicate a causal role for thiyl radicals in aging. *Exp. Gerontol.* **2011**, *46*, 164–169. [CrossRef]
24. Moosmann, B.; Schindeldecker, M.; Hajieva, P. Cysteine, glutathione and a new genetic code: Biochemical adaptations of the primordial cells that spread into open water and survived biospheric oxygenation. *Biol. Chem.* **2020**, *401*, 213–231. [CrossRef]
25. Fuhrmeister, J.; Tews, M.; Kromer, A.; Moosmann, B. Prooxidative toxicity and selenoprotein suppression by cerivastatin in muscle cells. *Toxicol. Lett.* **2012**, *215*, 219–227. [CrossRef]
26. Hughes, V.S.; Wiggins, J.M.; Siemann, D.W. Tumor oxygenation and cancer therapy-then and now. *Br. J. Radiol.* **2019**, *92*, 20170955. [CrossRef]
27. Brizel, D.M.; Scully, S.P.; Harrelson, J.M.; Layfield, L.J.; Bean, J.M.; Prosnitz, L.R.; Dewhirst, M.W. Tumor oxygenation predicts for the likelihood of distant metastases in human soft tissue sarcoma. *Cancer Res.* **1996**, *56*, 941–943.
28. Monteiro, A.R.; Hill, R.; Pilkington, G.J.; Madureira, P.A. The role of hypoxia in glioblastoma invasion. *Cells* **2017**, *6*, 45. [CrossRef]

29. Lohitesh, K.; Chowdhury, R.; Mukherjee, S. Resistance a major hindrance to chemotherapy in hepatocellular carcinoma: An insight. *Cancer Cell Int.* **2018**, *18*, 44. [CrossRef]
30. Pacifici, G.M.; Santerini, S.; Giuliani, L.; Rane, A. Thiol methyltransferase in humans: Development and tissue distribution. *Dev. Pharmacol. Ther.* **1991**, *17*, 8–15. [CrossRef]
31. Lipton, S.A. Pathologically activated therapeutics for neuroprotection. *Nat. Rev. Neurosci.* **2007**, *8*, 803–808. [CrossRef]
32. Negre-Salvayre, A.; Auge, N.; Ayala, V.; Basaga, H.; Boada, J.; Brenke, R.; Chapple, S.; Cohen, G.; Feher, J.; Grune, T.; et al. Pathological aspects of lipid peroxidation. *Free Radic. Res.* **2010**, *44*, 1125–1171. [CrossRef] [PubMed]
33. Tudek, B.; Zdżalik-Bielecka, D.; Tudek, A.; Kosicki, K.; Fabisiewicz, A.; Speina, E. Lipid peroxidation in face of DNA damage, DNA repair and other cellular processes. *Free Radic. Biol. Med.* **2017**, *107*, 77–89. [CrossRef] [PubMed]
34. Kunath, S.; Moosmann, B. What is the rate-limiting step towards aging? Chemical reaction kinetics might reconcile contradictory observations in experimental aging research. *Geroscience* **2020**, *42*, 857–866. [CrossRef] [PubMed]
35. Moosmann, B. Flux control in the aging cascade. *Aging* **2021**, *13*, 6233–6235. [CrossRef] [PubMed]
36. Moosmann, B.; Behl, C. Mitochondrially encoded cysteine predicts animal lifespan. *Aging Cell* **2008**, *7*, 32–46. [CrossRef]
37. Pradas, I.; Huynh, K.; Cabré, R.; Ayala, V.; Meikle, P.J.; Jové, M.; Pamplona, R. Lipidomics reveals a tissue-specific fingerprint. *Front. Physiol.* **2018**, *9*, 1165. [CrossRef]
38. Jové, M.; Mota-Martorell, N.; Pradas, I.; Galo-Licona, J.D.; Martín-Gari, M.; Obis, È.; Sol, J.; Pamplona, R. The lipidome fingerprint of longevity. *Molecules* **2020**, *25*, 4343. [CrossRef]
39. Bartoli, G.M.; Bartoli, S.; Galeotti, T.; Bertoli, E. Superoxide dismutase content and microsomal lipid composition of tumours with different growth rates. *Biochim. Biophys. Acta* **1980**, *620*, 205–211. [CrossRef]
40. Peck, B.; Schug, Z.T.; Zhang, Q.; Dankworth, B.; Jones, D.T.; Smethurst, E.; Patel, R.; Mason, S.; Jiang, M.; Saunders, R.; et al. Inhibition of fatty acid desaturation is detrimental to cancer cell survival in metabolically compromised environments. *Cancer Metab.* **2016**, *4*, 6. [CrossRef]
41. Szlasa, W.; Zendran, I.; Zalesińska, A.; Tarek, M.; Kulbacka, J. Lipid composition of the cancer cell membrane. *J. Bioenerg. Biomembr.* **2020**, *52*, 321–342. [CrossRef]
42. n-Dodecyl Mercaptan. *Safety Data Sheet, Version 4.14*; SDS Number 100000068622; Chevron Phillips Chemical: The Woodlands, TX, USA, 2019.
43. Fluorouracil Injection. *Safety Data Sheet, Version 1.1*; Pfizer: New York City, NY, USA, 2012.
44. Actinomycin, D. *Safety Data Sheet, Version 7.0*; Merck: Darmstadt, Germany, 2021.
45. Bekbulat, F.; Schmitt, D.; Feldmann, A.; Huesmann, H.; Eimer, S.; Juretschke, T.; Beli, P.; Behl, C.; Kern, A. RAB18 loss interferes with lipid droplet catabolism and provokes autophagy network adaptations. *J. Mol. Biol.* **2020**, *432*, 1216–1234. [CrossRef]
46. Hajieva, P.; Bayatti, N.; Granold, M.; Behl, C.; Moosmann, B. Membrane protein oxidation determines neuronal degeneration. *J. Neurochem.* **2015**, *133*, 352–367. [CrossRef]

molecules MDPI

Article

The Implication of Low Dose Dimethyl Sulfoxide on Mitochondrial Function and Oxidative Damage in Cultured Cardiac and Cancer Cells

Nonhlakanipho F. Sangweni [1,2,*], Phiwayinkosi V. Dludla [1], Nireshni Chellan [1,2], Lawrence Mabasa [1], Jyoti R. Sharma [1] and Rabia Johnson [1,2,*]

1 Biomedical Research and Innovation Platform (BRIP), South African Medical Research Council, Tygerberg 7505, South Africa; phiwayinkosi.dludla@mrc.ac.za (P.V.D.); nireshni.chellan@mrc.ac.za (N.C.); Lawrence.Mabasa@mrc.ac.za (L.M.); jyoti.sharma@mrc.ac.za (J.R.S.)

2 Centre for Cardio-Metabolic Research in Africa, Division of Medical Physiology, Faculty of Medicine and Health Sciences, Stellenbosch University, Tygerberg 7505, South Africa

* Correspondence: Nonhlakanipho.Sangweni@mrc.ac.za (N.F.S.); rabia.johnson@mrc.ac.za (R.J.)

Citation: Sangweni, N.F.; Dludla, P.V.; Chellan, N.; Mabasa, L.; Sharma, J.R.; Johnson, R. The Implication of Low Dose Dimethyl Sulfoxide on Mitochondrial Function and Oxidative Damage in Cultured Cardiac and Cancer Cells. *Molecules* **2021**, *26*, 7305. https://doi.org/10.3390/molecules26237305

Academic Editors: Višnja Stepanić, Marta Kučerová-Chlupáčová and Drazen Raucher

Received: 26 October 2021
Accepted: 28 November 2021
Published: 1 December 2021

Publisher's Note: MDPI stays neutral with regard to jurisdictional claims in published maps and institutional affiliations.

Abstract: Although numerous studies have demonstrated the biological and multifaceted nature of dimethyl sulfoxide (DMSO) across different in vitro models, the direct effect of "non-toxic" low DMSO doses on cardiac and cancer cells has not been clearly explored. In the present study, H9c2 cardiomyoblasts and MCF-7 breast cancer cells were treated with varying concentrations of DMSO (0.001–3.7%) for 6 days. Here, DMSO doses < 0.5% enhanced the cardiomyoblasts respiratory control ratio and cellular viability relative to the control cells. However, 3.7% DMSO exposure enhanced the rate of apoptosis, which was driven by mitochondrial dysfunction and oxidative stress in the cardiomyoblasts. Additionally, in the cancer cells, DMSO (≥0.009) led to a reduction in the cell's maximal respiratory capacity and ATP-linked respiration and turnover. As a result, the reduced bioenergetics accelerated ROS production whilst increasing early and late apoptosis in these cells. Surprisingly, 0.001% DMSO exposure led to a significant increase in the cancer cells proliferative activity. The latter, therefore, suggests that the use of DMSO, as a solvent or therapeutic compound, should be applied with caution in the cancer cells. Paradoxically, in the cardiomyoblasts, the application of DMSO (≤0.5%) demonstrated no cytotoxic or overt therapeutic benefits.

Keywords: mitochondria; bioenergetics; oxidative stress; apoptosis

1. Introduction

Dimethyl sulfoxide (DMSO) is a versatile compound that is extensively used as a solvent in pharmacology and toxicology to enhance drug delivery, and dissolve numerous drugs and herbal extracts. However, in recent years, increasing evidence has demonstrated that the amphiphilic nature of DMSO allows it to influence diverse biological and medical processes, such as disease pathology and intervention [1,2]. Certainly, the exploration of this amphiphilic solvent in clinical research and cell biology continues to highlight avenues to be meticulously investigated to broaden its use in biomedical science. Generally, the applied concentrations of DMSO are often unreported due to its obvious and frequent use [3]. This, coupled with its apparent low toxicity at concentrations less than 10% and its classification as a class 3 solvent, which is the safest category with low toxic potential, has led to its ubiquitous use and widespread application [3–5]. The influence that DMSO has on cellular mechanisms has been implicated in the modifications of essential cellular structures, such as proteins and DNA, and has been thus, studied for its involvement in cancer and cardiovascular diseases [6,7].

Given the lack of long-term effective therapies against such complications, it remains imperative to study the pharmacological actions of compounds such as DMSO, particularly, by targeting the implications of its non-toxic low doses. Previously, a dose of 0.1% DMSO was shown to induce epigenetic modifications, which impaired the expression of genes involved in cellular senescence and DNA repair in a 3D maturing cardiac model [3]. In contrast, 1% DMSO exposure significantly improved the nuclear morphology and antioxidant status of dermal fibroblasts [8]. This improvement was determined to be even higher than that observed in known antioxidants, such as N-acetylcysteine [8] Similarly, in colon cancer cells, low DMSO doses (0.1–1.5%) were shown to reduce the cellular levels of reactive oxygen species (ROS) and the cells proliferative activity relative to the control [9]. Literature has further shown that the addition of DMSO (1%) to culture medium at the formation stage of cardiomyocyte progenitor cells stimulates the differentiation of pluripotent stem cells (PSCs) into cardiomyocytes by a 1.5-fold increase [10]. Conversely, the addition of DMSO before the initiation of embryoid bodies is reported to suppress the differentiation of PSCs and therefore, the formation of cardiomyocytes [10]. Engineered cardiac tissue, which is generated from differentiated PSCs, are reported to have enhanced engraftment rates, as well as increased survival and progressive maturation of human cardiomyocytes [11]. In cancer research, the differentiated PSCs are currently used to develop cancer-based vaccines and have reportedly inhibited the formation of new tumors in 75% adenocarcinomas [12]. While it is evident that the changes in cellular processes following DMSO exposure in the PSCs appear to have some beneficial properties, the direct effect of DMSO on the cardiomyocytes and cancer cells has not been clearly explored. As such, we investigated the biological effect of "non-toxic" low DMSO doses in an in vitro model of cardiac and cancer cells. In this study, special attention was paid into understanding the effects of DMSO on the most studied cytotoxicity parameters, namely, oxidative stress, mitochondrial dysfunction, and resultant apoptosis.

2. Results

2.1. Cell Viability

In the present study, a range of DMSO concentrations (0.001–3.7%) were selected to determine and compare their cytotoxic profile in the H9c2 cardiomyoblasts and MCF-7 breast cancer cells after a 6-day treatment period. The results obtained revealed a concentration-dependent effect of DMSO, with the highest concentration being cytotoxic to both cell lines relative to the control group ($p < 0.001$). However, contrary to previous reports, a concentration of 0.5% DMSO had no significant effects on the growth and survival of both cell lines. Interestingly, the H9c2 cells treated with concentrations less than 0.5% DMSO presented with significantly augmented cellular viability relative to the control group ($p < 0.001$ and $p < 0.01$) after 6 days of treatment (Figure 1A). These findings were comparable with the effects observed in the viability of the MCF-7 cells (Figure 1B).

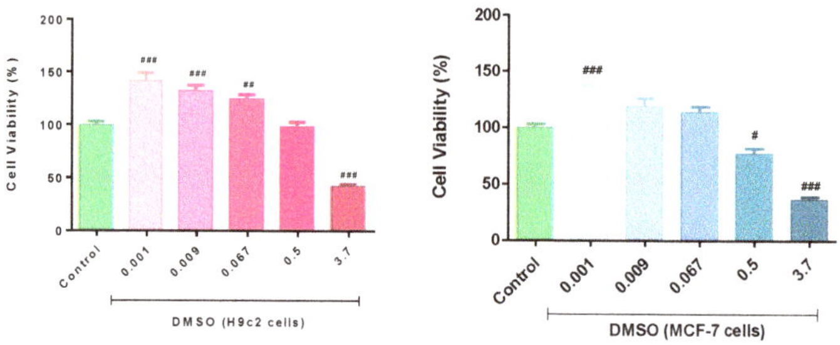

(**A**) (**B**)

Figure 1. The effect of dimethyl sulfoxide (DMSO) on cell viability. An overview of the cytotoxicity effects of DMSO on the (**A**) H9c2 cardiomyoblasts and (**B**) MCF-7 breast cancer cells. Briefly, H9c2 and MCF-7 cells were treated with varying DMSO concentrations (0.001, 0.009, 0.067, 0.5, and 3.7%) every second day for 6 days. Untreated cells served as the control. Data are presented as the mean ± SEM of 3 biological experiments with 5 technical repeats (n = 3). Significance is indicated as # $p < 0.05$, ## $p < 0.01$, and ### $p < 0.001$ versus the control.

2.1.1. Mitochondrial Bioenergetics

To determine the effect that DMSO has on the efficiency of mitochondrial oxidative phosphorylation and functionality, we quantified the parameters associated with native cellular respiration and respiration in the presence of known energy metabolism inhibitors. When comparing the energy phenotype inherent to the cardiomyoblasts and breast cancer cells, our data demonstrated that the MCF-7 cells were more energetic than the H9c2 cells, as they had a higher oxygen consumption rate (OCR) and glycolytic activity (Figure 2A). In contrast, the H9c2 cells had a more quiescent energy phenotype, with an inherently lower respiratory capacity than the MCF-7 cells (Figure 2A). In the case of DMSO, H9c2 cells treated with 3.7% DMSO presented with significantly impaired physiological mitochondrial OCR, as seen by the significant loss in the cell's basal respiratory capacity ($p < 0.001$) when compared to the control (Figure 2B,F). This loss was concomitant with the observed reduction in the cells extracellular acidification rate (ECAR) (Figure 2D). Likewise, MCF-7 cells treated with 3.7% DMSO presented with decreased OCR, ECAR, and basal respiratory activity, relative to the untreated cells (Figure 2C,E,G). Interestingly, DMSO (≤0.5%) had no significant effect on the cardiomyoblasts ATP turnover and spare respiratory capacity when compared to the control group (Figure 2J,N). In contrast, DMSO at 3.7% reduced the MCF-7 cells ATP linked respiration, ATP turnover, and maximal respiration relative to the control (Figure 2I,K,M).

Subsequently, we then quantified the respiratory flux ratios, which is an estimation of relative mitochondrial work and function. The results showed an increase in the H9c2 cells respiratory control ratio (RCR) following DMSO exposure with doses ≤ 0.5% (Table 1). We further observed a significant decrease in the cardiomyoblasts coupling efficiency ($p < 0.001$), relative to control, thus, indicating a significantly lower proportion of oxygen consumed to stimulate ATP production compared with that driving proton leak. Surprisingly, we observed no significant changes in the MCF-7 cells respiratory flux ratios following DMSO exposure with doses < 3.7% (Table 2).

Table 1. Respiratory flux ratios of DMSO treated H9c2 cells.

Flux Ratios (pmol/min/µg Protein)	Treatment					
	Control	0.001	0.009	0.067	0.5	3.7
State $_{apparent}$	3.6 ± 0.1	3.5 ± 0.1	3.5 ± 0.1	3.7 ± 0.2	3.4 ± 0.1	3.5 ± 0.1
Respiratory control ratio	6.96 ± 1.65	13.01 ± 0.79 $^{\#}$	15.14 ± 1.84 $^{\#\#}$	11.99 ± 1.24 $^{\#}$	11.38 ± 0.81 $^{\#}$	4,90 ± 1.20 $^{\#}$
Coupling efficiency	0.91 ± 0.01	0.85 ± 0.04	0.79 ± 0.03	0.82 ± 0.01	0.83 ± 0.02	0.64 ± 0.03 $^{\#\#\#}$

The state apparent, respiratory control ratio (RCR) and coupling efficiency of H9c2 cells in the presence or absence of DMSO were derived from the mitochondrial parameters presented in Figure 3A,B. Data represents mean ± SEM; $n = 6$. Significance is indicated as $^{\#}$ $p < 0.05$, $^{\#\#}$ $p < 0.01$, $^{\#\#\#}$ $p < 0.001$ versus the control.

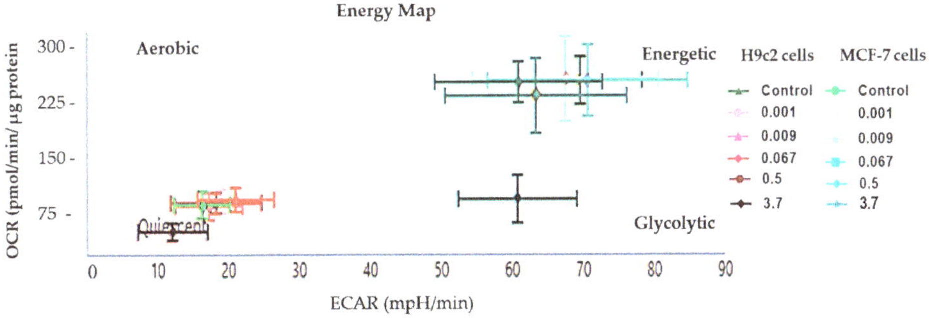

(A)

Figure 2. *Cont.*

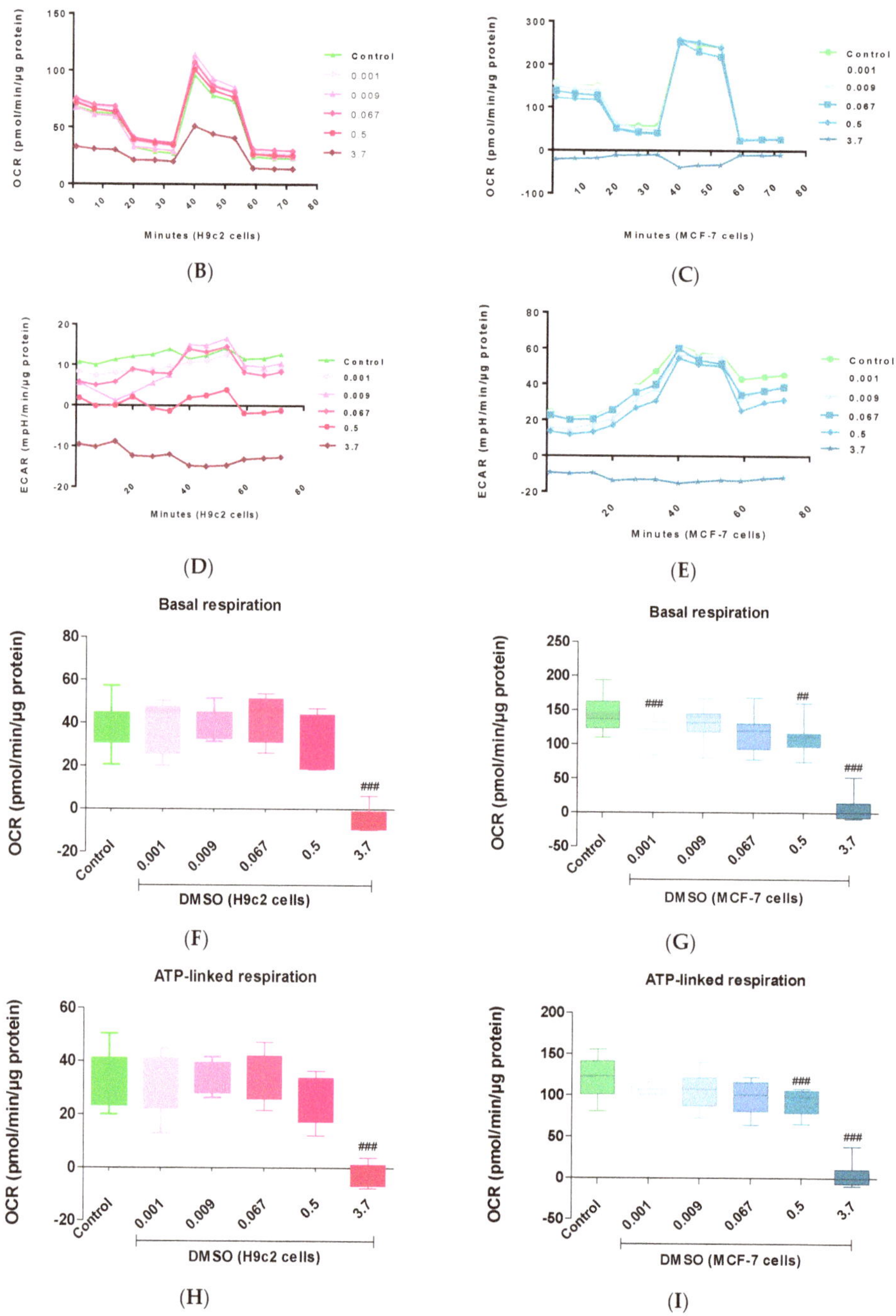

Figure 2. *Cont.*

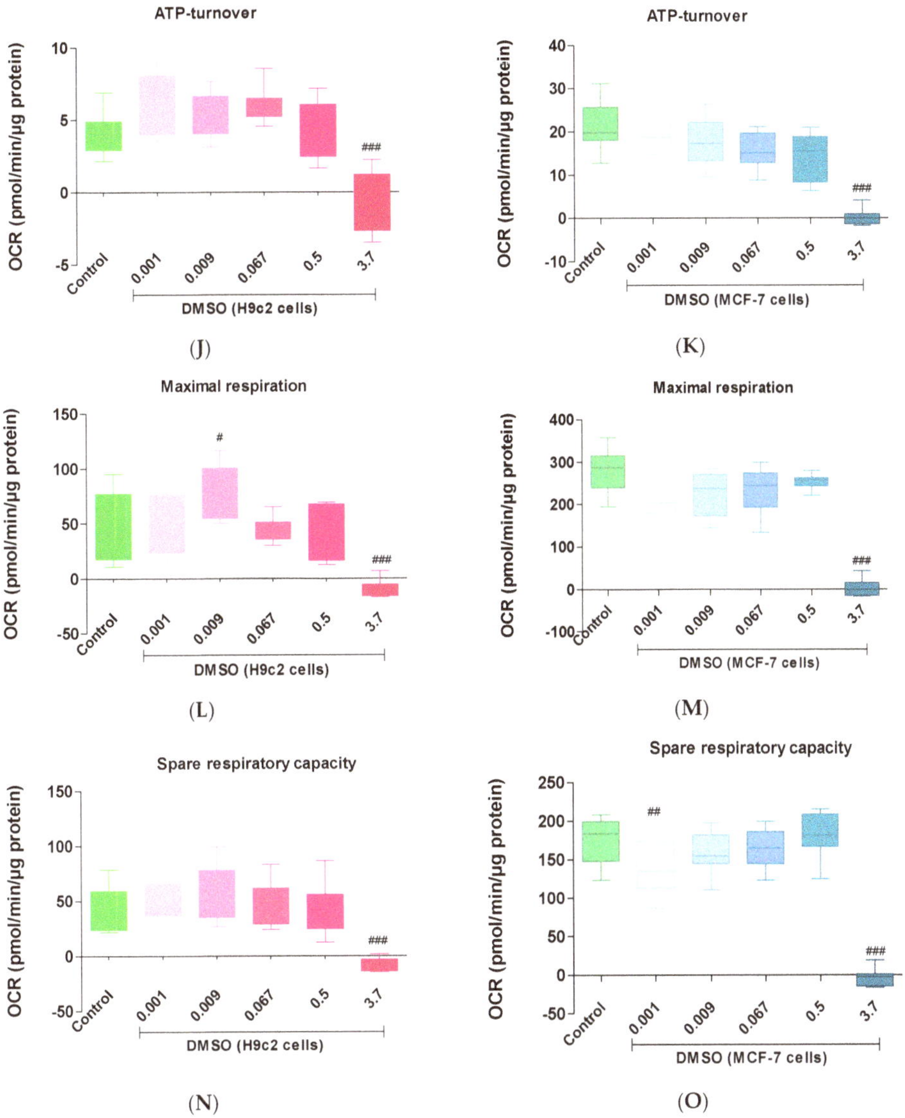

Figure 2. The effect of dimethyl sulfoxide (DMSO) on the mitochondrial bioenergetics in H9c2 cardiomyoblasts and MCF-7 breast cancer cells. (**A**) Energy phenotype, (**B**,**C**) Oxygen consumption rate (OCR), (**D**,**E**) extracellular acidification rate (**ECAR**), (**F**,**G**) basal respiration, (**H**,**I**) ATP-linked respiration, (**J**,**K**) ATP turnover, (**L**,**M**) maximal respiration and, (**N**,**O**) spare respiratory capacity. Both cell lines were treated every second day for 6 days with varying DMSO concentrations (0.001, 0.009, 0.067, 0.5, and 3.7%). Data are presented as the mean ± SEM of 3 biological experiments with 8 technical repeats ($n = 3$). Significance is indicated as $^{\#}$ $p < 0.05$, $^{\#\#}$ $p < 0.01$, and $^{\#\#\#}$ $p < 0.001$ versus the control.

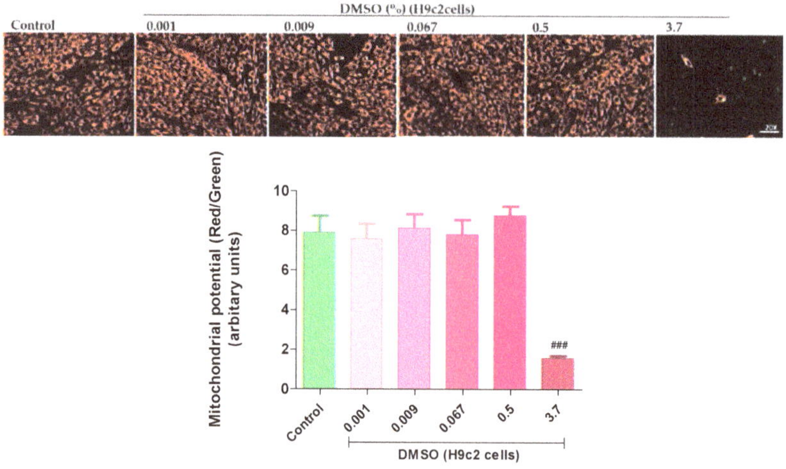

(A)

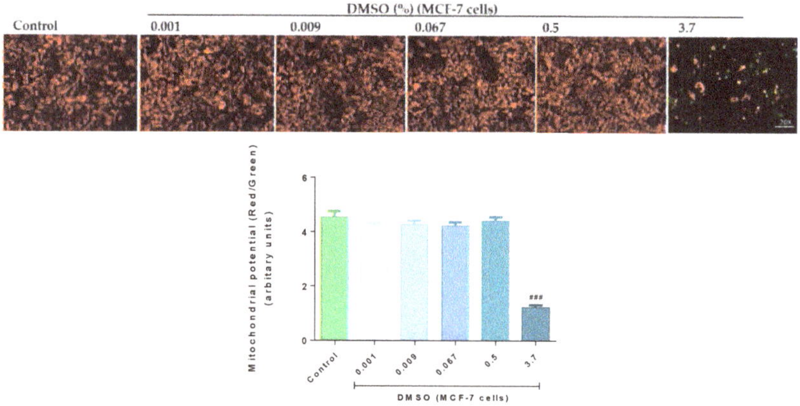

(B)

Figure 3. The effect of dimethyl sulfoxide (DMSO) the mitochondrial integrity of the (**A**) H9c2 cardiomyoblasts and (**B**) MCF-7 breast cancer cells. Cells were treated every second day for 6 days with varying DMSO concentrations (0.001, 0.009, 0.067, 0.5, and 3.7%). Data are presented as the mean ± SEM of 3 biological experiments with 5 technical repeats (n = 3). Significance is indicated as $^{###}$ $p < 0.001$ versus the control.

Table 2. Respiratory flux ratios of DMSO treated MCF-7 breast cancer cells.

Flux ratios (pmol/min/µg protein)	Treatment					
	Control	0.001	0.009	0.067	0.5	3.7
State $_{apparent}$	3.58 ± 0.02	3.57 ± 0.02	3.57 ± 0.02	3.61 ± 0.01	3.65 ± 0.01	3.58 ± 0.08
Respiratory control ratio	13.95 ± 1.04	12.12 ± 0.59	12.14 ± 0.56	14.66 ± 0.55 $^{\#}$	15.23 ± 0.97	9.26 ± 1.23 $^{\#}$
Coupling efficiency	0.87 ± 0.01	0.85 ± 0.01	0.86 ± 0.01	0.83 ± 0.02	0.84 ± 0.02	0.79 ± 0.03 $^{\#}$

The state apparent, respiratory control ratio (RCR) and coupling efficiency of MCF-7 cells in the presence or absence of DMSO were derived from the mitochondrial parameters presented in Figure 3A,B. Data represents mean ± SEM; n = 6. Significance is indicated as $^{\#}$ $p < 0.05$ versus the control.

2.1.2. Mitochondrial Membrane Potential

Mitochondrial damage is a prominent precursor of decreased cell viability. In the present study, H9c2 cells treated with 3.7% DMSO presented with a significant loss ($p < 0.001$) in cell viability (due to a loss in MMP) as seen in the JC-1 fluorescent images versus the control (Figure 3A). Similarly, chronic 3.7% DMSO exposure led to a significant loss ($p < 0.001$) in MMP and resultant mitochondrial deformities, as seen by the significant reduction in J-aggregate fluorescence intensity in the MCF-7 cells relative to the control group (Figure 3B). However, in both cell lines, the lower concentrations (0.001%, 0.009%, 0.067%, and 0.5%) had no effect on the structural and functional integrity of the mitochondria and were further found to be significantly less toxic than the higher 3.7% DMSO ($p < 0.001$) (Figure 3A,B).

2.1.3. Oxidative Stress

Impaired mitochondrial bioenergetics and the resultant loss in MMP can alter aerobic metabolism thus, stimulating increased ROS production. Here, the results demonstrated that intracellular ROS activity in the H9c2 cells exposed to 3.7% DMSO was significantly augmented when compared to the control group ($p < 0.05$) (Figure 4A). This increase in ROS was determined by the dramatic shift in DCFH-DA dye intensity, which is proportional to intracellular hydrogen peroxide concentrations. However, a steady reduction in ROS activity was observed when these cells were treated with DMSO doses $\leq 0.5\%$ (Figure 4A). Interestingly, a significant increase, in a dose-dependent manner, in intracellular ROS activity was observed in MCF-7 cells treated with either 0.009% ($p < 0.05$), 0.067 ($p < 0.05$), 0.5% ($p < 0.05$), or 3.7% ($p < 0.001$) DMSO, relative to the control group (Figure 4B). Although not significant when compared to the control, a dose of 0.001% DMSO also elevated ROS in these cells (Figure 4B).

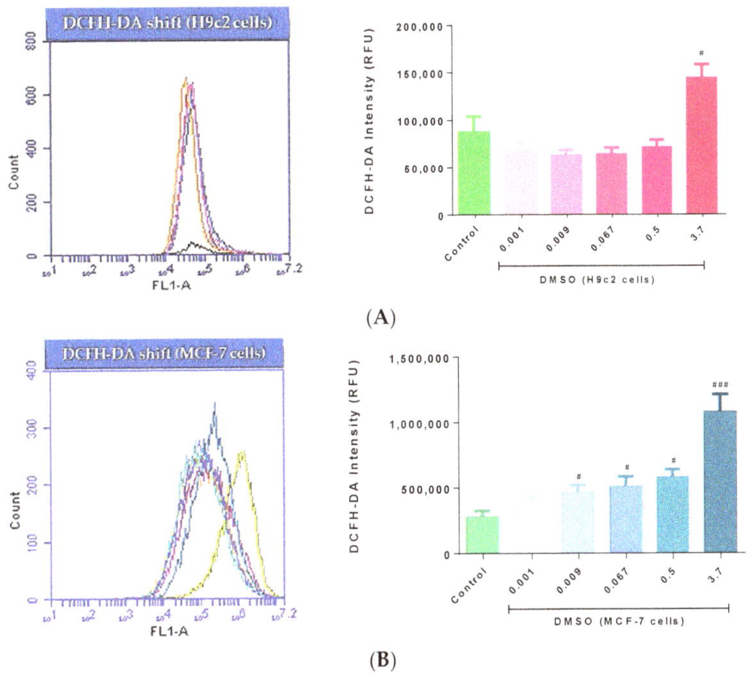

Figure 4. The effect of dimethyl sulfoxide (DMSO) on the production of reactive oxygen species (ROS) in (**A**) fluorescent shift of H9c2 cardiomyoblasts and (**B**) fluorescent shift of MCF-7 breast cancer cells. Cells were treated every second day for 6 days with varying DMSO concentrations (0.001, 0.009, 0.067, 0.5, and 3.7%). Data are presented as the mean ± SEM of 3 biological experiments with 3 technical repeats ($n = 3$). Significance is indicated as $^{\#}$ $p < 0.05$, $^{\#\#\#}$ $p < 0.001$ versus the control.

2.1.4. Apoptosis

The intricate role of DMSO-induced oxidative stress, which is often exacerbated by mitochondrial damage, accelerates cellular apoptosis. As such, the rate of cell death in H9c2 and MCF-7 cells was investigated following DMSO exposure. Here, only 3.7% DMSO treatment significantly increased early (lower right quadrant; $p < 0.001$) and late (upper right quadrant; $p < 0.001$) apoptosis, as could be seen by the significant reduction in the number of live (lower left quadrant) H9c2 cells ($p < 0.001$) relative to the control. The results further showed a significant increase in the number of necrotic (upper left quadrant) H9c2 cells when compared to the control (Figure 5A–E). In contrast, in the MCF-7 cells, DMSO exposure with doses $\leq$ 3.7% also led to a significant ($p < 0.05$) increase in early and late apoptosis, as demonstrated by enhanced annexin V positive cells (Figure 5F,H). The results further showed a significant increase in the number of cells going into late apoptosis in a dose dependent manner compared to the control (Figure 5I). While treatment with 3.7% DMSO led to a significant reduction in the number of live MCF-7 cells ($p < 0.001$), the results also demonstrated a significant increase in the rate of necrosis ($p < 0.001$) (Figure 5G,J).

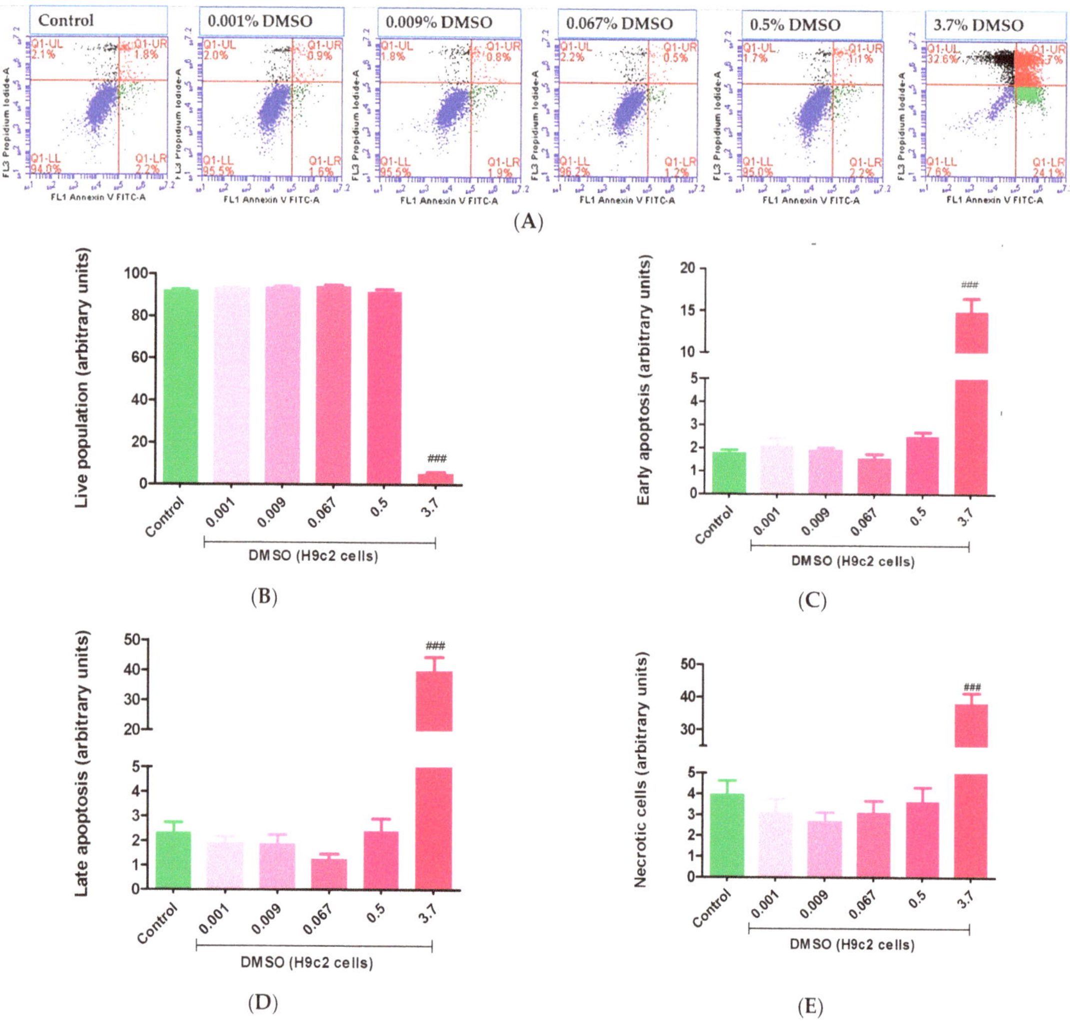

Figure 5. *Cont.*

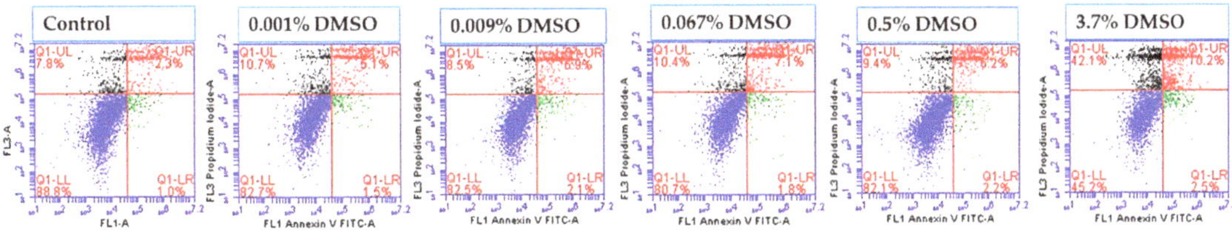

(F)

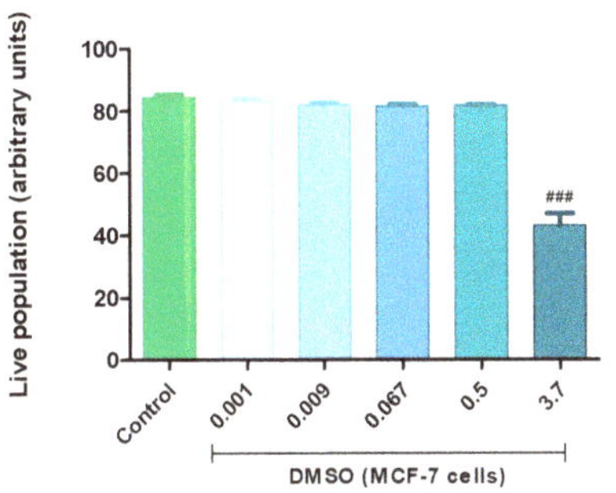

(G)

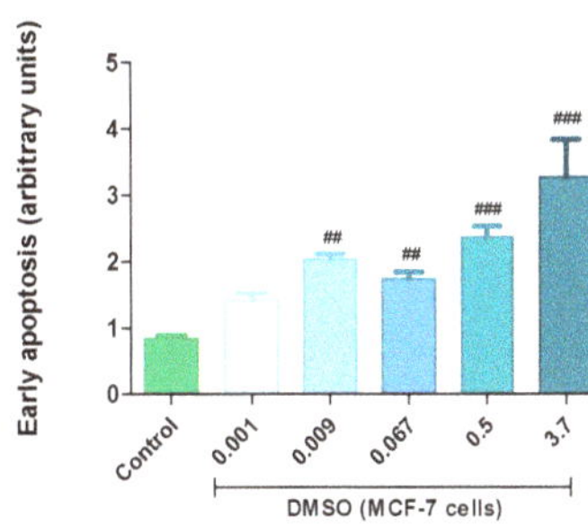

(H)

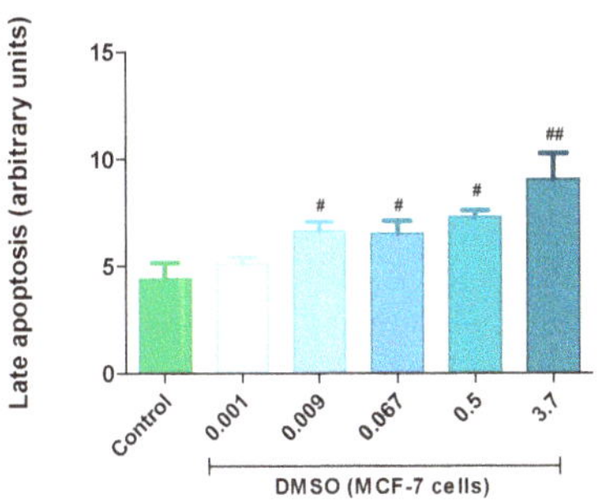

(I)

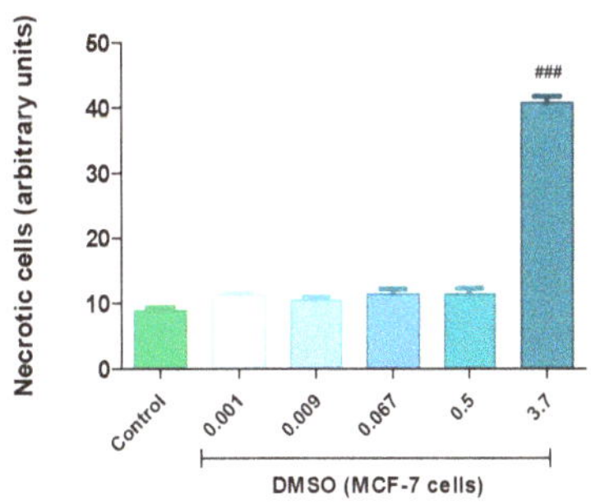

(J)

Figure 5. Dimethyl sulfoxide (DMSO)-induced apoptosis in the H9c2 cardiomyoblasts and MCF-7 breast cancer cells. (**A**) Flow cytometry scatter plot, (**B**) live population, (**C**) early apoptotic, (**D**) late apoptotic and (**E**) necrotic H9c2 cells. (**F**) Flow cytometry scatter plot, (**G**) live population, (**H**) early apoptotic, (**I**) late apoptotic and (**J**) necrotic MCF-7 cells. Cells were treated every second day for 6 days with varying DMSO concentrations (0.001, 0.009, 0.067, 0.5, and 3.7%). Data are presented as the mean ± SEM of 3 biological experiments with 3 technical repeats (n = 3). Data are presented as the mean ± SEM of 3 biological experiments with 3 technical repeats (n = 3). Significance is indicated as $^{\#}$ $p < 0.01$, $^{\#\#}$ $p < 0.01$, and $^{\#\#\#}$ $p < 0.001$ versus the control.

3. Discussion

In many disease models, uncontrolled production of ROS remains a central mechanism accountable for excess accumulation of damaged organelles, translating to tissue injury and other related deleterious effects [13]. Uncontrolled ROS is a known source for the depletion of intracellular antioxidant defense systems, a process implicated in the generation of oxidative stress. Within the cardiovascular system, besides being the precursor of plaque formation through its attack on circulating lipid products to cause endothelial dysfunction [14], oxidative stress can directly cause tissue injury, leading to cardiac fibrosis and conditions such as cardiomyopathies [15]. In cancer research, although the role of oxidative stress is controversial [16], cancer cells display an abnormal redox homeostasis, with very high ROS shown to be cytotoxic [17]. Thus, exploring the role of ROS in redox signaling and in tumor propagation, development, and metastasis, including uncovering associated cellular mechanisms remains relevant to discover effective therapies.

For example, while leading chemotherapeutic agents such as doxorubicin are effective against certain tumours, these drugs are known to present with undesirable side effects such as inducing cardiotoxicity when used for a prolonged period [18,19]. Certainly, current research [19,20], including that from our group [14], is targeting the use of antioxidant therapies to ameliorate oxidative stress-related detrimental effects consistent with the development of cancer and cardiac injury. Here, cultured cardiomyocytes and cancer cells were exposed to different doses of DMSO to understand their modulatory effect on oxidative stress-related parameters, such as mitochondrial function and apoptosis. This is especially important, since while low doses of DMSO can present with some antioxidant properties [21], this effect is known to vary with experimental models. Furthermore, as one of the commonly used solvents to deliver drugs or as a biological drug itself, it remains essential to broaden our understanding on the therapeutic properties of DMSO, especially its dose-dependent effect.

Both H9c2 and MCF-7 cells are widely used to investigate the therapeutic potential of drugs against oxidative stress mechanisms [14,22–26]. In the current study, these cells were exposed to DMSO doses, ranging from 0.001 to 3.7% for 6 days, before assessing endpoints such as cell viability, ROS production, efficiency of the mitochondrial respiratory capacity, changes in mitochondrial membrane potential, as well as markers of early and late apoptosis. Firstly, it was clear that DMSO doses $\leq$ 0.067% were not cytotoxic and, in fact, improved viability was observed for both cell lines. Alternatively, a dose of 0.5% had no effect on cell viability but, that of 3.7%, led to a significant reduction in cardiomyoblast viability. Conversely, both doses of 0.5 and 3.7% demonstrated signs of toxicity in the cancer cells. The next question was to determine the oxidative stress-related parameter that may explain improved cell viability with low DMSO doses ($\leq$0.067%), or the cytotoxicity seen with high doses (>0.5%) in both cell lines. As one of the major organelles that can determine cellular fate, special attention was paid into understanding how the various doses of DMSO affected mitochondrial respiration. This aspect remained vital to explore since, as part of being the cell's powerhouse, mitochondria are important for ATP production, controlling ROS production and antioxidant activity, as well as regulating apoptosis. The latter defines a highly ordered process that could be severely influenced by harmful stimuli, such as accelerated ROS activity and mitochondrial depolarization [27,28]. As a result, it has become necessary to collectively assess the therapeutic potential of any drug at the cellular level, by measuring its effect on cell viability, mitochondrial function, and apoptosis.

In the current study, we demonstrated that cardiomyoblasts have an inherently lower energy phenotype than breast cancer cells, which presented with a significantly higher oxygen consumption rate (OCR) and glycolytic profile. Here, the cardiomyoblasts displayed a physiologically lower respiratory capacity, which validated their quiescent energy phenotype. Upon DMSO exposure, with 0.001–0.067%, we observed a noticeable improvement in the cardiomyoblasts respiratory control ratio and ATP turnover, but not the breast cancer cells (Figure 3, Tables 1 and 2). Instead, although not significant, these doses appeared to reduce the cancer cells mitochondrial bioenergetics. Briefly, the respiratory control ratio, represented the maximum factorial increase in mitochondrial OCR that could be attained above the leak oxygen prerequisite when driving the conversion of ADP into ATP [29]. This clearly explained the potential capacity of low DMSO doses to improve cell viability and maintain mitochondrial function, whilst protecting against apoptosis in the cardiomyoblasts; an outcome that was not observed in the cancer cells. While these experimental results still need to be confirmed in vivo, it appears that low DMSO doses ($\leq$0.5%) may have had a positive effect on the cardiomyoblasts when considering processes that prevent oxidative damage, via improving mitochondrial respiration. Likewise, the findings of this study further demonstrated that the rate of apoptosis in the cardiomyoblasts treated with the "non-toxic" low DMSO doses ($\leq$0.5%) was comparable to the severity of apoptosis observed in the untreated control group. These findings were supported by the preserved live population and reduced number of early and late apoptotic cardiomyoblasts. In contrast, the data showed that 3.7% DMSO exposure presented with severe cardiomyoblast toxicity, as shown by the increased number of apoptotic and necrotic cells.

Interestingly, in the breast cancer cells, 3.7% DMSO exposure impaired the cells oxidative phosphorylative capacity, which was demonstrated by the cells reduced basal respiration, ATP-linked respiration, ATP turnover, and maximal respiration. The data further highlighted that DMSO doses $\leq$ 3.7% triggers oxidative stress, in a dose dependent manner, as determined by the significant increase in ROS activity in the cancer cells. Further justifying its potential exploration as a possible therapeutic compound, was the significant increase in the number of early and late apoptotic cancer cells following treatment with DMSO ($\leq$3.7%). Additionally, a much higher degree of necrosis was observed in these cells.

4. Materials and Methods

4.1. Reagents

Dimethyl sulfoxide (DMSO), Dulbecco's phosphate-buffered saline (DPBS), Dulbecco's Modified Eagle Medium (DMEM), tissue culture grade water, trypsin, and Hanks balanced salt solution (HBSS) were obtained from Lonza (Walkersville, MD, USA). Fluorescent probes, 5,5′, 6,6′-tetrachloro-1,1′, 3,3-tetraethylbenzimidazolyl-carbocyanine iodide (JC-1), and propidium iodide (PI) were obtained from Sigma-Aldrich (St Louis, MO, USA). Fetal Bovine Serum (FBS) was purchased from Thermo Fisher Scientific (Waltham, MA, USA). Annexin V and fluorescein conjugate (FITC annexin V) was purchased from Invitrogen (Carlsbad, CA, USA). The oxiselect ™ intracellular ROS Assay Kit (Green Fluorescence) was purchased from Cell Biolabs (San Diego, CA, USA). The XF Cell Mito Stress Kit was purchased from Agilent technologies (Santa Clara, CA, USA).

4.2. In Vitro Models

MCF-7 human breast cancer cells and H9c2 cardiomyoblasts were purchased from the American Type Culture Collection (ATCC, catalogue number HTB-22 and CRL-1446, respectively). Both cell lines were cultured in DMEM supplemented with 10% Fetal Bovine Serum (FBS, Thermo Fisher Scientific, Waltham, MA, USA) under standard tissue culture (TC) conditions (37 °C, 95% humidified air, and 5% CO_2). The cytotoxic threshold of DMSO was investigated on the MCF-7 and H9c2 cells, using a series of concentrations (0.001, 0.009, 0.067, 0.5, and 3.7%). These concentrations were randomly selected from a previously conducted preliminary study on healthy cardiomyoblasts (Figure S1, Supplementary Materials). Briefly, treatment was prepared in DMEM without phenol (Lonza, Walkersville, MD, USA), supplemented with 2% FBS. Prior to the experiments, the DMSO treatment was filter-sterilized using 0.22 µM syringe filter systems in a Class II Type A2 Biological Safety Cabinet. To assess the chronic effect of DMSO on the H9c2 and MCF-7 cells, the cells were treated every second day for 6 days, as demonstrated in Figure 6. Biochemical assays were conducted after terminating treatment.

4.3. Determination of Cell Viability

The effect of DMSO on mitochondrial activity, as a measure of cell viability, was assessed using the MTT (3-(4,5-dimethylthiazol-2-yl)-2,5-diphenyltetrazolium bromide) assay following an in-house protocol previously described [25]. Concisely, H9c2 and MCF-7 cells were seeded (0.8×10^5/well) in 96-well clear plates and then treated as described above. Treatment was terminated by washing the cells with 100 µL DPBS. Subsequently, the cells were exposed to 100 µL MTT solution (2 mg/mL) and then incubated for 1 h under standard TC conditions. Cell viability was quantified by measuring absorbance on the SpectraMax® i3x Multi-Mode Microplate Reader, at a wavelength of 570 nm.

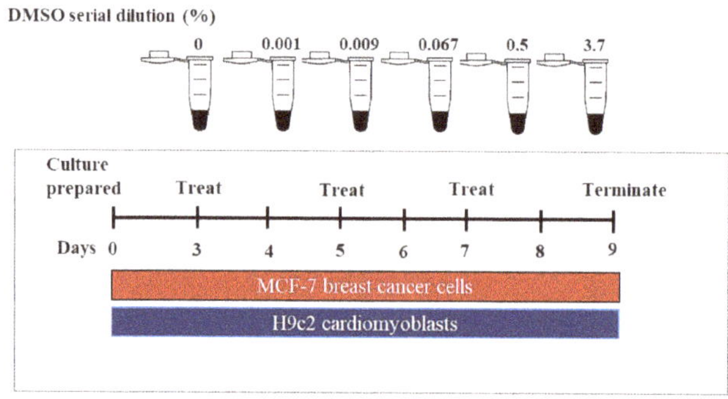

Figure 6. An in vitro experimental design of dimethyl sulfoxide (DMSO) exposure on H9c2 cardiomyoblasts and MCF-7 breast cancer cells. Concisely, cells were treated with varying DMSO concentrations (0.001 0.009, 0.067, 0.5, and 3.7%) for 6 days. Treatment was terminated on day 6 and biochemical analysis were conducted.

4.4. Assessment of Mitochondrial Respiratory Capacity

Respiratory parameters associated with mitochondrial bioenergetics were measured in intact breast cancer cells and cardiomyoblasts using the XF Cell Mito Stress Kit (Seahorse Bioscience, Billerica, MA, USA) according to the manufacturer's instruction, which is based on the sequential injection of oligomycin, carbonyl cyanide-4-(trifluoromethoxy) phenylhydrazone, rotenone, and antimycin, as per the manufacturer's instructions. Seeding densities for the H9c2 (1×10^4/well) and MCF-7 (5×10^3/well) cells were determined from a previous study [25]. Concisely, cells were seeded into XF96-well cell culture microplates (Seahorse Bioscience, Billerica, MA, USA) after which, treatment was initiated after 48 h as described above. In preparation of the assay, the cells were incubated for 1 h with the Seahorse base assay medium (supplemented with 2 mM glutamine, 10 mM glucose, and 1 mM pyruvate). Hereafter, mitochondrial oxygen consumption rate (OCR) and extracellular acidification rate (ECAR) were quantified on the intact live cells using the Seahorse XF96 extracellular flux analyser (Seahorse Bioscience, Billerica, MA, USA). The data were normalized to protein concentrations using the Bio-Rad DC Protein assay (Bio-Rad, Hercules, CA, USA), as per the manufacturer's instructions with the resultant absorbance read at 695 nm using the SpectraMax® i3x Multi-Mode Microplate Reader. The results were expressed as pmol/min/mg protein for the OCR, and mpH/min/mg protein for the ECAR.

4.5. Assessing Changes in Mitochondrial Membrane Potential ($\Delta\Psi m$)

The effect of DMSO on the mitochondrial integrity of the H9c2 and MCF-7 cells was assessed using the fluorescent JC-1 Assay Kit (Sigma-Aldrich, St Louis, MO, USA), as per the manufacturer's instructions. Briefly, cells seeded (0.8×10^5/well) in black clear bottom 96-well plates were stained with 100 μL JC-1 dye (8 μM) and then incubated for 45 min under standard TC conditions. Mitochondrial membrane potential (MMP) was then quantified by measuring the fluorescence intensity of J-aggregates at 590 nm and JC-1 monomers (at 529 nm using the SpectraMax® i3x Multi-Mode Microplate Reader.

4.6. Quantification of Reactive Oxygen Species (ROS)

The OxiSelect™ Intracellular ROS assay kit (Cell Biolabs, San Diego, CA, USA) was used to quantify DMSO-stimulated ROS production. Cells were seeded in 24-well plates (1×10^5 cells/well) and treated as described above. Once treatment with DMSO was terminated, cells were washed with pre-warmed DPBS before being stained with 100 μL/well 2′,7′dichlorofluorescin diacetate (DCFH-DA, 20 μM) dye and then incubated for 30 min. Following incubation, the DCFH-DA dye was aspirated, and the cells were washed with pre-warmed DPBS. Thereafter, cells were trypsinized (150 μL) for either

6 min for the H9c2 cells, or 5 min for the MCF-7 cells, in an incubator under standard TC conditions. Trypsinization was deactivated by the addition of 300 μL pre-warmed media (DMEM, supplemented with 10% FBS). The cell suspension was then collected into 2 mL Eppendorf tubes and centrifuged for 5 min at 1500 rpm for H9c2 cells, and 120 RCF for the MCF-7 cells. Hereafter, the cells were resuspended in 150 μL DPBS and then placed on ice. ROS activity was quantified using the BD Accuri C6 flow cytometer (BD Biosciences, Franklin Lakes, NJ, USA).

4.7. Apoptosis Assay

DMSO-induced apoptosis was quantified by staining cells with Annexin V-FITC (Invitrogen, Carlsbad, CA, USA) and propidium iodide (PI, Sigma-Aldrich, St. Louis, MO, USA) as previously described [25]. Briefly, H9c2 and MCF-7 cells were seeded in 24-well plates (1×10^5 cells/well) and treated as described above. The cells were trypsinized and collected into 2 mL Eppendorf tubes as previously reported [25]. The cells were then co-stained with 1.5 μL Annexin V and 1 μL PI (2 μg/mL) before being incubated in the dark for at least 10 min for the H9c2 cells, or 20 min for the MCF-7 cells. The rate of apoptosis was determined on the BD Accuri C6 flow cytometer (BD Biosciences) using the BD Accuri C6 Annexin V-FITC/PI template. Live, early, late apoptotic, as well as necrotic cells were quantified with the BD Accuri C6 software using the FITC signal detector FL1 (excitation = 488 nm; emission = 530 nm) for Annexin V positive cells and FL3 detector (excitation = 488 nm; emission = 670/LP) for PI positive cells.

4.8. Statistical Analysis

Data are represented as the mean ± standard error of the mean (SEM). Statistical comparisons between the control and different DMSO concentrations were performed using one-way analysis of variance (ANOVA), followed by a Tukey post hoc test and a student's *t*-test using GraphPad Prism software version 5.0 (GraphPad Software, Inc., La Jolla, San Diego, CA, USA). Differences were regarded significant at p values < 0.05.

5. Conclusions

While DMSO remains an extensively used vehicle control and widespread solvent in numerous research settings, it is evident that its influence on cellular biological processes requires further investigation. Our findings demonstrate that DMSO doses higher than 0.001%, but not more than 0.5%, can still be safely used in experimental set-ups involving the use of H9c2 cells. However, these doses were determined to have some adverse effects in the breast cancer cells, as they stimulated the cells proliferative activity. Considering this, it may not be reasonable to recommend DMSO as a potential therapeutic compound for cancer cells. In the same context, it is important to note that when used as a solvent, the low DMSO doses may influence the study outcome, especially in experimental models where the proliferative activity of the cells is reduced by low FBS (2%) concentrations and when the treatment duration exceeds 6 days. Nonetheless, the observed effects of DMSO on the cardiac and cancer cells advocates for further investigation to fully understand the cytotoxic and beneficial effects of the "non-toxic" low DMSO doses.

Supplementary Materials: The following are available online, Figure S1: Cytotoxicity screening of DMSO on H9c2 cells.

Author Contributions: Conceptualization: N.F.S. and N.C.; methodology: N.F.S.; formal analysis: N.F.S.; resources: R.J.; writing—original draft preparation: N.F.S., P.V.D. and R.J.; writing—review and editing: N.F.S., P.V.D., N.C., L.M., J.R.S. and R.J.; supervision: R.J. and L.M. funding acquisition: R.J. All authors have read and agreed to the published version of the manuscript.

Funding: This study was in part, funded by the South African Medical Research Council (SAMRC), through its division of the Biomedical Research and Innovation Platform (baseline funding), and the National Research Foundation (NRF) Thuthuka Programme (UID120812). The authors also acknowledge the financial support for Ms. Sangweni as a PhD student funded by the SAMRC through its Division of Research Capacity Development under the Internship Scholarship Programme. The content hereof is the sole responsibility of the authors and do not necessarily represent the official views of the SAMRC or the NRF.

Institutional Review Board Statement: Not applicable.

Informed Consent Statement: Not applicable.

Data Availability Statement: Data is contained within the article or supplementary material The data presented in this study are available within the article and in the supplementary material. Design of the study; in the collection, analyses, or interpretation of data; in the writing of the manuscript, or in the decision to publish the results.

Acknowledgments: The research reported in this publication was supported by the SAMRC through funding received from the South African National Treasury. We would also like to acknowledge BRIP/Technical Staff.

Conflicts of Interest: The authors declare no conflict of interest. The funders had no role in our data interpretation.

Sample Availability: Samples of the compounds are not available from the authors.

References

1. Moskot, M.; Jakóbkiewicz-Banecka, J.; Kloska, A.; Piotrowska, E.; Narajczyk, M.; Gabig-Cimińska, M. The Role of Dimethyl Sulfoxide (DMSO) in Gene Expression Modulation and Glycosaminoglycan Metabolism in Lysosomal Storage Disorders on an Example of Mucopolysaccharidosis. *Int. J. Mol. Sci.* **2019**, *20*, 304. [CrossRef] [PubMed]
2. Penazzi, L.; Lorengel, J.; Sündermann, F.; Golovyashkina, N.; Marre, S.; Mathis, C.M.; Lewejohann, L.; Brandt, R.; Bakota, L. DMSO modulates CNS function in a preclinical Alzheimer's disease model. *Neuropharmacology* **2017**, *113*, 434–444. [CrossRef]
3. Verheijen, M.; Lienhard, M.; Schrooders, Y.; Clayton, O.; Nudischer, R.; Boerno, S.; Timmermann, B.; Selevsek, N.; Schlapbach, R.; Gmuender, H.; et al. DMSO induces drastic changes in human cellular processes and epigenetic landscape in vitro. *Sci. Rep.* **2019**, *9*, 1–12. [CrossRef]
4. Food & Drug Administration, U.S, C. for D.E. and, 2020. Q3C Tables and List Rev. 4 [WWW Document]. U.S. Food and Drug Administration. Available online: https://www.fda.gov/regulatory-information/search-fda-guidance-documents/q3c-tables-and-list-rev-4 (accessed on 6 October 2021).
5. Brayton, C.F. Dimethyl sulfoxide (DMSO): A review. *Cornell Veter* **1986**, *76*, 61–90.
6. Marks, P.A.; Breslow, R. Dimethyl sulfoxide to vorinostat: Development of this histone deacetylase inhibitor as an anticancer drug. *Nat. Biotechnol.* **2007**, *25*, 84–90. [CrossRef]
7. Parisi, A.; Alfieri, A.; Mazzella, M.; Mazzella, A.; Scognamiglio, M.; Scognamiglio, G.; Mascolo, N.; Cicala, C. Protective Effect of Dimethyl sulfoxide on Acute Myocardial Infarction in Rats. *J. Cardiovasc. Pharmacol.* **2010**, *55*, 106–109. [CrossRef]
8. Akinci, B.; Sankella, S.; Gilpin, C.; Ozono, K.; Garg, A.; Agarwal, A.K. Progeroid syndrome patients with ZMPSTE24 deficiency could benefit when treated with rapamycin and dimethylsulfoxide. *Mol. Case Stud.* **2017**, *3*, a001339. [CrossRef]
9. Tunçer, S.; Gurbanov, R.; Sheraj, I.; Solel, E.; Esentürk, O.; Banerjee, S. Low dose dimethyl sulfoxide driven gross molecular changes have the potential to interfere with various cellular processes. *Sci. Rep.* **2018**, *8*, 14828. [CrossRef] [PubMed]
10. Budash, H.V.; Bilko, N.M. Embryonic and Induced Pluripotent Stem Cells and Their Differentiation in the Cardiomyocyte Direction in the Presence of Dimethyl Sulfoxide. *Cytol. Genet.* **2019**, *53*, 34–41. [CrossRef]
11. Riegler, J.; Tiburcy, M.; Ebert, A.D.; Tzatzalos, E.; Raaz, U.; Abilez, O.J.; Shen, Q.; Kooreman, N.G.; Neofytou, E.; Chen, V.C.; et al. Human Engineered Heart Muscles Engraft and Survive Long Term in a Rodent Myocardial Infarction Model. *Circ. Res.* **2015**, *117*, 720–730. [CrossRef]
12. Ouyang, X.; Liu, Y.; Zhou, Y.; Guo, J.; Wei, T.-T.; Liu, C.; Lee, B.; Chen, B.; Zhang, A.; Casey, K.M.; et al. Antitumor effects of iPSC-based cancer vaccine in pancreatic cancer. *Stem Cell Rep.* **2021**, *16*, 1468–1477. [CrossRef]
13. Liu, Z.; Ren, Z.; Zhang, J.; Chuang, C.-C.; Kandaswamy, E.; Zhou, T.; Zuo, L. Role of ROS and Nutritional Antioxidants in Human Diseases. *Front. Physiol.* **2018**, *9*, 477. [CrossRef] [PubMed]
14. Incalza, M.A.; D'Oria, R.; Natalicchio, A.; Perrini, S.; Laviola, L.; Giorgino, F. Oxidative stress and reactive oxygen species in endothelial dysfunction associated with cardiovascular and metabolic diseases. *Vasc. Pharmacol.* **2018**, *100*, 1–19. [CrossRef]
15. Zhang, D.; Li, Y.; Heims-Waldron, D.; Bezzerides, V.; Guatimosim, S.; Guo, Y.; Gu, F.; Zhou, P.; Lin, Z.; Ma, Q.; et al. Mitochondrial Cardiomyopathy Caused by Elevated Reactive Oxygen Species and Impaired Cardiomyocyte Proliferation. *Circ. Res.* **2018**, *122*, 74–87. [CrossRef]
16. Hayes, J.D.; Dinkova-Kostova, A.T.; Tew, K.D. Oxidative Stress in Cancer. *Cancer Cell* **2020**, *38*, 167–197. [CrossRef]
17. Reczek, C.R.; Birsoy, K.; Kong, H.; Martinez-Reyes, I.; Wang, T.; Gao, P.; Sabatini, D.M.; Chandel, N.S. A CRISPR screen identifies a pathway required for paraquat-induced cell death. *Nat. Chem. Biol.* **2017**, *13*, 1274–1279. [CrossRef]
18. Monahan, D.S.; Almas, T.; Wyile, R.; Cheema, F.H.; Duffy, G.P.; Hameed, A. Towards the use of localised delivery strategies to counteract cancer therapy–induced cardiotoxicities. *Drug Deliv. Transl. Res.* **2021**, *11*, 1924–1942. [CrossRef] [PubMed]

19. Sergazy, S.; Shulgau, Z.; Fedotovskikh, G.; Chulenbayeva, L.; Nurgozhina, A.; Nurgaziyev, M.; Krivyh, E.; Kamyshanskiy, Y.; Kushugulova, A.; Gulyayev, A.; et al. Cardioprotective effect of grape polyphenol extract against doxorubicin induced cardiotoxicity. *Sci. Rep.* **2020**, *10*, 1–12. [CrossRef] [PubMed]
20. Razavi-Azarkhiavi, K.; Iranshahy, M.; Sahebkar, A.; Shirani, K.; Karimi, G. The Protective Role of Phenolic Compounds Against Doxorubicin-induced Cardiotoxicity: A Comprehensive Review. *Nutr. Cancer* **2016**, *68*, 892–917. [CrossRef]
21. Dludla, P.V.; Nkambule, B.B.; Mazibuko-Mbeje, S.E.; Nyambuya, T.M.; Silvestri, S.; Orlando, P.; Mxinwa, V.; Louw, J.; Tiano, L. The impact of dimethyl sulfoxide on oxidative stress and cytotoxicity in various experimental models. In *Toxicology*; Academic Press: Amsterdam, The Netherlands, 2021; pp. 243–261.
22. Barteková, M.; Adameová, A.; Görbe, A.; Ferenczyová, K.; Pecháňová, O.; Lazou, A.; Dhalla, N.S.; Ferdinandy, P.; Giricz, Z. Natural and synthetic antioxidants targeting cardiac oxidative stress and redox signaling in cardiometabolic diseases. *Free. Radic. Biol. Med.* **2021**, *169*, 446–477. [CrossRef]
23. Sirangelo, I.; Sapio, L.; Ragone, A.; Naviglio, S.; Iannuzzi, C.; Barone, D.; Giordano, A.; Borriello, M. Vanillin Prevents Doxorubicin-Induced Apoptosis and Oxidative Stress in Rat H9c2 Cardiomyocytes. *Nutrients* **2020**, *12*, 2317. [CrossRef] [PubMed]
24. Sangweni, N.; Moremane, M.; Riedel, S.; Van Vuuren, D.; Huisamen, B.; Mabasa, L.; Barry, R.; Johnson, R. The Prophylactic Effect of Pinocembrin Against Doxorubicin-Induced Cardiotoxicity in an In Vitro H9c2 Cell Model. *Front. Pharmacol.* **2020**, *11*, 1172. [CrossRef] [PubMed]
25. Qiu, J.; Zhang, T.; Zhu, X.; Yang, C.; Wang, Y.; Zhou, N.; Ju, B.; Zhou, T.; Deng, G.; Qiu, C. Hyperoside Induces Breast Cancer Cells Apoptosis via ROS-Mediated NF-κB Signaling Pathway. *Int. J. Mol. Sci.* **2019**, *21*, 131. [CrossRef]
26. Shin, B.; Feser, R.; Nault, B.; Hunter, S.; Maiti, S.; Ugwuagbo, K.C.; Majumder, M. miR526b and miR655 Induce Oxidative Stress in Breast Cancer. *Int. J. Mol. Sci.* **2019**, *20*, 4039. [CrossRef] [PubMed]
27. Kang, M.-H.; Das, J.; Gurunathan, S.; Park, H.-W.; Song, H.; Park, C.; Kim, J.-H. The cytotoxic effects of dimethyl sulfoxide in mouse preimplantation embryos: A mechanistic study. *Theranostics* **2017**, *7*, 4735–4752. [CrossRef]
28. Yang, H.; Xie, Y.; Yang, D.; Ren, D. Oxidative stress-induced apoptosis in granulosa cells involves JNK, p53 and Puma. *Oncotarget* **2017**, *8*, 25310–25322. [CrossRef]
29. Harper, M.-E.; Green, K.; Brand, M.D. The Efficiency of Cellular Energy Transduction and Its Implications for Obesity. *Annu. Rev. Nutr.* **2008**, *28*, 13–33. [CrossRef]

molecules

MDPI

Article

Pharmaceutical Drug Metformin and MCL1 Inhibitor S63845 Exhibit Anticancer Activity in Myeloid Leukemia Cells via Redox Remodeling

Giedrė Valiulienė *, Aida Vitkevičienė, Giedrė Skliutė, Veronika Borutinskaitė and Rūta Navakauskienė

Department of Molecular Cell Biology, Institute of Biochemistry, Life Sciences Center, Vilnius University, Sauletekio av. 7, LT-01257 Vilnius, Lithuania; aida.vitkeviciene@gmc.vu.lt (A.V.); giedre.skliute@gmc.vu.lt (G.S.); veronika.borutinskaite@bchi.vu.lt (V.B.); ruta.navakauskiene@bchi.vu.lt (R.N.)
* Correspondence: giedre.valiuliene@bchi.vu.lt

Abstract: Metabolic landscape and sensitivity to apoptosis induction play a crucial role in acute myeloid leukemia (AML) resistance. Therefore, we investigated the effect of metformin, a medication that also acts as an inhibitor of oxidative phosphorylation (OXPHOS), and MCL-1 inhibitor S63845 in AML cell lines NB4, KG1 and chemoresistant KG1A cells. The impact of compounds was evaluated using fluorescence-based metabolic flux analysis, assessment of mitochondrial $\Delta\psi$ and cellular ROS, trypan blue exclusion, Annexin V-PI and XTT tests for cell death and cytotoxicity estimations, also RT-qPCR and Western blot for gene and protein expression. Treatment with metformin resulted in significant downregulation of OXPHOS; however, increase in glycolysis was observed in NB4 and KG1A cells. In contrast, treatment with S63845 slightly increased the rate of OXPHOS in KG1 and KG1A cells, although it profoundly diminished the rate of glycolysis. Generally, combined treatment had stronger inhibitory effects on cellular metabolism and ATP levels. Furthermore, results revealed that treatment with metformin, S63845 and their combinations induced apoptosis in AML cells. In addition, level of apoptotic cell death correlated with cellular ROS induction, as well as with downregulation of tumor suppressor protein MYC. In summary, we show that modulation of redox-stress could have a potential anticancer activity in AML cells.

Keywords: acute myeloid leukemia (AML); metformin; MCL-1 inhibitor S63845; reactive oxygen species (ROS)

Citation: Valiulienė, G.; Vitkevičienė, A.; Skliutė, G.; Borutinskaitė, V.; Navakauskienė, R. Pharmaceutical Drug Metformin and MCL1 Inhibitor S63845 Exhibit Anticancer Activity in Myeloid Leukemia Cells via Redox Remodeling. *Molecules* **2021**, *26*, 2303. https://doi.org/10.3390/molecules26082303

Academic Editors: Višnja Stepanić and Marta Kučerová-Chlupáčová

Received: 27 February 2021
Accepted: 13 April 2021
Published: 15 April 2021

1. Introduction

Acute myeloid leukemia (AML) is one of the most common forms of blood cancer in adults. Typically, the five-year survival rate for AML patients is ~29% [1]. The prognosis of specific AML cases varies widely as patient age and type of AML should be taken into account. For example, APL (acute promyelocytic leukemia) subtype of AML is mostly well managed, while others have a more complicated prognosis, as AML resistance and relapse still remain very serious issues [2]. Therefore, better treatment options in AML cases remain an urgent need.

A very interesting and innovative approach is a drug repositioning—when existing drugs are investigated for new therapeutic purposes. For example, metformin (*N,N*-dimethylbiguanide), which is applied widely to treat the symptoms of hyperglycemia in type 2 diabetes [3] and also used in the treatment of polycystic ovary syndrome [4], may be repurposed to manage cancer as well. Metformin was shown to have a cytostatic activity; it can strengthen chemotherapeutic effect [5]. In human breast cancer and osteosarcoma cells metformin was shown to induce the generation of reactive oxygen species (ROS) and partly elicit its anticancerous effect via redox remodeling [6,7]. Regarding leukemia research, metformin was demonstrated to sensitize AML cell lines to cytarabine [8]. Recently, we demonstrated that metformin in combination with cytarabine, and venetoclax also slightly inhibited AML patients' cell proliferation and profoundly reduced oxidative phosphorylation (OXPHOS) rate ex vivo [9]. It has been reported consistently that metformin elicits its effect on oxidative phosphorylation via inhibition of Complex I [10].

As mitochondrial oxidative phosphorylation is an essential factor in AML chemoresistance [11], targeting mitochondrial respiration combined with modulation of other cancer-associated pathways may be beneficial in AML treatment. Since the survival of cancer cells is heavily promoted by anti-apoptotic BCL-2 family members, such as BCL-2 (B cell lymphoma-2) and MCL-1 (myeloid cell leukemia-1), which prevent malignant cells from the mitochondrial cell death pathway [12], targeting both mitochondrial respiration and intrinsic apoptotic processes may suggest more potent therapeutic strategies. A recent study by Ewald et al. [13] revealed that MCL-1 may be a more universal therapeutic target in AML compared to BCL-2. S63845 (α(R)-[[(5S)-5-[3-chloro-2-methyl-4-[2-(4-methyl-1-piperazinyl)ethoxy]phenyl]-6-(5-fluoro-2-furanyl)thieno[2,3-d]pyrimidin-4-yl]oxy]-2-[[1-(2,2,2-trifluoroethyl)-1H-pyrazol-5-yl]methoxy]-benzenepropanoic acid) is an inhibitor of MCL-1, a pro-survival protein that is highly overexpressed in many cancer types and in AML as well [14,15]. Research has demonstrated that S63845 induced cell death in various AML cells [13]. In addition, AML cells that are in general resistant to venetoclax have been shown to be very sensitive to S63845 [16].

In the present study we aimed to test the effect of antidiabetic drug metformin and MCL-1 inhibitor S63845 in AML cell lines NB4 (APL subtype), KG1 and chemoresistant AML cell line KG1A. We were mainly interested in the potential of metabolic and redox modulation, as well as an effect on growth restriction. Therefore, we treated cells with metformin and S63845 alone or in combination. It should be emphasized that in this experimental set-up the supratherapeutic doses of metformin (10 mM) were used in order to alter the mitochondrial metabolism. After the treatments, we evaluated changes in cell proliferation and viability, metabolic activity, as well as ROS production and induction of apoptotic cell death.

2. Results

2.1. Metformin and S63845 Cytotoxicity and Effect on AML Cell Proliferation and Viability

After AML cell treatment with antidiabetic drug and oxidative phosphorylation inhibitor metformin at 10 mM concentration and treatment with various concentrations of MCL-1 inhibitor S63845 (25–2000 nM), as well as after their combined treatments, cell viability (Figure 1a) and proliferation (see Figure S1 in Appendix No. 1, Supplementary Materials) were assessed using the trypan blue exclusion test. Agent concentrations were chosen after testing a range of different concentrations (data not shown). We selected effective cell proliferation and survival-inhibiting concentrations, which did not show too high toxicity. In general, NB4, KG1 and KG1A cells were sensitive to applied treatments, as reduction in cell proliferation and cell viability was observed. Effect of 10 mM metformin was of comparable degree between all cell lines, though metformin had a slightly stronger effect on cell viability in NB4 cells compared to KG1A cells. However, evidently NB4, KG1 and KG1A cells were not equally sensitive to MCL-1 inhibitor S63845. For example, inhibition of cell proliferation (Figure S1 in Appendix No. 1, Supplementary Materials) in NB4 cells was achieved when using 25 nM of S63845. In order to reach a similar effect in KG1 cells we had to use 100 nM of S63845, whereas in KG1A cells only 1000 nM of S63845 could slow down the cell proliferation. Using XTT Cell Proliferation Assay Kit (ATTC) we also determined the antiproliferative IC_{50} values for metformin and S63845 (Figure 1b), which confirmed that metformin did have very similar cytotoxicity patterns in NB4, KG1 and KG1A cells (e.g., after 24 h treatments IC_{50} values ranged 40.0–46.3 mM). IC_{50} analysis results also verified that NB4 cells were the most sensitive for S63845 treatment, whereas for KG1A cells approx. 20-fold higher doses were needed to reach IC_{50} (after 24 h treatments IC_{50} was 90.2 nM and 1863.2 nM in NB4 and KG1A cells, respectively).

(a)

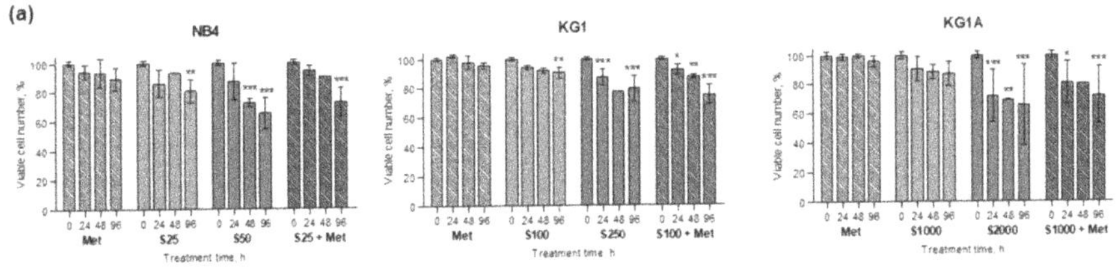

(b)

NB4	IC_{50} (24 h treatment)	IC_{50} (72 h treatment)
Metformin, mM	43.0 ± 1.5	3.6 ± 0.4
S63845, nM	90.2 ± 2.45	76.9 ± 2.4

KG1	IC_{50} (24 h treatment)	IC_{50} (72 h treatment)
Metformin, mM	40.0 ± 3.1	7.7 ± 0.9
S63845, nM	905.9 ± 82.8	879.2 ± 128.9

KG1A	IC_{50} (24 h treatment)	IC_{50} (72 h treatment)
Metformin, mM	46.3 ± 6.2	6.6 ± 3.1
S63845, nM	1863.2 ± 265.7	1324.5 ± 98.7

Figure 1. Effect of metformin and MCL-1 inhibitor S63845 on myeloid leukemia cell viability and induced cytotoxicity. (**a**) NB4, KG1 and KG1A cells were treated with 10 mM metformin and with different concentrations of MCL-1 inhibitor S63845 (S25–25 nM of S63845, etc.). Cell survival was evaluated by trypan blue exclusion test. Viability data were normalized to untreated controls. Results are mean ± S.D. ($n \geq 3$). (**b**) Antiproliferative IC_{50} values for metformin and S63845 were evaluated using XTT Cell Proliferation Assay Kit (ATTC). Average ± S.D. is presented ($n \geq 3$, except where columns are without error bars $n = 1$). Note: * denotes significant difference between treated vs. control cells with $p < 0.05$, ** denotes significant difference with $p < 0.01$, and *** denotes significant difference with $p < 0.005$, as evaluated using 1-way ANOVA with Dunnett post hoc test.

All around, it should be stressed that in NB4, KG1 and KG1A cells co-treatments with S63845 and metformin possibly had an additive effect on cell viability, as combined treatments reduced viable cell numbers more efficiently than S63845 and metformin acting alone. However, further analysis is needed in order to find the combination index (CI) values and confirm or reject this hypothesis.

2.2. Energetic Profile Regulation by Metformin and S63845 in AML Cells

Multiparametric analysis of cellular energy flux was performed using Extracellular Oxygen Consumption Assay together with Glycolysis Assay and Luminescent ATP Detection Assay (Figure 2a). A shift in the metabolic phenotype of NB4 cells (treated with 100 nM S63845, 10 mM metformin and combination of 100 nM S63845 + 10 mM metformin), KG1 cells (treated with 250 nM S63845, 10 mM metformin and combination of 250 nM S63845 + 10 mM metformin), as well as KG1A cells (treated with 1000 nM S63845, 10 mM metformin and combination of 1000 nM S63845 + 10 mM metformin) was evaluated upon 24- and 72-h-long treatments. As visible from the data, in all cell lines, treatment with metformin completely downregulated oxidative phosphorylation, as illustrated by the decrease in extracellular oxygen consumption rates by 91–97% (after 72 h treatment). It should be noticed that in NB4 and KG1A cells glycolysis rates were strongly boosted (up to 400–500% compared to control) after 24 h treatment with metformin. Accordingly, at this time point after treatment with metformin, no decline in ATP levels was registered in NB4 and KG1A cells. Related effects were also observed in NB4 and KG1A cells after 24-h-long combined treatments. However, after 72 h with metformin, as well as with the combined treatment, both NB4 and KG1A cells dropped in the levels of glycolysis rate as well as ATP quantity. In contrast to metformin, S63845, acting as a single agent in all cell lines that were tested, especially in KG1 and KG1A cells, diminished glycolysis, although it had no effect on oxidative phosphorylation rate or even slightly increased it. All in all, results of cellular energy flux analysis revealed that combined treatment with metformin and S63845, in comparison to metformin and S63845 acting alone, had a stronger inhibitory effect on AML cell oxidative phosphorylation and glycolysis rate and, consequently, on cellular ATP levels. The most robust response was registered in KG1A cells after 72 h exposure.

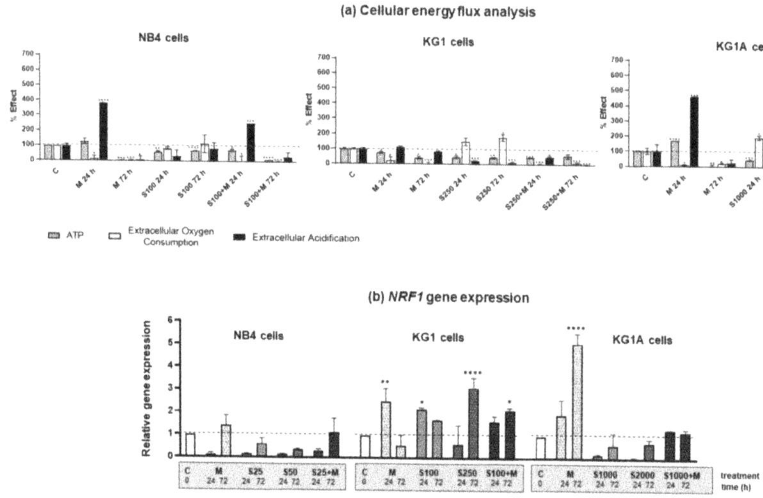

Figure 2. Cellular energy flux after treatment with metformin and MCL-1 inhibitor S63845. (**a**) Cellular energy flux for NB4 cells (treated with 100 nM S63845, 10 mM metformin and combination of 100 nM S63845 + 10 mM metformin), KG1 cells (treated with 250 nM S63845, 10 mM metformin and combination of 250 nM S63845 + 10 mM metformin), as well as KG1A cells (treated with 1000 nM S63845, 10 mM metformin and combination of 1000 nM S63845 + 10 mM metformin) shown as a percentage relative to untreated control cells. Comparative measurements after 24 and 72 h of incubation were taken with Extracellular Oxygen Consumption Assay (ab197243; seeded at 500,000 viable cells per well; white columns), Glycolysis Assay [Extracellular acidification] (ab197244; seeded at 500,000 viable cells per well; black columns) and Luminescent ATP Detection Assay Kit (ab113849; seeded at 75,000 viable cells per well; striped columns). (**b**) Gene expression changes of nuclear respiratory factor 1 (NRF1) after treatments were measured using RT-qPCR ΔΔCt method. HPRT1 gene expression was used for normalization; results are presented as relative changes in comparison to untreated cells; results are mean ± S.D. ($n = 3$). Note: * denotes significant difference between treated vs. control cells with $p < 0.05$, ** denotes significant difference with $p < 0.01$, *** denotes significant difference with $p < 0.005$ and **** denotes significant difference with $p < 0.0001$, as evaluated using 1-way ANOVA with Dunnett post hoc test.

In addition, we investigated the effect that treatments had on the expression of nuclear respirator factor 1 (NRF1) gene, which is a well-known transcription factor regulating oxidative phosphorylation, mitochondrial biogenesis and response to oxidative stress [17]. In general, results of RT-qPCR analysis demonstrated that treatments with MCL-1 inhibitor S63845 diminished NRF1 gene expression in NB4 and KG1A cells, whereas in KG1 cells upregulation of NRF1 gene expression was registered (Figure 2b). Tendency of metformin to induce overexpression of NRF1 gene was also observed, particularly in KG1 and KG1A cells, whereas combined treatments had an intermediate effect.

2.3. *Effect of Metformin and S63845 on AML Cell Redox Modulation*

Mitochondria have a major role in regulating cellular energetics, as well as activating the intrinsic pathway of controlled cell death. Therefore, we further evaluated metformin and S63845 activity on NB4, KG1 and KG1A cell mitochondrial potential and ROS production. Results of flow cytometric TMRE dye accumulation analysis (Figure 3a) revealed that in NB4 cells 24 h treatment with 10 mM metformin, as well as 24 h treatment with a combination of 25 nM S63845 + 10 mM metformin significantly increased Δψ (by approx. 60% compared to untreated control cells). However, opposite results were obtained in KG1 and KG1A cells, when metformin alone or together with S63845 reduced the mitochondrial membrane potential. It also should be stressed that MCL-1 inhibitor S63845, when used as a single agent, had no effect on Δψ. Nevertheless, after 72 h exposure both metformin alone and a combined treatment with S63845 diminished AML cell mitochondrial membrane potential.

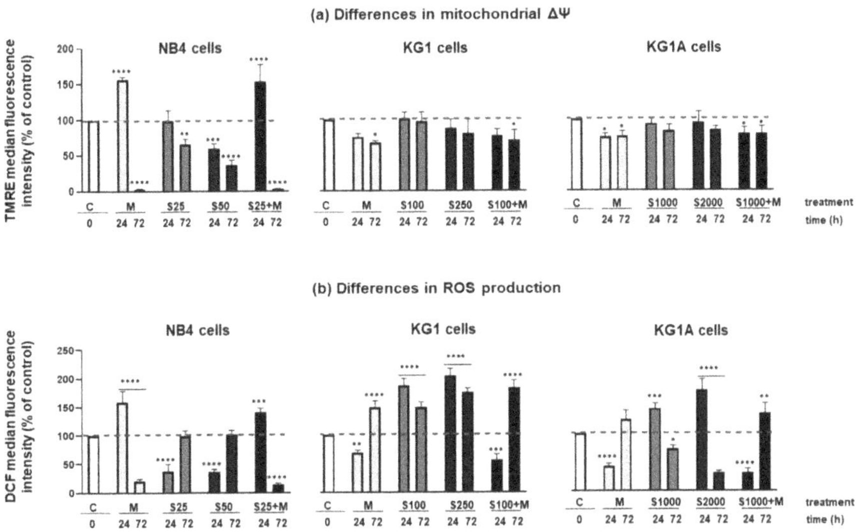

Figure 3. Effect of metformin and MCL-1 inhibitor S63845 on myeloid leukemia cell mitochondrial membrane potential and cellular reactive oxygen species production. For subsequent flow cytometric analysis NB4 cells were treated with 25 nM and 50 nM S63845, 10 mM metformin and combination of 25 nM S63845 + 10 mM metformin. KG1 cells were treated with 100 nM and 250 nM S63845, 10 mM metformin and combination of 100 nM S63845 + 10 mM metformin. KG1A cells were treated with 1000 nM and 2000 nM S63845, 10 mM metformin and combination of 1000 nM S63845 + 10 mM metformin. (**a**) Mitochondrial membrane potential ($\Delta\psi m$) of control (untreated) and treated NB4, KG1 and KG1A cells was evaluated using TMRE Mitochondrial membrane potential assay kit (ab113852) after 24 and 72 h of incubation. (**b**) Measurements of ROS production in control (untreated) and treated NB4, KG1 and KG1A cells were performed using Abcam DCFDA Cellular ROS detection assay kit (ab113851) after 24 and 72 h of incubation. Note: * denotes significant difference between treated vs. control cells with $p < 0.05$, ** denotes significant difference with $p < 0.01$, *** denotes significant difference with $p < 0.005$ and **** denotes significant difference with $p < 0.0001$, as evaluated using 1-way ANOVA with Dunnett post hoc test.

As shown in Figure 3b, metformin increased intracellular ROS production in NB4 cells, whereas in KG1 and KG1A cells the highest increase in ROS production was registered upon treatments with S63845. Therefore, the effects of metformin and MCL-1 inhibitor S63845 on antioxidant enzymes were investigated (Figure 4). We hypothesized that upregulation of intracellular ROS levels would result from the downregulation of gene expression of antioxidant enzymes, such as catalase 1 (CAT1), glutathione peroxidase (GPX) and thioredoxin system enzymes. Results showed that CAT1 gene expression was reduced only in KG1 cells, when treated with 100 nM S63845 (24 and 72 h), whereas treatment with metformin significantly upregulated expression of CAT1 in NB4 and KG1A cells. MCL-1 inhibitor S63845 also reduced GPX gene expression in KG1A cells after 24 h treatment, though, in general, 72-h-long treatments with metformin, S63845 and their combinations, tended to upregulate GPX expression in tested AML cells. However, our further analysis demonstrated that MCL-1 inhibitor S63845 had a significant effect on thioredoxin system regulation: in NB4 cells thioredoxin-1 (TXN) was downregulated 1.6-fold after treatment with 25 nM S63845 (24 and 72 h). In addition, the effect was much stronger when 25 nM S63845 was applied in combination with 10 mM metformin (expression was downregulated 4-fold). Supporting our hypothesis, inhibition of mitochondrial thioredoxin (TXN2) gene expression was also observed in KG1A cells upon treatments with S63845.

However, no reduction in gene expression of thioredoxin reductases 1 and 2 (TXNRD1 and TXNRD2) was detected, while after treatments their expression was evidently increased (see Figure S2 in Appendix No. 1, Supplementary Materials). Nonetheless, it should be noticed that gene expression of tumor suppressor thioredoxin interacting protein (TXNIP), which is a well-known inhibitor of thioredoxin-1 [18], was significantly upregulated after used treatments. For example, in NB4 cells gene expression of TXNIP was increased the most strongly after treatment with S63845, while in KG1A cells the most significant effect was obtained after treatment with metformin.

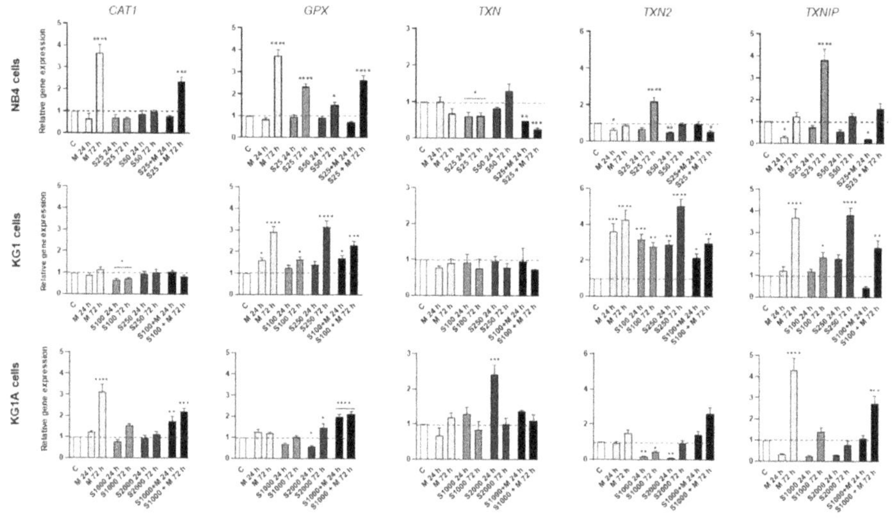

Figure 4. Metformin and MCL-1 inhibitor S63845 induced changes in the expression of antioxidant system genes. NB4 cells were treated with 25 nM and 50 nM S63845, 10 mM metformin and combination of 25 nM S63845 + 10 mM metformin. KG1 cells were treated with 100 nM and 250 nM S63845, 10 mM metformin and combination of 100 nM S63845 + 10 mM metformin. KG1A cells were treated with 1000 nM and 2000 nM S63845, 10 mM metformin and combination of 1000 nM S63845 + 10 mM metformin. Gene expression changes of antioxidant system genes *CAT1*, *GPX*, *TXN*, *TXN2* and *TXNIP* after 24 and 72 h of treatment were measured using RT-qPCR ΔΔCt method. *HPRT1* gene expression was used for normalization; results are presented as relative changes in comparison to untreated cells; results are mean ± S.D. (n = 3). Note: * denotes significant difference between treated vs. control cells with $p < 0.05$, ** denotes significant difference with $p < 0.01$, *** denotes significant difference with $p < 0.005$ and **** denotes significant difference with $p < 0.0001$, as evaluated using 1-way ANOVA with Dunnett post hoc test.

2.4. Treatment-Induced Apoptosis and Apoptosis-Related Gene and Protein Expression

The type of cell death induced in AML cells by treatment with metformin, MCL-1 inhibitor S63845 or their combination was determined by staining with Annexin V and Propidium Iodide (Figure 5). Analysis revealed that NB4, KG1 and KG1A cells treated with metformin at 10 mM concentration and with S63845 at selected concentrations (25 and 50 nM in NB4 cells, 100 and 250 nM in KG1 cells, as well as 1000 and 2000 nM in KG1A cells) induced cell apoptosis. In accordance with cell viability estimations (Figure 1), Annexin V and PI analysis confirmed that addition of metformin weakens the effect of S63845, especially in KG1 and KG1A cells. However, this phenomenon was not statistically significant (see the exact statistical data in Appendix No. 2, Supplementary Material). Interestingly, in KG1 and KG1A cells metformin upregulated gene expression of cyclin-dependent kinase inhibitor p21 (see Figure S3 in Appendix No. 1, Supplementary Materials), whereas, after treatment with combinations of metformin and S63845, such an effect on CDKN1A expression was weaker.

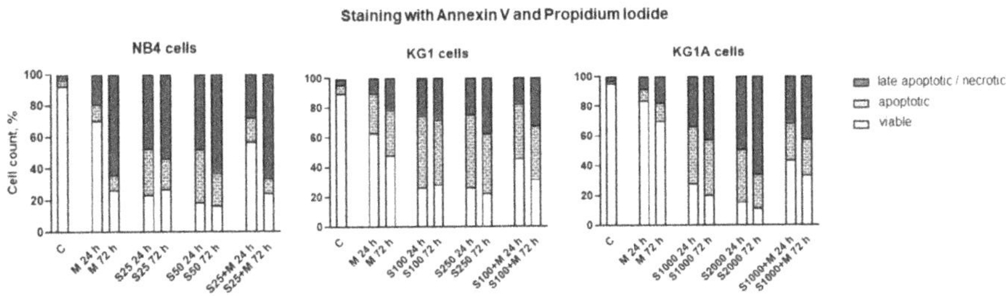

Figure 5. Metformin and MCL-1 inhibitor S63845 effect on cell apoptosis. Cell apoptosis of NB4 cells (treated with 25 nM and 50 nM S63845, 10 mM metformin and combination of 25 nM S63845 + 10 mM metformin), KG1 cells (treated with 100 nM and 250 nM S63845, 10 mM metformin and combination of 100 nM S63845 + 10 mM metformin), as well as KG1A cells (treated with 1000 nM and 2000 nM S63845, 10 mM metformin and combination of 1000 nM S63845 + 10 mM metformin) was evaluated by staining with Annexin V and Propidium Iodide after 24 and 72 h of treatment. Results are mean (n = 3; S.D. < ± 10%). Note: statistical analysis was performed using 2-way ANOVA with Tukey's multiple comparison test (due to the complexity of data, exact statistical analysis results are presented in Appendix No. 2, (Supplementary Materials).

To test whether metformin- and S63845-induced cell death was mediated by the regulation of oncogenic- and apoptosis-related factors that are linked to mitochondria, we further performed gene expression and Western blot analysis and measured the levels of BCL-2 (BCL-2), BCL-2-like-1 (BCL2L1), MCL-1 (MCL1) and MYC (MYC) (Figure 6a,b). Protein expression of MYC was downregulated time-dependently, especially strongly after incubation of cells with S63845 alone or in combination with metformin (in all cell lines that were tested, combined 24-h treatments with S63845 and metformin elicited stronger inhibitory effects on MYC protein levels compared to agents acting alone). However, we did not detect any significant decrease in MYC gene expression. Similarly, neither did we detect any significant downregulation of BCL2 and MCL1 gene expression. Nevertheless, BCL2L1 gene expression was significantly decreased in NB4 and KG1A cells after incubation with MCL-1 inhibitor S63845. Regarding protein level modulation, metformin had no effect or even lightly increased the levels of BCL-2, whereas treatment with S63845 had a reducing effect on NB4 and KG1 cells (however, in KG1A cells treatment with S63845 showed opposite results). It should be noted that in both NB4 and KG1 cells, combined treatments with S63845 and metformin had a stronger effect on BCL-2 protein level reduction than treatments with separate agents. Metformin also proved to be potent in MCL-1 level reduction, while treatment with S63845 evidently upregulated the levels of MCL-1. This indicated that S63845-induced apoptosis is not straightforwardly linked with the decrease in MCL-1 level, but rather with MCL-1 activity inhibition, as, in our study, increase in the levels of MCL-1 was concomitant with apoptotic cell death (Figure 5).

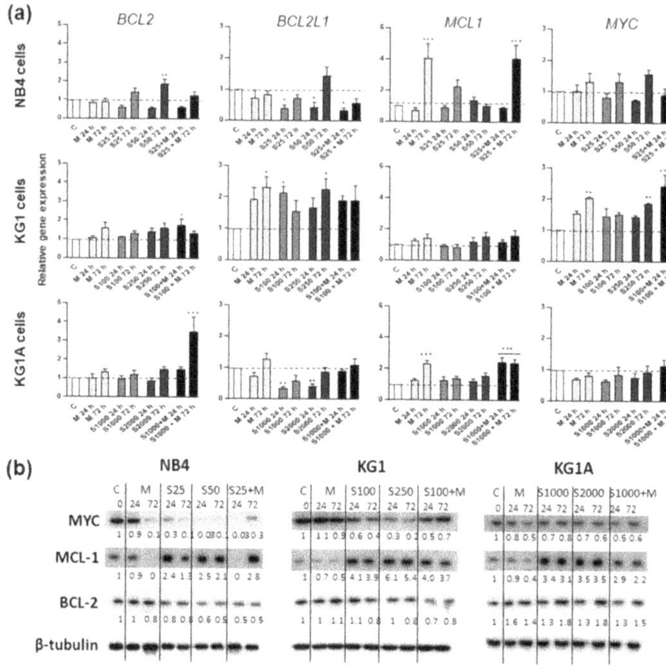

Figure 6. Anti-apoptotic gene and protein expression after treatment with metformin and MCL-1 inhibitor S63845. For subsequent analysis NB4 cells were treated with 25 nM and 50 nM S63845, 10 mM metformin and combination of 25 nM S63845 + 10 mM metformin. KG1 cells were treated with 100 nM and 250 nM S63845, 10 mM metformin and combination of 100 nM S63845 + 10 mM metformin. KG1A cells were treated with 1000 nM and 2000 nM S63845, 10 mM metformin and combination of 1000 nM S63845 + 10 mM metformin. C denotes control, untreated, cells. (**a**) Gene expression changes of apoptosis-suppressing genes *BCL2*, *BCL2L1*, *MCL1* and *MYC* after 24 and 72 h of treatment were measured using RT-qPCR ΔΔCt method. *HPRT1* gene expression was used for normalization; results are presented as relative changes in comparison to untreated cells; results are mean ± S.D. ($n = 3$). (**b**) Protein level changes after cell treatment with S63845 and metformin were assessed by immunoblot. β-tubulin was used as a loading control. Protein band intensity was measured using ImageJ software; band intensity results were normalized according to β-tubulin from the same membrane ($n \geq 2$, representative results are displayed). Note: * denotes significant difference between treated vs. control cells with $p < 0.05$, ** denotes significant difference with $p < 0.01$, *** denotes significant difference with $p < 0.005$, as evaluated using 1-way ANOVA with Dunnett post hoc test.

3. Discussion

Mitochondria-associated processes such as oxidative phosphorylation, production of reactive oxygen species and regulation of the intrinsic death pathway have a major role in the progression and survival of a plethora of cancer types, not excluding AML. Therefore, in this study we investigated the effect of the antidiabetic drug metformin, which is also an inhibitor of oxidative phosphorylation [19], alone and in combination with S63845, the inhibitor of anti-apoptotic protein MCL-1. We tested these agents on APL subtype cell line NB4, AML cell line KG1 and its chemoresistant counterpart KG1A cells.

We showed that metformin's effect on cell proliferation was of comparable degree in all cell lines that were tested. However, differences were detected in cell viability, as 10 mM metformin decreased viable cell percentage the most strongly in NB4 cells, while the weakest impact was observed in chemoresistant KG1A cells (Figures 1a and 5). In addition, downregulation of CDKN1A gene expression in NB4 cells was also registered after treatment with metformin, while in KG1 and KG1A cells metformin increased CDKN1A gene expression after prolonged exposure (Figure S3 in Appendix No. 1, Supplementary Materials). These observations coincide with cell viability and apoptosis evaluation data, as it is widely accepted that p21 can play the inhibitory role in apoptosis [20]. At this point it is very important to stress that co-treatment with metformin restrained S63845 effect on apoptosis induction, while, in general, S63845-induced level of apoptotic cell death in KG1 and KG1A cells was higher compared to the effect of metformin. However, results of cell viability analysis (Figure 1a) revealed that combined treatments with metformin and S63845 had a stronger effect, which was the most evident in KG1 and KG1A cells. Based on results obtained in our study, we hypothesize that apoptotic cell death of AML cells after treatments with either metformin, S63845 or their combination could be elicited, at least partly, via ROS level modulation. Previously it was thought that overproduction of ROS can only promote carcinogenesis and resistance to therapy, whereas currently it is assumed to have a split role—induction of ROS over the certain threshold can also lead to cancer cell death [21]. Our research demonstrated that in NB4 cells 10 mM metformin upregulated cellular ROS production by 1.6-fold, whereas in KG1A cells ROS accumulation was more pronounced upon treatments with S63845 (increased approx. 1.8-fold) (Figure 3b). Li et al. [7] showed that treatment with metformin can induce heightened ROS production in osteosarcoma cells. However, at the same time they did observe the reduction in mitochondrial membrane potential, while in our study, upregulation of ROS in NB4 cells after 24 h treatment was accompanied by the increase in $\Delta\psi$. This seems credible, as, in general, increased mitochondrial membrane potential is associated with higher ROS production rates [22]. Of course, the mitochondrial pathway is not the only pathway that can generate ROS in myeloid leukemia cells, as NADPH oxidase is a critical enzyme in these cells, and its activation can lead to ROS accumulation and consequently it can potentiate apoptotic cell death [23]. It is plausible that MCL-1 inhibitor S63845 upregulated ROS generation in KG1 and KG1A cells mainly by acting via NADPH oxidase activity rather than through mitochondrial electron transport chain, as there were no significant changes in mitochondrial membrane potential detected (Figure 3a). However, we did notice a mild increase in the oxidative phosphorylation rate in KG1 and KG1A cells after treatments with S63845 (Figure 2a), which could be regarded as a compensatory mechanism due to reduction in glycolysis. The exact mechanism explaining the effect of S63845 on KG1 and KG1A cell glycolysis inhibition remains elusive and further studies are needed. Nevertheless, research by other authors has revealed that targeting MCL-1 indeed affects carbohydrate metabolism [24]. In addition, it should be emphasized that results of our study demonstrated that combined treatments, in comparison to metformin and S63845 acting alone, had more profound effects on inhibition of AML cell oxidative phosphorylation and glycolysis, as well as on reduction of ATP levels (Figure 2a).

Furthermore, in our study, treatment with MCL-1 inhibitor S63845 profoundly increased the levels of MCL-1 protein in all tested AML cells (Figure 6b). In addition, the increase of MCL-1 protein in cancer cells after treatment with S63845 has been demonstrated by other authors [15]. Their research revealed that binding of S63845 to MCL-1 disrupts MCL-1 interaction with pro-apoptotic proteins BAX/BAK, and thereafter indeed induces apoptosis. However, the increase in MCL-1 protein levels was associated with protein half-life extension [15]. In our study, increase in MCL-1 levels by treatment with S63845 was accompanied by the reduction in oncogenic protein MYC (Figure 6b). The strongest effect on MYC downregulation was observed in NB4 and KG1A cells, though in all cell lines that were tested, co-treatment with S63845 elevated the activity of metformin on MYC downregulation, as observed after 24 h combined treatments. Furthermore, co-treatments of NB4 and KG1 cells with S63845 and metformin had a stronger effect on BCL-2 protein level reduction than treatments with separate agents. These findings illustrated that the combination of MCL-1 inhibitor S63845 and metformin indeed has a repressive effect on NB4 and KG1 cells pro-survival proteins.

It is worth mentioning that in NB4 cells treatment with MCL-1 inhibitor S63845 statistically significantly lowered mitochondrial membrane potential and reduced production of cellular ROS (Figure 3a,b). Such a phenomenon could also be regarded as an apoptosis-promoting factor, as prolonged low values of $\Delta\psi$ are also threatening due to the cells' inability to make enough ATP, as well as a reduced amount of ROS that could lead to the other kind of stress, the so-called "reductive stress" [25]. Therefore, drug-induced modulation of mitochondrial $\Delta\psi$ and ROS production could possibly play a major role in the anticancer activity of metformin and S63845 in AML cells. Obviously, we should admit that the concentration of metformin that was used in these experiments (10 mM) is too high for the clinical setting. Therefore, future studies will be necessary in order to test the potential of metformin and its combination with S63845 in AML treatment in vivo.

4. Materials and Methods

4.1. Cell Cultivation and Treatment

NB4 cell line (DSMZ, Braunschweig, Germany) was cultured in RPMI 1640 medium supplemented with 10% fetal bovine serum, 100 U/mL penicillin and 100 µg/mL streptomycin (Gibco, Carlsbad, CA, USA); KG1 and KG1A cell lines (ATCC, Manassas, VA, USA) were cultured in IMDM medium supplemented with 20% fetal bovine serum, 100 U/mL penicillin and 100 µg/mL streptomycin (Gibco, Carlsbad, CA, USA) at 37 °C in a humidified 5% CO_2 atmosphere. For the treatment with metformin (Cayman Chemical Company, Ann Arbor, MI, USA) and S63845 (Cayman Chemical Company, Ann Arbor, MI, USA) cell seeding density was 0.5×10^6 cells/mL. Cell proliferation and survival were assessed by trypan blue exclusion test using a hemocytometer. In short, 1 part of 0.2% trypan blue and 1 part of the cell suspension were mixed (100 µL of each). A drop of the mixture was applied to the hemocytometer (Neubauer-improved, 0.1 mm depth of chamber; Paul Marienfeld GmbH & Co. KG, Lauda-Königshofen, Germany) and manual counting using a binocular microscope was performed within 1 min. Unstained (viable) and stained (nonviable) cells were counted separately. Two technical replicates were performed. To obtain the total number of viable cells per ml of aliquot, the average count of viable cells from four hemocytometer fields (a square size 1 mm^2) were multiplied by 2 (dilution factor) and once again multiplied by 10^4. The percentage of viable cells was counted using the formula:

$$\text{viable cells } (\%) = \frac{\text{total number of viable cells per mL of aliquot}}{\text{total number of cells per mL of aliquot}} \times 100 \quad (1)$$

Antiproliferative IC_{50} values were evaluated using XTT Cell Proliferation Assay Kit (ATTC, Manassas, VA, USA).

4.2. Apoptosis Evaluation

For the detection of apoptosis, ApoFlowEx® FITC Kit (Exbio, Vestec, Czech Republic) was used. This assay detects viable, early apoptotic, and late apoptotic or necrotic cells depending on how cells are stained by Annexin V-FITC and Propidium Iodide. Stained cells were analyzed on the Millipore Guava® easyCyte 8HT flow cytometer with InCyte 2.2.2 software.

4.3. Cellular Energy Flux Measurement

Cell extracellular oxygen consumption and extracellular acidification were measured using Extracellular Oxygen Consumption Assay Kit (Abcam, Cambridge, UK) and Glycolysis Assay Kit (Abcam, Cambridge, UK) according to the manufacturer's instructions. Cells were treated with MCL1 inhibitor S63845 and metformin for 24 h and 72 h before measurements. Cells were harvested, washed, and seeded at 5×10^5 viable cells/well in a 96-well plate, detection reagents were added and measurements performed in a plate reader. The aforementioned assays were validated using glucose oxidase as a positive control (Sigma-Aldrich, St. Louis, MO, USA). Cellular ATP concentration was evaluated using Luminescent ATP Detection Assay Kit (Abcam, Cambridge, UK) according to the manufacturer's instructions. ATP standards provided in the kit were used in order to test the validity of the assay. Cells were seeded at 7.5×10^4 viable cells/well in a white 96-well plate, detergent solution was added and incubated for 5 min to lyse cells and stabilize ATP, then substrate solution was added and the plate was stored in the dark until analysis of luminescence. For cell metabolic activity measurements the Thermo Scientific Varioskan® plate reader (Thermo Fisher Scientific, Waltham, MA, USA) was used.

4.4. Assessment of Mitochondrial Membrane Potential

For TMRE Mitochondrial Membrane Potential Assay (Abcam, Cambridge, UK), 1×10^5 of NB4, KG1 and KG1A cells were resuspended in PBS with 0.2% BSA and incubated with 400 nM TMRE for 30 min at 37 °C. Samples were analyzed with Millipore Guava® easyCyte 8HT flow cytometer, using the InCyte 2.2.2 software. Ten thousand events were collected for each sample.

4.5. Cellular ROS Measurement

Levels of cellular reactive oxygen species (ROS) of NB4, KG1 and KG1A cells were determined using DCFDA Cellular ROS Detection Assay Kit (Abcam, Cambridge, UK). All procedures were carried out according to the manufacturer's instructions. In brief, 2.5×10^4 cells/sample were incubated with 25 μM of 2′,7′–dichlorofluorescin diacetate (DCFDA) for 30 min at 37 °C and then analyzed using Millipore Guava® easyCyte 8HT flow cytometer with InCyte 2.2.2 software. Ten thousand events were collected for each sample.

4.6. Gene Expression Analysis by RT-qPCR

Total RNA was purified using TRI Reagent (Zymo, Irvine, CA, USA). Traces of DNA in RNA preparations were removed using DNAse I, Amplification Grade (ThermoFisher Scientific, Waltham, MA, USA). cDNA was synthesized using SensiFAST™ cDNA Synthesis Kit (Bioline, Memphis, TN, USA) and qPCR was performed using SensiFAST™ SYBR® No-ROX Kit (Bioline) on the RotorGene 6000 system (Corbett Life Science, QIAGEN, Hilden, Germany). Primer sequences (Metabion international AG, Planegg/Steinkirchen, Germany) are presented in Table S1 (see Table S1 in Appendix No. 1, Supplementary Materials). mRNA levels were normalized to HPRT1 expression. Relative gene expression was calculated using $\Delta\Delta$Ct method. Data are expressed as mean ±standard deviation (S.D.)

4.7. Immunoblotting

Cell lysates were prepared as described previously [26] and fractionated in 7.5–15% SDS-PAGE gradient electrophoresis gel. Proteins were transferred on PVDF membrane and specific proteins were detected using antibodies against BCL-2 (Cell Signalling Technology, Danvers, MA, USA), MCL-1 (Proteintech, Rosemont, IL, USA), MYC (Novus Biologicals, Centennial, CO, USA) and β-tubulin (Abcam, Cambridge, UK). β-tubulin was used as a loading control. Chemiluminescent detection was carried out using WesternBright ECL (Advansta, San Jose, CA, USA) on ChemiDoc™ XRS+ System (BIORAD, Hercules, CA, USA). Quantitative evaluation was performed using ImageJ software (1.48v).

4.8. Statistical Analysis

Unless otherwise specified, all experiments were repeated at least three times. Data were expressed as mean values with S.D. One-way ANOVA with Dunnett post hoc test and two-way ANOVA with Tukey's multiple comparison test in GraphPad Prism software (8.0.1) were used for statistical analysis. Significance was set at $p \leq 0.05$ (*), $p \leq 0.005$ (**), $p \leq 0.0005$ (***) and $p \leq 0.0001$ (****).

5. Conclusions

Overall, our study demonstrated that combined treatment with metformin and S63845, in comparison to metformin and S63845 acting alone, had a stronger inhibitory effect on AML cell oxidative phosphorylation and glycolysis rate and consequently on cellular ATP levels. In addition, treatment-induced apoptotic cell death was concomitant with changes in levels of cellular ROS. Therefore, such modulation of cellular energetics and redox status might be beneficial in targeting chemoresistant AML. However, future studies are needed in order to verify these results in more clinically appropriate settings.

Supplementary Materials: The following are available online. Table S1: Primers used for RT-qPCR analysis. Figure S1: Cell proliferation and viability assessment; Figure S2: Metformin and MCL-1 inhibitor S63845 induced changes in the expression of antioxidant system genes; Figure S3: Metformin and MCL-1 inhibitor S63845 effect on CDKN1A gene expression.

Author Contributions: G.V., Conceptualization, investigation, formal analysis, data curation, visualization, writing—original draft. A.V., conceptualization, investigation, data curation, visualization, writing—review and editing. G.S., investigation, formal analysis, data curation, visualization. V.B., contribution to the interpretation of data, approval of the final version of the manuscript. R.N., design of research study, conceptualization, experiments conception, funding acquisition, writing—review and editing. All authors have read and agreed to the published version of the manuscript.

Funding: This work was supported by the Research Council of Lithuania (grant number S-SEN-20-2).

Institutional Review Board Statement: Not applicable.

Informed Consent Statement: Not applicable.

Data Availability Statement: The data presented in this study are available in Supplementary Material.

Conflicts of Interest: The authors declare that there is no conflict of interest regarding the publication of this article.

Sample Availability: The data sets (data in native formats) used to support the findings of this study are available from the corresponding author upon request.

References

1. Howlader, N.; Noone, A.M.; Krapcho, M.; Miller, D.; Brest, A.; Yu, M.; Ruhl, J.; Tatalovich, Z.; Mariotto, A.; Lewis, D.R.; et al. (Eds.) *SEER Cancer Statistics Review, 1975–2017*; Based on November 2019 SEER Data Submission, Posted to the SEER Web Site; National Cancer Institute: Bethesda, MD, USA, 2020. Available online: https://seer.cancer.gov/csr/1975_2017/ (accessed on 18 January 2021).
2. Hackl, H.; Astanina, K.; Wieser, R. Molecular and genetic alterations associated with therapy resistance and relapse of acute myeloid leukemia. *J. Hematol. Oncol.* **2017**, *10*, 1–16. [CrossRef] [PubMed]
3. Baker, C.; Retzik-Stahr, C.; Singh, V.; Plomondon, R.; Anderson, V.; Rasouli, N. Should metformin remain the first-line therapy for treatment of type 2 diabetes? *Ther. Adv. Endocrinol. Metab.* **2021**, *12*, 2042018820980225. [CrossRef] [PubMed]
4. Fruzzetti, F.; Perini, D.; Russo, M.; Bucci, F.; Gadducci, A. Comparison of two insulin sensitizers, metformin and myo-inositol, in women with polycystic ovary syndrome (PCOS). *Gynecol. Endocrinol.* **2017**, *33*, 39–42. [CrossRef] [PubMed]
5. Schein, C.H. Repurposing approved drugs for cancer therapy. *Br. Med Bull.* **2021**, *137*, 13–27. [CrossRef] [PubMed]
6. Gao, Z.-Y.; Liu, Z.; Bi, M.-H.; Zhang, J.-J.; Han, Z.-Q.; Han, X.; Wang, H.-Y.; Sun, G.-P.; Liu, H. Metformin induces apoptosis via a mitochondria-mediated pathway in human breast cancer cells in vitro. *Exp. Ther. Med.* **2016**, *11*, 1700–1706. [CrossRef]
7. Li, B.; Zhou, P.; Xu, K.; Chen, T.; Jiao, J.; Wei, H.; Yang, X.; Xu, W.; Wan, W.; Xiao, J. Metformin induces cell cycle arrest, apoptosis and autophagy through ROS/JNK signaling pathway in human osteosarcoma. *Int. J. Biol. Sci.* **2020**, *16*, 74–84. [CrossRef]
8. Yuan, F.; Cheng, C.; Xiao, F.; Liu, H.; Cao, S.; Zhou, G. Inhibition of mTORC1/P70S6K pathway by Metformin synergistically sensitizes Acute Myeloid Leukemia to Ara-C. *Life Sci.* **2020**, *243*, 117276. [CrossRef]
9. Vitkevičienė, A.; Janulis, V.; Žučenka, A.; Borutinskaitė, V.; Kaupinis, A.; Valius, M.; Griškevičius, L.; Navakauskienė, R. Oxidative phosphorylation inhibition induces anticancerous changes in therapy-resistant–acute myeloid leukemia patient cells. *Mol. Carcinog.* **2019**, *58*, 2008–2016. [CrossRef]
10. Fontaine, E. Metformin-Induced Mitochondrial Complex I Inhibition: Facts, Uncertainties, and Consequences. *Front. Endocrinol.* **2018**, *9*, 753. [CrossRef]
11. Carter, J.L.; Hege, K.; Kalpage, H.A.; Edwards, H.; Hüttemann, M.; Taub, J.W.; Ge, Y. Targeting mitochondrial respiration for the treatment of acute myeloid leukemia. *Biochem. Pharmacol.* **2020**, *182*, 114253. [CrossRef]
12. Maji, S.; Panda, S.; Samal, S.K.; Shriwas, O.; Rath, R.; Pellecchia, M.; Emdad, L.; Das, S.K.; Fisher, P.B.; Dash, R. Bcl-2 Antiapoptotic Family Proteins and Chemoresistance in Cancer. *Adv. Cancer Res.* **2018**, *137*, 37–75. [CrossRef]
13. Ewald, L.; Dittmann, J.; Vogler, M.; Fulda, S. Side-by-side comparison of BH3-mimetics identifies MCL-1 as a key therapeutic target in AML. *Cell Death Dis.* **2019**, *10*, 917. [CrossRef]

14. Glaser, S.P.; Lee, E.F.; Trounson, E.; Bouillet, P.; Wei, A.; Fairlie, W.D.; Izon, D.J.; Zuber, J.; Rappaport, A.R.; Herold, M.J.; et al. Anti-apoptotic Mcl-1 is essential for the development and sustained growth of acute myeloid leukemia. *Genes Dev.* **2012**, *26*, 120–125. [CrossRef]
15. Kotschy, A.; Szlavik, Z.; Murray, J.; Davidson, J.; Maragno, A.L.; Le Toumelin-Braizat, G.; Chanrion, M.; Kelly, G.L.; Gong, J.-N.; Moujalled, D.M.; et al. The MCL1 inhibitor S63845 is tolerable and effective in diverse cancer models. *Nature* **2016**, *538*, 477–482. [CrossRef]
16. Hormi, M.; Birsen, R.; Belhadj, M.; Huynh, T.; Aguilar, L.C.; Grignano, E.; Haddaoui, L.; Guillonneau, F.; Mayeux, P.; Hunault, M.; et al. Pairing MCL-1 inhibition with venetoclax improves therapeutic efficiency of BH3-mimetics in AML. *Eur. J. Haematol.* **2020**, *105*, 588–596. [CrossRef]
17. Yuan, J.; Zhang, S.; Zhang, Y. Nrf1 is paved as a new strategic avenue to prevent and treat cancer, neurodegenerative and other diseases. *Toxicol. Appl. Pharmacol.* **2018**, *360*, 273–283. [CrossRef]
18. Jia, J.-J.; Geng, W.-S.; Wang, Z.-Q.; Chen, L.; Zeng, X.-S. The role of thioredoxin system in cancer: Strategy for cancer therapy. *Cancer Chemother. Pharmacol.* **2019**, *84*, 453–470. [CrossRef]
19. Pecinová, A.; Brázdová, A.; Drahota, Z.; Houštěk, J.; Mráček, T. Mitochondrial targets of metformin—Are they physiologically relevant? *BioFactors* **2019**, *45*, 703–711. [CrossRef]
20. Karimian, A.; Ahmadi, Y.; Yousefi, B. Multiple functions of p21 in cell cycle, apoptosis and transcriptional regulation after DNA damage. *DNA Repair* **2016**, *42*, 63–71. [CrossRef]
21. Snezhkina, A.V.; Kudryavtseva, A.V.; Kardymon, O.L.; Savvateeva, M.V.; Melnikova, N.V.; Krasnov, G.S.; Dmitriev, A.A. ROS Generation and Antioxidant Defense Systems in Normal and Malignant Cells. *Oxidative Med. Cell. Longev.* **2019**, *2019*, 1–17. [CrossRef]
22. Nicholls, D.G. Mitochondrial membrane potential and aging. *Aging Cell* **2003**, *3*, 35–40. [CrossRef]
23. Khan, N.H.; Sexton, K.J.; Grimm, M.J.; Segal, B.H.; E Vigil, C. NADPH Oxidase Regulates Cytarabine-Induced Apoptotic Death in Acute Myeloid Leukemia Cells. *Blood* **2011**, *118*, 4258. [CrossRef]
24. Carter, B.Z.; Mak, P.Y.; Tao, W.; Warmoes, M.; Lorenzi, P.L.; Mak, D.; Ruvolo, V.; Tan, L.; Cidado, J.; Drew, L.; et al. Targeting MCL-1 dysregulates cell metabolism and leukemia-stroma interactions and resensitizes acute myeloid leukemia to BCL-2 inhibition. *Haematologica* **2020**. [CrossRef]
25. Zorova, L.D.; Popkov, V.A.; Plotnikov, E.Y.; Silachev, D.N.; Pevzner, I.B.; Jankauskas, S.S.; Babenko, V.A.; Zorov, S.D.; Balakireva, A.V.; Juhaszova, M.; et al. Mitochondrial membrane potential. *Anal. Biochem.* **2018**, *552*, 50–59. [CrossRef]
26. Savickiene, J.; Treigyte, G.; Valiuliene, G.; Stirblyte, I.; Navakauskiene, R. Epigenetic and molecular mechanisms underlying the antileukemic activity of the histone deacetylase inhibitor belinostat in human acute promyelocytic leukemia cells. *Anti-Cancer Drugs* **2014**, *25*, 938–949. [CrossRef]

molecules

MDPI

Article

Influence of Oxidative Stress on Time-Resolved Oxygen Detection by $[Ru(Phen)_3]^{2+}$ In Vivo and In Vitro

Veronika Huntosova [1,*], Denis Horvath [1], Robert Seliga [1] and Georges Wagnieres [2]

1 Center for Interdisciplinary Biosciences, Technology and Innovation Park, P.J. Safarik University in Kosice, Jesenna 5, 041 54 Kosice, Slovakia; denis.horvath@upjs.sk (D.H.); robert.seliga@upjs.sk (R.S.)
2 Laboratory for Functional and Metabolic Imaging, Institute of Physics, Swiss Federal Institute of Technology in Lausanne (EPFL), Station 6, Batiment de Chimie, 1015 Lausanne, Switzerland; georges.wagnieres@epfl.ch
* Correspondence: veronika.huntosova@upjs.sk; Tel.: +421-55-243-2243

Abstract: Detection of tissue and cell oxygenation is of high importance in fundamental biological and in many medical applications, particularly for monitoring dysfunction in the early stages of cancer. Measurements of the luminescence lifetimes of molecular probes offer a very promising and non-invasive approach to estimate tissue and cell oxygenation in vivo and in vitro. We optimized the evaluation of oxygen detection in vivo by $[Ru(Phen)_3]^{2+}$ in the chicken embryo chorioallantoic membrane model. Its luminescence lifetimes measured in the CAM were analyzed through hierarchical clustering. The detection of the tissue oxygenation at the oxidative stress conditions is still challenging. We applied simultaneous time-resolved recording of the mitochondrial probe MitoTracker™ OrangeCMTMRos fluorescence and $[Ru(Phen)_3]^{2+}$ phosphorescence imaging in the intact cell without affecting the sensitivities of these molecular probes. $[Ru(Phen)_3]^{2+}$ was demonstrated to be suitable for in vitro detection of oxygen under various stress factors that mimic oxidative stress: other molecular sensors, H_2O_2, and curcumin-mediated photodynamic therapy in glioma cancer cells. Low phototoxicities of the molecular probes were finally observed. Our study offers a high potential for the application and generalization of tissue oxygenation as an innovative approach based on the similarities between interdependent biological influences. It is particularly suitable for therapeutic approaches targeting metabolic alterations as well as oxygen, glucose, or lipid deprivation.

Keywords: oxygen detection; dendrogram analysis; hierarchical clustering; cancer cells; luminescence lifetime; hydrogen peroxide; oxidative stress; photodynamic therapy; time-resolved imaging

Citation: Huntosova, V.; Horvath, D.; Seliga, R.; Wagnieres, G. Influence of Oxidative Stress on Time-Resolved Oxygen Detection by $[Ru(Phen)_3]^{2+}$ In Vivo and In Vitro. *Molecules* **2021**, *26*, 485. https://doi.org/10.3390/molecules26020485

Academic Editors: Višnja Stepanić and Marta Kučerová-Chlupáčová

Received: 9 November 2020
Accepted: 14 January 2021
Published: 18 January 2021

Publisher's Note: MDPI stays neutral with regard to jurisdictional claims in published maps and institutional affiliations.

1. Introduction

Tissue oxygenation is closely connected with the cell's metabolic activity. In particular, oxygen supply reflects the pathophysiology of cancer cells. Several approaches to assess the tissue oxygenation were developed. One minimally invasive method to assess the level of oxygen is the time-resolved detection of oxygen-sensitive probes [1–3].

Dichlorotris(1,10-phenanthroline)-ruthenium(II) hydrate ($[Ru(Phen)_3]^{2+}$) is an interesting molecule emitting a bright luminescence. This relatively small molecule (700 Da) is sensitive to oxygen in solutions, in cells, and in vivo [4–6]. The luminescence lifetime of $[Ru(Phen)_3]^{2+}$ is quenched by oxygen, a mechanism that is governed by the Stern-Volmer relation [3]. This hydrophilic molecule is biocompatible and has low phototoxicity when it is localized in the extracellular space [3,7,8]. It takes a long time (24 h) for $[Ru(Phen)_3]^{2+}$ to accumulate in cells [7]. However, straight nuclear localization of $[Ru(Phen)_3]^{2+}$ was observed in damaged cells due to the increased permeability of their membrane [7,9]. Interaction with DNA increases the $[Ru(Phen)_3]^{2+}$ luminescence intensity as well as its luminescence lifetimes [10,11]. Therefore, $[Ru(Phen)_3]^{2+}$ is considered as an interesting oxygen-sensitive molecule, mainly to probe the oxygen level in extracellular compartments and solutions.

The concentration of oxygen inside and outside blood vessels can vary from place to place, and it can cause inaccuracies in determining the oxygen content. Recent trends suggest that a consistent interpretation of the measurements can be achieved not only by statistical evaluation but particularly by data classification. In general, it can be considered as a form of unsupervised learning [12,13] that is a form of learning leading to self-organized data clustering without previous labeling of the collected data by human experts. It is also important for the approach we present in this article to have a visualization that helps to understand the data. A specific application we present in this work is the use of clustering to facilitate the interpretation of $[Ru(Phen)_3]^{2+}$ luminescence lifetime data obtained in a pre-clinical in vivo model, the chicken embryo chorioallantoic membrane (CAM). We used hierarchical cluster analysis, which, in our specific conditions, involves the processing of data that reflect fluctuating oxygen levels, including fluctuations of the $[Ru(Phen)_3]^{2+}$ luminescence lifetimes, similarity aspects $[Ru(Phen)_3]^{2+}$ luminescence lifetime statistics, as well as spatial heterogeneities of the $[Ru(Phen)_3]^{2+}$ luminescence lifetimes induced by vascular heterogeneity.

Imbalance in oxygen supply and consumption may induce oxidative stress resulting from the production of reactive oxygen species (ROS) in cells. ROS are created as inherent products of oxygen metabolism in cells. Generally, the level of oxidative stress in cells is related to the amount of protein, lipids and DNA damages, and antioxidant status [14]. Recently, we have employed time-resolved imaging based on MitoTracker^{TM} Orange CMTMRos (MTO) to evaluate oxidative stress-induced in mitochondria of cancer cells [15]. This approach allows us to detect low-level oxidative stress-induced in cells. MTO is a mitochondrial probe which, besides oxidative stress visualization, enables to assess the integrity of mitochondria and to measure the mitochondrial membrane potential [16].

It is time and cost-saving, and very informative to perform multimodal visualization and simultaneous detection of several parameters, such as the level of oxygen and oxidative stress, in the studied system. Several approaches based on the intensity and time-resolved microscopy were applied for dual molecular sensors detection of oxidative stress, mitochondria integrity, thiophenols, and oxygen consumption by genetically encoded photosensitizers (PSs) [17–23]. Recently, metal complexes were developed to detect oxygen consumption, enabling mitochondria targeting and tracking morphological changes of mitochondria [24–26].

It is of high importance to eliminate the photodamages generated during oxygen and oxidative stress detection. Virtually, all fluorescence/phosphorescence molecular probes are more or less potent PSs. Their fluorescence and phosphorescence are often excited with laser light at wavelengths corresponding to maxima of the probe's absorption. Excited probes/molecules stay in their triplet states during times ranging typically between 100 ns and 100 μs. These relatively long times are sufficient to enable an interaction of the probes with other molecules and biomolecules like proteins and lipids. Moreover, in the presence of molecular oxygen, such photoreaction can lead to oxidation and/or peroxidation of proteins and lipids. The initiation of these photoreactions strongly depends on the probes/molecule concentration and of the fluence rate generated by the light source. The photoreaction in cells may produce oxidative stress, ROS, and global photodamages in cells, which can lead to a decrease of their proliferation. The mechanisms described above are involved in photodynamic therapy (PDT), a treatment modality, in which light, PSs, and oxygen interplay together [27]. Ideal PS for PDT should be selective enough to generate photodamages only in the lesions to be treated. In contrast to PSs, ideal oxygen and oxidative stress-sensitive fluorescence and phosphorescence probes should be minimally phototoxic to the probed tissues.

The spectral properties of MTO and $[Ru(Phen)_3]^{2+}$ interfere with spectral properties of clinically applied PSs [27]. To demonstrate the sensitivity of MTO and $[Ru(Phen)_3]^{2+}$ during PDT, we have chosen curcumin with promising spectral and PDT properties [28,29]. Curcumin is a natural polyphenolic compound extracted from Curcuma longa which increases or decreases oxidative stress according to the condition in which it is present [30].

The phototoxicity of curcumin is ascribed to singlet oxygen and superoxide radicals production [28,29]. However, singlet oxygen scavenging by curcumin was reported as well [31]. Due to curcumin spectral properties and its biological activity, curcumin represents an interesting molecule in PDT.

In the present work, the main aim was to optimize the evaluation of oxygen detection in vivo by $[Ru(Phen)_3]^{2+}$. This is why we decided to apply the cluster analysis of the $[Ru(Phen)_3]^{2+}$ luminescence lifetimes measured in the CAM for this reason. The detection of the tissue oxygenation at an oxidative stress condition is still challenging. Therefore, we have suggested applying time-resolved recording of MTO-FLIM and $[Ru(Phen)_3]^{2+}$-PLIM in the intact cell without disturbing the sensitivities of these molecular probes. The aim was also to demonstrate the stability of $Ru(Phen)_3]^{2+}$ as a molecular probe that can be used in vitro for oxygen detection in the presence of various stress factors mimicking the oxidative stress: other molecular sensors, H_2O_2, and the irradiation as a form of photodynamic therapy in cancer cells.

2. Results and Discussion

2.1. Finding the Limitations of $[Ru(Phen)_3]^{2+}$ Oxygen Sensitivity in CAM

The luminescence properties of $[Ru(Phen)_3]^{2+}$ were previously described [3,7,8,10,32]. As a small molecule, $[Ru(Phen)_3]^{2+}$ leaks out of the vessels and remains in the interstitial space of blood vessel cells [3]. Representative biodistribution of $[Ru(Phen)_3]^{2+}$ in the CAM at 1 and 20 min after its administration is presented in Figure 1b. While bright luminescence of $[Ru(Phen)_3]^{2+}$ was detected inside the vessels at 1 min post-administration, the majority of $[Ru(Phen)_3]^{2+}$ luminescence was observed out of the vessels at 20 min.

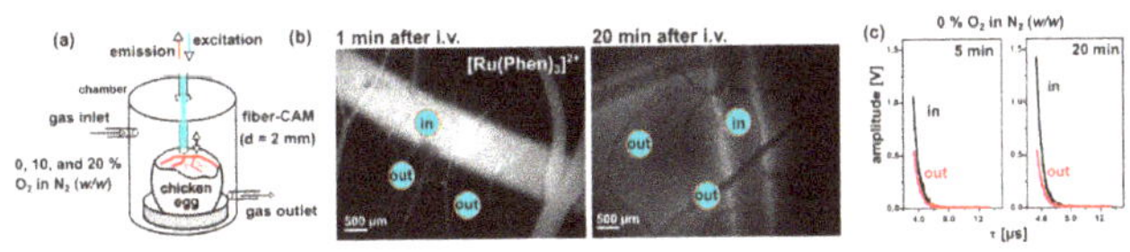

Figure 1. (**a**) Illustrative scheme of the gas chamber used to perform $[Ru(Phen)_3]^{2+}$ luminescence lifetime measurements in the CAM. (**b**) Fluorescence images of the CAM recorded 1 and 20 min after intravenous (iv) administration of $[Ru(Phen)_3]^{2+}$ (10 mg/kg of b.w.). (**c**) Illustrative outputs of $[Ru(Phen)_3]^{2+}$ (1 mg/kg of b.w.) luminescence lifetime detection at 0% O_2 in N_2 (*w*/*w*) atmosphere. The decays were detected in the blood vessels and out of the blood vessels as demonstrated in the fluorescence images.

The luminescence decays of $[Ru(Phen)_3]^{2+}$ presented a typical monoexponential character (Figure 1c). Its lifetime rapidly changed with the absence of atmospheric oxygen. The $[Ru(Phen)_3]^{2+}$ luminescence lifetime sensitivity to oxygen in solutions is quite unambiguous [32]. Since the CAM is a living organism, it is more complex and heterogeneous than cells monolayers. The ideal mono-exponential decay observed in homogeneous solutions is, therefore, no longer observed in the CAM. This can be explained by the complexity and multi-component character of this living system. While the CAM membrane plexus, composed of endodermal cells, represents a static component, the dynamic blood flow is present in the whole system.

Placement of the CAM into an atmosphere with low oxygen leads to significant changes in the oxygenation of the extracellular/interstitial space. However, the transport of oxygen at the cellular level and in the vessels is not a simple process. On the other hand, we had to keep the embryos alive during the measurements. For this reason, the chicken embryo kept residual oxygen supply as it can be recognized from Table 1.

Table 1. Additional information on descriptive statistics of $[Ru(Phen)_3]^{2+}$ luminescence lifetimes corresponding to different experimental conditions in the CAM. To clarify the meaning of the symbols used in the table headings, please refer to their meanings below in Equations (1)–(3). The color code is the same as in dendrograms in Figures 3 and 4.

Regime Data Set	rel_{Md}	rel_{OI}	Mean τ [ns]	% O_2 in N_2 Calculated Equation (1)
0% O_2 in N_2, 5 min, in	0.029		710.8	10.80
0% O_2 in N_2, 5 min, out	0.009	0.032	733.6	9.83
0% O_2 in N_2, 10 min, in	0.013		743.1	9.44
0% O_2 in N_2, 10 min, out	0.024	0.033	768.4	8.46
0% O_2 in N_2, 20 min, in	−0.006		774.3	8.24
0% O_2 in N_2, 20 min, out	0.030	0.028	796.6	7.44
10% O_2 in N_2, 5 min, in	0.005		626.5	14.99
10% O_2 in N_2, 5 min, out	−0.013	0.067	670.4	12.67
10% O_2 in N_2, 10 min, in	0.000		616.6	15.55
10% O_2 in N_2, 10 min, out	−0.018	0.103	683.3	12.05
10% O_2 in N_2, 20 min, in	−0.001		679.6	12.23
10% O_2 in N_2, 20 min, out	−0.004	0.002	681.1	12.15
20% O_2 in N_2, 5 min, in	0.012		609.0	16.00
20% O_2 in N_2, 5 min, out	−0.005	0.039	633.3	14.61
20% O_2 in N_2, 10 min, in	0.008		624.9	15.08
20% O_2 in N_2, 10 min, out	0.012	0.041	651.3	13.64
20% O_2 in N_2, 20 min, in	0.005		644.6	13.99
20% O_2 in N_2, 20 min, out	−0.011	0.006	648.5	13.79

Due to the non-Gaussian character of the $[Ru(Phen)_3]^{2+}$ luminescence lifetime (values detected in different CAMs) distributions, the data treatment could not be carried out by obvious reduction to arithmetic central values (mean values). Therefore, several central trend indicators were provided, as part of the R-programming environment [33]. The oxygenation values in Table 1 were estimated from Equation (1) (see below) that represents a calibration curve obtained from our previous study of $[Ru(Phen)_3]^{2+}$ in an isotonic solution of 0.9% NaCl [32]. The $[Ru(Phen)_3]^{2+}$ luminescence lifetimes measured at different regimes (0%, 10%, and 20% O_2 in N_2 for 5, 10, and 20 min) were converted to% oxygen by the Stern-Volmer relation as follows:

$$\%\ O_2 \text{ in } N_2 = (1/\text{Mean } \tau - 919{,}155.47)/45{,}157.84. \quad (1)$$

The 'in' values of oxygenation, which correspond to $[Ru(Phen)_3]^{2+}$ luminescence lifetimes inside the vessels, are higher than the 'out' values measured in the less vascularized areas. Moreover, the values detected during the first 5 min, which correspond to intravascular $[Ru(Phen)_3]^{2+}$, were also higher than at 10 and 20 min post-administration. We can see, that the % O_2 in N_2 (from 12 to 16) measured in the CAM at 20% and 10% O_2 in N_2 applied externally were not significantly different. At the contrary, 0% O_2 in N_2 regime (from 7 to 11) were significantly different from both 20 and 10% O_2 in N_2 regimes (*** $p < 0.001$).

A statistical overview of the measurements collected under defined conditions is shown in Figure 2. This figure explicitly shows a structured six-dimensional summary(.) output of the program R [33], as previously described in the methods section. We can see that the use of this representation also reveals some equivalence with $[Ru(Phen)_3]^{2+}$ luminescence lifetime distribution characteristics. Initial findings revealed remarkable dynamic differences in statistics of 'out' and 'in' cases, as well as significant variations between results obtained with early, intermediate, and late administration. Let us first look at the values of the arithmetical means. These values can be arranged into the chronological layout of Table 1. Within this table, we specified the column where the generalized mean values of the $[Ru(Phen)_3]^{2+}$ luminescence lifetimes (τ) are transformed to the relative quantity.

$$rel_{Md} = 2(\text{Median } \tau - \text{Mean } \tau)/(\text{Median } \tau + \text{Mean } \tau). \tag{2}$$

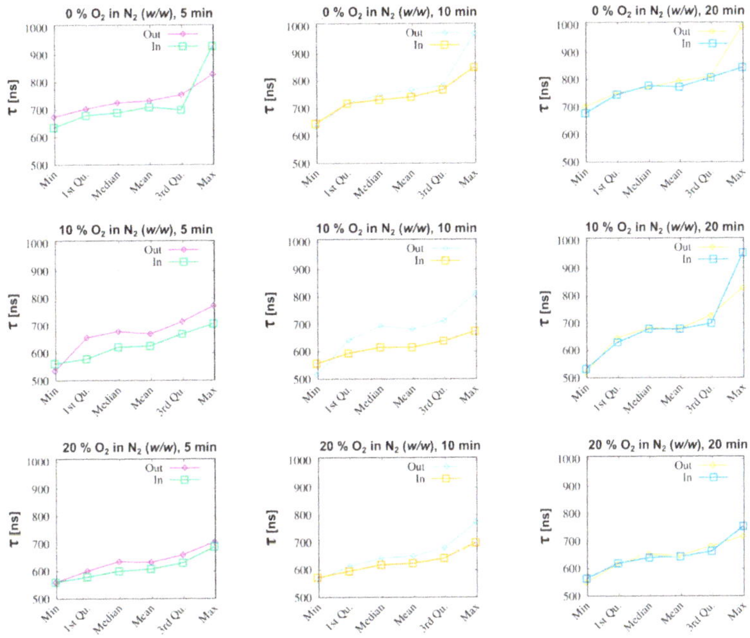

Figure 2. Statistical summaries of the $[Ru(Phen)_3]^{2+}$ luminescence lifetime measurements analysis in the CAM at 5, 10, and 20 min after iv administration and in atmospheres of 0%, 10%, and 20% O_2 in N_2 (w/w).

Here Median τ is a symbol for the median calculated from the data sets of luminescence decay times. More precisely, these are the values obtained by exponential regression. As a basis for comparison, we use the arithmetic mean of luminescence lifetime values denoted as Mean τ. For such averages, the zero of rel_{Md} indicates Gaussian behavior. From a practical point of view, obviously, it is important to know how the rel_{Md} differs from zero. An application example arose in the process of identifying 10% of O_2 in N_2 as the anomalous condition. Further, rel_{Md} (consider 0.029, 0.024, 0.03 to be representatively high values) becomes even more pronounced at 0% of O_2 in N_2.

For a more detailed interpretation, let's also consider the 'out-in' sensitivity indicator:

$$rel_{OI} = 2(\text{Mean}_{out}\ \tau - \text{Mean}_{in}\ \tau)/(\text{Mean}_{out}\ \tau + \text{Mean}_{in}\ \tau). \tag{3}$$

The indices 'in' and 'out' mean the selection of data under specific conditions. The relation is performed to conduct a statistical characterization of the corresponding 'in', 'out' pairs at given administration times under fixed O_2% in N_2 conditions. Table 1 shows a variation of a few percent with the outlier value of $rel_{OI} = 0.103$.

Hypoxic conditions applied to certain cells of the CAM may induce the fatal effect. This is of importance since the luminescence lifetime of $[Ru(Phen)_3]^{2+}$ in the presence of DNA is longer than in aqueous solution, and comparable with the values obtained in the absence of oxygen [10,11]. CAM cells suffering from acute hypoxia are more likely to become necrotic [34]. Cell membranes and nuclear envelopes of these cells can be more permeable for $[Ru(Phen)_3]^{2+}$ molecules. We hypothesize that this effect can explain the highest sensitivity to the oxygen level in the case of 10% O_2 in N_2 applied for 10 min to the CAM. Nevertheless, certain cells are not affected by these conditions. It is expected that more cells should suffer from hypoxia at the 0% O_2 in N_2 regimes. However, the $[Ru(Phen)_3]^{2+}$ luminescence lifetimes distribution is less dispersed. This identifiable anomaly is consistent with the results presented in Figure 2 which are highlighting the differences between the 'in' and 'out' curves that relate to the $[Ru(Phen)_3]^{2+}$ luminescence lifetime statistics.

Analysis of data based on multidimensional statistical inputs requires visual comparison and additional interpretation. The implementation of hierarchical processing in this project is new, but more natural because the clustering provides advanced tools to achieve interpretable and even visual consequence. In the present work, we also emphasize that clustering does not takes place once, but rather as an iterative mechanism with feedback.

Figure 3 shows the results of the cluster analysis based on luminescence lifetimes obtained under different experimental conditions. The specific arrangement of the results and conditions is such that, while (a) focuses on the 'in' cases, panel (b) monitors the 'out' scenarios. The color and the corresponding contents reflect the differences between panels (a) and (b). In panel (a) the structure of the dendrogram is more fragmented, while (b) is more ordered. This can be explained by the experimental control (high tissue homogeneity) of the 'out' cases, whereas the 'in' case data also reflect more complex physiological internal processes.

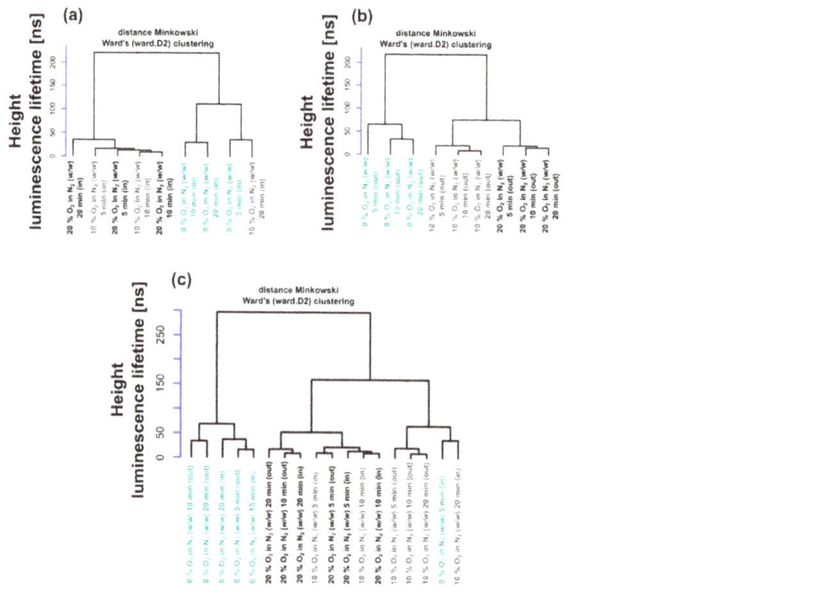

Figure 3. Dendrograms of $[Ru(Phen)_3]^{2+}$ luminescence lifetime detected (**a**) in the blood vessels and (**b**) out of the blood vessels of the CAM at 5, 10, and 20 min after iv administration and with O_2 in N_2 (w/w) atmosphere ranging between 0% and 20%. (**c**) The dendrogram, which is derived from the whole dataset showing mutual relationships.

The results obtained with 10% O_2 in N_2 are the most fragmented, as shown in the overall statistics in panel (c). This is in agreement with the aforementioned anomaly analysis expressed by means of the rel_{OI} indicator. The cophenetic distance in the case of the panel (c) also clearly shows that while 10% and 20% O_2 in N_2 are statistically close ($[Ru(Phen)_3]^{2+}$ luminescence lifetime height 150 ns, highest square arch), the case of 0% O_2 in N_2 dataset cases is a significant outlier ($[Ru(Phen)_3]^{2+}$ luminescence lifetime height = 270 ns).

However, one might ask where the roots of mismatch and fragmentation are within the 'in' cases. The answer is indicated by the three upper panels of Figure 4, which were generated by the separate processing of ('in', 0%), ('in', 10%), and ('in', 20%) subsets grouped from the entire 'in' statistics. The administration time (read the dendrogram from left to right) of the legs (branches) changed significantly from the 0% sequence (5 min, 10 min, and 20 min) to (20 min, 5 min, and 10 min) in the cases of 10% and 20% O_2 in N_2. This means that conditions corresponding to the sensitive branch with a duration of 20 min behave as if it more closely mimics the distance conditions of 'in' case. Another perspective of data perception is established in the last three 'in' dendrograms where the selection of data for clustering is carried out according to the time after administration. Maintaining the canonical order (0%, 10%, and 20%) is a surprising result of these experiments. Here, too, an unusual dendrogram height shift occurs at 20 min.

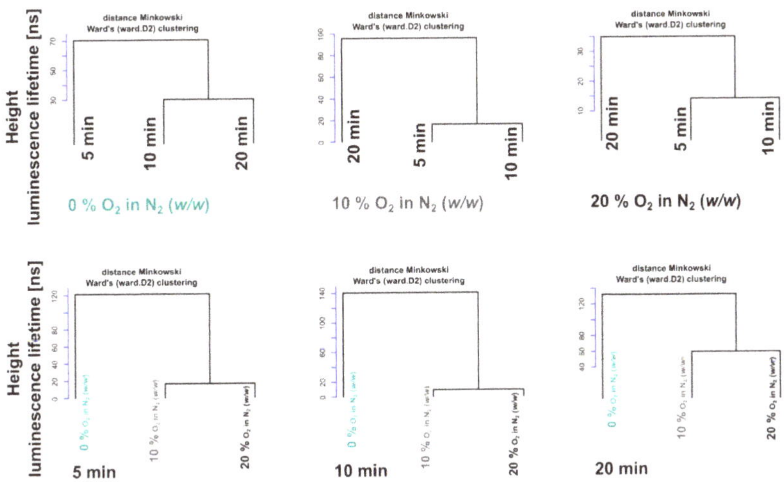

Figure 4. System of sub-dendrograms obtained from $[Ru(Phen)_3]^{2+}$ luminescence lifetime datasets detected in the blood vessels of the CAM at 5, 10, and 20 min after i.v. administration and 0–20% O_2 in N_2 (*w*/*w*) atmosphere. The lifetime data set visualization is divided into an upper and a lower part. Each part consists of three dendrograms, which are composed of separately processed data (time-dependent, O_2 in N_2 concentration-dependent). While the first three dendrograms are constructed according to the similarity of the data at given concentrations, the other three are constructed at given detection times.

We recommend reading Figure 3 panels in two directions. One can proceed from the details of the panels (a) and (b), and connect them in the direction of the composite panel (c) as the results. On the other hand, there is a top-down, reductionist way of interpreting panel (c) with less detailed visual representations of (a) and (b). Figure 4 provides alternative and complementary methods for processing and understanding the results.

We know from previous examples as well as from practical experience that because of their flexibility, straightforwardness, and controllability, the dendrogram and cluster methods could still compete with many other forms of machine learning and data processing. They have become known and used not only in population genetics but also in many other bio-motivated fields, including brain sciences [35].

We assumed that the change of oxygen atmosphere from normoxia to hypoxia in the gas chamber may have induced oxidative stress. The level of oxidative stress in CAM cells may then locally influence $[Ru(Phen)_3]^{2+}$ luminescence lifetime detection.

In the next step, we have evaluated the influence of the oxidative stress on the $[Ru(Phen)_3]^{2+}$ luminescence lifetime detection in glioma cells.

2.2. Localization of $[Ru(Phen)_3]^{2+}$ in Glioma Cells and Oxidative Stress

We have used a unique FLIM/PLIM system to see how $[Ru(Phen)_3]^{2+}$ localization influences the $[Ru(Phen)_3]^{2+}$ luminescence lifetime. Illustrative PLIM images of $[Ru(Phen)_3]^{2+}$ in the extracellular and intracellular spaces are presented in Figure 5. U87 MG cells incubated during 1 h with 200 μM $[Ru(Phen)_3]^{2+}$ are recognized as the dark areas in the intensity and PLIM images. Extracellular $[Ru(Phen)_3]^{2+}$ luminescence lifetime was ≈650 ns (according to Equation (3) it corresponds to 13.7% of O_2 in N_2). $[Ru(Phen)_3]^{2+}$ was not recognized in the cytoplasm. However, the subcellular vesicles loaded with $[Ru(Phen)_3]^{2+}$ were identified at longer (24 h) incubations time. We have previously demonstrated that $[Ru(Phen)_3]^{2+}$ crossed the plasma membrane via endocytosis and maintained localized in the endocytotic vesicles, and peroxisomes [7]. The intracellular lifetimes of $[Ru(Phen)_3]^{2+}$ were longer (≈800 ns and 7.3% of O_2 in N_2) than for extracellularly measured. Global lifetimes, i.e., the extracellular plus intracellular components, as represented by luminescence lifetime distribution histograms (Figure 5), fall in the interval ranging between 600 and 900 ns (16.5% and 4.2% of O_2 in N_2). The increase of the intracellularly localized $[Ru(Phen)_3]^{2+}$ luminescence lifetime reflects lower oxygenation within the cells. However, the luminescence lifetime also may dependents on other environmental factors that could explain the longer luminescence lifetimes.

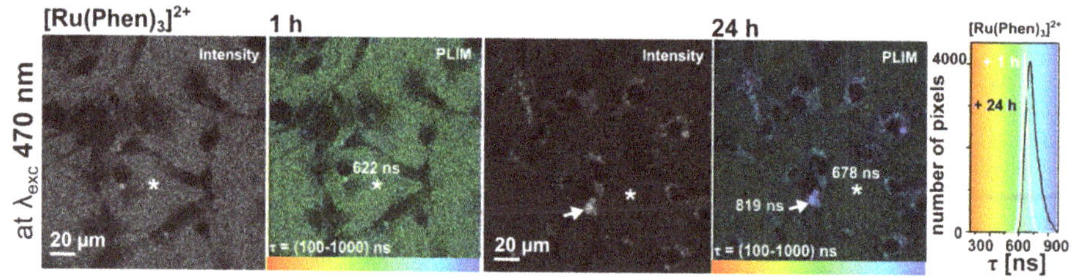

Figure 5. Four images on the left: Illustrative intensity and PLIM images obtained with 200 μM $[Ru(Phen)_3]^{2+}$ applied in the culture medium of U87 MG cells for 1 and 24 h. Right: $[Ru(Phen)_3]^{2+}$ luminescence lifetime distribution histograms (1 h—white, 24 h black). The luminescence lifetimes are color-coded (100 ns—red, 1000 ns—blue). The samples were excited with laser light at 470 nm. The white arrows point to intracellular and asterisks to extracellular localization of $[Ru(Phen)_3]^{2+}$.

$[Ru(Phen)_3]^{2+}$ can be considered as a metalotoxin. However, it has a low toxicity and phototoxicity in comparison with porphyrins, as we have already demonstrated [3,7,8]. The presence of metalotoxin in cells may induce oxidative stress. One of the most sensitive organelles to oxidative stress in cells are mitochondria. Representative confocal fluorescence images of CellROX green, a probe sensitive to oxidative stress induction in mitochondria, are presented in Figure 6. We have selected this ROS unspecific probe due to its spectral properties and subcellular localization which are different from those of $[Ru(Phen)_3]^{2+}$ (Figure 6b). Comparison of CellROX green localization with rhodamine 123 (a mitochondrial probe sensitive to mitochondria membrane potential), and $[Ru(Phen)_3]^{2+}$ is presented in Figure S1 in Supplementary Materials. Intact mitochondria in U87 MG cells were detected in the presence of $[Ru(Phen)_3]^{2+}$. Application of light with the microscopy detection system-induced photodamages in cells, as revealed by the destabilization of the mitochondria membrane potential (tubular structures cannot be recognized, see Figure S2 in Supplementary Materials) and an increase in oxidative stress (CellROX localization in the nucleus), as can be seen in Figure 6c. The localization of $[Ru(Phen)_3]^{2+}$ before and after illumination did not change (Figure 6b,c).

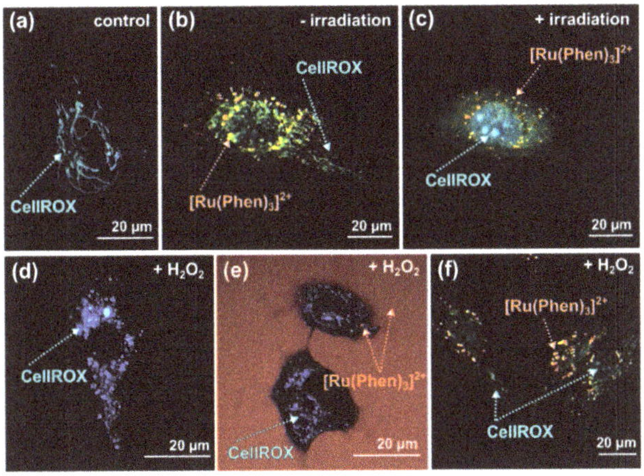

Figure 6. Illustrative confocal microscopy images of U87 MG cells stained with (**a**–**f**) 1 µM (30 min) CellROX®Green (cyan and blue) and 200 µM $[Ru(Phen)_3]^{2+}$ (orange and red). These dyes were applied in the U87 MG cells culture medium for (**e**) 1 and (**b**,**c**,**f**) 24 h. (**c**) The cells were irradiated with microscopy (detection time less than 2 min, 2% of the power of the laser: 405 and 488 nm)—the second scan of cells stained with $[Ru(Phen)_3]^{2+}$ and CellROX®Green. The photoreaction induced oxidative stress. (**d**–**f**) Extracellular oxidative stress was induced by 200 µM H_2O_2.

To see the stability of $[Ru(Phen)_3]^{2+}$ distribution and luminescence at external induction of oxidative stress, we have applied H_2O_2 into the cell culture. The presence of 200 µM H_2O_2 in U87 MG resulted in significant reduction of intracellular catalase and superoxide dismutase 1 (see Figure S3 in the Supplementary Materials). Zhang et al. observed that apoptosis triggered by H_2O_2 mediated oxidative stress in hepatoma cells involved decreasing of catalase and superoxide dismutase activity [36]. It is in agreement with our findings. In our study, we have applied H_2O_2 for short time (<1 h) to induce acute stress, and prevent its complete metabolism. We have previously demonstrated by MitoSOX^{TM} Red [15] that superoxide production increased in U87 MG cells treated with H_2O_2. Besides, application of H_2O_2 resulted in significant ROS production extracellularly and intracellularly detected by DCFDA/H2DCFDA assay (see Figure S4 in Supplementary Materials), and further caused significant lipid peroxidation inside the cells (see Figure S5 in Supplementary Materials).

The extracellular application of H_2O_2 (without molecular probes) induced dissipation of mitochondrial membrane potential, and fission of mitochondria in U87 MG cells (Figure 6d and Figure S2). The same effect as in the absence of $[Ru(Phen)_3]^{2+}$ (fission of mitochondria) was observed on mitochondria in the presence of $[Ru(Phen)_3]^{2+}$ and H_2O_2 keeping in the cell culture medium (Figure 6e). $[Ru(Phen)_3]^{2+}$ distribution and luminescence were not changed.

A similar experiment was performed in cells that were incubated for 24 h with $[Ru(Phen)_3]^{2+}$, at that time it crossed plasma membrane via endocytosis [7]. In this case, $[Ru(Phen)_3]^{2+}$ was removed from the cell culture medium prior observation. The expected fission of mitochondria induced by H_2O_2 was less effective (Figure 6f). Indeed, oxidative stress induced by H_2O_2 caused partial relocalization of CellROX into the nucleus. The number of $[Ru(Phen)_3]^{2+}$ loaded vesicular structures previously identified as the peroxisomes [7] were localized nearby tubular structured mitochondria (see Figures S1 and S6 in Supplementary Materials), where it may attenuate the oxidative stress. This finding strongly suggests the antioxidants-like effects of $[Ru(Phen)_3]^{2+}$. The partial localization of CellROX in the nuclei, on the other hand, suggests that oxidative stress level increased in these cells due to H_2O_2.

2.3. $[Ru(Phen)_3]^{2+}$ Luminescence Lifetime Stability during Oxidative Stress in Solution

We have demonstrated that the localization of $[Ru(Phen)_3]^{2+}$ in cells was the same in the presence and absence of H_2O_2. We assumed that the $[Ru(Phen)_3]^{2+}$ luminescence lifetime was more sensitive to this molecule than its intensity. For this reason, we have conducted $[Ru(Phen)_3]^{2+}$ luminescence lifetime measurements in the solutions of H_2O_2. Figure 7a demonstrates unaffected absorption spectra of $[Ru(Phen)_3]^{2+}$ in the presence and absence of H_2O_2. The decays of $[Ru(Phen)_3]^{2+}$ luminescence lifetimes are plotted in Figure 7b. The extracted values of $[Ru(Phen)_3]^{2+}$ lifetimes were found to be significantly different in the presence of low concentrated (200 nM) H_2O_2 (Figure 7c). A small decrease of the $[Ru(Phen)_3]^{2+}$ lifetimes could be connected with the decomposition of H_2O_2 with time. The oxygen developed in this reaction can induce quenching of the $[Ru(Phen)_3]^{2+}$ luminescence lifetime. The highest concentration of H_2O_2 did not induce further shortening of the lifetimes. The deviations of the luminescence lifetime values induced by H_2O_2 were small, i.e., within the range (560–575 ns). Estimated values of applied O_2 in N_2 (19–18%) were higher in solutions in comparison to the % of O_2 in N_2 detected in the CAM. This suggests that the effect of H_2O_2 on the $[Ru(Phen)_3]^{2+}$ luminescence lifetime is minor.

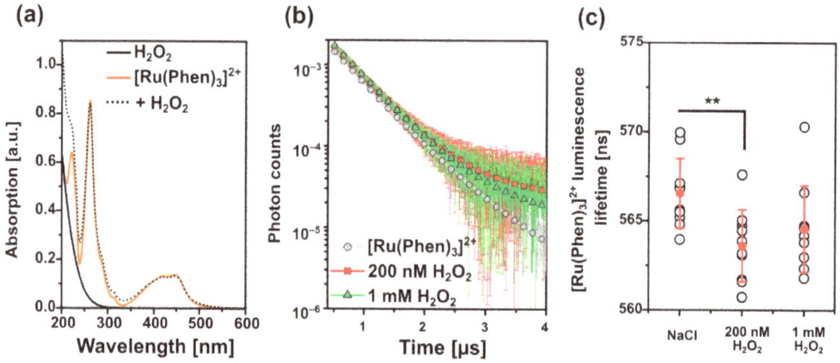

Figure 7. (**a**) Absorption spectra of 1 mM H_2O_2 (black), 10 µM $[Ru(Phen)_3]^{2+}$ (orange), and $[Ru(Phen)_3]^{2+}$ with 1 mM H_2O_2 (dotted black) 0.9% NaCl solutions. (**b**) 10 µM $[Ru(Phen)_3]^{2+}$ luminescence lifetime decays in 0.9% NaCl physiological solution (grey circles) and the presence of 200 nM (red squares) and 1 mM H_2O_2 (green triangles). (**c**) $[Ru(Phen)_3]^{2+}$ luminescence lifetime values detected at the conditions mentioned in (**b**). The mean values of the decays and the lifetimes are the averages of ten measurements with the appropriated standard deviations as the error bars. The level of significance was assessed with the Student *t*-test: ** $p < 0.01$.

2.4. MTO Fluorescence and $[Ru(Phen)_3]^{2+}$ Luminescence Lifetimes Sensitivity during Oxidative Stress Induced in Glioma Cells

In our previous study, we have demonstrated that oxidative stress induced in cancer cells fragmented MTO such a way that one fraction was bound to DNA in the nucleus resulting in an increase of MTO fluorescence lifetime [15]. Interestingly, MTO fluorescence lifetime increased in mitochondria in which oxidative stress level increased. For this reason, we have selected MTO to detect oxidative stress levels and $[Ru(Phen)_3]^{2+}$ to detect oxygenation in cells with a unique time-resolved microscopy system. With this setup, excitation and emission of MTO and $[Ru(Phen)_3]^{2+}$ were clearly discriminated, i.e., no signals of MTO was detected in PLIM, whereas no $[Ru(Phen)_3]^{2+}$ luminescence was detected in the MTO FLIM channel.

Oxidative stress was induced externally by H_2O_2. Figure 8 shows FLIM images of MTO fluorescence lifetimes and PLIM images of $[Ru(Phen)_3]^{2+}$ luminescence lifetimes in U87MG cells in the presence and absence of H_2O_2. We can recognize the increase of MTO fluorescence lifetime (blue-colored) in the presence of H_2O_2 (Figure 8a). Two main regions of interest (ROI) in cells were recognized. One ROI is in the perinuclear area (in,

≈1500 ps) with long MTO fluorescence lifetimes, whereas the second ROI is in the periphery, nearby the plasma membrane (out, ≈1200 ps) with shorter MTO fluorescence lifetimes (Figure 8b,c). As the intracellular H_2O_2 concentration increased, the MTO delocalizes into the nucleus and the MTO fluorescence lifetime dramatically increased in this region (≈1900 ps).

PLIM measurements of $[Ru(Phen)_3]^{2+}$ were performed within 1 h after its administration, i.e., when $[Ru(Phen)_3]^{2+}$ localized extracellularly. Dark areas in images showing the cell localization of $[Ru(Phen)_3]^{2+}$ in PLIM can be recognized in Figure 8a. Application of H_2O_2 resulted in a decrease of the extracellular $[Ru(Phen)_3]^{2+}$ luminescence lifetimes from ≈650 ns (13.7% of O_2 in N_2) to ≈400 ns (35% of O_2 in N_2), as presented in Figure 8b. As observed in solutions, this effect could be caused by the H_2O_2 decomposition. The PLIM detection at longer incubation time (1 h) after the H_2O_2 application developed a bimodal character of $[Ru(Phen)_3]^{2+}$ luminescence lifetimes (see the distribution histograms in Figure 8c). The shorter lifetimes < 600 ns (>16.5% of O_2 in N_2) were detected in the extracellular area, whereas the longer lifetime > 600 ns (<16.5% of O_2 in N_2) were detected in cells. It should be noted that the nuclei of the cells were brightly labeled with $[Ru(Phen)_3]^{2+}$ and with the longest luminescence lifetimes. This can be explained by the interaction of $[Ru(Phen)_3]^{2+}$ with cellular DNA. Indeed, Komor et al. described that the luminescence of $[Ru(Phen)_3]^{2+}$ increased in the presence of DNA [10,11]. We assume that the H_2O_2 applied destabilized the plasma membrane, which became permeable to $[Ru(Phen)_3]^{2+}$ molecules. These results suggest that the evaluation of $[Ru(Phen)_3]^{2+}$ lifetimes and subsequent determination of oxygenation level strongly depend on cell fitness.

One original aspect of our work was to apply simultaneously MTO and $[Ru(Phen)_3]^{2+}$ in cell culture media and to record FLIM and PLIM images from the same cells to monitor oxidative stress level and oxygenation changes. MTO FLIM and $[Ru(Phen)_3]^{2+}$ PLIM images were detected in the absence and presence of H_2O_2. No detectable FLIM and PLIM signals were observed in U87 MG cells in the absence of MTO and $[Ru(Phen)_3]^{2+}$ (Figure 9a). As can be seen in Figure 9, we have obtained results comparable to those observed when MTO and $[Ru(Phen)_3]^{2+}$ were applied independently. The MTO fluorescence lifetimes increased and the nucleus localized MTO appeared after H_2O_2 application. The $[Ru(Phen)_3]^{2+}$ luminescence lifetimes decreased shortly after H_2O_2 application and increased with time and nuclear distribution (Figure 9b,c). It should be noted that the oxidative stress level evidenced by the MTO fluorescence lifetime increased in the presence of $[Ru(Phen)_3]^{2+}$ in the perinuclear area. Besides, the $[Ru(Phen)_3]^{2+}$ luminescence lifetime slightly increased in the presence of MTO. Therefore, the mutual influence of $[Ru(Phen)_3]^{2+}$ and MTO should be taken into consideration when oxidative stress level and oxygenation are estimated simultaneously.

2.5. MTO Fluorescence and $[Ru(Phen)_3]^{2+}$ Luminescence Lifetimes Changes Observed during PDT Induced by Curcumin and Blue Light in Glioma Cells

In the following, we aimed to apply the simultaneous FLIM/PLIM detection of MTO and $[Ru(Phen)_3]^{2+}$ after intracellular induction of oxidative stress by PDT. We have tested the hypothesis to detect oxygen consumption in the extracellular medium (with $[Ru(Phen)_3]^{2+}$), and oxidative stress (with MTO) induction within the cells by PDT.

The absorption curves of $[Ru(Phen)_3]^{2+}$, MTO, and curcumin are partially overlapping. Therefore, intensity-based techniques (e.g., fluorescence microscopy) present obstacles to discriminate their emissions. In our laboratory, we utilized a time-resolved microscopy system that enabled us to detect fluorescence (FLIM) and phosphorescence (PLIM) lifetimes from sample (cell) after its excitation with a white femtosecond laser. Thanks to this system we could differentiate by time-resolved spectroscopy the $[Ru(Phen)_3]^{2+}$, MTO, and curcumin localization in the same cell. Moreover, the lifetimes of MTO-FLIM and $[Ru(Phen)_3]^{2+}$ provided information regarding the oxidative stress level and oxygenation of the sample [3,15].

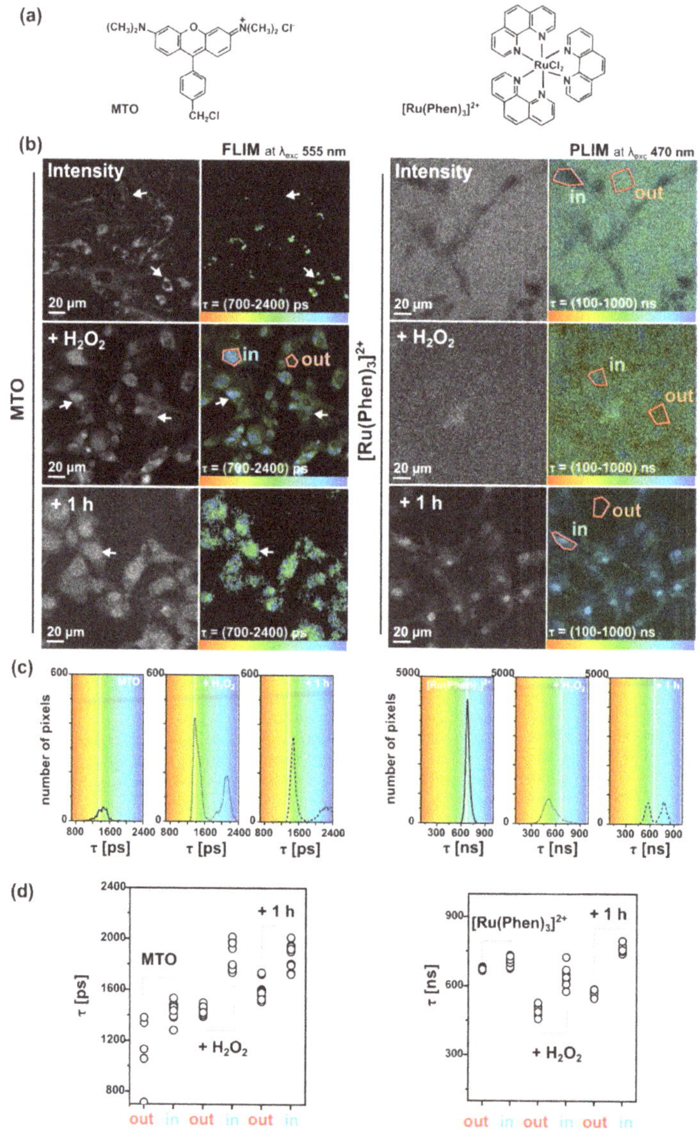

Figure 8. (**a**) Chemical structures of MTO (left) and [Ru(Phen)$_3$]$^{2+}$ (right). (**b**) Representative FLIM and PLIM images of MTO (left) and [Ru(Phen)$_3$]$^{2+}$ (right) individually applied in the medium of U87 MG cells for 30 min in the absence and presence of 200 µM H_2O_2 (shortly and 1 h after H_2O_2 administration). White arrows point to MTO mitochondrial and nuclear localization. The MTO fluorescence lifetimes ROI was selected in the perinuclear/nuclear area (in) and far from this area (out). The [Ru(Phen)$_3$]$^{2+}$ luminescence lifetimes ROI were selected in the intracellular (in) and extracellular area (out). (**c**) MTO and [Ru(Phen)$_3$]$^{2+}$ luminescence lifetime distribution histograms. The luminescence lifetimes are color-coded (minima—red, maxima—blue). Gray-scaled images represent the luminescence intensity of MTO and [Ru(Phen)$_3$]$^{2+}$. The samples were excited with laser light at 555 nm (MTO) and 470 nm ([Ru(Phen)$_3$]$^{2+}$). (**d**) MTO and [Ru(Phen)$_3$]$^{2+}$ luminescence lifetime values detected at the conditions mentioned in (**b**). At least ten representative ROI was selected.

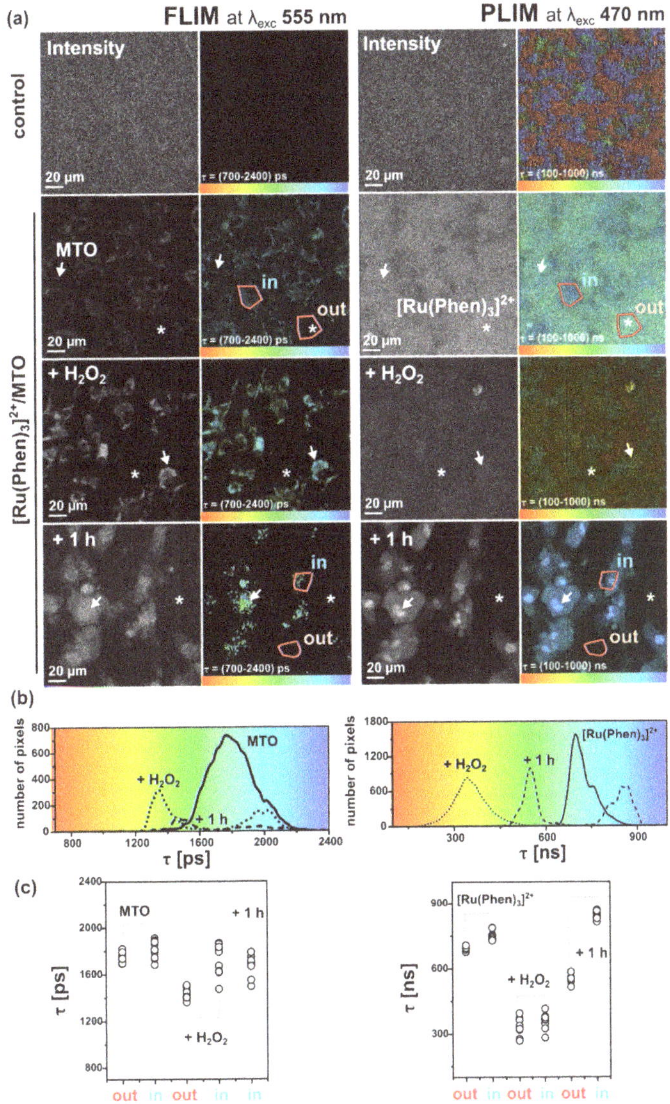

Figure 9. (**a**) Representative FLIM and PLIM images of MTO (left) and $[Ru(Phen)_3]^{2+}$ (right) both applied in the medium of U87 MG cells for 30 min in the absence and presence of 200 μM H_2O_2 (shortly and 1 h after H_2O_2 administration). The cells without staining were collected in the FLIM and PLIM channels as well. White arrows point to MTO mitochondrial and nuclear localization. The MTO fluorescence lifetimes ROI was selected in the perinuclear/nuclear area (in) and far from this area (out). The $[Ru(Phen)_3]^{2+}$ luminescence lifetimes ROI were selected in the intracellular (in) and extracellular (white asterisk) area (out). (**b**) MTO and $[Ru(Phen)_3]^{2+}$ luminescence lifetime distribution histograms. The luminescence lifetimes are color-coded (minima—red, maxima—blue). Gray-scaled images represent the luminescence intensity of MTO and $[Ru(Phen)_3]^{2+}$. The samples were excited with laser light at 555 nm (MTO) and 470 nm ($[Ru(Phen)_3]^{2+}$). (**c**) MTO and $[Ru(Phen)_3]^{2+}$ luminescence lifetimes detected at the conditions mentioned in (**a**). At least ten representative ROIs were selected.

As mentioned in the introduction, curcumin can be a scavenger of singlet oxygen but, under irradiation with blue light, it can produce superoxide radicals, singlet oxygen and other reactive oxygen species [29–31]. A combination of dual-detection (MTO-$[Ru(Phen)_3]^{2+}$) could reveal both, oxidative stress and antioxidant effect of curcumin during PDT in the same cells.

In the present study, we have incubated U87 MG cells with 10 µM curcumin during 1 h. Representative fluorescence intensity images of curcumin, MTO, and $[Ru(Phen)_3]^{2+}$ are presented in Figure 10. We have observed an intracellular localization of curcumin, whereas MTO localizes in mitochondria, and $[Ru(Phen)_3]^{2+}$ in the extracellular space. Irradiation (detection time less than 3 min, laser was set at 2% of the total power of the laser: 405, 488, and 555 nm) coming from the microscopy system during fluorescence imaging triggered photoreactions generated oxidative stress production, photobleaching of curcumin fluorescence, and nuclear localization of MTO (Figure 10c). The intracellular localization of curcumin was not specifically in mitochondria. This is in agreement with the observations performed by Sala de Oyanguren et al. who demonstrated specific localization of curcumin in the endoplasmic reticulum and described its role in the process of autophagy and apoptosis [37].

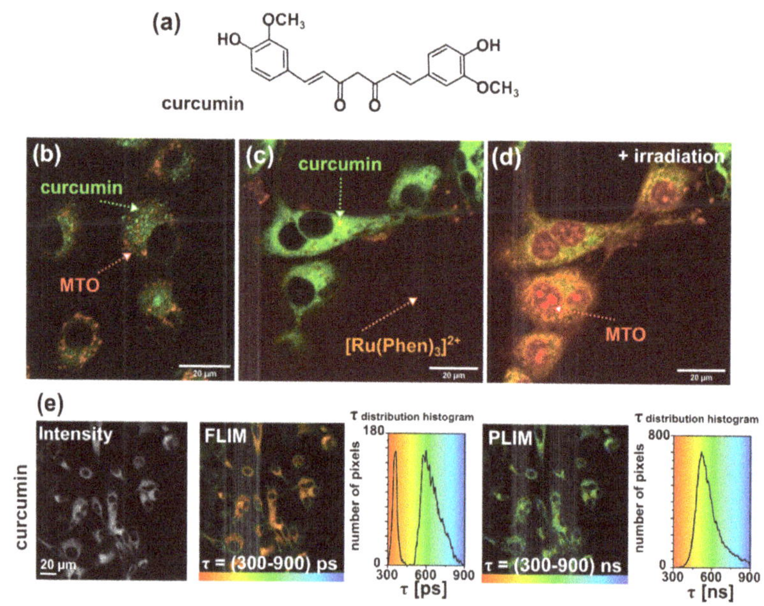

Figure 10. (**a**) Chemical structure of curcumin. Illustrative confocal microscopy fluorescence intensity images of U87 MG cells stained with (**b**) 10 µM curcumin (green) and 400 nM MTO (red), (**c**) curcumin, MTO (localized in the mitochondria), and 200 µM $[Ru(Phen)_3]^{2+}$ (red, extracellularly localized) first scan and (**d**) second scan that induced oxidative stress (MTO is localized in the nuclei). (**e**) FLIM and PLIM images of curcumin in U87 MG cells and corresponding fluorescence and phosphorescence lifetimes histograms. The lifetimes are color-coded (minima—red, maxima—blue). The gray-scaled image represents the fluorescence and phosphorescence intensity of curcumin.

The fluorescence lifetime of curcumin is short (≈hundreds of ps), as reported by Khopde et al. in different solvent [38]. FLIM and PLIM images of curcumin were detected in the same spectral window as $[Ru(Phen)_3]^{2+}$. Fluorescence lifetime histograms of curcumin have a bimodal character (Figure 10e). Short components (≈350 ps) were localized in the perinuclear area, whereas longer fluorescence lifetimes of curcumin (≈580 ps) were observed in the periphery near the plasma membrane. In comparison to fluorescence, very little is known about curcumin phosphorescence. In the nineties, Chignell et al. described the photobleaching of curcumin and reported that its phosphorescence depends on the excitation wavelength [29]. A broad phosphorescence spectrum > 650 nm was observed at 77 K and 282 nm excitation by this group. In the present study, the phosphorescence lifetime of curcumin was observed in the same PLIM detection window as it was for $[Ru(Phen)_3]^{2+}$ (Figure 10d). Phosphorescence lifetimes of ≈590 ns were detected within U87 MG cells. To the best of our knowledge, this is the first time that curcumin phosphorescence imaging is demonstrated. We should note, that possible other product of curcumin degradation in the cytoplasm could be detected in PLIM mode. This effect should be verified in the future study. Very often the PS is degraded during PDT due to oxidative stress [39]. For this reason, the curcumin PLIM, in the present study, may be considered as a marker for PDT efficacy in cells. With this regard, curcumin degradation should result in decrease of its phosphorescence.

In the following, curcumin was irradiated in cells with blue light emitted by a diode (463 ± 10 nm and 100 $\mu W/cm^2$ irradiance) for 30 and 90 min. Our time-resolved microscopy system detected the FLIM image of MTO and PLIM images of curcumin and $[Ru(Phen)_3]^{2+}$ in the same cell. Irradiation of cells in the presence of $[Ru(Phen)_3]^{2+}$ only and in combination with MTO in the absence of curcumin did not result in significantly different changes in FLIM/PLIM images (see Figures S7 and S8 in Supplementary Materials). Indeed, irradiation of cells in the presence of $[Ru(Phen)_3]^{2+}$ resulted in significant ROS formation, and detected with H2DCFDA sensor in the extracellular area (see Figure S4 in Supplementary Materials). It should be noted that higher light dose may induce certain damages as it was observed by confocal fluorescence microscopy. However, the presence of MTO during irradiation resulted to photodamage of the cells that affected FLIM/PLIM detection. The FLIM and PLIM images of MTO, curcumin, and $[Ru(Phen)_3]^{2+}$ are presented in Figure 11. We can see that the presence of curcumin in U87 MG cells without irradiation increases the MTO fluorescence lifetimes (>2000 ps). This suggests that curcumin increases the oxidative stress in cells. The luminescence lifetimes of $[Ru(Phen)_3]^{2+}$ were not modified by the presence of curcumin (≈700 ns; 11.28% of O_2 in N_2). The histograms of curcumin phosphorescence lifetimes revealed values as short as ≈400 ns.

Irradiation dramatically changed the MTO and curcumin luminescence lifetime images. Indeed, the MTO fluorescence lifetimes decreased below 1400 ps after 30 min irradiation with blue light. Longer irradiation induced an increase of MTO fluorescence lifetimes. This suggests that the oxidative stress firstly decreased and later increased in cells along the irradiation time. We can recognize that the phosphorescence of curcumin was quenched by the 30 min irradiation with blue light, which led to a slight increase of the $[Ru(Phen)_3]^{2+}$ luminescence lifetime from 631 to 658 ns (from 14.7% to 13.3% of O_2 in N_2). These changes can be seen in the histograms of $[Ru(Phen)_3]^{2+}$ and curcumin phosphorescence lifetimes (Figure 11). It could be that these changes are due to a slight oxygen consumption. On the contrary, the longest 90 min irradiation resulted in decreasing $[Ru(Phen)_3]^{2+}$ luminescence lifetimes down to 613 ns (15.7% of O_2 in N_2). It should be noted that the irradiation of cells during FLIM/PLIM detection did not induce any photodamage as it was the case for the confocal fluorescence microscopy.

One can reasonably assume that the antioxidant effects of irradiated curcumin prevail for short irradiation times. The photobleaching of curcumin probably induced oxygen consumption ($[Ru(Phen)_3]^{2+}$ luminescence lifetimes increasing). Moreover, degraded curcumin fractions, active as antioxidants, reduced the oxidative stress level in cells (MTO fluorescence lifetime decreasing).

Long-time irradiation of U87 MG cells probably reversed the curcumin antioxidant effect to photodamage induction. The oxidative stress level (MTO fluorescence lifetimes) increased in those cells. The cells became leakier for $[Ru(Phen)_3]^{2+}$ molecules, which were internalized in the cells. However, we could not distinguish between the signal from $[Ru(Phen)_3]^{2+}$ and possible photoproducts of curcumin that may remain in cells after irradiation and create bias in the $[Ru(Phen)_3]^{2+}$ luminescence lifetime determination.

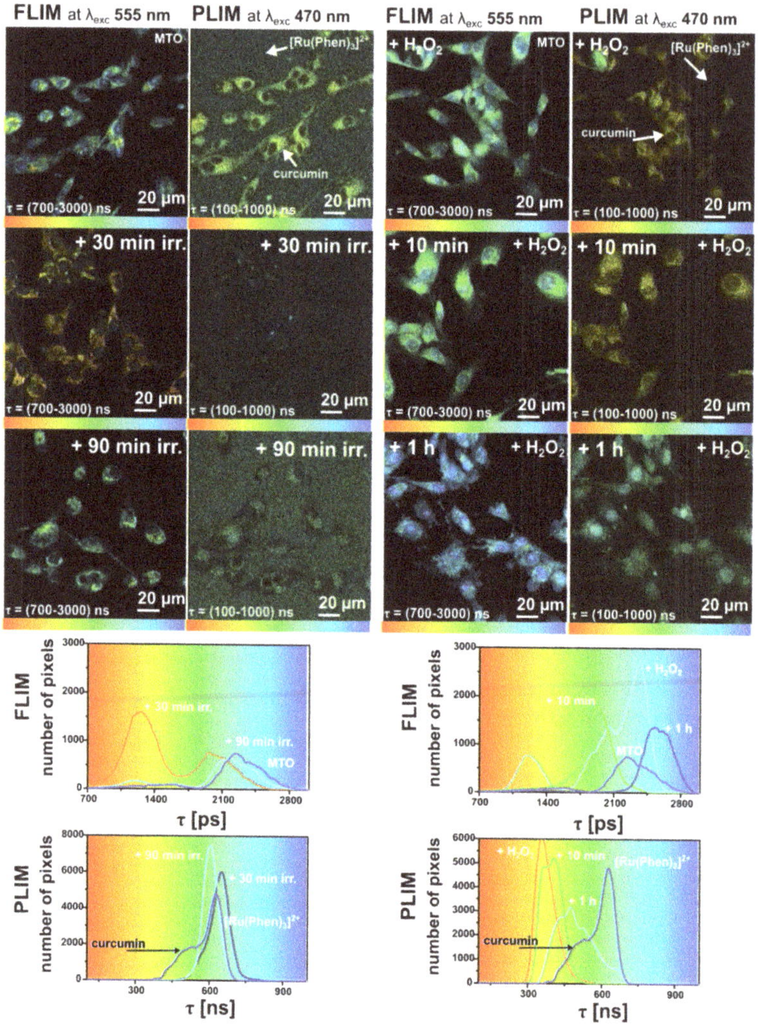

Figure 11. Illustrative FLIM images of MTO and PLIM images of $[Ru(Phen)_3]^{2+}$ both applied in the medium of U87 MG cells for 30 min in the presence of 10 μM curcumin. The phosphorescence of curcumin (localized only in cells) was detected in the same channel as $[Ru(Phen)_3]^{2+}$. The cells were irradiated 30 and 90 min with blue light, 463 ± 10 nm with an irradiance of 100 μW/cm^2 (six images in the upper left). Alternatively, the cells were stressed (six images in the upper right) with 200 μM H_2O_2 (10 min and 1 h after H_2O_2 administration). FLIM and PLIM luminescence lifetime histograms are plotted for both conditions (below the images). The luminescence lifetimes are color-coded (minima—red, maxima—blue).

As we have demonstrated above (Section 2.2), extracellular H_2O_2 induced significant oxidative stress in cells. In one of our previous study, we have demonstrated that this effect can be reduced by catalase (antioxidant) application [15]. One can assume that curcumin-PDT may induce H_2O_2 production in cells and subsequently increase intracellular catalase concentration. For this reason, the oxidative stress was induced extracellularly by H_2O_2 in U87 MG cells treated with curcumin (Figure 11). Indeed, we have first observed a decrease of oxidative stress (decreased MTO fluorescence lifetimes), and a decrease of

$[Ru(Phen)_3]^{2+}$ luminescence lifetimes. This is consistent with the results obtained in the absence of curcumin (Figures 8 and 9). Interestingly, short phosphorescence lifetimes of curcumin were detected in cells. As expected, an increase of MTO fluorescence and $[Ru(Phen)_3]^{2+}$ luminescence lifetimes were observed 1 h after H_2O_2 administration. These results suggest that the H_2O_2 concentration was too high to be reduced by curcumin.

Although we have demonstrated that curcumin possesses antioxidant properties this compound is not as antioxidating as catalase. On the other hand, curcumin can be degraded with H_2O_2. The resulting products of this reaction could result in their interaction with curcumin, and decrease its phosphorescence lifetime. Low concentrations of curcumin at 1 h after H_2O_2 administration disable the antioxidant effect of curcumin and could be another reason for higher for the oxidative stress.

Photodamages induced in cells by curcumin differ from massive injury caused by H_2O_2. For this reason, we have conducted an assessment of the metabolic activity and phototoxicity in U87 MG cells in the presence of all studied molecules before and after irradiation with the blue light. Those results also define the limits of simultaneous MTO and $[Ru(Phen)_3]^{2+}$ luminescence lifetimes detection with minimal phototoxicity.

Two irradiation times were applied: 5 min and 30 min. We have chosen to irradiate cells for 5 min to evaluate if the irradiation during PLIM/FLIM measurement may induce photodamages in the cells. The evaluation was performed in living cells 3 h after irradiation with the WST-8 kit (Figure 12a,b). This kit is biocompatible for living cells and does not requires DMSO administration. Standard MTT-assays were performed 24 h after cell irradiations (Figure 12c,d). It should be noted, that the absorption of WST-8 overlaps with that of $[Ru(Phen)_3]^{2+}$, MTO, and curcumin. For this reason, a level of significance was estimated within the studied sets. A significant decrease of metabolic activity was found 30 min and 3 h after the irradiations in U87 MG cells treated with 10 µM curcumin, and in the presence of MTO (Figure 12a). The metabolic activity of those cells was remarkably suppressed 24 h after the irradiation (Figure 12c). Presence of $[Ru(Phen)_3]^{2+}$ in cells amplified the photoeffect (Figure 12b,d). However, the viability of these cells did not drop below 60%. These results suggest that the antioxidizing activity of curcumin is more important than the apoptotic one. Our results suggest that a simultaneous detection of $[Ru(Phen)_3]^{2+}$ and MTO can be performed without photodamages if the irradiation from excitation laser maintain less than 30 min when the light dose is equal to 100 µW/cm^2.

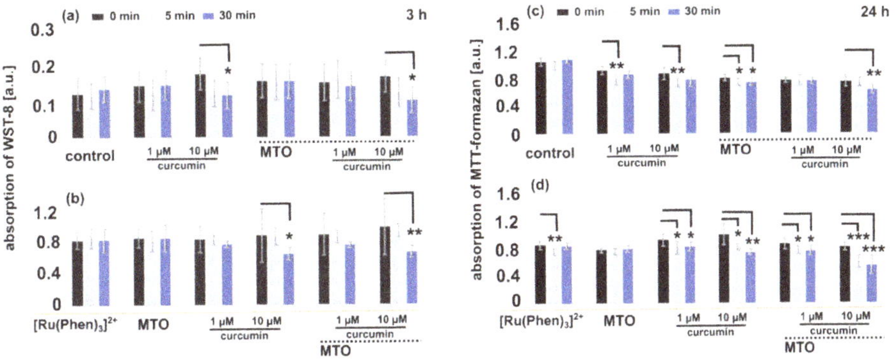

Figure 12. Metabolic activity test of U87 MG cells performed with (**a**,**b**) WST-8 and (**c**,**d**) MTT-formazan 3 and 24 h after the irradiation, respectively. The cells were pre-treated during 1 h with 1 and 10 µM curcumin, 400 nM MTO, and 200 µM $[Ru(Phen)_3]^{2+}$, individually or in combinations, as represented by histograms. The histograms present the results obtained in the (**a**,**c**) absence, and (**b**,**d**) presence of $[Ru(Phen)_3]^{2+}$. The cells were irradiated with blue light (463 ± 10 nm; 100 µW/cm^2) for 5 and 30 min. The average values presented correspond to the means of two independent measurements. Error bars: standard deviations from the mean values. The level of significance was estimated by the Student *t*-test: * $p < 0.05$, ** $p < 0.01$, and *** $p < 0.001$.

3. Materials and Methods

3.1. CAM Model Preparation and Luminescence Lifetime Detection

Fertilized chicken eggs (Animalco AG, Staufen, Switzerland) were incubated in an automatic turning incubator (FIEM snc, Buttigliera d'Asti AT, Italy) with the blunt end up for 3 days (37 °C, 65% humidity, 155.4 mmHg atmospheric oxygen pressure). On the 3rd embryo development day (EDD) a small hole (3 mm in diameter) was perforated in the shell and covered by tape (Scotch® Magic™, St. Paul, MN, USA). Eggs were then returned into the incubator with the blunt end down in a static position until a measurement. The hole in the shell was enlarged (2.5 cm in diameter) at EDD 11.

The chicken embryo chorioallantoic membrane (CAM) was placed under an epi-fluorescence microscope (Nikon Eclipse E 600 FN, Nikon, Tokyo, Japan) to visualize and measure luminophores. A physiological solution of 0.9% NaCl (Braun Melsungen AG, Melsungen, Germany) containing 10 mg/kg of body weight (b.w.) of dichlorotris(1,10-phenanthroline)-ruthenium(II) hydrate ($[Ru(Phen)_3]^{2+}$, 98% purity powder, Sigma-Aldrich, St. Louis, MO, USA) was administered intravenously (iv) with 20 µL aliquots into the main vein of the CAM. The $[Ru(Phen)_3]^{2+}$ luminescence was detected with a digital scientific camera (PCO.1300, PCO Imaging, Kelheim, Germany) under a Hg-arc lamp (HBO 103 W/2, Osram, Munich, Germany) excitation at 470 ± 20 nm. The emission was separated using a 505 nm dichromatic mirror, and a long-pass emission filter at 520 nm. A low magnification objective (4×/0.13, Plan Fluor ∞/−, Nikon, Tokyo, Japan) was used to visualize the $[Ru(Phen)_3]^{2+}$ biodistribution. Image analysis was performed with the Image J software (National Institutes of Health, Bethesda, MD, USA).

The luminescence lifetime of 1 mg/kg of b.w. $[Ru(Phen)_3]^{2+}$ was measured using a dedicated optical fiber-based, time-resolved spectrometer previously described [3,7]. A nitrogen laser-pumped tunable dye (Coumarin 102) laser emitting at 470 nm (<10 ns pulse duration, 10 Hz repetition range) was coupled into a single optical fiber (500 µm diameter) to probe the CAM vessels. The same fiber was used to collect the $[Ru(Phen)_3]^{2+}$ luminescence filtered by a 660–735 nm emission filter (HQ 700/75 M), and detected by a gateable photomultiplier. Autofluorescence was subtracted. Luminescence decays were measured at different locations on the CAM as presented in Figure 1: in the vessels ('in') and out of the vessels ('out'). The eggs were placed into a gas chamber and subjected to 0, 10, and 20% O_2 in N_2 (w/w) that correspond to pO_2: 0, 74, and 155.4 mmHg. Pure nitrogen gas (Carbagas, Muri bei Bern, Switzerland) was mixed by the BRICK-gas mixer (Life Imaging Services GmbH, Basel, Switzerland) with air to reach gas mixtures. The $[Ru(Phen)_3]^{2+}$ luminescence lifetimes were measured at 5, 10, and 20 min after its administration and gas application. Eggs were kept at 37 °C. Ten eggs were measured per condition.

3.2. Luminescence Lifetime Determination

Although the clustering problem is the subject of an extensive literature with stringent arguments, we will briefly deal with only a few of its aspects that are relevant for the experimental data application in question. Generally, the clustering approach can be useful in scientific areas where data occur in clumps, which were referred to as clusters. In short, our study can be viewed as a technique where the decays of luminescence intensity are derived (see Figures 1c and 13a) to provide data inputs that the clustering handles further. The next step introduces the set of lifetimes, which is assumed to consist of a system of all lifetimes corresponding to the same conditions of observation. Using approaches known as descriptive statistics, we can better understand the properties of a set of observations. As a next step, we used the summary (.) function in R [33]. The six components: minimum, 1st quartile, arithmetic mean, median, 3rd quartile, and maximum can be calculated for the set of lifetimes corresponding to the given experimental conditions (see Figures 2 and 13b). The components are forming six-dimensional vectors which can be compared in terms of distances or distance-based measures. More specifically, we used Ward's agglomerative clustering algorithm, which is implemented in R (open-source software for statistical calculations) along with a standard hclust (.) function. The effects were discussed in connection with the Minkowski-type of the distance matrix (in the case of $p = 1$) leading to Ward's technique [40]. For specificity, we state that using R defined by the Ward.D2 method, the distance matrix was transformed into a system of dendrograms (Figure 3). The cophenetic distance between two objects in the dendrogram is the distance of the objects—vectors that are members of the same cluster. Climbing up the dendrogram tree is described by the Height, which is the normalized lifetime difference referring to the pair of experimental conditions associated with the dendrogram legs.

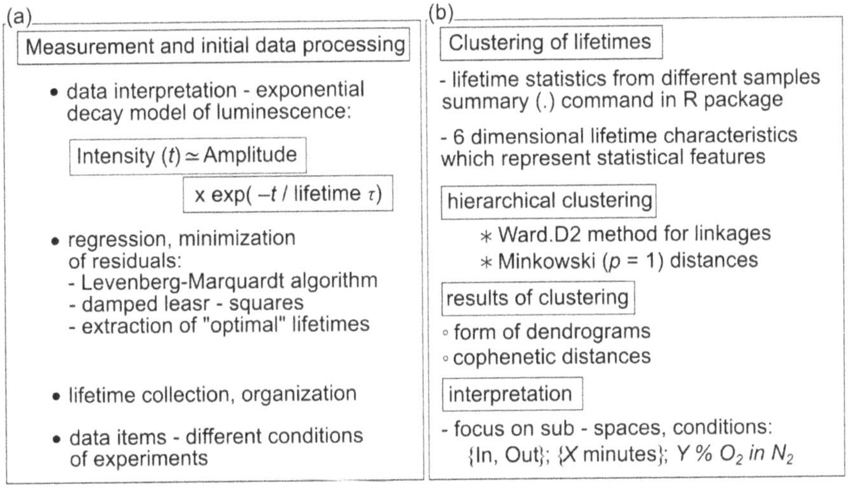

Figure 13. Schematic diagrams of (**a**) measurement and initial data processing, and (**b**) clustering of lifetimes.

3.3. Luminescence Properties of [Ru(Phen)$_3$]$^{2+}$ in Solutions

The [Ru(Phen)$_3$]$^{2+}$ UV-Vis absorption spectra in the presence and absence of 1 mM H_2O_2 (VWR, Paris, France) and H_2O_2 alone diluted in distilled water were detected with an absorption spectrometer (UV-2401 PE, Shimadzu, Sydney, Australia) in the spectral range 200–600 nm with 1 nm step.

The luminescence decays of 10 µM [Ru(Phen)$_3$]$^{2+}$ in the 0.9% NaCl solutions in the absence and presence of 200 nM and 1 mM H_2O_2 were detected as previously described [32]. The samples were excited at 476 nm by CW laser (90C FreD, Coherent, Santa Clara, CA, USA). The laser beam passed via an acousto-optical modulator (AOM, 1205C, Isomet, Springfield, VA, USA) operated in switching ON/OFF mode. It was driven by electrical pulses from a delay generator (20 µs laser pulse, 2 kHz repetition rate). Gas flow (20% O_2 in N_2 (w/w)) and temperature (25 °C) controllers were embedded in a quartz cuvette. The phosphorescence signal was purified by a long-pass filter (>500 nm) and measured with an avalanche photo-diode (APD, APD110A2, Thorlabs, North Newton, NJ, USA). A mono-exponential fit was applied to derive the [Ru(Phen)$_3$]$^{2+}$ luminescence lifetime. The average value was estimated from 10 measurements.

3.4. Cell Culture Preparation and Confocal Fluorescence Imaging

The U87 MG human glioma cells (Cells Lines Services, Eppelheim, Germany) were grown in cell culture medium Dulbecco's modified Eagle medium (D-MEM, Gibco-Invitrogen, Life Technologies Ltd., Paisley, UK) supplemented with 10% fetal bovine serum (FBS), L-glutamine (862 mg/L), sodium pyruvate (110 mg/L), glucose (4500 mg/L) and penicillin/streptomycin (1% w/w), all from Gibco-Invitrogen, Life Technologies Ltd., Paisley, UK. The cells were incubated in the dark at 37 °C, 5% CO_2, and 80% humidified atmosphere. Cells were seeded into the glass-bottom embedded Petri dishes (SPL Confocal dish PS/glass hole, 35 × 10 mm^2) for microscopy at density 10^4 cells per petri dish.

The cells were labeled with 1 μM CellROX® Green (CellROX, Life Technologies™, Carlsbad, CA, USA) and 400 nM MitoTracker™ Orange CMTMRos (MTO, ThermoFisher Scientific, Waltham, MA, USA) to detect oxidative stress and mitochondrial integrity. Cells were treated with 200 μM $[Ru(Phen)_3]^{2+}$ and 10 μM curcumin (Sigma-Aldrich, Darmstadt, Germany) for 1 and 24 h. Fluorescence images were collected with a confocal fluorescence microscope system (LSM 700, Zeiss, Oberkochen, Germany), 63X oil immersion objective (NA 1.46, Zeiss), and a CCD camera (AxioCam HRm, Zeiss). The samples were excited with 405, 488, and 555 nm lasers. The emission was detected as follows: CellROX and curcumin 488 nm/500–550 nm, $[Ru(Phen)_3]^{2+}$ 405 nm/<580 nm, MTO 555 nm/<580 nm. The fluorescence images were analyzed with Zen 2011 software (Zeiss). The oxidative stress was induced by 200 μM H_2O_2 administration and by light from the microscope system during detection. The experiments were performed in triplicates. Cells with $[Ru(Phen)_3]^{2+}$ only and in combination with MTO were irradiated during 30 min with blue light as well.

Viable mitochondria were also stained with 5 μM Rhodamine 123 (Rh123, Sigma-Aldrich, Darmstadt, Germany) for 30 min. The samples were excited with 488 nm laser and the fluorescence was detected in the spectral range 490–530 nm. The nuclei were stained with 10 μg/mL Hoechst 33258 (ThermoFisher Scientific, Waltham, MA, USA) during 30 min. The samples were excited with 405 nm laser and the fluorescence was detected in the spectral range 410–480 nm.

Cellular ROS were determined by 2′,7′-dichlorofluorescin diacetate (DCFDA/H2DCFDA) assay kit (ab113851, Abcam, Cambridge, UK) in cells before and after irradiation, in the presence of $[Ru(Phen)_3]^{2+}$, and 200 μM H_2O_2. The samples were excited with 488 nm laser and the fluorescence was detected in the spectral range 490–530 nm. The quantification of H2DCFDA fluorescence intensity was performed with ImageJ. The mean values of fluorescence intensities (8 images) in the extracellular area and the intracellular area were plotted in histograms. Error bars represent standard deviations. The level of significant difference from the control was calculated with one-way ANOVA: * $p < 0.05$, ** $p < 0.01$, *** $p < 0.001$.

Lipid peroxidation was estimated with the lipid peroxidation assay kit (ab243377, Abcam) in cells in the absence and presence of 200 μM H_2O_2. The samples were excited with 488 nm laser and the fluorescence was detected in the spectral ranges 490–530 nm (green channel) and 560–630 nm (red channel). Lipid peroxidation is represented with increasing of green fluorescence.

Quantification of lipid peroxidation sensor fluorescence was performed with a 96-well plate fluorescence reader (GloMax TM-Multi1Detection system with Instinct Software, Madison, WI, USA) with blue (excitation at 490 nm, green emission at 510–570 nm) and green (excitation at 525 nm, red emission at 580–640 nm) filters. Cells were seeded in the wells of 96-well plate at the density 9×10^3 cells per well and treated with H_2O_2 and lipid peroxidation sensor similarly as for microscopy. The mean values of the fluorescence (three measurements) are presented as the histograms. Error bars represent standard deviations. The level of significant difference was calculated with one-way ANOVA: * $p < 0.05$, ** $p < 0.01$.

3.5. Metabolic Activity of U87 MG Cells before and after Irradiation

The U87 MG cells were treated with 400 nM MTO, 200 μM $[Ru(Phen)_3]^{2+}$, 10 μM curcumin, and their combinations for 1 h. After 1 h the treated cells were kept in dark and irradiated for 5 and 20 min with home-made blue LED-based irradiation platform at 463 ± 10 nm and 100 $\mu W/cm^2$ light dose rate. Metabolic activity was assessed by cell counting Kit-8 (WST-8, Merck, Darmstadt, Germany) at 3rd hour after the irradiation in living cells according to supplier protocol. The absorption of 96-well plates with treated cells was detected with the reader (GloMax TM-Multi1Detection system with Instinct Software, Madison, WI, USA) at 450 nm. The 3-(4,5-dimethylthiazol-2-yl)-2,5-diphenyltetrazolium bromide (MTT, Sigma-Aldrich, Darmstadt, Germany) was applied to cells 24 h after the irradiation according to supplier protocol. The purple crystals of formazan were dissolved in 100% dimethyl sulfoxide (DMSO, Sigma-Aldrich, Darmstadt, Germany) and measured with the absorption reader at 560 nm.

3.6. Fluorescence and Phosphorescence Lifetime Imaging of U87 MG Cells

An inverted fluorescence microscopy system (Zeiss AxioObserver Z1, Zeiss) equipped with 40× water immersion objective (NA = 1.2, Zeiss) and connected to a fluorescence lifetime imaging system (FLIM, DSC-120 Dual Channel Confocal Scanning system, Becker & Hickl GmbH, Berlin, Germany) was used for time-resolved fluorescence microscopy. The 400 nM MTO was detected in the fluorescence lifetime imaging (FLIM) mode. The samples were excited with a pulsed NKT-Super-K Extreme laser (40 MHz, 5 ps pulse, NKT Photonics, Birkerød, Denmark) at 555 nm (2% of total power). The MTO fluorescence lifetimes were detected at the 25 ns time range with an HPM-100-50 hybrid detector (Becker & Hickl GmbH). The 200 μM $[Ru(Phen)_3]^{2+}$ and 10 μM curcumin were detected in the phosphorescence lifetime imaging (PLIM) mode. Samples were excited with the same laser as for MTO at 470 nm (10% of total power) and a scan rate of 10.23 ms line time, 2.62 s/frame. The $[Ru(Phen)_3]^{2+}$ and curcumin phosphorescence lifetimes were detected at the 25.6 μs time range with the same HPM-100-50 hybrid detector. The light path was filtered with LP 488 and BP 624 ± 20 nm (NT67-035, EDMUND Optics, Barrington, NJ, USA). Time decays were analyzed with SPC image analysis software (Becker & Hickl GmbH). The quality of the fits was graphically checked by plotting the residuals and $\chi^2 \sim 1$. The experiments were performed in triplicates. Different regions of interest (ROI) were selected. The positions of ROI were depicted in figures. The FLIM of MTO was fitted with a mono-exponential function. The PLIM of $[Ru(Phen)_3]^{2+}$ were fitted with bi-exponential function and t2 was fixed at 20,000 ns (a probability of the occurrence was below 1). FLIM and PLIM were detected in living cells treated 1 h with MTO, curcumin, and $[Ru(Phen)_3]^{2+}$. All probes were maintained in the media during detection. Intracellular $[Ru(Phen)_3]^{2+}$ detection was performed 24 h after the treatment. Oxidative stress in cells was induced by 200 μM H_2O_2 and by irradiation with blue LED light (463 ± 10 nm and 100 $\mu W/cm^2$ light dose rate) for 30 and 90 min. Cells with $[Ru(Phen)_3]^{2+}$ only and in combination with MTO were 30 min irradiated with blue light as well. The detection was performed shortly after the irradiation. The luminescence signal and lifetime of $[Ru(Phen)_3]^{2+}$ was not detected in the MTO fluorescence detection window, and the phosphorescence of MTO in $[Ru(Phen)_3]^{2+}$ window was eliminated. Curcumin fluorescence was not detected in the MTO fluorescence window. Curcumin phosphorescence was present in $[Ru(Phen)_3]^{2+}$ detection window, however, the localization of this signal was clearly defined only within the cells.

3.7. Western Blot Analysis

The U87 MG human glioma cells were seeded in 25 cm^2 flasks at the density 5×10^5 cells. Cells were treated for 30 min with 200 μM H_2O_2 before lysis in radioimmunoprecipitation (RIPA) buffer (150 mM sodium chloride, 1% Triton X-100, 0.5% sodium deoxycholate, 0.1% sodium dodecyl sulphate, 50 mM Tris, pH 8). Western blot analysis was performed similarly as described in [15]. Oxidative stress defense cocktail (ab179843, Abcam, Cambridge, UK) was applied to estimate the expression level of catalase, superoxide dismutase 1, thioredoxin, and smooth muscle actin in cells. Anti-β-actin antibody (ab8227, Abcam, Cambridge, UK) was determined as a housing protein. The WesternBreeze chromogenic kit anti- rabbit was purchased from ThermoFisher Scientific. The image analysis of proteins on the membrane was performed with ImageJ software. Optical densities (O.D.) of the bands were detected with ImageJ and analyzed, the normalized O.D. values are the mean values from 4 measurements and are plotted in histograms (down). Error bars represent standard deviations. The level of significant difference was calculated with one-way ANOVA: * $p < 0.05$, ** $p < 0.01$.

4. Conclusions

The present manuscript brings a new point of view to interpret the data obtained by time-resolved measurement of the $[Ru(Phen)_3]^{2+}$ luminescence lifetimes in cells. We have shown that hierarchical clustering, a key element in the processing of data sets reflecting the $[Ru(Phen)_3]^{2+}$ oxygen sensitivity, offers many opportunities to carry out the analysis. Significant relationships in the tissue oxygenation level detections were revealed by specific differences between dendrograms. This study enabled to show that $[Ru(Phen)_3]^{2+}$ molecules distributed in the extravascular space were more sensitive to external changes of the oxygen level, the dendrograms unambiguously segmented the data of high and low oxygenated tissues (in space and time). We have assumed that the oxygen sensitivity of $[Ru(Phen)_3]^{2+}$ luminescence lifetime is influenced by oxidative stress induced in cells. For this reason, and for the simplicity, $[Ru(Phen)_3]^{2+}$ and MTO were co-administered and their luminescence and fluorescence lifetimes were detected with the unique fluorescence and phosphorescence lifetime imaging microscope to register the changes in the oxygenation and oxidative stress in glioma cells.

With our new approach, we have demonstrated that $[Ru(Phen)_3]^{2+}$ can be applied in combination with MTO to detect extracellular oxygen and intracellular oxidative stress levels. This combination works well when oxidative stress is induced externally, for example when H_2O_2 is used for this purpose. However, PDT which is known to consumed oxygen intracellularly did not induce variations of the $[Ru(Phen)_3]^{2+}$ luminescence lifetime in cell culture media in our studied system. On the other hand, MTO sensing revealed both, antioxidant and oxidative stress production activity of curcumin-mediated PDT.

We have demonstrated that the photo-toxicity of the combination of MTO and $[Ru(Phen)_3]^{2+}$ was low and that the detection of oxygenation and oxidative stress can be performed without cell photodamage when such measurements are performed up to 1 h after the administration of these dyes when the excitation is at 463 $\pm$ 10 nm with an irradiance of 100 $\mu W/cm^2$ (30 min).

It would be interesting to see what would happen if this approach is applied to cancer cell spheroids as a model of tumor. One can expect that $[Ru(Phen)_3]^{2+}$ will localize in the interstitial space, hence probing the oxygen level in spheroids instead of in the cells monolayers. Another advantage of the small 3D system is that the diffusion of oxygen and its consumption will be better controlled in the small volume of the spheroid in comparison with the CAM. Using optimized PS and oxygen-sensitive probes will probably enable to improve the imaging in real-time of the photo-destruction, oxygen consumption, and oxidative stress induction during PDT.

Supplementary Materials: The following are available online, Figure S1: Decomposition of confocal microscopy images of U87 MG cells stained with CellROX®Green, rhodamine 123, and $[Ru(Phen)_3]^{2+}$; Figure S2: Mitochondria of U87 MG cells in the presence and absence of H_2O_2; Figure S3: Western blot analysis of oxidative stress defense proteins: catalase, superoxide dismutase 1, and thioredoxin in U87 MG cells in the absence and presence H_2O_2; Figure S4: Cellular ROS visualized with DCFDA/H2DCFDA assay in U87 MG cells in the presence of $[Ru(Phen)_3]^{2+}$, light and H_2O_2; Figure S5: Lipid peroxidation visualized in U87 MG cells in the absence and presence H_2O_2; Figure S6: Mitochondria of U87 MG cells in the presence of $[Ru(Phen)_3]^{2+}$ after irradiation; Figure S7: PLIM images of $[Ru(Phen)_3]^{2+}$ in U87 MG cells after irradiation; Figure S8: FLIM and PLIM images of MTO and $[Ru(Phen)_3]^{2+}$ before and after irradiation of U87 MG.

Author Contributions: Conceptualization, V.H. and D.H., and G.W.; methodology, V.H., D.H., and G.W.; software, D.H.; validation, R.S., and V.H.; investigation, V.H., D.H., R.S., and G.W.; data curation, V.H., and D.H.; writing—original draft preparation, V.H., D.H., R.S., and G.W.; visualization, V.H., and D.H. All authors have read and agreed to the published version of the manuscript.

Funding: This research was funded by the Ministry of education, science, research and sport of the Slovak Republic, grant number VEGA 1/0156/18, and VEGA 1/0421/18, the Swiss National Science Foundation (project 315230 185262/1) and by the Slovak Research and development agency, grant number APVV-15-0485. This publication is the result of the project implementation: Open scientific community for modern interdisciplinary research in medicine (Acronym: OPENMED), ITMS2014+: 313011V455 supported by the Operational Programme Integrated Infrastructure, funded by the ERDF.

Data Availability Statement: Data contained in this paper are available from the authors.

Conflicts of Interest: The authors declare no conflict of interest.

Sample Availability: Samples of the compounds are commercially available.

References

1. Stepinac, T.K.; Chamot, S.K.; Rungger-Brandle, E.; Ferrez, P.; Munoz, J.L.; van den Bergh, H.; Riva, C.E.; Pournaras, C.J.; Wagnieres, G.A. Light-induced retinal vascular damage by Pd-porphyrin luminescent oxygen probes. *Investig. Ophthalmol. Vis. Sci.* **2005**, *46*, 956–966. [CrossRef] [PubMed]
2. Piffaretti, F.; Santhakumar, K.; Forte, E.; van den Bergh, H.E.; Wagnieres, G.A. Optical fiber-based setup for in vivo measurement of the delayed fluorescence lifetime of oxygen sensors. *J. Biomed. Opt.* **2011**, *16*, 069801, Erratum in **2011**, *16*, 037005. [CrossRef]
3. Huntosova, V.; Gay, S.; Nowak-Sliwinska, P.; Rajendran, S.K.; Zellweger, M.; van den Bergh, H.; Wagnieres, G. In vivo measurement of tissue oxygenation by time-resolved luminescence spectroscopy: Advantageous properties of dichlorotris(1, 10-phenanthroline)-ruthenium(II) hydrate. *J. Biomed. Opt.* **2014**, *19*, 077004. [CrossRef] [PubMed]
4. O'Neal, D.P.; Meledeo, M.A.; Davis, J.R.; Ibey, B.L.; Gant, V.A.; Pishko, M.V.; Cote, G.L. Oxygen sensor based on the fluorescence quenching of a ruthenium complex immobilized in a biocompatible poly(ethylene glycol) hydrogel. *IEEE Sens. J.* **2004**, *4*, 728–734. [CrossRef]
5. Grist, S.M.; Chrostowski, L.; Cheung, K.C. Optical Oxygen Sensors for Applications in Microfluidic Cell Culture. *Sensors* **2010**, *10*, 9286–9316. [CrossRef] [PubMed]
6. Baleizao, C.; Nagl, S.; Schaferling, M.; Berberan-Santos, M.N.; Wolfbeis, O.S. Dual fluorescence sensor for trace oxygen and temperature with unmatched range and sensitivity. *Anal. Chem.* **2008**, *80*, 6449–6457. [CrossRef]
7. Huntosova, V.; Stroffekova, K.; Wagnieres, G.; Novotova, M.; Nichtova, Z.; Miskovsky, P. Endosomes: Guardians against [Ru(Phen)(3)](2+) photo-action in endothelial cells during in vivo pO(2) detection? *Metallomics* **2014**, *6*, 2279–2289. [CrossRef]
8. Huntosova, V.; Gerelli, E.; Horvath, D.; Wagnieres, G. Measurement of pO(2) by luminescence lifetime spectroscopy: A comparative study of the phototoxicity and sensitivity of [Ru(Phen)(3)](2+) and PdTCPP in vivo. *J. Biophotonics* **2017**, *10*, 708–717. [CrossRef]
9. Zarebski, M.; Kordon, M.; Dobrucki, J.W. Photosensitized Damage Inflicted on Plasma Membranes of Live Cells by An Extracellular Generator of Singlet Oxygen-A Linear Dependence of A Lethal Dose on Light Intensity. *Photochem. Photobiol.* **2014**, *90*, 709–715. [CrossRef]
10. Belej, D.; Jurasekova, Z.; Nemergut, M.; Wagnieres, G.; Cura, D.J.; Huntosova, V. Negligible interaction of [Ru(Phen)(3)](2+) with human serum albumin makes it promising for a reliable in vivo assessment of the tissue oxygenation. *J. Inorg. Biochem.* **2017**, *174*, 37–44. [CrossRef]
11. Komor, A.C.; Barton, J.K. The path for metal complexes to a DNA target. *Chem. Commun.* **2013**, *49*, 3617–3630. [CrossRef] [PubMed]
12. Hinton, G.; Sejnowski, T.J. *Unsupervised Learning: Foundations of Neural Computation*; The MIT Press: Cambridge, MA, USA, 1999. [CrossRef]
13. Albalate, A.; Minker, W. (Eds.) *Semi-Supervised and Unsupervised Machine Learning: Novel Strategies*; WILEY: Hoboken, NJ, USA, 2011.
14. Katerji, M.; Filippova, M.; Duerksen-Hughe, P. Approaches and Methods to Measure Oxidative Stress in Clinical Samples: Research Applications in the Cancer Field. *Oxid. Med. Cell Longev.* **2019**, *2019*, 1–29. [CrossRef] [PubMed]
15. Tomkova, S.; Misuth, M.; Lenkayska, L.; Miskovsky, P.; Huntosova, V. In vitro identification of mitochondrial oxidative stress production by time-resolved fluorescence imaging of glioma cells. *BBA-Mol. Cell Res.* **2018**, *1865*, 616–628. [CrossRef] [PubMed]
16. Scorrano, L.; Petronilli, V.; Colonna, R.; Di Lisa, F.; Bernardi, P. Chloromethyltetramethylrosamine (Mitotracker Orange (TM)) induces the mitochondrial permeability transition and inhibits respiratory complex I—Implications for the mechanism of cytochrome c release. *J. Biol. Chem.* **1999**, *274*, 24657–24663. [CrossRef] [PubMed]
17. Szilagyi, G.; Simon, L.; Koska, P.; Telek, G.; Nagy, Z. Visualization of mitochondrial membrane potential and reactive oxygen species via double staining. *Neurosci. Lett.* **2006**, *399*, 206–209. [CrossRef] [PubMed]
18. Liu, Q.M.; Li, A.Y.; Li, X.K.; Li, B.; Zhang, Y.H.; Li, J.; Guo, Y. Selective visualization of live-cell mitochondria' thiophenols and their induced oxidative stress process by a rationally designed rhodol-based fluorescent probe. *Sens. Actuators B Chem.* **2019**, *283*, 820–830. [CrossRef]
19. Formella, I.; Svahn, A.J.; Radford, R.A.W.; Don, E.K.; Cole, N.J.; Hogan, A.; Lee, A.; Chung, R.S.; Morsch, M. Real-time visualization of oxidative stress-mediated neurodegeneration of individual spinal motor neurons in vivo. *Redox Biol.* **2018**, *19*, 226–234. [CrossRef] [PubMed]

20. Alavi, M.V.; Chiang, W.C.; Kroeger, H.; Yasumura, D.; Matthes, M.T.; Iwawaki, T.; LaVail, M.M.; Gould, D.B.; Lin, J.H. In Vivo Visualization of Endoplasmic Reticulum Stress in the Retina Using the ERAI Reporter Mouse. *Investig. Ophthalmol. Vis. Sci.* **2015**, *56*, 6961–6970. [CrossRef]
21. Choi, H.; Yang, Z.L.; Weisshaar, J.C. Single-cell, real-time detection of oxidative stress induced in Escherichia coli by the antimicrobial peptide CM15. *Proc. Natl. Acad. Sci. USA* **2015**, *112*, E303–E310. [CrossRef]
22. Maharjan, S.; Oku, M.; Tsuda, M.; Hoseki, J.; Sakai, Y. Mitochondrial impairment triggers cytosolic oxidative stress and cell death following proteasome inhibition. *Sci. Rep.* **2014**, *4*, 5896. [CrossRef]
23. Balke, J.; Volz, P.; Neumann, F.; Brodwolf, R.; Wolf, A.; Pischon, H.; Radbruch, M.; Mundhenk, L.; Gruber, A.D.; Ma, N.; et al. Visualizing Oxidative Cellular Stress Induced by Nanoparticles in the Subcytotoxic Range Using Fluorescence Lifetime Imaging. *Small* **2018**, *14*, 1800310. [CrossRef] [PubMed]
24. Yang, J.; Cao, Q.; Zhang, H.; Hao, L.; Zhou, D.; Gan, Z.; Li, Z.; Tong, Y.X.; Ji, L.N.; Mao, Z.W. Targeted reversal and phosphorescence lifetime imaging of cancer cell metabolism via a theranostic rhenium(I)-DCA conjugate. *Biomaterials* **2018**, *176*, 94–105. [CrossRef] [PubMed]
25. Cao, Q.; Zhou, D.J.; Pan, Z.Y.; Yang, G.G.; Zhang, H.; Ji, L.N.; Mao, Z.W. CAIXplatins: Highly Potent Platinum(IV) Prodrugs Selective Against Carbonic Anhydrase IX for the Treatment of Hypoxic Tumors. *Angew. Chem. Int. Ed.* **2020**, *59*, 18556–18562. [CrossRef] [PubMed]
26. Yang, J.; Zhao, J.X.; Cao, Q.; Hao, L.; Zhou, D.; Gan, Z.; Ji, L.N.; Mao, Z.W. Simultaneously Inducing and Tracking Cancer Cell Metabolism Repression by Mitochondria-Immobilized Rhenium(I) Complex. *ACS Appl. Mater. Interfaces* **2017**, *9*, 13900–13912. [CrossRef] [PubMed]
27. Agostinis, P.; Berg, K.; Cengel, K.A.; Foster, T.H.; Girotti, A.W.; Gollnick, S.O.; Hahn, S.M.; Hamblin, M.R.; Juzeniene, A.; Kessel, D.; et al. Photodynamic Therapy of Cancer: An Update. *CA Cancer J. Clin.* **2011**, *61*, 250–281. [CrossRef] [PubMed]
28. Shen, L.; Ji, H.F.; Zhang, H.Y. A TD-DFT study on triplet excited-state properties of curcumin and its implications in elucidating the photo sensitizing mechanisms of the pigment. *Chem. Phys. Lett.* **2005**, *409*, 300–303. [CrossRef]
29. Chignell, C.F.; Bilski, P.; Reszka, K.J.; Motten, A.G.; Sik, R.H.; Dahl, T.A. Spectral and Photochemical Properties of Curcumin. *Photochem. Photobiol.* **1994**, *59*, 295–302. [CrossRef]
30. Klinger, N.V.; Mittal, S. Therapeutic Potential of Curcumin for the Treatment of Brain Tumors. *Oxid. Med. Cell. Longev.* **2016**, *2016*, 1–14. [CrossRef]
31. Das, K.C.; Das, C.K. Curcumin (diferuloylmethane), a singlet oxygen (O-1(2)) quencher. *Biochem. Biophys. Res. Commun.* **2002**, *295*, 62–66. [CrossRef]
32. Varchola, J.; Huntosova, V.; Jancura, D.; Wagnieres, G.; Miskovsky, P.; Bano, G. Temperature and oxygen-concentration dependence of singlet oxygen production by RuPhen as induced by quasi-continuous excitation. *Photochem. Photobiol. Sci.* **2014**, *13*, 1781–1787. [CrossRef]
33. R Documentation. Summary(.) Function in R. Available online: https://www.rdocumentation.org/packages/base/versions/3.6.2/topics/summary (accessed on 1 July 2020).
34. Warren, D.R.; Partridge, M. The role of necrosis, acute hypoxia and chronic hypoxia in F-18-FMISO PET image contrast: A computational modelling study. *Phys. Med. Biol.* **2016**, *61*, 8596–8624. [CrossRef] [PubMed]
35. Wang, Y.L.; Msghina, M.; Li, T.Q. Studying Sub-Dendrograms of Resting-State Functional Networks with Voxel-Wise Hierarchical Clustering. *Front. Hum. Neurosci.* **2016**, *10*, 75. [CrossRef] [PubMed]
36. Zhang, X.; Wang, L.; Lu, H.; Zong, Z.; Chen, Z.; Li, Y.; Luo, X.; Li, Y. Preservation of hydrogen peroxide-induced oxidative damage in HepG-2 cells by rice protein hydrolysates pretreated with electron beams. *Sci. Rep.* **2020**, *10*, 8415. [CrossRef] [PubMed]
37. de Oyanguren, F.J.S.; Rainey, N.E.; Moustapha, A.; Saric, A.; Sureau, F.; O'Connor, J.E.; Petit, P.X. Highlighting Curcumin-Induced Crosstalk between Autophagy and Apoptosis as Supported by Its Specific Subcellular Localization. *Cells* **2020**, *9*, 361. [CrossRef]
38. Khopde, S.M.; Priyadarsini, K.; Palit, D.K.; Mukherjee, T. Effect of solvent on the excited-state photophysical properties of curcumin. *Photochem. Photobiol.* **2000**, *72*, 625–631. [CrossRef]
39. Huntosova, V.; Gerelli, E.; Zellweger, M.; Wagnieres, G. Effect of PpIX photoproducts formation on pO(2) measurement by time-resolved delayed fluorescence spectroscopy of PpIX in solution and in vivo. *J. Photochem. Photobiol. B* **2016**, *164*, 49–56. [CrossRef]
40. Lee, A.; Willcox, B. Minkowski Generalizations of Ward's Method in Hierarchical Clustering. *J. Classif.* **2014**, *31*, 194–218. [CrossRef]

molecules

MDPI

Article

Intermediate Detection in the Casiopeina–Cysteine Interaction Ending in the Disulfide Bond Formation and Copper Reduction

Lillian G. Ramírez-Palma [1,2,†], Adrián Espinoza-Guillén [3,†], Fabiola Nieto-Camacho [3], Alexis E. López-Guerra [3], Virginia Gómez-Vidales [1], Fernando Cortés-Guzmán [1,2,*] and Lena Ruiz-Azuara [3,*]

1 Instituto de Química, Universidad Nacional Autónoma de México, Av. Universidad 3000, C. U., México City 04510, Mexico; lila.gis.rp@gmail.com (L.G.R.-P.); gomvidal@unam.mx (V.G.-V.)

2 Centro Conjunto de Investigación en Química Sustentable UAEMex-UNAM, Carretera Toluca-Atlacomulco km 14.5, Toluca 50200, Mexico

3 Facultad de Química, Universidad Nacional Autónoma de México, Av. Universidad 3000, C. U., México City 04510, Mexico; adrianeg24@gmail.com (A.E.-G.); fabille.nc@gmail.com (F.N.-C.); eduga_9@live.com.mx (A.E.L.-G.)

* Correspondence: fercor@unam.mx (F.C.-G.); lenar701@gmail.com (L.R.-A.)

† These authors contributed equally to this work.

Abstract: A strategy to improve the cancer therapies involves agents that cause the depletion of the endogenous antioxidant glutathione (GSH), increasing its efflux out of cells and inducing apoptosis in tumoral cells due to the presence of reactive oxygen species. It has been shown that Casiopeina copper complexes caused a dramatic intracellular GSH drop, forming disulfide bonds and reducing Cu^{II} to Cu^{I}. Herein, through the determination of the $[Cu^{II}]$–SH bond before reduction, we present evidence of the adduct between cysteine and one Casiopeina as an intermediate in the cystine formation and as a model to understand the anticancer activity of copper complexes. Evidence of such an intermediate has never been presented before.

Keywords: copper complexes; glutathione depletion; cysteine; Casiopeina; intermediate; copper reduction

Citation: Ramírez-Palma, L.G.; Espinoza-Guillén, A.; Nieto-Camacho, F.; López-Guerra, A.E.; Gómez-Vidales, V.; Cortés-Guzmán, F.; Ruiz-Azuara, L. Intermediate Detection in the Casiopeina–Cysteine Interaction Ending in the Disulfide Bond Formation and Copper Reduction. *Molecules* **2021**, *26*, 5729. https://doi.org/10.3390/molecules26195729

Academic Editors: Višnja Stepanić, Marta Kučerová-Chlupáčová and Helen Osborn

Received: 10 July 2021
Accepted: 14 September 2021
Published: 22 September 2021

1. Introduction

The study of the oxidation of endogenous thiolated compounds, such as glutathione (GSH) and cysteine (Cys), is essential to understand the oxidative stress within the cells [1]. Glutathione (γ-L-glutamyl-L-cysteinyl-glycine, GSH) is a tripeptide that participates in redox processes into the cells, where the Cys residues of GSH are easily oxidized to disulfide (GSSG) [2,3]. It also participates in cancer cell protection against xenobiotics, ionizing radiations, and oxidative stress. Its oxidation favors the opening of the mitochondrial permeability transition pore complex, facilitating the release of death-related molecular signals [4,5]. A strategy to improve cancer therapies' efficacy should involve cytosolic and mitochondrial GSH depletion through an increase of GSH efflux out of cells [4]. Kachadourian and coworkers tested, in human lung cancer cells (H157 and A549), one copper complex of the Casiopeina® family, showing that it induced a dramatic drop in the intracellular levels of GSH (Figure 1A) [6]. Additionally, there are reports of GSH depletion produced by copper complexes on cervix HeLa and neuroblastoma CHP-212. GSSG/GSH and cystine/cysteine's redox potentials are very similar, −263 and −220 mV vs. normal hydrogen electrode (NHE) respectively [7,8], with glutathione being a slightly better reducing agent (Figure 1C). The oxidation by copper(II) ions of cysteine-containing peptides such as glutathione and cysteine has been studied to understand this biometal's role in oxidative stress processes [9–12]. The copper favors the oxidation of thiols [11,13], where Cu^{II} reacts with thiols to form $[Cu^{II}]$–thiol adducts depending on their molar ratio. The $[Cu^{II}]$–thiol complex is reduced to Cu^{I}, and in turn, thiols are oxidized to the corresponding radicals. The CasIII-ia ([Cu(4,4′-dimethyl-2,2′-bipyridine)(acetylacetonate)]NO_3*$(H_2O)_2$,

Figure 1B) is a copper(II) complex from the Casiopeina family with a potential of 62 mV vs. NHE [14], which could mediate the oxidation of biological thiols (Figure 1C). Recently, a paper [15] demonstrated the formation of $O_2{\cdot}^-$ when bisdiimine copper(II) chelates were reduced by ascorbate. However, the pathway by which Cu^{II} is reduced to Cu^{I} has not yet been described; therefore, in this work, we present a strategy to understand this reduction process and determine the possible intermediate.

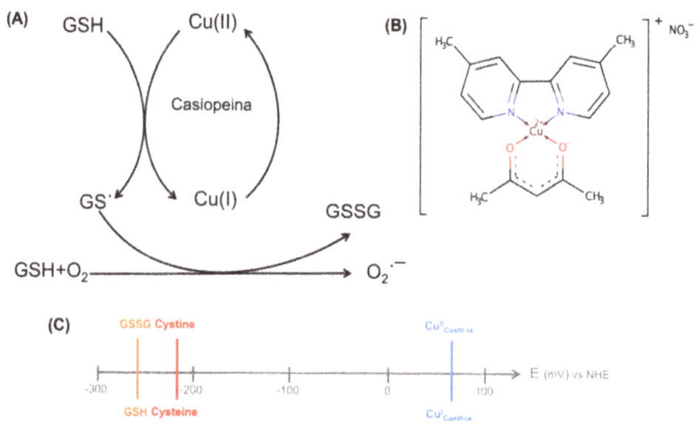

Figure 1. (**A**) The hypothesis of the interaction between the Casiopeina family and a thiolated redactor [6]. (**B**) CasIII-ia structure. (**C**) The redox potential of thiol-containing biomolecules and CasIII-ia.

These observations suggest the importance of copper complexes, such as the Casiopeinas, to oxidate thiol residues, inducing mitochondrial damage [6]. From the above considerations, in the present work, we study the specific interaction between cysteine and CasIII-ia to understand the pathway leading to a final reduction of the copper metal center and the disulfide bond formation. We focused on Cys, the reactive site of GSH, because it reacts slower than GSH in solution, and standard analytic techniques can detect its reactive intermediates. Additionally, we used mechanochemical methods to reduce the oxidation reaction rate. The family of Casiopeínas® [16] was designed considering three elements: copper(II) as a central metal and two mixed ligands with several substituents, providing to copper the possibility of producing cytotoxicity through different mechanisms of action [17]. Ligands modify the cationic copper's transport properties [18], the electronic properties of the central metal [19], and the molecular recognition of the complex [20]. The Casiopeina activity goes along with the cytotoxic effects, such as the generation of reactive oxygen species (ROS) [21,22], which can damage cellular components through oxidation and alter the oxidation-reduction balance cell or interfere with the mechanisms of cellular signaling related to the redox state [23]. There are reports about the antiproliferative and antineoplastic activities over murine and xenografted human tumors [24,25]. On the other hand, survival was evaluated in murine models: L1210 (leukemia), S180 (sarcoma), and B16 (melanoma) [26]. In non-tumor cells, the mean inhibitory concentration of this compound was 4.7, determined in lymphocytes, approximately 250 times higher than that observed in tumor lines, suggesting a selectivity towards tumoral cells [27]. Additionally, pharmacokinetic analyses performed with rat urine [28] and dog blood [29] samples have proved a high elimination rate of the Casiopeinas. All these results indicate an increase in the in vivo activity of the compound CasIII-ia, concerning the cisplatin activity as a positive control, evaluated in the same system. Our group has investigated the relationship between the features of the metal complex and their activity. A QSAR study showed that the half-wave potential and aromatic ring in the molecule are relevant for the compounds' action [14].

Other derived models can predict mixed chelate copper complexes' degree of activity based on the chemical correlation between structure, EPR, and electrochemical behavior, supported by DFT calculations [19]. We also developed a regression model to reproduce the antiproliferative activity involving the atomic delocalization and dipole moment changes within the ligands' C-N bonds [30]. These bonds are also determinant for the recognition site of copper complexes by the DNA backbone [20].

2. Results and Discussion

In solution, the endogenous reductant species, such as GSH, react very fast with Casiopeinas to be detected by standard analytical methods. For this reason, we decided to focus on the reactive site of GSH and Cys, and perform the reaction process in the solid state to observe the steps occurring in the Cys oxidation process. It has been reported that the Cys redox reaction barrier is smaller than the GSH one and also that the redox process is faster than any ligand exchange. In this way, Cys is an acceptable model of GSH for the process catalyzed by the CasIII-ia.

The CasIII-ia was prepared by the patent procedure [31–33]. To prove the inclusion of Cys in the copper coordination sphere and determine the features of the $[Cu^{II}]$–Cys interaction, we performed powder X-ray diffraction (PXRD), electronic spectra (UV-Vis-NIR), infrared spectroscopy (FTIR), electron paramagnetic resonance spectroscopy (EPR), and mass spectra-direct analysis in real-time (MS-DART) analyses. To understand the structural evolution of the $[Cu^{II}]$–Cys adduct, we carried out a DFT computational analysis for the cooper reduction process.

The reaction between equimolar solutions of Cys (colorless) and CasIII-ia (blue) showed color changes from brown/green to blue color (see the Movie S1 in the Supplementary Materials), and the same observations were reported by Seko et al. [34] and Ugone et al. [35]. EPR spectra of fresh mixtures of two CasIII-ia and Cys stoichiometric solutions were recorded at 77 K. Figure 2 shows the copper electronic environment change of CasIII-ia when it interacts with Cys. Therefore, it is proposed that CasIII-ia generates a $[Cu^{II}]$–Cys adduct before its reduction. However, the reaction's rapid kinetics avoids studying the $[Cu^{II}]$–Cys adduct in solution using this technique. Then, we opted to change the conditions to reduce the reaction rate with a mechanochemical solid-state approach. CasIII-ia and Cys were milled until a homogenous solid mixture was obtained, EtOH was added, and was mixed until dryness. During the mechanical process, it was possible to observe a color change, from Pantone 2139c to Pantone 289c (Figure 3), that can be associated with Cys' coordination to CasIII-ia. It was impossible to isolate and purify the reaction intermediates; therefore, we used several techniques to identify them within the reaction mixture.

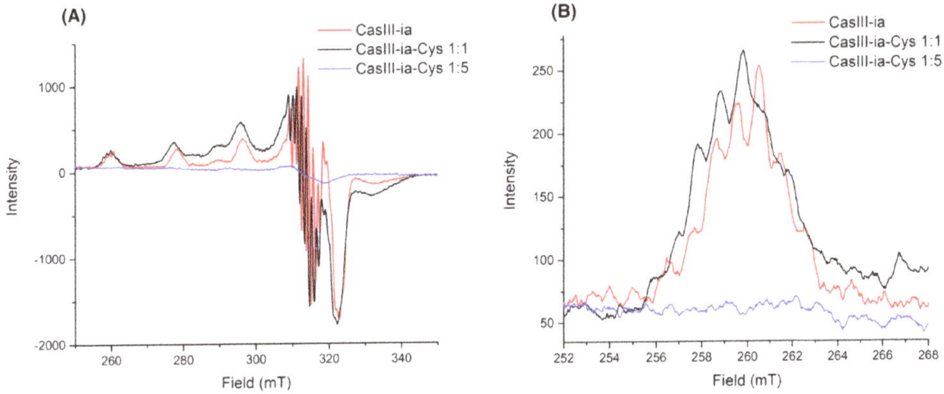

Figure 2. Experimental EPR spectra for CasIII-ia (1 mM) in red, CasIII-ia–Cys (1:1) in black, and CasIII-ia–Cys (1:5) in blue, in MeOH/H_2O (1:1) frozen solution at 77 K. (**A**) From 250 to 350 mT, and (**B**) from 252 to 268 mT.

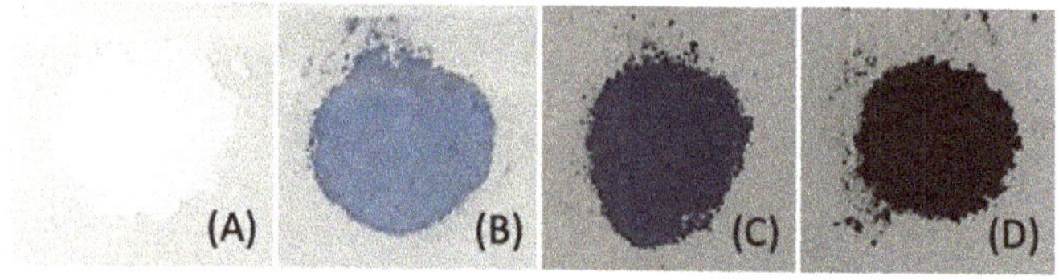

Figure 3. Reactant and intermediates of the mechanochemical solid-state reaction process: (**A**) cysteine (white), (**B**) CasIII-ia (Pantone 2139c), (**C**) intermediate 1 (Pantone 289c), and (**D**) intermediate 2 (Pantone 532c).

First, we analyzed the powder X-Ray diffraction patterns of the reactants and the reaction mixture. The crystal square pyramid structure of CasIII-ia has been reported in a previous description (CCDC 1440021). Supplementary Figure S1 shows the diffraction pattern of the reaction mixture and the two reactants. It is possible to observe that the crystalline arrangement is conserved. Some remnant signals can be associated with the Cys and the CasIII-ia; however, some signals are no longer present, such as the 7.9 intense signals in 2θ. On the other hand, new signals are presented, such as 11.24, 12.12, 17.18, 25.94, 32.58, 33.02, 37.92, and 38.58 (see Supplementary Materials for details). These peaks confirm a new species that still have copper(II) as a metal center.

The second evidence of the presence of a new copper(II) complex is the UV-Vis spectrum of the reaction mixture, as shown in Figure 4. The reaction mixture spectrum resembles that for CasIII-ia–Cys, with maxima at 395 and 598 nm. The latter can be associated with the electronic transitions of an elongated D_{4h} octahedral structure. The simulated spectra show that the signals of two possible CasIII-ia–Cys adduct arrangements, octahedral (axial Cys) and square planar pyramid (equatorial Cys), are very close to the experimental ones, 626 and 663 nm, respectively. The deconvolution of the two observed signals agrees with three theoretical excited states of octahedral and square planar pyramid geometries. These signals are mainly related to the transition to SOMO or LUMO molecular orbitals (for octahedral geometry: SOMO-20 → SOMO, SOMO-16 → SOMO, and SOMO-16 → LUMO; for square planar pyramid: SOMO-18 → SOMO, SOMO-18 → LUMO, and SOMO-15 → LUMO). Details are provided in the Supplementary Materials.

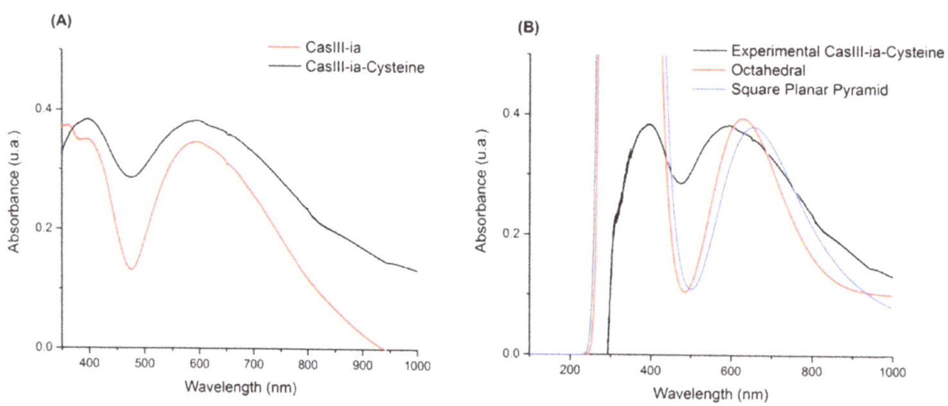

Figure 4. (**A**) UV-Vis spectra for CasIII-ia (red) and the reaction between CasIII-ia and cysteine (black). (**B**) UV-Vis spectra comparison between experimental reaction (black) and calculated geometries for the CasIII-ia–Cysteine system: octahedral (red) and square planar pyramid (blue).

The third evidence of the interaction between CasIII-ia and Cys is the EPR data. Figure 5 shows the experimental spectra and fitting models of CasIII-ia (A) and CasIII-ia–Cys (B). Table 1 presents experimental and computational values for the g and A tensors of the spectra in Figure 5. We compared the observed spectra with DFT simulations of octahedral and square planar pyramid structures. CasIII-ia shows an axial symmetry spectrum associated with a square planar pyramid geometry. The reaction product is a 1:1 CasIII-ia/CasIII-ia–Cys mixture in the solid state. The linear combination of the axial and isotropic EPR profiles of CasIII-ia and CasIII-ia–Cys reproduce the experimental spectrum. The axial symmetry can be associated with geometry with a pseudo-Janh-Teller effect, where a Cys occupies the axial position of CasIII-ia. The difference between the experimental and theoretical g values ranges from 0.004 to 0.10, and for A values, from 25 to 98 MHz. These differences agree with that obtained in other reports [36,37]. For the CasIII-ia–Cys, we compared the experimental data with two different geometries: octahedral and square planar pyramid. The difference between the experimental and theoretical data is 0.004 and 0.030 for g_{iso}, respectively. The A_{iso} value differences are 56 and 25 MHz, respectively. The best fit for g_{iso} corresponds to octahedral geometry, while for A_{iso} it is given by the square planar pyramid.

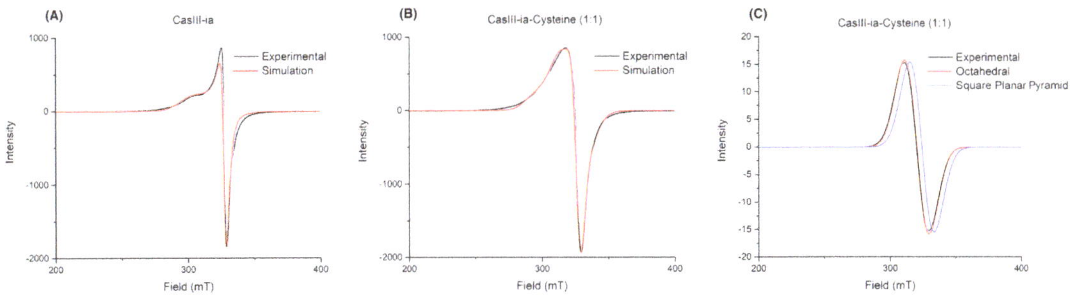

Figure 5. EPR spectra in the solid-state for (**A**) CasIII-ia and (**B**) CasIII-ia with cysteine reaction, and (**C**) comparison between experimental reaction and calculated geometries for the CasIII-ia–Cysteine system.

Table 1. Experimental and computational values for the g tensor and the A tensor (in MHz) parameters of spectra A and B in Figure 5.

	CasIII-ia (A)		
	Experimental		**Computational**
$g_{xx} = g_{yy}$	2.0767		2.0466
g_{zz}	2.2517		2.1458
$A_{xx} = A_{yy}$	7.47		105.26
A_{zz}	117.4		−185.99
	CasIII-ia–Cysteine 1:1 (B)		
	Experimental	**Octahedral**	**Square Planar Pyramid**
$g_{xx} = g_{yy} = g_{zz}$	2.10557	2.1010	2.0755
$A_{xx} = A_{yy} = A_{zz}$	48.797	104.57	23.85

The fourth evidence of the Cu–S interaction is the I.R. signals presented in the Supplementary Materials. Since cysteine is a molecule of biological interest, its solid-state vibrational spectra have been extensively studied, considering the polymorphisms that it can show [38] and the I.R. modifications with different protonation modes [39]. The Cys I.R. signals S-H (1063 cm^{-1}), C-S (692 cm^{-1}), and C-N (291 cm^{-1}) [38], agreeing with theoretical frequencies (1056.73, 669.99, and 278 cm^{-1}, respectively), yield the most significant changes when interacting with the copper complex. In the case of CasIII-ia, the Cu-O and C-N bonds' I.R. signals appear at 596 and 294 cm^{-1} (603.88 and 279.73 cm^{-1} theoretical values), respectively. The 1:1 solid reaction mixture, with drops of EtOH, produces an adduct which presents I.R. signal variations as evidence of the CasIII-ia–Cys interaction (Figure 6). Based on computational information (179 and 194 cm^{-1} for octahedral and SPP theoretical values, respectively), we can assign the frequency at 187 cm^{-1} to the Cu–S interaction between CasIII-ia and Cys. It is possible to note signals associated with the metal–ligand interaction within CasIII-ia, such as Cu–O (594 cm^{-1}) and Cu–N (289 cm^{-1}). Cu–N weakened after Cys coordination from 294 to 289 cm^{-1}. The weakening of the C-S bond is shown by reducing the frequency from 692 to 685 cm^{-1}. One can observe a small S-H at 1063 cm^{-1} as evidence of unreacted Cys.

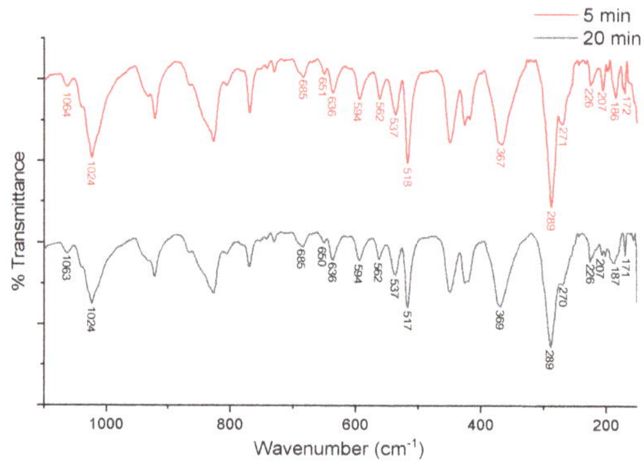

Figure 6. ATR-FTIR spectrum of CasIII-ia–Cys (1:1) mixture at 5 (red) and 20 (black) minutes.

After NaOH was added in an equimolar amount, S-H frequency, at 1063 cm^{-1}, disappeared, but the C-S and Cu-O vibrations remain. We confirmed that Cu–S interaction is associated with the 187 cm^{-1} frequency. These signals correlate with those exhibited by octahedral theoretical geometry. Details are provided in the Supplementary Materials.

Our last evidence is from the direct analysis in real-time mass spectrometry (DART-MS). Under positive ionization of the DART technique, it has been reported that the amino-acid presents protonation, radical, and adduct formation [40]. In the Cys case, it is possible to observe $[Cys + H]^+$ = 122 m/Z and [2Cys + H + H] = 243 m/Z. The mechanochemical solid-state mixture was analyzed using DART-MS, whose spectrum is presented in Figure 7. It is possible to observe the signal associated with dimethylbipyridine (m/Z = 185.1), Cu(acetylacetonate)$_2$ (m/Z = 262), and CasIII-ia (m/Z = 346). At 200 °C, we found a signal of m/Z = 423 corresponding to the CasIII-ia–Cys adduct, with a loss of carboxylic group. It has been reported that Cys loses CO in situations where the sulfur atom is involved in a bond or a strong interaction. The fragmentation pattern of Cys by DART-MS and cystine by TANDEM-MS both present the [Cistina-H_2O-CO + H] ion with m/Z = 195 when the CO loss can be observed [41,42]. The presence of the signals at 346 and 348 for CasIII-ia, at 262 and 264 for Cu(acetylacetonate)$_2$, and at 422.99 and 424.98 for the adduct agree with the isotopic distribution for ^{63}Cu and ^{65}Cu observed in a copper(II) species. The abundance of each peak can be related to the stability of the analyzed species [43]. In this way, the adduct is an unstable species compared with the CasIII-ia.

To understand the pathway leading to a final reduction of the copper metal center, we calculated the structures and their dynamics. The two possible molecular arrangements of the CasIII-ia–Cys adduct are the octahedral and the square planar pyramid (Figure 8). In the former, the sulfur atom occupies an axial coordination position at 2.82 Å Cu–S distance. The latter presents the sulfur atom located in an equatorial position, at 2.37 Å Cu–S distance. The square planar pyramid is 4.9 kcal/mol more stable than octahedral geometry.

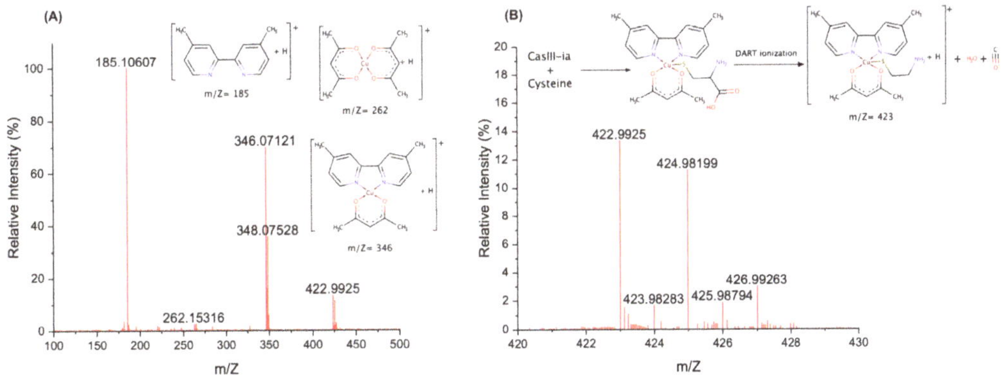

Figure 7. DART spectra of the mechanochemical solid-state mixture CasIII-ia–Cys at 200 °C. (**A**) From 100 to 500 m/Z, and (**B**) from 420 to 430 m/Z.

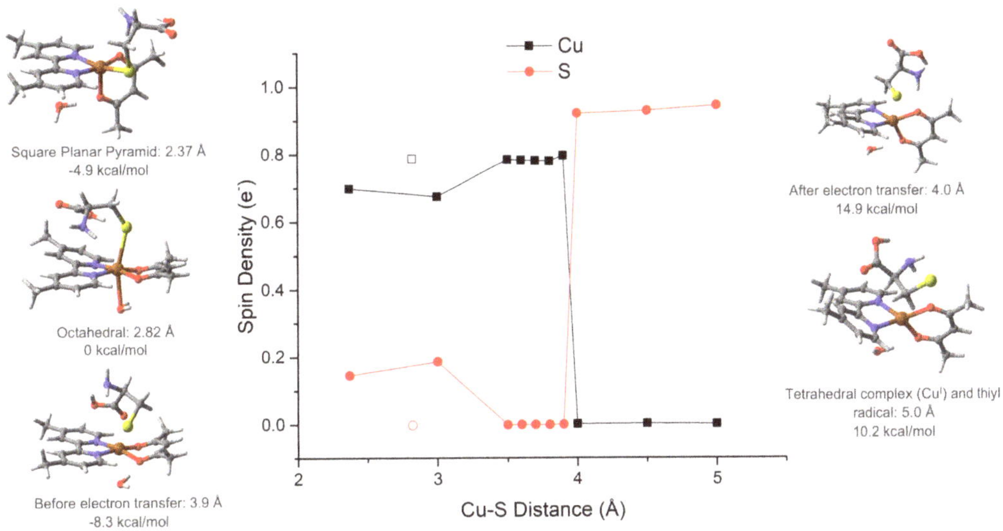

Figure 8. From square planar pyramid Cu^{II} complex to tetrahedral Cu^{I} complex, electron transfer path. Spin density values for copper atom in black and for sulfur atom in red.

The reductive process's reaction path begins from the Cys coordination to CasIII-ia in the square planar pyramid geometry, and it is presented in Figure 8, along with the atomic spin population. From the initial 2.37 Å, the Cu–S distance increases until 3.9 Å, where the electron is transferred from sulfur to the copper atom. It is possible to observe a spin population change from 0.78 to 0.001 e^- in the copper atom, which is transferred to the sulfur atom, which shows an increase from 0.00 to 0.94 e^-. In this process, the copper valence shell changes, and then the geometry of the complex becomes tetrahedral. Our group has previously reported the spin change effect on the metal valence shell and thus on the complex structure [44,45]. From this point, the thiyl radical is free to participate in the following reaction to form the disulfide bond. Details are provided in the Supplementary Materials.

3. Materials and Methods

3.1. Chemicals

All reagents: acetylacetone (acacH) (Sigma-Aldrich, St. Louis, MO, USA), $Cu(NO_3)_2\cdot 2.5H_2O$(Sigma Aldrich, St. Louis, MO, USA), and 4,4′-dimethyl-2,2′-bipyridine (dmbpy) (Sigma Aldrich, St. Louis, MO, USA), as the organic solvents, were used without further purification. L-cysteine (Cys) (Sigma Aldrich, St. Louis, MO, USA) was also used without further purification. The elemental analysis of the white crystalline powder for $C_3H_7O_2S$ was %C 29.82 (29.73), %H 5.92(5.82), %N 11.70 (11.56), %S 26.97 (26.46) ((#) calculated values). The far FTIR-ATR spectrum of a deep white powder of Cys showed characteristic bands at 1614 cm^{-1} CO_2, 1063 cm^{-1} S-H, 692 cm^{-1} C-S, and 637 cm^{-1} CH-CO_2.

3.2. CasIII-ia Synthesis

[Cu(4,4′-dimethyl-2,2′-bipyridine)(acetylacetonate)]NO_3*$(H_2O)_2$ (CasIII-ia) CAS [223930-33-4], the copper(II) complex, was prepared following the reported patent [31–33]. The complex was isolated on MeOH/H_2O solution, and a blue crystalline powder was obtained. The elemental analysis of the blue powder for $CuC_{17}H_{19}N_3O_5$*$(H_2O)_2$ was %C 44.26 (45.89), %H 4.79(5.21), and %N 9.45 (9.44) ((#) calculated values). The far FTIR-ATR spectrum of a deep blue powder of CasIII-ia showed characteristic bands at 1616 cm^{-1} C=O (acac), 1373 cm^{-1} N-O (nitrate), 596 cm^{-1} Cu-O, and 294 cm^{-1} Cu-N.

3.3. Solid-State Reaction

CasIII-ia (30 mg, 0.0674 mmol) and cysteine (8.17 mg, 0.0674 mmol) were milled until a homogenous solid was obtained. EtOH (400 μL) was added and mixed until dryness.

3.4. Measurements

Powder X-ray diffraction (PXRD) data were collected under ambient conditions on a Rigaku ULTIMA IV diffractometer operated at 160 W (40 kV, 40 mA) for Cu Kα1 (λ = 1.5406 Å).

Electron paramagnetic resonance spectroscopy (EPR) measurements were carried out in a JEOL JES-TE300 spectrometer operated at X-Band mode at a microwave frequency of 9.4 GHz and center field of 300 mT. Solid-state measurements were performed at room temperature, where the samples were placed in a quartz cell. The acquisition and manipulation of spectra were performed using the ES-IPRIT/TE program. The g and hyperfine tensors were determined by fitting the powder spectra using the EasySpin [46] simulation package (Version 5.2.28, easyspin.org, (accessed on 6 May 2020)) for MATLAB R2019b.

The solid-state electronic spectra (UV-Vis-NIR) for the samples were measured over the range 40,000–5000 cm^{-1} by the diffuse reflectance method on a Cary-5000 Varian spectrophotometer at room temperature.

The near-FTIR attenuated total reflectance (ATR) spectra were obtained over the range 4000–250 cm^{-1} on a Thermo Fisher Scientific Nicolet IS-50 spectrophotometer. The samples were examined as solid. The middle-FTIR spectra were obtained over the range 4000–400 cm^{-1} on a Nicolet spectrophotometer Nicolet AVATAR 320. The samples were analyzed as KBr disk.

The MS-DART spectra were acquired with a JEOL AccuTOF JMS-T100LC spectrometer. The samples were examined as solid. The values of the signals are expressed in mass/charge units (m/Z), followed by the relative intensity with reference to a 100% base peak.

The elemental analysis (EA) was carried out using The PerkinElmer® 2400 Series II CHNS/O Elemental Analyzer.

3.5. Computational Details

All structures were fully optimized at the DFT level in gas phase, with the m05-2x functional and the LanL2DZ basis set, as implemented in Gaussian 09 software [47]. Then, we performed frequency calculations to verify the equilibrium states and to obtain infrared spectra. We used the TD-DFT CAM-B3LYP/SDD theoretical level for the UV-Vis spectra, with the SMD solvation model, with water as the solvent. For the electron transfer studies, we used the Quantum Theory of Atoms In Molecules [48], using the set of molecular orbitals of each molecule to compute the atomic properties of the electron density with the AIMAll software [49]. EPR parameters and g and A tensors of the optimized structures were calculated with ORCA software [50], using the B3LYP functional and the def2-SVP basis set.

The initial complex presented a square planar geometry with a water molecule in one axial site, as in the reported crystallographic geometry [51]. Then, a cysteine molecule was linked to the complex by the vacant axial site, yielding an octahedral geometry. A base in the environment removes the -SH proton, expelling the water molecule from the opposite side. We detected the electron transfer by the complex change to a tetrahedral geometry, which is the geometry preferred by a copper atom with a 1+ oxidation state.

4. Conclusions

A strategy to improve cancer therapies' efficacy should involve cytosolic and mitochondrial GSH depletion through an increase of GSH efflux out of cells. There are reports of GSH depletion produced by copper complexes of the Casiopeina family. In this work, we presented experimental evidence of the formation of an adduct between cysteine and a Casiopeina complex. This adduct shows equilibrium between octahedral and square planar pyramid structures. From this equilibrium, it is possible to identify an electron transfer path when the Cu–S distance increases to 3.9 Å, which produces a thiyl radical and a reduced tetrahedral copper(I) complex. The proposed mechanism is presented in Figure 9.

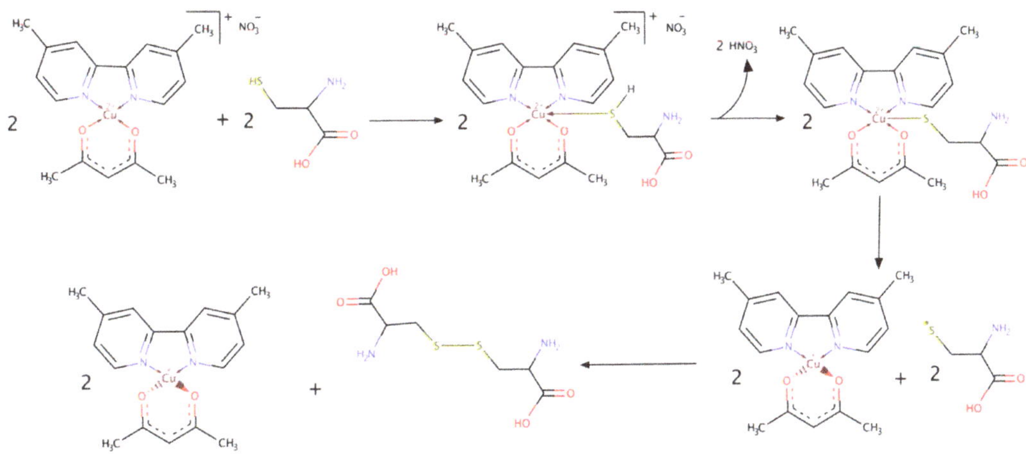

Figure 9. Proposed mechanism for the interaction between CasIII-ia and cysteine.

Supplementary Materials: The following are available online, Video S1: Reaction in solution. Experimental Details: Figure S1: Powder X-ray diffraction, Table S1: Powder X-ray diffraction (PXRD) data, Table S2: Experimental EPR parameters, Figure S2: Details of the CasIII-ia:Cysteine (1:1) EPR simulation, Figure S3: Deconvolved UV-Vis spectra, Table S3: Comparison between deconvolved and computational values for UV-Vis transitions, Figures S4–S7: Details of ATR-FTIR spectra, Table S4: Main absorption infrared signals, Figures S8 and S9: MS-DART spectra for reactants, computational calculations: optimized geometries, Figures S10–S14: UV-Vis spectra, Figures S15–S17: Infrared spectra, Tables S5–S8: Infrared signals, Table S9: Calculated EPR parameters, Figure S18: Spin density changes, Table S10: Spin density values for Cu and S atoms, Figure S19: Spin density changes in Cu-S bond distance scan, Figure S20: Geometries in the Cu–S bond distance scan, Figures S21 and S22: Copper Atomic graphs.

Author Contributions: L.G.R.-P.: methodology, formal analysis, investigation, data curation, writing—original draft preparation, and writing—review and editing; A.E.-G.: methodology, formal analysis, investigation, data curation, writing—original draft preparation, and writing—review and editing; F.N.-C.: investigation; A.E.L.-G.: investigation; V.G.-V.: methodology; F.C.-G.: conceptualization, writing—original draft preparation, writing—review and editing, supervision, project administration, and funding acquisition; L.R.-A.: conceptualization, writing—original draft preparation, writing—review and editing, supervision, project administration, and funding acquisition. All authors have read and agreed to the published version of the manuscript.

Funding: This research was funded by UNAM (PAPIIT-IN212520, PAPIIT IN218013, and PAIP 5000-9047) and CONACYT (CB179119). F.N.-C. and A.E.L.-G. thank SNI (grant "Ayudantes de Investigador Nacional SNI III/Emérito") and L.G.R.-P. thanks CONACyT for the fellowship (308338). We also thank DGTIC-UNAM (LANCAD-UNAM-DGTIC-194) for supercomputer time and USAII-UNAM for elemental analysis.

Institutional Review Board Statement: Not applicable.

Informed Consent Statement: Not applicable.

Data Availability Statement: Data is contained within the article and Supplementary Material.

Acknowledgments: The authors thank I.A. Ibarra-Alvarado and E. Sánchez-González for PXRD determinations, M. C. García for MS-DART determinations, and R. Patiño Maya for I.R. determinations.

Conflicts of Interest: The authors declare no conflict of interest.

Sample Availability: Samples of the compounds are available from the authors.

References

1. Baba, S.P.; Bhatnagar, A. Role of thiols in oxidative stress. *Curr. Opin. Toxicol.* **2018**, *7*, 133–139. [CrossRef] [PubMed]
2. Aquilano, K.; Baldelli, S.; Ciriolo, M.R. Glutathione: New roles in redox signaling for an old antioxidant. *Front. Pharmacol.* **2014**, *5*, 196. [CrossRef] [PubMed]
3. Ulrich, K.; Jakob, U. The role of thiols in antioxidant systems. *Free Radic. Biol. Med.* **2019**, *140*, 14–27. [CrossRef]
4. Ortega, A.L.; Mena, S.; Estrela, J.M. Glutathione in Cancer Cell Death. *Cancers* **2011**, *3*, 1285–1310. [CrossRef] [PubMed]
5. Bansal, A.; Simon, M.C. Glutathione metabolism in cancer progression and treatment resistance. *J. Cell Biol.* **2018**, *217*, 2291–2298. [CrossRef] [PubMed]
6. Kachadourian, R.; Brechbuhl, H.M.; Ruiz-Azuara, L.; Gracia-Mora, I.; Day, B.J. Casiopeína IIgly-induced oxidative stress and mitochondrial dysfunction in human lung cancer A549 and H157 cells. *Toxicology* **2010**, *268*, 176–183. [CrossRef] [PubMed]
7. Millis, K.K.; Weaver, K.H.; Rabenstein, D.L. Oxidation/Reduction Potential of Glutathione. *J. Org. Chem.* **1993**, *58*, 4144–4146. [CrossRef]
8. Jocelyn, P.C. The Standard Redox Potential of Cysteine-Cystine from the Thiol-Disulphide Exchange Reaction with Glutathione and Lipoic Acid. *Eur. J. Biochem.* **1967**, *2*, 327–331. [CrossRef]
9. Ngamchuea, K.; Batchelor-McAuley, C.; Compton, R. The Copper(II)-Catalyzed Oxidation of Glutathione. *Chem.—Eur. J.* **2016**, *22*, 15937–15944. [CrossRef]
10. Kachur, A.V.; Koch, C.J.; Biaglow, J.E. Mechanism of copper-catalyzed autoxidation of cysteine. *Free Radic. Res.* **1999**, *31*, 23–34. [CrossRef]
11. Prudent, M.; Girault, H.H. The role of copper in cysteine oxidation: Study of intra- and inter-molecular reactions in mass spectrometry. *Metallomics* **2009**, *1*, 157–165. [CrossRef]
12. Carrasco-Pozo, C.; Aliaga, M.E.; Olea-Azar, C.; Speisky, H. Double edge redox-implications for the interaction between endogenous thiols and copper ions: In vitro studies. *Bioorg. Med. Chem.* **2008**, *16*, 9795–9803. [CrossRef] [PubMed]
13. Smith, R.C.; Reed, V.D.; Hill, W.E. Oxidation Of Thiols By Copper(II). *Phosphorus. Sulfur. Silicon Relat. Elem.* **1994**, *90*, 147–154. [CrossRef]
14. Bravo-Gómez, M.E.; García-Ramos, J.C.; Gracia-Mora, I.; Ruiz-Azuara, L. Antiproliferative activity and QSAR study of copper(II) mixed chelate [Cu(N–N)(acetylacetonato)]NO3 and [Cu(N–N)(glycinato)]NO3 complexes, (Casiopeínas®). *J. Inorg. Biochem.* **2009**, *103*, 299–309. [CrossRef]
15. Santoro, A.; Calvo, J.S.; Peris-Díaz, M.D.; Krężel, A.; Meloni, G.; Faller, P. The Glutathione/Metallothionein System Challenges the Design of Efficient O2-Activating Copper Complexes. *Angew. Chem.-Int. Ed.* **2020**, *59*, 7830–7835. [CrossRef] [PubMed]
16. Ruiz-Azuara, L. MéxicoTítulo de Marca: Casiopeína. Reg. 407543 SECOFI, 1992.
17. Arredondo, M.; Núñez, M.T. Iron and copper metabolism. *Mol. Asp. Med.* **2005**, *26*, 313–327. [CrossRef]
18. Bravo-Gómez, M.E.; Dávila-Manzanilla, S.; Flood-Garibay, J.; Muciño-Hermández, M.Á.; Mendoza, Á.; García-Ramos, J.C.; Moreno-Esparza, R.; Ruiz-Azuara, L. Secondary Ligand Effects on the Cytotoxicity of Several Casiopeína's Group II Compounds. *J. Mex. Chem. Soc.* **2012**, *56*, 85–92. [CrossRef]

19. García-Ramos, J.C.; Galindo-Murillo, R.; Tovar-Tovar, A.; Alonso-Saenz, A.L.; Gómez-Vidales, V.; Flores-Álamo, M.; Ortiz-Frade, L.; Cortes-Guzmán, F.; Moreno-Esparza, R.; Campero, A.; et al. The π-Back-Bonding Modulation and Its Impact in the Electronic Properties of CuII Antineoplastic Compounds: An Experimental and Theoretical Study. *Chem.—Eur. J.* **2014**, *20*, 13730–13741. [CrossRef]
20. Galindo-Murillo, R.; Ruiz-Azuara, L.; Moreno-Esparza, R.; Cortés-Guzmán, F. Molecular recognition between DNA and a copper-based anticancer complex. *Phys. Chem. Chem. Phys.* **2012**, *14*, 15539–15546. [CrossRef]
21. Klaunig, J.E.; Kamendulis, L.M. The Role of Oxidative Stress in Carcinogenesis. *Annu. Rev. Pharmacol. Toxicol.* **2004**, *44*, 239–267. [CrossRef]
22. Gaetke, L.M.; Chow, C.K. Copper toxicity, oxidative stress, and antioxidant nutrients. *Toxicology* **2003**, *189*, 147–163. [CrossRef]
23. Huang, R.; Wallqvist, A.; Covell, D.G. Anticancer metal compounds in NCI's tumor-screening database: Putative mode of action. *Biochem. Pharmacol.* **2005**, *69*, 1009–1039. [CrossRef]
24. Carvallo-Chaigneau, F.; Trejo-Solís, C.; Gómez-Ruiz, C.; Rodríguez-Aguilera, E.; Macías-Rosales, L.; Cortés-Barberena, E.; Cedillo-Peláez, C.; Gracia-Mora, I.; Ruiz-Azuara, L.; Madrid-Marina, V.; et al. Casiopeina III-ia induces apoptosis in HCT-15 cells in vitro through caspase-dependent mechanisms and has antitumor effect in vivo. *BioMetals* **2008**, *21*, 17–28. [CrossRef] [PubMed]
25. Bravo-Gómez, M.E.; Hernández de la Paz, A.L.; Gracia-Mora, I. Antineoplastic evaluation of two mixed chelate copper complexes (Casiopeínas®) in HCT-15 xenograft model. *J. Mex. Chem. Soc.* **2013**, *57*, 205–211. [CrossRef]
26. Ruiz-Azuara, L.; Bravo-Gómez, M.E. Copper Compounds in Cancer Chemotherapy. *Curr. Med. Chem.* **2010**, *17*, 3606–3615. [CrossRef]
27. García-Ramos, J.C.; Gutiérrez, A.G.; Vázquez-Aguirre, A.; Toledano-Magaña, Y.; Alonso-Sáenz, A.L.; Gómez-Vidales, V.; Flores-Alamo, M.; Mejía, C.; Ruiz-Azuara, L. The mitochondrial apoptotic pathway is induced by Cu(II) antineoplastic compounds (Casiopeínas®) in SK-N-SH neuroblastoma cells after short exposure times. *BioMetals* **2017**, *30*, 43–58. [CrossRef]
28. Vértiz, G.; García-Ortuño, L.E.; Bernal, J.P.; Bravo-Gómez, M.E.; Lounejeva, E.; Huerta, A.; Ruiz-Azuara, L. Pharmacokinetics and hematotoxicity of a novel copper-based anticancer agent: Casiopeina III-Ea, after a single intravenous dose in rats. *Fundam. Clin. Pharmacol.* **2014**, *28*, 78–87. [CrossRef]
29. Cañas-Alonso, R.C.; Fuentes-Noriega, I.; Ruiz-Azuara, L. Pharmacokinetics of Casiopeína IIgly in beagle dog: A copper based compound with antineoplastic activity. *J. Bioanal. Biomed.* **2010**, *2*, 28–34. [CrossRef]
30. Ramírez-Palma, L.G.; García-Jacas, C.R.; García-Ramos, J.C.; Almada-Monter, R.; Galindo-Murillo, R.; Cortés-Guzmán, F. Pharmacophoric sites of anticancer metal complexes located using quantum topological atomic descriptors. *J. Mol. Struct.* **2020**, *1204*, 127480. [CrossRef]
31. Ruiz-Azuara, L. Process to Obtain New Mixed Copper Aminoacidate from Methylate Phenanthroline Complexes to Be Used as Anticancerigenic Agents. U.S. Patent 5,576,326, 19 November 1996.
32. Ruiz-Azuara, L. Procedimiento para la Obtención de Complejos Metálicos como Agentes Anticancerígenos. Tipo II. Patente de in-Vención SECOFI 18801, 1994.
33. Ruiz-Azuara, L. Composición Parental de Casiopeína y usos de la Misma. IMPI No. Solicitud MX/a/2017/016444, 2017.
34. Seko, H.; Tsuge, K.; Igashira-Kamiyama, A.; Kawamoto, T.; Konno, T. Autoxidation of thiol-containing amino acid to its disulfide derivative that links two copper(II) centers: The important role of auxiliary ligand. *Chem. Commun.* **2010**, *46*, 1962–1964. [CrossRef]
35. Ugone, V.; Pisanu, F.; Sanna, D.; Garribba, E. Interaction of the potent antitumoral compounds Casiopeinas®with blood serum and cellular bioligands. *J. Inorg. Biochem.* **2021**, *224*, 111566–111576. [CrossRef]
36. Rivillas-Acevedo, L.; Grande-Aztatzi, R.; Lomelí, I.; García, J.E.; Barrios, E.; Teloxa, S.; Vela, A.; Quintanar, L. Spectroscopic and Electronic Structure Studies of Copper(II) Binding to His111 in the Human Prion Protein Fragment 106-115: Evaluating the Role of Protons and Methionine Residues. *Inorg. Chem.* **2011**, *50*, 1956–1972. [CrossRef] [PubMed]
37. Ames, W.M.; Larsen, S.C. DFT calculations of the EPR parameters for Cu(II) DETA imidazole complexes. *Phys. Chem. Chem. Phys.* **2009**, *11*, 8266–8274. [CrossRef]
38. Parker, S.F. Assignment of the vibrational spectrum of L-cysteine. *Chem. Phys.* **2013**, *424*, 75–79. [CrossRef]
39. Gaillard, T.; Trivella, A.; Stote, R.H.; Hellwig, P. Far infrared spectra of solid state L-serine, L-threonine, L-cysteine, and L-methionine in different protonation states. *Spectrochim. Acta-Part A Mol. Biomol. Spectrosc.* **2015**, *150*, 301–307. [CrossRef] [PubMed]
40. Sekimoto, K.; Sakakura, M.; Kawamukai, T.; Hike, H.; Shiota, T.; Usui, F.; Bando, Y.; Takayama, M. Ionization characteristics of amino acids in direct analysis in real time mass spectrometry. *Analyst* **2014**, *139*, 2589–2599. [CrossRef] [PubMed]
41. Zhang, P.; Chan, W.; Ang, I.L.; Wei, R.; Lam, M.M.T.; Lei, K.M.K.; Poon, T.C.W. Revisiting Fragmentation Reactions of Protonated α-Amino Acids by High-Resolution Electrospray Ionization Tandem Mass Spectrometry with Collision-Induced Dissociation. *Sci. Rep.* **2019**, *9*, 6453. [CrossRef] [PubMed]
42. Zhang, P.; Chan, W.; Ang, I.L.; Wei, R.; Lam, M.M.T.; Lei, K.M.K.; Poon, T.C.W. Gas-Phase Fragmentation Reactions of Protonated Cystine using High-Resolution Tandem Mass Spectrometry. *Molecules* **2019**, *24*, 747. [CrossRef]
43. Nicolescu, T.O. Interpretation of Mass Spectra. In *Mass Spectrometry*; Aliofkhazraei, M., Ed.; In Tech Open: Rijeka, Croatia, 2017; pp. 30–40.

44. Gutiérrez-Arzaluz, L.; Ramírez-Palma, D.I.; Ramírez-Palma, L.G.; Barquera-Lozada, J.E.; Peon, J.; Cortés-Guzmán, F. Origin of the Photoinduced Geometrical Change of Copper(I) Complexes from the Quantum Chemical Topology View. *Chem.—Eur. J.* **2019**, *25*, 775–784. [CrossRef]
45. Ramirez-Palma, D.I.; Cortes-Guzman, F. From Linnett-Gillespie Model to the Polarization of the Spin Valence Shells of Metals in Complexes. *Phys. Chem. Chem. Phys.* **2020**, *22*, 24201–24212. [CrossRef]
46. Stoll, S.; Schweiger, A. EasySpin, a comprehensive software package for spectral simulation and analysis in EPR. *J. Magn. Reson.* **2006**, *178*, 42–55. [CrossRef] [PubMed]
47. Frisch, M.J.; Trucks, G.W.; Schlegel, H.B.; Scuseria, G.E.; Robb, M.A.; Cheeseman, J.R.; Scalmani, G.; Barone, V.; Mennucci, B.; Petersson, G.A.; et al. *Gaussian 09 (Revision E.01)*; Gaussian Inc.: Wallingford, CT, USA, 2009.
48. Bader, R. *Atoms in Molecules. A Quantum Theory*, 1st ed.; Oxford University Press: New York, NY, USA, 1990.
49. Keith, T.A. AIMAll (Version 16.01.09), TK Gristmill Software, Overland Park, KS, USA, 2016. Available online: aim.tkgristmill.com (accessed on 23 June 2021).
50. Neese, F. The ORCA program system. *Wiley Interdiscip. Rev. Comput. Mol. Sci.* **2012**, *2*, 73–78. [CrossRef]
51. García-Ramos, J.C.; Tovar-Tovar, A.; Hernández-Lima, J.; Cortés-Guzmán, F.; Moreno-Esparza, R.; Ruiz-Azuara, L. A new kind of intermolecular stacking interaction between copper (II) mixed chelate complex (Casiopeína III-ia) and adenine. *Polyhedron* **2011**, *30*, 2697–2703. [CrossRef]

molecules

MDPI

Article

Antioxidant Activity of Bioactive Peptide Fractions from Germinated Soybeans Conjugated to Fe_3O_4 Nanoparticles by the Ugi Multicomponent Reaction

Yarelys Elena Augusto-Jimenez [1,*], Marcela González-Montoya [2], Dany Naranjo-Feliciano [3], Daniel Uribe-Ramírez [2], Eliseo Cristiani-Urbina [2], Carlos Díaz-Águila [4], Hernani Yee-Madeira [1] and Rosalva Mora-Escobedo [2,*]

[1] Instituto Politécnico Nacional–ESFM, U.P.A.L.M., San Pedro Zacatenco, Ciudad de Mexico 07738, Mexico; hernaniyee@gmail.com

[2] Instituto Politécnico Nacional–ENCB, U.P.A.L.M., San Pedro Zacatenco, Ciudad de Mexico 07738, Mexico; mrcela_mecs@hotmail.com (M.G.-M.); daniel.uriberam@gmail.com (D.U.-R.); ecristiani@ipn.mx (E.C.-U.)

[3] Centro Nacional de Sanidad Agropecuaria, San Jose de las Lajas 32700, Mayabeque, Cuba; danynaranjofeliciano@gmail.com

[4] Centro de Biomateriales, Universidad de La Habana, Plaza de la Revolucion 10400, La Habana, Cuba; cdiaz@biomat.uh.cu

* Correspondence: syleray91@gmail.com (Y.E.A.-J.); rosalmorae@gmail.com (R.M.-E.)

Abstract: The conjugation of biomolecules to magnetic nanoparticles has emerged as promising approach in biomedicine as the treatment of several diseases, such as cancer. In this study, conjugation of bioactive peptide fractions from germinated soybeans to magnetite nanoparticles was achieved. Different fractions of germinated soybean peptides (>10 kDa and 5–10 kDa) were for the first time conjugated to previously coated magnetite nanoparticles (with 3-aminopropyltriethoxysilane (APTES) and sodium citrate) by the Ugi four-component reaction. The crystallinity of the nanoparticles was corroborated by X-ray diffraction, while the particle size was determined by scanning transmission electron microscopy. The analyses were carried out using infrared and ultraviolet–visible spectroscopy, dynamic light scattering, and thermogravimetry, which confirmed the coating and functionalization of the magnetite nanoparticles and conjugation of different peptide fractions on their surfaces. The antioxidant activity of the conjugates was determined by the reducing power and hydroxyl radical scavenging activity. The nanoparticles synthesized represent promising materials, as they have found applications in bionanotechnology for enhanced treatment of diseases, such as cancer, due to a higher antioxidant capacity than that of fractions without conjugation. The highest antioxidant capacity was observed for a >10 kDa peptide fraction conjugated to the magnetite nanoparticles coated with APTES.

Keywords: soybean; peptides; antioxidant capacity; magnetite nanoparticles; conjugation; multicomponent reaction

Citation: Augusto-Jimenez, Y.E.; González-Montoya, M.; Naranjo-Feliciano, D.; Uribe-Ramírez, D.; Cristiani-Urbina, E.; Díaz-Águila, C.; Yee-Madeira, H.; Mora-Escobedo, R. Antioxidant Activity of Bioactive Peptide Fractions from Germinated Soybeans Conjugated to Fe_3O_4 Nanoparticles by the Ugi Multicomponent Reaction. *Molecules* **2021**, *26*, 5726. https://doi.org/10.3390/molecules26195726

Academic Editors: Višnja Stepanić and Marta Kučerová-Chlupáčová

Received: 22 July 2021
Accepted: 13 September 2021
Published: 22 September 2021

1. Introduction

The main challenges for magnetic nanoparticles (MNPs) used in biomedicine are related to their tendency to agglomerate and lack of biocompatibility [1]. Regarding biocompatibility, iron oxide nanoparticles (NPs) maghemite (γ-Fe_2O_3) and magnetite (Fe_3O_4) are advantageous for in vivo applications. Unlike other materials with good magnetic behavior, iron cell homeostasis is well-controlled by absorption, excretion, and storage processes. Even excess iron is efficiently removed from the body [2]. Specially, different works show that the superparamagnetic behavior of magnetite NPs is closely related to their nanometric size [3]. This property is essential for the in vivo applications of the material as it ensures that no magnetization remains in the system after stopping the action of an external magnetic field. Moreover, uncoated magnetite NPs tend to decrease the surface free energy by forming stable aggregates under physiological conditions [4]. In this regard, it has been observed that the coupling of different types of biomolecules to Fe_3O_4 NPs increases the stability of the system while directing it towards the desired biological target. Thereby, magnetite has been coupled to polymers and different anticancer drugs, with prominent applications in biomedicine, nanoscience, and nanotechnology [5–7]. These conjugates are of interest mainly because the benefits of the Fe_3O_4 NPs and biomolecules are combined in the same system.

Peptide–Fe_3O_4 conjugates are promising for biotechnological applications [8–10], particularly because peptides from natural sources and synthetic derivatives have been identified as promising for the treatment of diseases, such as cancer [11,12]. From the conjugation of peptides to Fe_3O_4 NPs, one could not only selectively transport drugs towards biological targets, but also treat different diseases using magnetic hyperthermia [13–16]. Despite the advances in the development of methodologies for the conjugation of peptides to magnetite NPs, the current protocols are still limited by long and complex reaction steps. In most cases, they are based on classical peptide coupling between peptides and MNPs functionalized with carboxylic acid groups [17]. In some studies, surface modification of MNPs has been performed with ligands suitable for their subsequent coupling to peptides through click chemistry [14]. This restricts and makes the application of peptide–Fe_3O_4 conjugates in biomedical investigations more expensive.

Several procedures for the combination of complex molecules based on multicomponent reactions have been reported [18,19]. Multicomponent protocols have a high potential owing to their high chemical efficiency, simple execution, and structural diversity at low synthesis costs [20]. One of the most used is the Ugi four-component (U-4C) reaction, in which the condensation of an oxo derivative, primary amine, carboxylic acid, and isonitrile leads to the formation of a dipeptoid in a single reaction step. The N-substituted amide in peptoids derived from the U-4C reaction is a structural motif of biological relevance because it prevents proteolytic degradation and limits the conformational orientation of peptide structures [21]. However, to the best of our knowledge, this type of reaction has not been used for the conjugation of bioactive peptides to MNPs.

On the other hand, several studies demonstrated the important role of oxidative stress and presence of different reactive oxygen and nitrogen species (ROS and RNS, respectively) in numerous noncommunicable diseases, such as cardiovascular diseases, diabetes, atherosclerosis, arthritis, and cancer [22]. These diseases lead to approximately 41 million deaths per year, equivalent to 71% of all deaths globally [23]. Some peptides and biomimetic peptides inhibit the generation of ROS and other free radicals [24,25]. Therefore, the search for antioxidant peptide structures that can prevent oxidative stress and its associated adverse effects is of increasing interest. In this sense, González-Montoya et al. evaluated the anti-inflammatory, antioxidant, and antiproliferative activities of three peptide fractions obtained from germinated soybeans [26,27]. These peptide fractions exhibited exceptional antioxidant and antiproliferative activities in breast and cervical cancer cell lines. Motivated by these studies, we present the synthesis, characterization, and biological evaluation of conjugates of Fe_3O_4 NPs with bioactive peptide fractions from germinated soybeans. To the best of our knowledge, these bioconjugates are reported for the first time. A multicomponent procedure is also used for the first time in the conjugation of bioactive peptide fractions from germinated soybeans to magnetite NPs.

2. Results and Discussion

A procedure for the conjugation of bioactive peptide fractions to magnetite NPs was developed (Scheme 1). The methodology comprised three fundamental parts: (1) synthesis of Fe_3O_4 NPs, (2) functionalization of Fe_3O_4 NPs in the form of either carboxylic acid or amine, and (3) conjugation using the U-4C reaction of the peptide fractions extracted from germinated soybeans to the functionalized Fe_3O_4 NPs. Once the desired conjugates were obtained, the antioxidant activity of both the peptide–Fe_3O_4 conjugates as well as the peptide fractions and coated magnetite were determined.

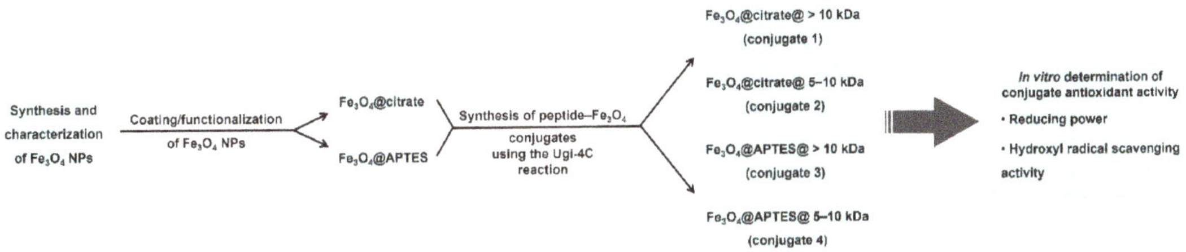

Scheme 1. Experimental workflow.

2.1. Characterization of Fe_3O_4

The X-ray diffraction (XRD) pattern of the synthesized magnetite is shown in Figure 1 (black profile). The main peaks associated with the characteristic spinel structure of magnetite were observed at 18.3, 30.2, 35.6, 43.3, 53.6, 57.3, 62.9, 74.5, and 90.6°. These peaks are indexed to the (111), (220), (311), (400), (422), (511), (440), (533), and (731) hkl planes of the Joint Committee on Powder Diffraction Standards (JCPDS) 19629 magnetite standard pattern, respectively. The estimated Fe_3O_4 cell parameter was 8.3824 Å and the average crystallite size calculated using the Debye–Scherrer equation (Equation (1)) was 17 nm. This last value allowed confirming the nanometric size of the obtained magnetite.

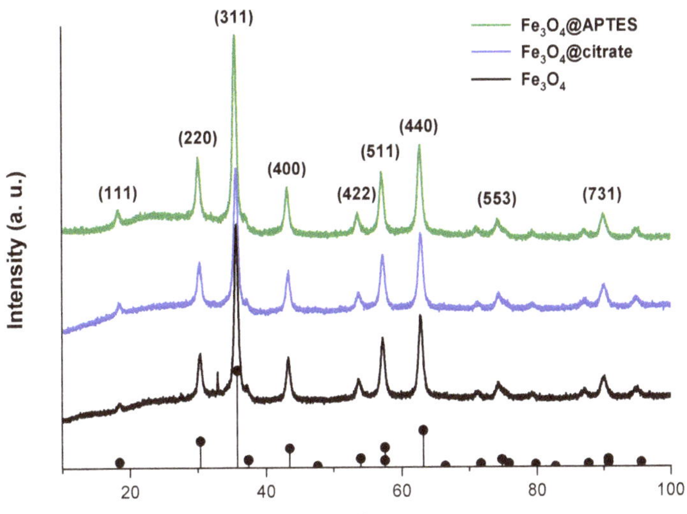

Figure 1. XRD patterns of the uncoated magnetite (Fe_3O_4) and magnetite samples coated with sodium citrate (Fe_3O_4@citrate) and APTES (Fe_3O_4@APTES).

Additional information about the size and morphology of Fe_3O_4 NPs was obtained using scanning transmission electron microscopy (STEM) (Figure 2). The particle size determined was 13 ± 5 nm. The fact that this value is similar to the calculated crystallite size suggests that the system was monocrystalline. In the STEM images, it is observed that uncoated Fe_3O_4 NPs have spherical morphology and are agglomerated, most likely because of their small sizes.

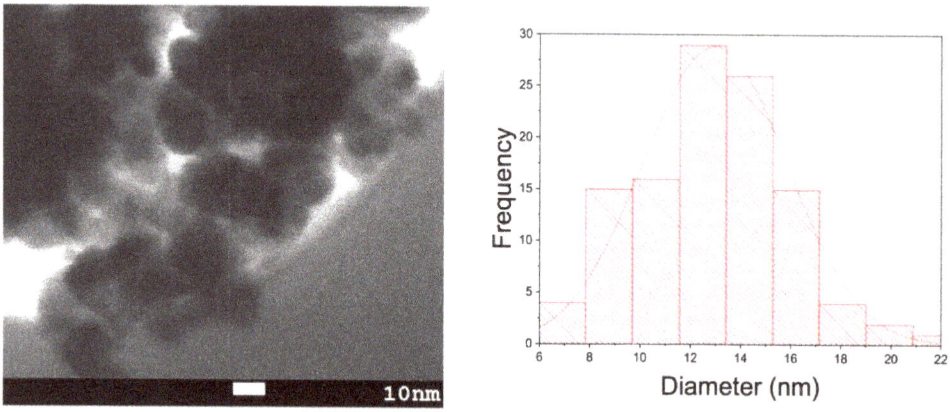

Figure 2. Microscopy image of uncoated Fe_3O_4 with the corresponding histogram.

The synthesis of Fe_3O_4 was also corroborated by infrared (IR) spectroscopy (Figure 3, black spectrum). From the assignment of the absorption bands observed in the IR spectra obtained, it was possible to analyze the respective functional groups of the molecules present in each of the synthesized materials. Thus, a broad band centered at 3252 cm^{-1}, corresponding to the overlapping valence vibrations of the N–H and O–H bonds, was observed. It was assigned to NH_4^+ and OH^- ions and adsorbed water molecules on the surface of MNPs. In addition, the band at 1628 cm^{-1} can be attributed to the bending of the NH_4^+ group on the magnetite surface. Finally, at 573 cm^{-1}, a band assigned to the valence vibration of the Fe–O bond of Fe_3O_4 was observed.

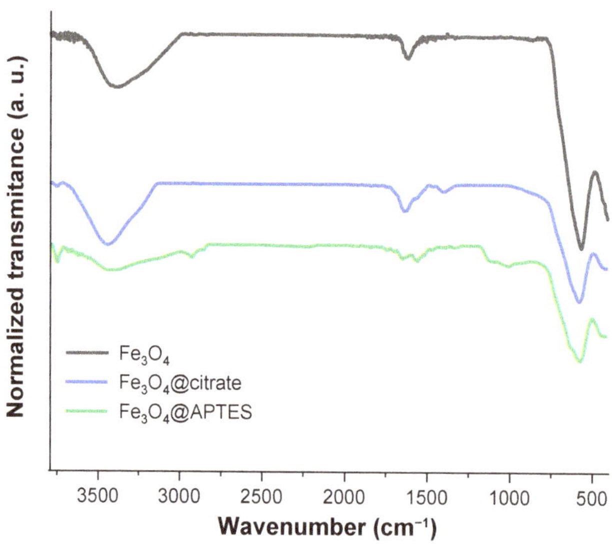

Figure 3. IR spectra of the uncoated magnetite (Fe_3O_4) and magnetite samples coated with sodium citrate (Fe_3O_4@citrate) and APTES (Fe_3O_4@APTES).

Finally, the hydrodynamic diameter of Fe_3O_4 was determined using dynamic light scattering (DLS) measurements. The value obtained (180 $\pm$ 70 nm) was used as a reference for comparison to the coated magnetite samples.

2.2. Characterization of the Functionalized Fe_3O_4

As mentioned in the Introduction, though magnetite is a biocompatible material, its stabilization in a physiological environment is only possible after coating [2,4]. In addition, functionalization enables the conjugation with therapeutic biomolecules and contributes to achieving effective directionality of magnetite nanoparticles towards the site of interest inside the body [7]. In this work, it was decided to carry out the coating/functionalization of the Fe_3O_4 NPs in a single step. Thus, from the use of sodium citrate and 3-aminopropyltriethoxysilane (APTES), it was possible to functionalize the magnetite surface in the form of carboxylic acid and amine, respectively. Both functional groups are required for subsequent conjugation to the peptide fractions from the Ugi 4C reaction. Each of the variants made is described below.

2.2.1. Characterization of Fe_3O_4@citrate

Carboxylate is one of the functional groups that most strongly binds to the surface of magnetite NPs [28]. Sodium citrate (or the corresponding acid), a small biocompatible molecule, has three carboxylate groups in its structure. Only one or two of the carboxylate functions of sodium citrate have been found to coordinate to the magnetite surface, allowing at least one of them to be exposed to the medium [29]. This effect is very convenient for the application of the magnetic system since it protects the nanoparticulate nucleus from the possible formation of aggregates while providing it with hydrophilicity and functionalizing it for subsequent derivatizations and couplings. Here, after the synthesis of Fe_3O_4 NPs coated with sodium citrate (Fe_3O_4@citrate), the resulting product was characterized.

The Fourier-transform infrared (FTIR) spectroscopy analysis of the MNPs coated with sodium citrate suggested that the desired product was obtained (Figure 3, blue spectrum). The three bands at 3435, 1629, and 1398 cm^{-1} indicate carboxylate in the sample. These signals correspond to the valence vibrations of the O–H, C=O, and COO– bonds, respectively. The band corresponding to Fe_3O_4 remained at 537 cm^{-1}. In addition, the XRD patterns of the starting magnetite (Figure 1, black profile) and that coated with citrate (Figure 1, blue profile) were similar, which suggests that the crystallinity of the material was not affected by the coating. However, the calculated cell parameter was 8.3643 Å. The difference with respect to the cell parameter of the uncoated magnetite indicates a strong interaction between citrate atoms and magnetite, which results in the contraction of the unit cell of the starting material [30]. The average crystallite size determined using the Debye–Scherrer equation (Equation (1)) was 16 nm. This value is similar to the particle size estimated using STEM (13 ± 3 nm). The similarity between these results suggests that the compound was monocrystalline. Figure 4a shows a microscopy image and obtained size distribution with the corresponding histogram. The obtained NPs had spherical morphology with some aggregation because of their small sizes.

Another evidence of the presence of citrate on the surface of the MNPs is the increase in the hydrodynamic diameter (284 ± 140 nm) relative to that of the uncoated MNPs. As some of the polar groups of citrates (carboxylate and hydroxyl) are dangling towards the surrounding medium, a larger hydration layer is generated. Finally, the weight percentage of the coating material on magnetite, with respect to the uncoated Fe_3O_4, was determined using thermogravimetric analysis (TGA). The weight loss was 2.21%, confirming that magnetite was coated with an organic material.

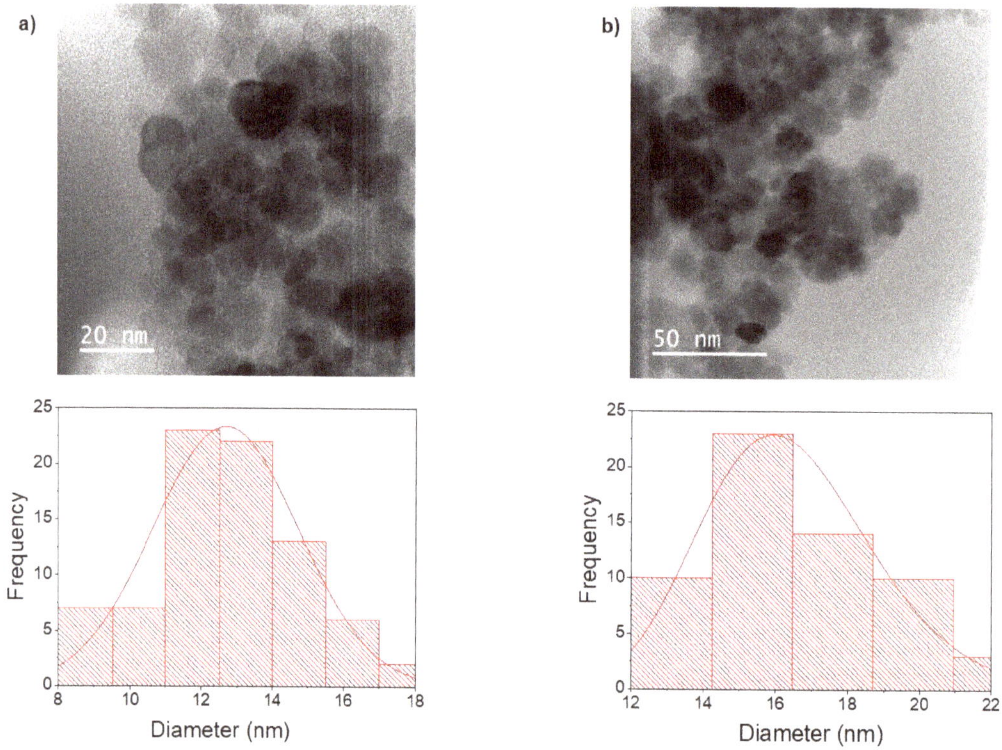

Figure 4. Microscopy images of (**a**) Fe_3O_4@citrate and (**b**) Fe_3O_4@APTES with the corresponding histograms.

2.2.2. Characterization of Fe_3O_4@APTES

The functionalization of the Fe_3O_4 NPs surface in the amine form was carried out from a silylation reaction with the use of APTES. This molecule is the most widely used alkoxysilane for coating magnetite NPs since it provides the highest density of amine groups on the surface of the magnetic core [31].

The IR spectrum of Fe_3O_4@APTES showed characteristic bands that confirmed the presence of the expected APTES coating (Figure 3, green spectrum). Thus, a wide band, centered at 3399 cm^{-1}, was observed, attributed to the overlapping valence vibrations of the O–H and N–H bonds of the hydroxyl and amine groups in APTES. The bands corresponding to the antisymmetric and symmetric vibrations of the APTES methylene groups appeared at 2928 and 2842 cm^{-1}, respectively. The bands corresponding to the vibrations antisymmetric and symmetric of the NH_2 group in APTES were observed around 1642 and 1549 cm^{-1}, respectively. The presence of silanol groups (Si–OH) and siloxanes (Si–O–Si) on the surface of the MNPs was verified by the bands of the valence vibrations of the Si–O–Si and Si–OH bonds at 1396 and 1015 cm^{-1}, respectively. Finally, the Fe_3O_4 band was maintained at 573 cm^{-1}, suggesting the presence of magnetite in the conjugate. Moreover, no differences were observed between the XRD patterns of the coated and uncoated magnetite samples (Figure 1, green and black profile, respectively), which confirms that the crystallinity of the material was not affected by the coating with APTES.

The calculated cell parameter of Fe_3O_4@APTES was 8.3744 Å, slightly lower than that determined for the uncoated magnetite, as the Fe and O atoms are closer to each other in the unit cell owing to the APTES molecules [30]. The crystallite size of the APTES-coated NPs was 17 nm, which confirms that the compound maintained its nanoscale dimensions. This value is very similar to the particle size obtained using STEM (16 $\pm$ 3 nm), which suggests that the system was monocrystalline. The STEM images (Figure 4b) show that the coated NPs have spherical morphology and are agglomerated because of their small sizes. The weight percentage of the APTES-coated Fe_3O_4 was determined using TGA. A weight loss of 5.22% with respect to the uncoated Fe_3O_4 was obtained, which confirmed the coating on the magnetite surface.

2.3. Determination of the Degree of Substitution of Functionalized MNPs

The degree of substitution on the magnetic oxide was assessed, i.e., the amount of free functional groups on the surface of Fe_3O_4. For this purpose, the adsorption capacity of Cu (II) on the functionalized MNPs was determined using a spectrophotometric analysis. The results are presented in Table 1. To assess the molar quantity of free carboxylate groups on the MNPs coated with sodium citrate, it was considered that 2 M of carboxylate is necessary to complex 1 M of Cu^{2+}. In the case of MNPs coated with APTES, it was assumed that the Cu^{2+} ions would form a complex with tetrahedral geometry with free amine groups on the magnetite [32]. Using this procedure, approximate molar quantities of free carboxylates or amine groups per gram of Fe_3O_4 were obtained.

Table 1. Determination of the degree of substitution on the MNPs using the bicinchoninate method.

Particle	t (h)	Cu^{2+} Adsorption Capacity (mg Cu^{2+}/g MNPs)	Amount (mmol of Free Groups/g MNPs)
Fe_3O_4@citrate	0	0	0
Fe_3O_4@citrate	1	14	0.44
Fe_3O_4@citrate	24	15	0.47
Fe_3O_4@APTES	0	0	0
Fe_3O_4@APTES	1	12	0.76
Fe_3O_4@APTES	24	12	0.76

The calculations for the subsequent synthesis step were carried out using the average value of the loading of functional groups on the Fe_3O_4 surface.

2.4. Characterization of Peptide–Fe_3O_4 Conjugates

The main objective of this study was to develop a simple procedure for the conjugation of bioactive peptide fractions extracted from germinated soybeans to magnetite NPs. To this end, the U-4C reaction was chosen because it occurs in a single step and can be easily implemented [33,34]. The results for the remaining peptide concentrations are shown in Figure 5. There were no statistically significant differences in the percentage of free peptides at 4 and 8 h in either variant for the same peptide fraction. This suggests that the reaction ended after 4 h. However, the difference between the percentages of conjugation when the fraction >10 kDa was used (approximately 60%) and that at 5–10 kDa (approximately 92%) was remarkable. This was expected considering the smaller size of the peptides contained in the latter, which would facilitate the multicomponent process. On the other hand, the yields of the developed conjugation process can be considered good in all the cases, considering that the magnetic component of the reaction is solid-support-dispersed, while the other reactants are dissolved in the medium [35].

The conjugates were characterized using FTIR spectroscopy (Figure 6). In all the cases, the bands that confirmed the presence of carbonyl groups in each of the structures on the MNPs were observed at approximately 1620 cm^{-1}. In addition, the band at 573 cm^{-1} was retained, which corroborated the presence of Fe_3O_4.

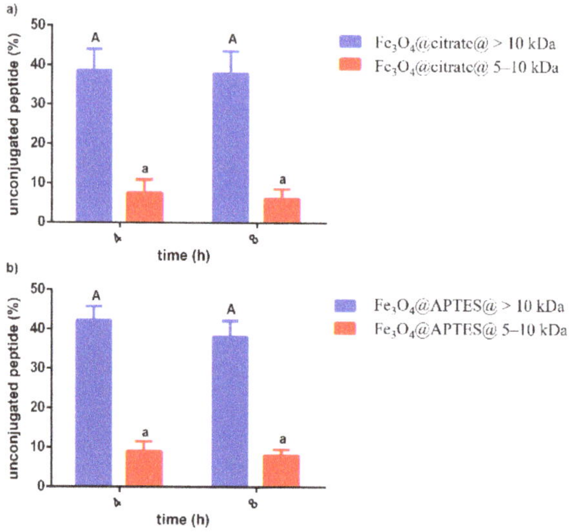

Figure 5. Percentages of unconjugated peptides to coated Fe_3O_4: (**a**) Fe_3O_4@citrate@ > 10 kDa, Fe_3O_4@citrate@ > 5–10 kDa; (**b**) Fe_3O_4@APTES@ 10 kDa, Fe_3O_4@APTES@ 5–10 kDa. The determinations were carried out in triplicate. The results are expressed as the means ± standard deviation. Different letters indicate statistical difference between the samples for each reaction time ($p \leq 0.05$).

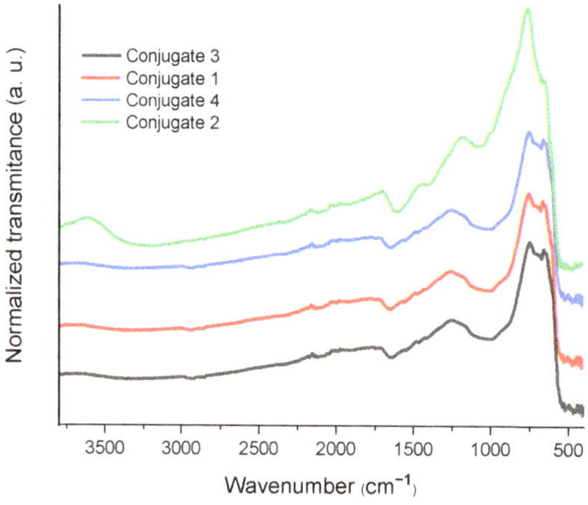

Figure 6. FTIR spectra of the final conjugates: Fe_3O_4@citrate@ > 10 kDa (conjugate 1), Fe_3O_4@citrate@ 5–10 kDa (conjugate 2), Fe_3O_4@APTES@ > 10 kDa (conjugate 3), and Fe_3O_4@APTES@ 5–10 kDa (conjugate 4).

2.5. Antioxidant Activities of the Conjugates

The association between the generation of ROS and the development of chronic degenerative diseases has been reported [27]. Furthermore, in cellular metabolism, oxidative compounds are produced constantly, so it is important to counteract these oxidizing spices to maintain a balance of intra- and extracellular homeostasis. Hence, the antioxidant capacities of the conjugates obtained in this study were evaluated. The determinations were made by verifying their reducing power (RP) and hydroxyl radical (OH·) scavenger abilities. The results presented below (Figures 7 and 8) correspond to the tests carried out at a concentration of 15 mg/mL, at which a stronger effect was observed in the determinations.

All the results were compared to those obtained for the nonconjugated peptide fractions and for the coated magnetite (Fe_3O_4@APTES and Fe_3O_4@citrate).

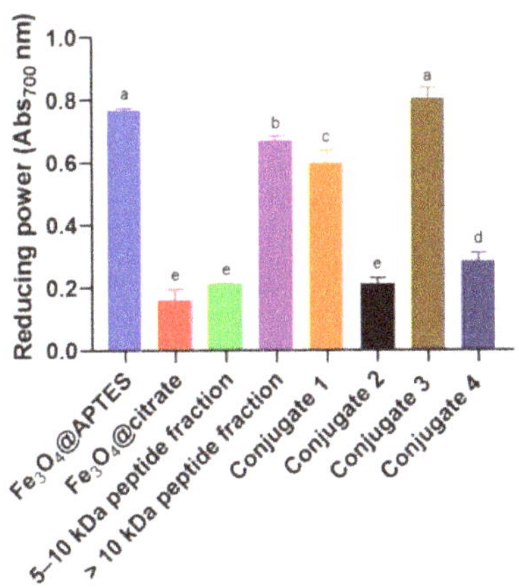

Figure 7. Average results of the RPs of coated magnetite (Fe_3O_4@APTES and Fe_3O_4@citrate) and of the conjugates and nonconjugated peptide fractions at 15 mg/mL: Fe_3O_4@citrate@ > 10 kDa (conjugate 1), Fe_3O_4@citrate@ 5–10 kDa (conjugate 2), Fe_3O_4@APTES@ > 10 kDa (conjugate 3), Fe_3O_4@APTES@ 5–10 kDa (conjugate 4), and nonconjugated peptide fractions (>10 kDa and 5–10 kDa peptide fractions). The determinations were carried out in triplicate. The results are expressed as the means ± standard deviation. Different letters indicate statistical difference between the samples ($p \leq 0.05$).

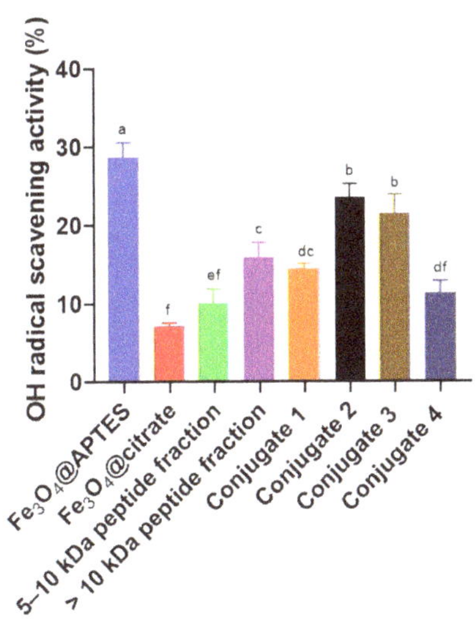

Figure 8. Average results of the OH· scavenging activities of coated magnetite (Fe_3O_4@APTES and Fe_3O_4@citrate) and of the conjugates and nonconjugated peptide fractions at 15 mg/mL: Fe_3O_4@citrate@ > 10 kDa (conjugate 1), Fe_3O_4@citrate@ 5–10 kDa (conjugate 2), Fe_3O_4@APTES@ > 10 kDa (conjugate 3), Fe_3O_4@APTES@ 5–10 kDa (conjugate 4), and nonconjugated peptide fractions (>10 kDa and 5–10 kDa peptide fractions). The determinations were carried out in triplicate. The results are expressed as the means ± standard deviation. Different letters indicate statistical differences between the samples ($p \leq 0.05$).

The RPs of the coated magnetite, as well as those of the conjugates and nonconjugated peptide fractions (Figure 7), were expressed as absorbances, where 1 represents the highest value. The RP of a sample refers to its ability to act as a proton acceptor (or electron donor) in an oxidation–reduction reaction and, therefore, the number of basic groups in the conjugates could be related to the obtained RP. In this sense, Fe_3O_4@APTES, with free amine groups on the magnetite surface, showed the highest RP value. Regarding the conjugates, those in which magnetite was conjugated to the fractions containing the peptides with masses >10 kDa (conjugates 1 and 3) had the highest RP values, with statistically significant differences between them and the conjugates 2 and 4. This result agrees with the study by González-Montoya et al. [27], who observed that the >10 kDa (RP = 0.7) fraction had a higher RP than the 5–10 kDa fraction (RP = 0.3). This could be attributed to the higher amount of basic amino acids in the >10 kDa fraction. A synergistic effect was observed when conjugating the peptide fractions to Fe_3O_4@APTES (conjugate 3 with respect to the >10 kDa peptide fraction), surely, due to the contribution of the amine groups that remained free, unreacted during the multicomponent reaction, on the surface of the magnetite coated with the APTES. It was also observed that the conjugation of the >10 kDa peptide fraction to magnetite NPs coated with sodium citrate (conjugate 1) slightly decreased the RP with respect to that of the peptide fraction, whilst the RP of the 5–10 kDa peptide fraction conjugated to Fe_3O_4@citrate (conjugate 2) did not exhibit statistically significant differences with respect to the nonconjugated fraction. In this way, conjugate 3 represents an opportunity for future evaluations of biological activities related with oxidative stress in chronic diseases.

The hydroxyl radical (OH·) is one of the most reactive free radicals and is generated by the Fenton reaction in cells [27]. This radical can be transformed into a superoxide anion and hydrogen peroxide in the presence of metal ions such as copper and iron. The OH· scavenging activities of the coated magnetite (Fe_3O_4@APTES and Fe_3O_4@citrate), as well as those of the conjugates and nonconjugated peptide fractions, are shown in Figure 8. According to the data, the nature of the groups exposed to the medium on each sample determines the properties of the compound. In the case of conjugates with the fraction > 10 kDa (conjugates 1 and 3), the activity is higher when it is used as an acidic component of the multicomponent reaction (conjugate 3). Therefore, in the final compound, the peptides would mostly expose their amine groups (which are apparently involved in the measured activity) to the environment. In the conjugates with the 5–10 kDa fraction (conjugates 2 and 4), the opposite effect was observed. The peptides contained in this fraction exerted a stronger OH· scavenging effect by exposing their acidic groups (conjugate 2). Remarkably, the OH· scavenging activities of conjugates 2 and 3 were significantly higher than that of the nonconjugated fractions. This result could be due to a synergistic effect of the respective carboxylic acid (from Fe_3O_4@citrate) and amine (from Fe_3O_4@APTES) groups in the corresponding final conjugates.

These results suggest that conjugates 2 and 3 could be a used as antioxidant conjugates and evaluated in the future in chronic diseases related with oxidative stress, such as cancer since it has been reported that antioxidant activity can modulate the activity of key proteins involved in the control of cell cycle progression and may influence the expression of many associated genes [36].

3. Materials and Methods

3.1. Materials

Ferrous chloride tetrahydrate ($FeCl_2 \cdot 4H_2O$, PA), sodium citrate dihydrate ($Na_3C_6H_5O_7 \cdot 2H_2O$, ACS >99%), APTES (97%), paraformaldehyde, 1-pentynylisonitrile, trichloroacetic acid (TCA), phenanthroline, and ferrous sulphate ($FeSO_4$) were purchased from Sigma-Aldrich, Mexico City, Mexico. Ferric chloride hexahydrate ($FeCl_3 \cdot 6H_2O$, 99.0%) and the ammonium hydroxide solution (NH_4OH, 28.4%) were purchased from J.T. Baker, Mexico City, Mexico. Hydrochloric acid (HCl, PA) and potassium ferricyanide ($K_3[Fe(CN)_6]$) were purchased from Fermont, Monterrey, Mexico, and hydrogen peroxide (H_2O_2) from Reasol, Mexico city, Mexico. The peptide fractions were supplied by Laboratorio de Bioquímica de la Nutrición (ENCB-IPN), Mexico City, Mexico.

3.2. *Synthesis of Fe_3O_4 NPs*

The NPs were synthesized using the coprecipitation method [37]. A mixture of acidic solutions (HCl, 2 M) of the $FeCl_2 \cdot 2H_2O$ (5 mL, 5 M) and $FeCl_3 \cdot 6H_2O$ (20 mL, 2.5 M) salts was added to an aqueous solution of NH_3 (250 mL, 0.7 M) contained in a three-neck balloon using a peristaltic pump. All the solutions were bubbled with N_2 (g) before the synthesis. The reaction mixture was kept at room temperature under an N_2 (g) atmosphere under mechanical stirring (500 rpm) and ultrasound-treated for 25 min. After the reaction time, the obtained MNPs were separated using a neodymium magnet. Subsequently, they were washed with 100 mL of deionized water and separated again to obtain a black powder corresponding to Fe_3O_4, which was dried in a desiccator containing activated silica for 12 h.

3.3. *Coating/Functionalization of Fe_3O_4 NPs*

3.3.1. Fe_3O_4@citrate

The functionalization of the MNPs as carboxylic acids was carried out by coating Fe_3O_4 with sodium citrate. Fe_3O_4 (500 mg) was added to an aqueous solution of $Na_3C_6H_5O_7 \cdot 2H_2O$ (50 mL, 1 mM). The reaction mixture was mechanically stirred for 4 h at room temperature, and then adjusted to pH = 9 using an aqueous solution of NH_3 (0.7 M). The obtained precipitate was magnetically separated and washed twice with 50 mL of deionized water. Finally, it was dried (as explained for the synthesis of the Fe_3O_4 NPs), and Fe_3O_4@citrate was obtained in the form of a black powder.

3.3.2. Fe_3O_4@APTES

Additionally, the functionalization of the MNPs as amines was carried out by modifying the surface of Fe_3O_4 with APTES [31]. Briefly, 500 mg of Fe_3O_4 were redispersed in a mixture of ethanol (80 mL) and deionized water (40 mL). The dispersion was adjusted to pH = 10 with an aqueous solution of NH_3 (0.7 M) and homogenized in an ultrasonic bath. Subsequently, an alcoholic solution of APTES (5 mL, 10%) was dripped to the reaction mixture under constant stirring. After 24 h, the precipitate was separated with a neodymium magnet and washed twice with 50 mL of ethanol. Finally, the product was dried according to the procedure described above to obtain Fe_3O_4@APTES in the form of a black powder.

3.4. *Cu(II) Adsorption Capacity of Functionalized NPs*

Once the functionalization of MNPs was completed, the degree of substitution on the magnetic oxide was calculated. The determination could be carried out using the bicinchoninate method from the quantification of Cu(II) ions remaining in aqueous dispersions [38]. An aqueous solution of $CuSO_4 \cdot 5H_2O$ (35.5 mL, 100 mg/L) was prepared in an Erlenmeyer flask. Subsequently, 5 mL of the solution were collected, and a bicinchoninic acid kit was added. The absorbance was measured using an ultraviolet (UV)–visible (Vis) spectrophotometer at 560 nm (time: 0 h), and then the concentration was determined. The functionalized MNP sample (0.03 g) was added to the remaining solution (30.5 mL) and the pH was adjusted to 5 with HCl (0.1 M). The reaction mixture was mechanically stirred with a glass-lined magnetic stirrer at 140 rpm for 1 h. The magnetic system was then separated with a neodymium magnet. The supernatant (0.25 mL) was then collected, filtered, and diluted with 4.75 mL of distilled water, and a bicinchoninic acid kit was added. The absorbance was measured at 560 nm (time: 1 h), and then the concentration was determined. The last absorbance measurement was performed 24 h after the functionalized MNPs were in contact with $CuSO_4 \cdot 5H_2O$ (time: 24 h). Each measurement was performed in duplicate.

3.5. Synthesis of Peptide–Fe_3O_4 Conjugates by the U-4C Reaction

To conjugate the peptide fractions to Fe_3O_4 using the U-4C reaction, the amino component (the peptide fractions or Fe_3O_4@APTES) and four equivalents (eq) of paraformaldehyde were reacted at room temperature for 24 h in the smallest required volume of phosphate-buffered saline (PBS) (pH = 7.4). The acid component (Fe_3O_4@citrate or peptide fractions) was then added; 10 min later, 1-pentynylisonitrile (4 eq) was added. The reactions were monitored at intervals of 4 h over 8 h and carried out in quintuplicate. Finally, the products were washed twice with 1 mL of distilled water and dried, as stated for the synthesis of Fe_3O_4 NPs.

3.6. Extension of the Conjugation

The amount of peptide remaining after conjugation to MNPs by the U-4C reaction was determined according to the Bradford assay [39]. The supernatants of each reaction (10 µL) were deposited with 100 µL of the Bradford reagent in a 96-well microplate. The readings were carried out at a wavelength of 595 nm. The remaining peptide concentration values were determined using a calibration curve of bovine serum albumin (BSA).

3.7. NP Characterizations Using X-ray Diffraction, Fourier-Transform Infrared Spectroscopy, Scanning Transmission Electron Microscopy, Thermogravimetric Analysis, Dynamic Light Scattering, and Spectrophotometric Analysis

XRD measurements were performed using a Bruker D8 Advance diffractometer. Cu $K\alpha$ (λ = 1.54183 Å) incident radiation was used. The measurements were carried out in the range of 5–100° with an increment of 0.02° and scan speed of 10 s. The cell parameters were calculated using the Powdercell 2.4 software. The average crystallite size was determined using the Debye–Scherrer equation (Equation (1))

$$D = \frac{\kappa\lambda}{\beta\cos\theta}. \quad (1)$$

where D is the crystallite size, κ is the Debye–Scherrer constant (0.89), λ is the operation wavelength, β is the average width of the most intense peak, and θ is the Bragg angle.

FTIR spectra were recorded using an Equinox 55 Bruker spectrometer in the absorbance measurement range of 4000 to 400 cm^{-1}. TGA measurements were performed using a Netzsch STA 409 PC Luxx thermal analyzer coupled to the FTIR equipment. Approximately 10 mg of the sample were heated from 25 to 1000 °C and analyzed at the heating flux of 20 °C/min in an Ar atmosphere. DLS measurements were carried out using a Zetasizer Nano sampler (Malvern Instruments, Cambridge, UK) equipped with a He–Ne laser (λ = 633 nm). The measurements were carried out in square polystyrene cells at 25 °C in the range of 0.3–10 µm using H_2O as a dispersion medium. The absorbance performance was assessed using two different types of equipment. To determine the adsorption capacity of Cu^{2+} ions, the absorbance values were obtained at 560 nm using a Genesys 10 UV spectrophotometer (Thermo Electron Corporation, Madison, WI, USA). During the experiment with the peptide fractions, the absorbances were measured using a microplate reader (Thermo Scientific Multiskan GO, Waltham, MA, USA). The readings were carried out at 536, 595, and 700 nm depending on the system and properties to be determined. The morphology and average size of the NPs were analyzed using STEM images. The measurements were carried out using a JEM-ARM200CF electronic microscope with the resolution of 80 pm for the employed mode. The acceleration voltage was 200 keV. The analysis of the STEM images to determine the NP average size was carried out with the ImageJ (Origin(Pro) Corporation, Northampton, MS, USA), and the histograms were obtained using Origin v9.0, (OriginLab Corporation, Northampton, MS, USA).

3.8. *In Vitro Determination of the Antioxidant Activity of the Samples*

3.8.1. Reducing Power (RP)

The RP of the conjugates was determined according to the method described by Oyaizu et al. [40]. It is based on the reduction of ferricyanide $[Fe(CN)_6]^{-3}$ to the ferrocyanide $[Fe(CN)_6]^{-4}$ anion. Owing to this reduction, Prussian blue was formed in the presence of Fe^{3+}. PBS (0.2 M, pH = 6.6; 50 μL) and 50 μL of $K_3[Fe(CN)_6]$ (1%) were added to 20 μL of the samples at the concentrations of 15, 7.5, 3.75, and 1.88 mg/mL in one 96-well microplate. The mixture was incubated at 40 °C for 20 min. Subsequently, 50 μL of TCA (10%) and 10 μL of $FeCl_3$ (0.1%) were added and incubated for 10 min at 40 °C. The absorbance was read at 700 nm using a microplate reader.

3.8.2. Hydroxyl Radical Scavenging Activity

The percentage hydroxyl radical (OH·) scavenging capacity of the conjugates was determined according to the method developed by Lin et al. [41]. In this procedure, OH· is generated using the Fenton reaction. Thus, 20 μL of each sample at the concentrations of 15, 7.5, 3.75, and 1.88 mg/mL were placed in a 96-well microplate. Subsequently, 50 μL of 1,10-phenanthroline and $FeSO_4$ (3 mM) were added. To start the reaction, 50 μL of H_2O_2 were added and incubated at 37 °C for 60 min. The absorbance was measured using a microplate reader at 536 nm.

3.9. *Statistical Analysis*

The statistical analysis of the data obtained during the determination of the extension of conjugation and the antioxidant activity tests was performed using GraphPad Prism v6.0, La Jolla, CA, USA. A one-way analysis of variance (ANOVA) and Tukey's test were performed for multiple comparisons. The statistical significance was set at $p \leq 0.05$.

4. Conclusions

The conjugation of peptide fractions from germinated soybeans to Fe_3O_4 NPs was achieved for the first time using a multicomponent protocol. The U-4C reaction was an efficient and versatile procedure for the conjugation of bioactive peptide fractions to Fe_3O_4 NPs. The experiments on the RP and hydroxyl radical scavenging activity demonstrated that the antioxidant capacity of the peptide fractions was maintained or increased after the conjugation. The only exception was the conjugation of the >10 kDa peptide fraction to the magnetite NPs coated with sodium citrate (conjugate 1). Its RP decreased slightly with respect to the nonconjugated peptide fraction. Fe_3O_4@APTES@ > 10 kDa showed the highest antioxidant capacity. The obtained results suggest a synergistic effect of the conjugation of peptide fractions from germinated soybeans to the coated Fe_3O_4 NPs by the U-4C reaction.

Author Contributions: Conceptualization, H.Y.-M., R.M.-E., Y.E.A.-J. and C.D.-Á.; Methodology, Y.E.A.-J., M.G.-M., D.U.-R., E.C.-U. and D.N.-F.; Validation, H.Y.-M. and R.M.-E.; Formal analysis, Y.E.A.-J.; Investigation, Y.E.A.-J.; Resources, R.M.-E. and H.Y.-M.; Writing, Y.E.A.-J. and R.M.-E.; Supervision, R.M.-E., H.Y.-M., C.D.-Á. and E.C.-U.; Funding acquisition, R.M.-E. and H.Y.-M. All authors have read and agreed to the published version of the manuscript.

Funding: This research was supported by grants from Instituto Politécnico Nacional (SIP:20181718 project) and CONACyT (PY-SEP-CONACyT 242860 project). The work was also supported with a PhD scholarship to A.T.M. from CONACyT.

Institutional Review Board Statement: Not applicable.

Informed Consent Statement: Not applicable.

Data Availability Statement: Data of this research is available from the authors.

Acknowledgments: Y.E.A.J. acknowledges CONACyT for scholarship No. 622043 and H.Y.M. E.C.U. and R.M.E. acknowledge the economic support from EDI, COFAA, and SNI. The authors acknowledge CNMN-IPN and Liliana Alamilla-Beltrán (from ENCB-IPN) for the experimental support provided. Y.E.A.J. is thankful to Eduardo Gonzalez-Martinez and Gerardo-Manuel Ojeda-Carralero for their valuable suggestions.

Conflicts of Interest: The authors declare no conflict of interest.

Sample Availability: Samples of the compounds Fe_3O_4@APTES@ > 10 kDa are available from the authors.

References

1. Mahmoudi, M.; Simchi, A.; Imani, M. Recent advances in surface engineering of superparamagnetic iron oxide nanoparticles for biomedical applications. *J. Iran. Chem. Soc.* **2010**, *7*, S1–S27. [CrossRef]
2. Corchero, J.L.; Villaverde, A. Biomedical applications of distally controlled magnetic nanoparticles. *Trends Biotechnol.* **2009**, *27*, 468–476. [CrossRef]
3. Li, L.; Jiang, W.; Luo, K.; Song, H.; Lan, F.; Wu, Y.; Gu, Z. Superparamagnetic iron oxide nanoparticles as MRI contrast agents for non-invasive stem cell labeling and tracking. *Theranostics* **2013**, *3*, 595–615. [CrossRef]
4. Kuznetsov, O.A.; Brusentsov, N.A.; Kuznetsov, A.A.; Yurchenko, N.Y.; Osipov, N.E.; Bayburtskiy, F.S. Correlation of the coagulation rates and toxicity of biocompatible ferromagnetic microparticles. *J. Mag. Magn. Mater.* **1999**, *194*, 83–89. [CrossRef]
5. Gómez-Pérez, A.; González-Martínez, E.; Díaz-Águila, C.R.; González-Martínez, D.A.; González-Ruiz, G.; García-Artalejo, A.; Yee-Madeira, H. Chitosan-coated magnetic iron oxide nanoparticles for DNA and rhEGF separation. *Coll. Surf. A: Physic. Eng. Aspects* **2020**, *591*, 124500. [CrossRef]
6. González-Martínez, E.; Gómez-Pérez, A.; González-Martínez, D.A.; Díaz-Águila, C.R.; Cristiani-Urbina, E.; Uribe-Ramírez, D.; Yee-Madeira, H. Chitosan-coated magnetic nanoparticles; exploring their potentialities for DNA and Cu(II) recovery. *Inor. Nano-Met. Chem.* **2021**, *51*, 1098–1107. [CrossRef]
7. Kogan, M.J.; Olmedo, I.; Hosta, L.; R Guerrero, A.; Cruz, L.J.; Albericio, F. Peptides and metallic nanoparticles for biomedical applications. *Nanomedicine* **2007**, *2*, 287–306. [CrossRef]
8. Zhang, W.; Taheri-Ledari, R.; Hajizadeh, Z.; Zolfaghari, E.; Ahghari, M.R.; Maleki, A.; Hamblin, M.R.; Tian, Y. Enhanced activity of vancomycin by encapsulation in hybrid magnetic nanoparticles conjugated to a cell-penetrating peptide. *Nanoscale* **2020**, *12*, 3855–3870. [CrossRef]
9. Scarberry, K.E.; Dickerson, E.B.; McDonald, J.F.; Zhang, Z.J. Magnetic nanoparticle—peptide conjugates for in vitro and in vivo targeting and extraction of cancer cells. *J. Am. Chem. Soc.* **2008**, *130*, 10258–10262. [CrossRef]
10. Nosrati, H.; Tarantash, M.; Bochani, S.; Charmi, J.; Bagheri, Z.; Fridoni, M.; Abdollahifar, M.-A.; Davaran, S.; Danafar, H.; Kheiri Manjili, H. Glutathione (GSH) peptide conjugated magnetic nanoparticles as blood–brain barrier shuttle for MRI-monitored brain delivery of paclitaxel. *ACS Biom. Sci. Eng.* **2019**, *5*, 1677–1685. [CrossRef]
11. Liscano, Y.; Oñate-Garzón, J.; Delgado, J.P. Peptides with dual antimicrobial–anticancer activity: Strategies to overcome peptide limitations and rational design of anticancer peptides. *Molecules* **2020**, *25*, 4245. [CrossRef]
12. Gaspar, D.; Veiga, A.S.; Castanho, M.A. From antimicrobial to anticancer peptides. A review. *Front. Microbiol* **2013**, *4*, 294. [CrossRef]
13. Jeong, W.-j.; Bu, J.; Kubiatowicz, L.J.; Chen, S.S.; Kim, Y.; Hong, S. Peptide–nanoparticle conjugates: A next generation of diagnostic and therapeutic platforms? *Nano Conver.* **2018**, *5*, 38. [CrossRef]
14. Arriortua, O.K.; Garaio, E.; de la Parte, B.H.; Insausti, M.; Lezama, L.; Plazaola, F.; García, J.A.; Aizpurua, J.M.; Sagartzazu, M.; Irazola, M. Antitumor magnetic hyperthermia induced by RGD-functionalized Fe3O4 nanoparticles, in an experimental model of colorectal liver metastases. *Beilstein, J. Nano.* **2016**, *7*, 1532–1542. [CrossRef] [PubMed]
15. Yen, S.; Padmanabhan, P.; Selvan, S. Multifunctional iron oxide nanoparticles for diagnostics, therapy and macromolecule delivery. *Theranostics* **2013**, *3*, 986–1003. [CrossRef] [PubMed]
16. Wadghiri, Y.Z.; Sigurdsson, E.M.; Sadowski, M.; Elliott, J.I.; Li, Y.; Scholtzova, H.; Tang, C.Y.; Aguinaldo, G.; Pappolla, M.; Duff, K. Detection of Alzheimer's amyloid in transgenic mice using magnetic resonance microimaging. *Magn. Reson. Med.* **2003**, *50*, 293–302. [CrossRef] [PubMed]
17. Arosio, D.; Manzoni, L.; Araldi, E.M.; Scolastico, C. Cyclic RGD functionalized gold nanoparticles for tumor targeting. *Bioconjugate Chem.* **2011**, *22*, 664–672. [CrossRef]
18. Ugi, I.; Werner, B.; Dömling, A. The chemistry of isocyanides, their multicomponent reactions and their libraries. *Molecules* **2003**, *8*, 53–66. [CrossRef]
19. Weber, L.; Illgen, K.; Almstetter, M. Discovery of new multi component reactions with combinatorial methods. *Synlett* **1999**, *3*, 366–374. [CrossRef]
20. Dömling, A.; Ugi, I. Multicomponent reactions with isocyanides. *Angew. Chem. Inter. Ed.* **2000**, *39*, 3168–3210. [CrossRef]
21. Rivera, D.G.; Vasco, A.V.; Echemendía, R.; Concepción, O.; Pérez, C.S.; Gavín, J.A.; Wessjohann, L.A. Frontispiece: A ulticomponent conjugation strategy to unique N-steroidal peptides: First evidence of the steroidal nucleus as a β-Turn inducer in acyclic peptides. *Chem. Eur. J.* **2014**, *20*, 13150–13161. [CrossRef] [PubMed]
22. Sarmadi, B.H.; Ismail, A. Antioxidative peptides from food proteins: A review. *Peptides* **2010**, *31*, 1949–1956. [CrossRef] [PubMed]
23. Branca, F.; Lartey, A.; Oenema, S.; Aguayo, V.; Stordalen, G.A.; Richardson, R.; Arvelo, M.; Afshin, A. Transforming the food system to fight non-communicable diseases. *BMJ* **2019**, *364*, l296. [CrossRef] [PubMed]

24. Zhang, C.; Zhou, Y.; Yang, G.-Y.; Li, S. Biomimetic peptides protect cells from oxidative stress. *Am. J. Trans. Res.* **2017**, *9*, 5518.
25. Elias, R.J.; Kellerby, S.S.; Decker, E.A. Antioxidant activity of proteins and peptides. *C. Food Sci. Nutr.* **2008**, *48*, 430–441. [CrossRef]
26. González-Montoya, M.; Hernández-Ledesma, B.; Silván, J.M.; Mora-Escobedo, R.; Martínez-Villaluenga, C. Peptides derived from in vitro gastrointestinal digestion of germinated soybean proteins inhibit human colon cancer cells proliferation and inflammation. *Food Chem.* **2018**, *242*, 75–82. [CrossRef]
27. González-Montoya, M.; Ramón-Gallegos, E.; Robles-Ramírez, M.C.; Mora-Escobedo, R. Evaluation of the antioxidant and antiproliferative effects of three peptide fractions of germinated soybeans on breast and cervical cancer cell lines. *PFHN* **2016**, *71*, 368–374.
28. Weissleder, R.; Ross, B.; Rehemtulla, A.; Gambhir, S. General principles of molecular imaging. In *Molecular Imaging: Principles and Practice*; PMPH: Shelton, CT, USA, 2010; pp. 1–9.
29. Li, L.; Mak, K.; Leung, C.W.; Chan, K.; Chan, W.; Zhong, W.; Pong, P. Effect of synthesis conditions on the properties of citric-acid coated iron oxide nanoparticles. *Microelectron. Eng.* **2013**, *110*, 329–334. [CrossRef]
30. Freire, T.; Dutra, L.M.; Queiroz, D.; Ricardo, N.; Barreto, K.; Denardin, J.; Wurm, F.R.; Sousa, C.; Correia, A.; de Lima-Neto, P. Fast ultrasound assisted synthesis of chitosan-based magnetite nanocomposites as a modified electrode sensor. *Carb. Polym.* **2016**, *151*, 760–769. [CrossRef]
31. Bini, R.A.; Marques, R.F.C.; Santos, F.J.; Chaker, J.A.; Jafelicci Jr, M. Synthesis and functionalization of magnetite nanoparticles with different amino-functional alkoxysilanes. *J. Magnet. Magn. Mat.* **2012**, *324*, 534–539. [CrossRef]
32. Basolo, F.; Johnson, R.; Busch, R.H. *Química Compuestos de Coordinación*; Reverté: Barcelona, Spain, 1980; pp. 1–178.
33. El Kaim, L.; Grimaud, L. Beyond the Ugi reaction: Less conventional interactions between isocyanides and iminium species. *Tetrahedron* **2009**, *65*, 2153–2171. [CrossRef]
34. Reguera, L.; Méndez, Y.; Humpierre, A.R.; Valdés, O.; Rivera, D.G. Multicomponent reactions in ligation and bioconjugation chemistry. *Acc. Chem. Res.* **2018**, *51*, 1475–1486. [CrossRef] [PubMed]
35. Morales, F.E.; Garay, H.E.; Muñoz, D.F.; Augusto, Y.E.; Otero-González, A.J.; Reyes Acosta, O.; Rivera, D.G. Aminocatalysis-mediated on-resin ugi reactions: Application in the solid-phase synthesis of n-substituted and tetrazolo lipopeptides and peptidosteroids. *Org. Lett.* **2015**, *17*, 2728–2731. [CrossRef] [PubMed]
36. Vega, C.; Delgado-Olivares, L.; Morales González, J.A.; García, E.A.; Villagomez Ibarra, R.J.; Moreno, E.R.; Gutiérrez, M.S.; Sumaya Martínez, M.T.; Clara, Z.P.; Ramos, Z.C. The role of natural antioxidants in cancer disease, oxidative stress and chronic degenerative diseases. In *Oxidative Stress and Chronic Degenerative Diseases—A Role for Antioxidants*; Morales-Gonzalez, J.A., Ed.; IntechOpen: London, UK, 2013; pp. 391–418.
37. Laurent, S.; Forge, D.; Port, M.; Roch, A.; Robic, C.; Vander Elst, L.; Muller, R.N. Magnetic iron oxide nanoparticles: Synthesis, stabilization, vectorization, physicochemical characterizations, and biological applications. *Chem. Rev.* **2008**, *108*, 2064–2110. [CrossRef]
38. Hach Company. DR/890 Colimeter Procedures Manual. Available online: http://www.hach.com (accessed on 15 February 2020).
39. Bradford, M.M. A rapid and sensitive method for the quantitation of microgram quantities of protein utilizing the principle of protein-dye binding. *Anal. Biochem.* **1976**, *72*, 248–254. [CrossRef]
40. Oyaizu, M. Studies on products of browning reaction. *Jpn. J. Nutr. Deiet.* **1986**, *44*, 307–315. [CrossRef]
41. Lin, S.-Y.; Wang, C.-C.; Lu, Y.-L.; Wu, W.-C.; Hou, W.-C. Antioxidant, anti-semicarbazide-sensitive amine oxidase, and anti-hypertensive activities of geraniin isolated from Phyllanthus urinaria. *Food Chem. Toxic.* **2008**, *46*, 2485–2492. [CrossRef] [PubMed]

molecules

MDPI

Article

Effects of Combination Treatments with Astaxanthin-Loaded Microparticles and Pentoxifylline on Intracellular ROS and Radiosensitivity of J774A.1 Macrophages

Eleonora Binatti [1,*], Gianni Zoccatelli [1,2], Francesca Zanoni [2], Giulia Donà [2], Federica Mainente [1] and Roberto Chignola [1]

1 Department of Biotechnology, University of Verona, Strada Le Grazie 15-CV1, I-37134 Verona, Italy; gianni.zoccatelli@univr.it (G.Z.); federica.mainente@univr.it (F.M.); roberto.chignola@univr.it (R.C.)
2 Sphera Encapsulation S.r.l., Strada Le Grazie 15-CV1, I-37134 Verona, Italy; zanoni@spheraencapsulation.com (F.Z.); lab@spheraencapsulation.com (G.D.)
* Correspondence: eleonora.binatti@univr.it

Abstract: Radiation-induced fibrosis (RIF) is a serious, yet incurable, complication of external beam radiation therapy for the treatment of cancer. Macrophages are key cellular actors in RIF because of their ability to produce reactive oxidants, such as reactive oxygen species (ROS) and inflammatory cytokines that, in turn, are the drivers of pro-fibrotic pathways. In a previous work, we showed that phagocytosis could be exploited to deliver the potent natural antioxidant astaxanthin specifically to macrophages. For this purpose, astaxanthin encapsulated into μm-sized protein particles could specifically target macrophages that can uptake the particles by phagocytosis. In these cells, astaxanthin microparticles significantly reduced intracellular ROS levels and the secretion of bioactive TGFβ and increased cell survival after radiation treatments. Here we show that pentoxifylline, a drug currently used for the treatment of muscle pain resulting from peripheral artery disease, amplifies the effects of astaxanthin microparticles on J774A.1 macrophages. Combination treatments with pentoxifylline and encapsulated astaxanthin might reduce the risk of RIF in cancer patients.

Keywords: pentoxifylline; astaxanthin; particle encapsulation; oxidative stress; ionizing radiations; macrophages; radiation-induced fibrosis

Citation: Binatti, E.; Zoccatelli, G.; Zanoni, F.; Donà, G.; Mainente, F.; Chignola, R. Effects of Combination Treatments with Astaxanthin-Loaded Microparticles and Pentoxifylline on Intracellular ROS and Radiosensitivity of J774A.1 Macrophages. *Molecules* **2021**, *26*, 5152. https://doi.org/10.3390/molecules26175152

Academic Editors: Višnja Stepanić and Marta Kučerová-Chlupáčová

Received: 26 July 2021
Accepted: 24 August 2021
Published: 25 August 2021

1. Introduction

Radiation-induced fibrosis (RIF) is the most common long-term complication in cancer radiotherapy, and it affects a significant proportion of the ~60% cancer patients that receive radiation as part of their therapeutic regimens [1]. The onset of RIF occurs between 1–4 months and 1–2 years after radiotherapy and it progresses in severity with time [2]. RIF does not regress spontaneously, and to date, there are no effective treatments for RIF prevention and/or treatment [3]. Investigations with patients and experimental animals have demonstrated a significant reduction of RIF with antioxidant treatments administered after radiation [4]. The pathogenesis of RIF is still not fully understood, although it is known that phagocytic cells, such as macrophages, play a key role in RIF for their ability to produce oxidant chemical species, in particular reactive oxygen species (ROS), and inflammatory molecules that in turn trigger the fibrotic pathways in injured tissues [5].

Pentoxifylline (3,7-dimethyl-1-(5-oxohexyl)purine-2,6-dione, PTX) is a derivative of xanthine. PTX was approved by FDA in 1984 for the treatment of muscle pain resulting from peripheral artery disease, but it is also used off-label for the treatment of leg ulcers [6]. Indeed, PTX is known to improve the rheological properties of blood and wound healing, and to have anti-inflammatory and antioxidant activity [7].

PTX has also been investigated as a radioprotector in vitro and in vivo because it can increase blood flow and tissue oxygenation upon oral administration [8]. It has provided satisfactory results in terms of RIF resolution but only at high doses (i.e., 800 mg/day over a 6-month treatment period [8]) that are not tolerated by patients and that can cause, among the side effects, serious gastrointestinal disorders [8,9]. Thus PTX alone cannot be used to cure RIF. Vitamin E is a well-known antioxidant, and it can effectively scavenge reactive oxygen species produced by radiation treatments. Moreover, its deficiency is associated with dysfunction of connective tissue repair [9]. When administered orally to patients its effects on RIF were only modest, but given in combination with PTX tablets, significantly reduced fibrosis and related clinical symptoms such as skin tightness, dyspnea, and pain [2].

Astaxanthin (ASX) is a xanthophyll ketocarotenoid with an antioxidant potential 100 times higher than vitamin E [10]. We previously explored in vitro whether ASX could be effective against RIF [11]. ASX was encapsulated into μm-sized protein particles that can only be taken up by cells through phagocytosis, and in this way, we could target them specifically to macrophages [11]. In these cells, ASX microparticles significantly reduced intracellular ROS levels and, importantly, the secretion of bioactive transforming growth factor β (TGFβ) [11]. The role of this pro-inflammatory cytokine is central to fibrosis because it promotes fibroblasts recruitment, and local deposition of extracellular matrix components [12]. In addition, it inhibits the expression of antioxidant enzymes in target cells, further contributing to intracellular ROS accumulation [12].

Given these encouraging results and the outcomes of combination treatments with PTX and vitamin E discussed above, we explored whether PTX could also enhance the effects of ASX when given in combination to macrophages. We found that PTX could indeed improve the antioxidant and radio-protective activities of ASX microparticles.

2. Results

Full details concerning the synthesis of ASX-loaded microparticles along with their complete biochemical and biological characterization have been given in reference [11]. In this work we used the same preparations and in this section we describe new data concerning PTX and its interactions with ASX microparticles.

2.1. Cytotoxicity of PTX on J774A.1 Macrophages

In our previous work, we showed that ASX microparticles were not cytotoxic for J774A.1 macrophage cells [11]. To the best of our knowledge no information is available on the safety of PTX in this cell system. We therefore assayed whether PTX could alter the growth kinetics of these cells. Intracellular ATP content was measured at different time points of the growth curve obtained with independent cell populations continuously exposed to different PTX concentrations, and the results are shown in Figure 1.

The data show that PTX, at all assayed concentrations, was not cytotoxic for J774A.1 cells.

2.2. PTX Does Not Alter Phagocytosis of ASX-Loaded Microparticles

Central to our treatment strategy is the idea that antioxidant molecules can be delivered specifically to macrophages (and other phagocytes) if loaded into particles of appropriate size that can be taken up by these cells through phagocytosis (see also our analyses carried out by confocal microscopy and reported in [11]). PTX has been reported to inhibit the phagocytosis induced by latex particles (~0.8 μm diameter) in monocytes and polymorphonuclear leukocytes in a dose-dependent manner. The critical PTX concentration at which the inhibitory activity became significant was 10 μg/mL [13]. We therefore assayed whether a slightly lower PTX concentration, namely 8 μg/mL, could inhibit phagocytosis of ASX microparticles in J774A.1 cells.

Figure 2 shows that 8 μg/mL PTX did not inhibit phagocytosis of ASX microparticles in J774A.1 macrophages.

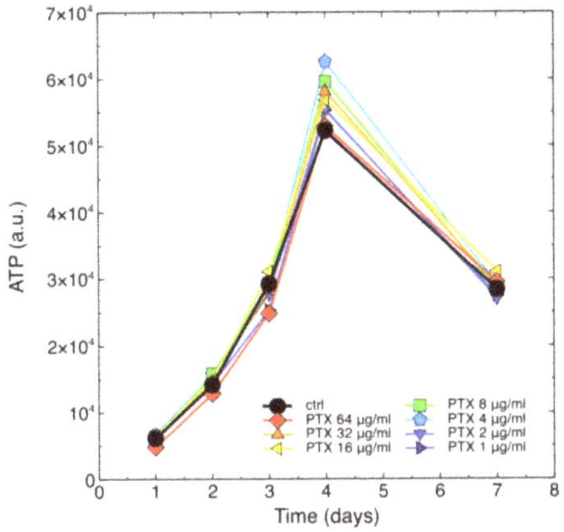

Figure 1. Growth kinetics of J774A.1 cells exposed to different concentrations of PTX. The drug was administered to the cells at the indicated final concentrations at the beginning of the growth assays, and ATP was then measured. Cell populations were also left untreated as controls (black circles marked as ctrl). The error bars correspond to calculated standard errors and are not clearly visible in this figure because they are masked by the symbols (min. and max. coefficient of variation, CV = 100∗SE/mean, 0.75% and 6.0%, respectively).

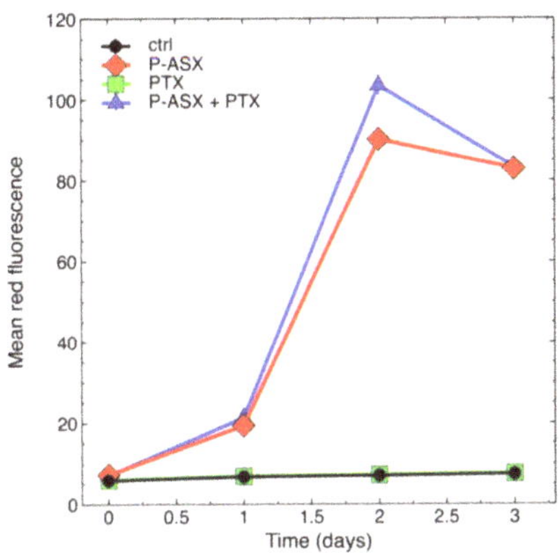

Figure 2. Phagocytosis kinetics of ASX microparticles. The cells were exposed to ASX microparticles (P-ASX) and were either treated or left untreated with PTX at a final concentration of 8 μg/mL. Phagocytosis was measured by flow cytometry thanks to the fluorescence emission of the ASX-containing oleoresin encapsulated into the protein particles. In this figure cell autofluorescence of control untreated cells (ctrl) or of cells treated with PTX only is also shown. Values are given as mean ± SE. The error bars correspond to calculated standard errors and are not clearly visible in this figure because they are masked by the symbols (min. and max. coefficient of variation, CV=100∗SE/mean, 0.45% and 5.1%, respectively).

2.3. Effects of PTX and ASX Microparticles on Intracellular ROS Levels

PTX has been reported to have antioxidant activity [7], and more importantly to have a synergic effect with other antioxidants, such as vitamin E, in combined treatments [14]. We previously showed that ASX microparticles could significantly reduce intracellular ROS levels in J774A.1 cells treated with H_2O_2 to induce oxidative stress [11], and we therefore wondered whether PTX could also improve the effects of ASX particles. Figure 3 shows that the intracellular ROS levels in J774A.1 macrophages decreased when PTX and ASX microparticles were given alone or in combination.

Table 1. Two-ways ANOVA analysis of the antioxidant effects of PTX and ASX microparticles given alone or in combination to J774A.1 cells.

Treatment	df [1]	SS	MS	*F*	*p*
P-ASX	1	97.4	97.4	381.9	1.8×10^{-10}
PTX	1	122.1	122.1	478.9	4.9×10^{-11}
PTX and ASX	1	41.2	41.2	161.4	2.6×10^{-8}
Error	12	3.1	0.255		
Total	15	263.8			

[1] df, degrees of freedom; SS, sum of squares; MS, mean squares.

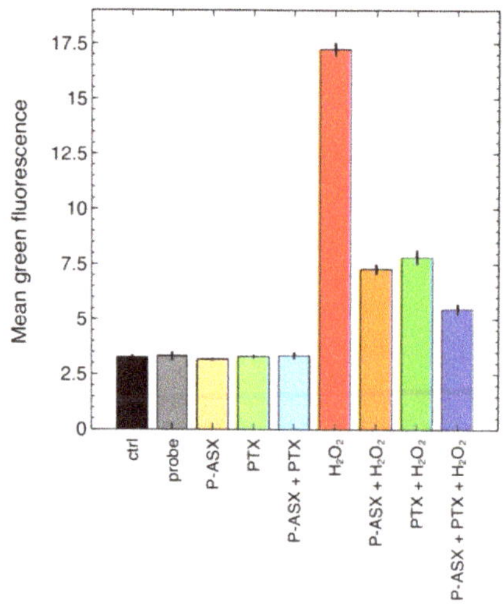

Figure 3. Effects of PTX and ASX-loaded microparticles on intracellular ROS as evaluated by flow cytometry with the DCF-DA probe. Cell fluorescence was almost equal for untreated cells (cell autofluorescence, ctrl) or cells treated with DCF-DA (probe), ASX microparticles (P-ASX), PTX or with P-ASX and PTX. Cell fluorescence significantly increased in cells loaded with DFC-DA and treated with H_2O_2, but treatments with PTX and ASX microparticles given alone or in combination significantly suppressed the effects of H_2O_2 (see Figure 4 and Table 1 for statistical details). Data are the mean ± SE of 4 independent samples.

Two-way ANOVA analyses indicated that PTX and ASX microparticles significantly interacted to reduce the intracellular ROS levels induced by hydrogen peroxide treatment in J774A.1 cells (Figure 4 and Table 1).

2.4. *Radical-Scavenging Activity of PTX and ASX-Containing Extract in a Cell-Free Assay*

Data in Figures 3 and 4 and Table 1 indicate that PTX is an intracellular ROS scavenger. We further investigated its activity in a well-controlled cell-free assay, i.e., ABTS assay, in the attempt to explore whether the drug could react with free radicals on its own. As the control we used ASX-containing extracts that have an acknowledged antioxidant activity [16]. This result is somewhat expected since in a previous study PTX showed poor scavenging activity using DPPH assay [17]. This behavior is probably due to the lack of hydroxyl groups on PTX structure that underpin the scavenging activity of other potent antioxidants, such as vitamin E and phenolic compounds [18].

Figure 5 shows that PTX had no radical-scavenging activity at all assayed concentrations.

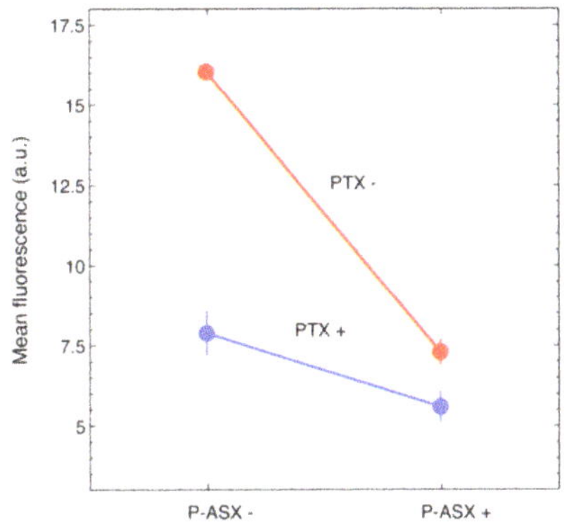

Figure 4. Graphical representation of the synergic effect of PTX and ASX microparticles on intracellular ROS reduction in J774A.1 cells. The two lines would be expected to be parallel if there were no synergic effects [15]. Formally, synergism is demonstrated by two-ways ANOVA statistical analysis (see Table 1).

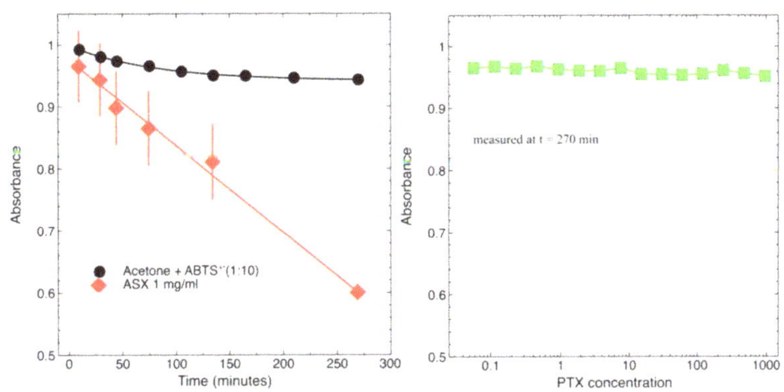

(**a**) (**b**)

Figure 5. Antioxidant activity of PTX and ASX extracts in a cell free assay. (**a**) Antioxidant kinetics of ASX extracts compared to the spontaneous decay of ABTS cation radicals ($ABTS^{+}$) in acetone solvent. (**b**) Antioxidant kinetics of PTX overlapped those of $ABTS^{+\cdot}$ in acetone at all assayed concentrations. For the sake of clarity, in this panel we show only the final time-point, i.e., the measurements carried out after 270 min for all PTX concentrations.

We therefore concluded that the effects observed for PTX on intracellular ROS levels likely depend on the ability of the drug to activate cellular detoxifying pathways (see also Section 3 for further discussion on this point).

2.5. Effects of PTX and ASX Microparticles on Irradiated Cells

Ionizing radiations initiate oxidative stress in biological samples through water radiolysis that ultimately leads to cell death [19]. Since PTX and ASX microparticles cooperated to significantly reduce intracellular ROS (Figures 3 and 4 and Table 1) we asked whether they could also increase the survival of irradiated cell samples when given alone or in combination to cells before radiation treatments.

The clonogenic assay is commonly performed in radiobiology to quantify the radiosensitivity of cell samples. This test is based on the ability of cells to grow in isolation and form colonies [20], but in spite of several attempts we could not obtain well defined colonies with J774A.1 cells (see also reference [11] for further discussion on this point). We therefore resorted to a recently developed method that translates into experiments a probabilistic model of cell survival after radiotherapy and that is based on Poisson statistics [11,21]. The model is described by Equation (1) (see Section 4.8), and the final goal is to estimate the model parameter $S(D)$ by nonlinear fitting of Equation (1) to experimental data. This parameter defines the survival probability of irradiated clonogenic cells. However, to univocally estimate its values an independent determination of the multiplicative parameter ε is required [11,21]. Parameter ε, on its turn, defines the clonogenic potential of non-irradiated cells and it can be precisely estimated using limiting dilution assays [11,21].

Figure 6 shows the results of our experiments with J774A.1 macrophages. ASX microparticles and PTX, given alone or in combination, did not alter significantly the clonogenic potential of the J774A.1 cells (parameter ε) if compared to that of control untreated cells, but significantly increase the survival of irradiated cells (parameter $S(D)$). The survival of irradiated cells increased progressively when the cells were treated with ASX microparticles, PTX or ASX particles in combination with PTX.

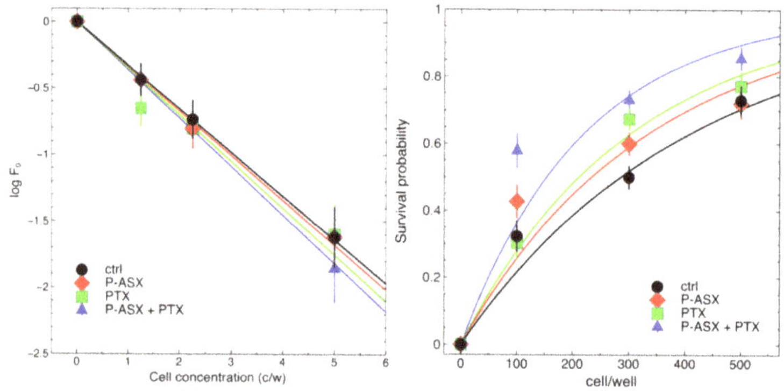

Figure 6. Effects of PTX and ASX microparticles on irradiated J774A.1 macrophages. (**a**) Limiting dilution assays were performed with non-irradiated cells to estimate parameter ε in Equation (1) (see the main text for details). Cells were seeded at each indicated density into the wells of 96-well culture plates and treated with ASX microparticles (P-ASX) and PTX, either alone or in combination (P-ASX + PTX), or left untreated (ctrl). F_0 is the fraction of cells where no proliferation was observed at the end of the observation period (~30 days). When the cells are randomly and independently distributed, F_0 is expected to obey Poisson statistics with parameter ε. Linear fitting of log-transformed data provided the following estimates: $\varepsilon_{ctrl} = 0.328 \pm 0.0325$, $\varepsilon_{P\text{-}ASX} = 0.336 \pm 0.036$, $\varepsilon_{PTX} = 0.351 \pm 0.036$, $\varepsilon_{P\text{-}ASX+PTX} = 0.364 \pm 0.038$. The differences in the estimated parameter values were not statistically significant (ANOVA test, $p = 0.9$). (**b**) Survival probability of independent cell populations treated with ASX microparticles (P-ASX), PTX, P-ASX in combination with PTX (P-ASX + PTX) or left untreated (ctrl) and then irradiated with a dose of 4 Gy γ-rays. Nonlinear fits with Equation (1) allowed to estimate the fraction of the cells surviving radiation treatments (see also the main text for details). The results are: $S(D)_{ctrl} = 0.0071 \pm 0.0005$ ($\chi^2/df = 1.5$), $S(D)_{P\text{-}ASX} = 0.0086 \pm 0.0006$ ($\chi^2/df = 1.2$), $S(D)_{PTX} = 0.0095 \pm 0.0007$ ($\chi^2/df = 1.7$), $S(D)_{P\text{-}ASX+PTX} = 0.0131 \pm 0.001$ ($\chi^2/df = 2.1$). All observed differences in estimated parameter values were statistically significant as evaluated by ANOVA ($p = 1.5 \times 10^{-12}$) followed by Tukey post-hoc test.

3. Discussion

Oxidative stress is involved in several diseases such as cancer, Alzheimer's disease, Parkinson's disease, atherosclerosis, heart failure, fibrosis and, in particular, RIF. It is a systemic pathologic condition caused by altered ROS accumulation in cells and tissues and hence by imbalance between ROS production and elimination through the activation of specific detoxifying mechanisms [22]. Ionizing radiations produce high amounts of ROS in both intracellular and extracellular compartments through water radiolysis [19]. This causes DNA lesions and other life-threatening damages not only to cancerous cells but also to irradiated and non-irradiated normal cells. It is nowadays accepted that radiation generates "danger" signals that propagate from irradiated to non-irradiated cells (the so-called off-target effects). Such molecular signals include ROS, other reactive molecular species, cytokines, ATP and extracellular DNA [23]. These signals, in turn, activate immune cells that, similar to macrophages, can further contribute to ROS production and can initiate and sustain the inflammatory cascade [23]. If uncontrolled, inflammation can cause tissue remodeling and dysfunction [24].

For example, the TGFβ cytokine promotes the recruitment of fibroblasts and local deposition of extracellular matrix components and it represents a key molecular actor in RIF [12].

RIF manifests in ~1/4 of all cancer patients that undergo radiotherapy with clinical consequences that impact the quality of their life [1]. Our previous study showed that ASX microparticles could not only reduce intracellular ROS levels in macrophages but also inhibited the secretion of bioactive TGFβ by these cells [11], indicating—although within the limitations of a pre-clinical in vitro study—that the treatment strategy developed to target specifically phagocytic cells could indeed be effective.

The results of the present study show that the effects of ASX microparticles can be further potentiated by combination treatment with PTX. PTX was nottoxic for the cells, did not interfere with phagocytosis of ASX microparticles, but at the same time could significantly reduce the oxidative stress in J774A.1 macrophages. Cell-free assays, however, clearly showed that PTX had no direct antioxidant activity on its own. It has been reported that PTX may have indirect antioxidant effects in neutrophils where it may reduce the superoxide production via NADPH oxidase [7]. PTX has also been shown to contribute to the maintenance of GSH levels, mitochondrial viability and in general to have protective effects against malathion-induced oxidative damage to rat brain mitochondria in vivo [25]. The antioxidant activity of PTX, therefore, appears to depend on its ability to modulate intracellular detoxifying pathways. Our results are in line with this interpretation of the molecular mechanisms of action of PTX. Importantly, the drug showed synergistic antioxidant effects with ASX microparticles in J774A.1 cells and significantly contributed to increasing the survival of irradiated macrophages. These findings collectively show that both direct and indirect mechanisms can be activated to restore intracellular ROS levels and protect cells from oxidative injury.

PTX has also a well-acknowledged anti-inflammatory activity, since it can inhibit TNFα production and signaling [8]. Given in combination with ASX microparticles, that on their own inhibit active TGFβ release by targeted macrophages, might result in more effective treatments against inflammation and fibrosis. Overall, our previous and present observations justify further investigations with in vitro and in vivo models of fibrosis.

4. Materials and Methods

4.1. Pentoxifylline and Astaxanthin Loaded Microparticles

Pentoxifylline P1784-100G powder was provided by Sigma-Aldrich (St. Louis, MO, USA). Astaxanthin Oleoresin (ASTAPure® 10% Oleoresin from Haematococcus pluvialis) was provided by Algatech (Ketura, Israel).

The microencapsulation of astaxanthin oleoresin into the whey protein isolate shell was performed by emulsification and solvent evaporation technique described in the PCT/IB2019/05991 application. Chemical and physical properties (e.g., distribution of particles size, storage stability, absorption spectra) of astaxanthin loaded microparticles were given in our previous work [11]. The median particle size was ~2.5 μm.

The concentration of ASX microparticles was expressed as μg of dried powder/mL of cell growth medium. The mass contribution of ASX into the microparticles is approximately 2.9%. Taking into account that the molecular weight of ASX is 596.8 g/mol this corresponds to a concentration of 48.6 nM ASX (μg/mL) dried particles.

When not specified, in all assays the final concentration of PTX was 8 μg/mL (28.7 mM).

4.2. Cells and Cell Culture

J774A.1 macrophages were obtained from the European Collection of Authenticated Cell Cultures (ECACC, Salisbury, UK; ECACC number 91051511). Murine macrophages were cultured in RPMI 1640 medium supplemented with foetal bovine serum (FBS), 10 mg/mL gentamicin (Biochrom) and 2 mM glutamine (Sigma-Aldrich, St. Louis, MO, USA) at 37 °C in a humidified 5% CO_2 atmosphere.

Cell morphology was routinely checked using an Evos (AMG, Life Technology) digital inverted microscope and an Olympus IX51 (Olympus Corporation, Corporate Parkway Center Valley, PA, USA) inverted microscope.

4.3. Cytotoxicity Assay

J774A.1 cells were seeded at 5000 cells/well in 96-wells culture plates in 200 μL RPMI medium containing different concentrations of PTX, from 1 μg/mL to 64 μg/mL. At each time point after treatments the intracellular ATP content was quantified using the Cell Titer Glo® Luminescent Cell Viability Assay (Promega, Milan, Italy) following the manufacturer's specification. Luminescence was measured with an FLX800 Microplate reader (FLX800, Bio-Tek Instruments, Bad Friedrichshall, Germany). All measurements were carried out at least in 4 replicates.

4.4. Flow Cytometry

A Guava easyCyte 5 flow cytometer (Merck Millipore, Billerica, MA, USA) was used. The instrument is equipped with a 488 nm, 20 mW, blue laser light. Light scattering is measured by means of a forward scatter (FSC) photodiode and a side scatter (SSC) photomultiplier. Three fluorescence channels, green, yellow, and red, allow to collect cell-associated fluorescence at the same time, thanks to the following filters: green, 525/30 filter; yellow, 583/26; red, 680/30. Instrument calibration was routinely carried out using the Guava EasyCheck kit (Merck Millipore, Billerica, MA, USA) following the manufacturer's instructions. We routinely collected at least 5000 gated events for analysis.

4.5. Phagocytosis Kinetics

J774A.1 cells were seeded at 20,000 cells/well in six-wells plates. Cells were incubated at 37 °C with 56 μg/mL of ASX microparticles for different days, and at each time point, phagocytosis of the microparticles was analyzed by flow cytometry. In fact, ASX oleoresin emits red fluorescence around 600 nm when excited by a blue laser light. In parallel assays, the cells were also treated with 8 μg/mL PTX.

4.6. Intracellular ROS Detection

The cell-permeable and H_2O_2-sensitive 2′,7′-Dichlorofluorescein diacetate (DCF-DA) fluorescent probe (Sigma-Aldrich, St. Louis, MO, USA) was used to measure intracellular ROS levels. When the diacetate group is cleaved by intracellular esterases the probe is retained into the cells.

Macrophages were seeded in six-well plates at 60,000 cell/wells in 3 mL of growth medium and treated with 56 mg/mL ASX particles, 8 μg/mL PTX or both for 5 h at 37 °C to allow phagocytosis. The cells were then washed twice with PBS 1×, and the medium was replaced with 500 μL of 50 μM DCF-DA. The cells were further incubated for 30 min. After washings with PBS 1× the cells were incubated for 30 min at 37 °C with 3 mL of a solution containing 2.5 mL of growth medium and 0.5 mL of 0.1% *w*/*w* hydrogen peroxide in PBS (0.017% H_2O_2 final concentration) to increase intracellular ROS levels and simulate oxidative stress. The fluorescence of the DCF-DA probe was then measured by flow cytometry.

4.7. Free Radical Scavenging: ABTS Test

The free radical scavenging capacity of PTX was also studied using the ABTS radical cation de-colorization assay which is based on the reduction of $ABTS^{+\bullet}$ radicals by antioxidants [17].

ABTS (Sigma-Aldrich, St. Louis, MO, USA) was dissolved in PBS to reach a final concentration of 7.4 mM. ABTS radical cation ($ABTS^{+\bullet}$) was produced by a chemical reaction of ABTS with 2.6 mM potassium persulfate (Sigma-Aldrich St. Louis, MO, USA). The reaction was carried out overnight, in the dark, and at room temperature. The $ABTS^{+\bullet}$ solution was then diluted in methanol to reach an absorbance of 0.75 at 734 nm. A BioTek PowerWave HT, microplate spectrophotometer (BioTek Instruments, Inc., Winoosky, VT, USA) was used.

Different concentrations of PTX dissolved in acetone (Sigma-Aldrich St. Louis, MO, USA) (20 μL) were allowed to react with 200 μL of $ABTS^{+\bullet}$ solution into the wells of a 96-well microplate kept in the dark. Absorbance kinetics were measured at room temperature starting at 5 min after initial mixing. All solutions were used the same day of their preparation, and all determinations were carried out in triplicate.

4.8. Irradiation of Cell Samples

The cells were seeded at different densities into the wells of 96-well culture plates, 1 plate for each assayed cell density, and treated with 56 μg/mL of ASX microparticles and/or with 8 μg/mL PTX. After 24 h cells were irradiated with a dose of 4 Gy using a Gammacell40 irradiator (Atomic Energy of Canada Limited, Kanata, ON, Canada) equipped with a ^{137}Cs source. The dose rate was 0.6654 Gy/min and the measured uniformity was ±1.3% over the entire sample chamber. Both parameters are monitored by the Radiation Protection Service of the University of Verona.

After ~20 days surviving cell populations were determined by careful microscopic inspection and counted. The number of wells containing survivors divided by the total number of seeded wells is an experimental estimate of the overall survival probability P of the cells at a given radiation dose, and this probability is expected to vary as the function of cell density μ as follows [11,21]:

$$P = 1 - e^{-S(D)\,\varepsilon\,\mu} \tag{1}$$

where $S(D)$ is the survival probability of the cells with self-renewing potential (i.e., clonogens) irradiated with a dose D, and ε is the fraction of clonogens in a population of μ cells on average. To quantify the effects of radiation the parameter $S(D)$ was estimated by nonlinear fitting of experimental data with Equation (1) knowing the fraction of clonogens ε in the cell population. This fraction was determined in independent limiting dilution assays as described [11,21].

4.9. Statistics

All assays were carried out in 3–4 replicates and repeated at least three times with different cell batches. Data were expressed as mean ± SE, where SE is the standard error of the mean. Statistical analyses and nonlinear regression were carried out using the software Mathematica (v.12, Wolfram Research Inc., Champaign, IL, USA). The reduced χ^2, i.e., χ^2/df where df is the number of degrees of freedom, was used to determine the goodness of the nonlinear fits.

Author Contributions: Conceptualization, E.B., R.C. and G.Z.; investigation, E.B.; methodology, E.B., F.Z., G.Z. and R.C.; resources, F.M. and G.D.; funding acquisition, G.Z.; supervision, R.C. and G.Z.; writing—original draft preparation, E.B. and R.C.; visualization, E.B. and R.C.; writing—review and editing, E.B., G.Z., F.Z., G.D., F.M. and R.C. All authors have read and agreed to the published version of the manuscript.

Funding: This research received no external funding.

Institutional Review Board Statement: Not applicable.

Informed Consent Statement: Not applicable.

Data Availability Statement: Data are contained within the article.

Conflicts of Interest: The authors declare no conflict of interest.

Sample Availability: Samples of the compounds are available from the authors.

References

1. Hall, S.; Rudrawar, S.; Zunk, M.; Bernaitis, N.; Arora, D.; McDermott, C.M.; Anoopkumar-dukie, S. Protection against radiotherapy-induced toxicity. *Antioxidants* **2016**, *5*, 22. [CrossRef]
2. Chiao, T.B.; Lee, A.J. Role of pentoxifylline and vitamin E in attenuation of radiation-induced fibrosis. *Ann. Pharmacother.* **2005**, *39*, 516–522. [CrossRef] [PubMed]
3. Delanian, S.; Lefaix, J. Current management for late normal tissue injury: Radiation-induced fibrosis and necrosis. *Semin. Radiat. Oncol.* **2006**, *17*, 99–107. [CrossRef] [PubMed]
4. Delanian, S.; Lefaix, J. The radiation-induced fibroatrophic process: Therapeutic perspective via the antioxidant pathway. *Radiother. Oncol.* **2004**, *73*, 119–131. [CrossRef]
5. Lech, M.; Anders, H. Macrophages and fibrosis: How resident and infiltrating mononuclear phagocytes orchestrate all phases of tissue injury and repair. *Biochim. Biophys. Acta* **2012**, *1832*, 989–997. [CrossRef] [PubMed]
6. Pareek, P.; Samdariya, S.; Sharma, A.; Gupta, N.; Shekhar, S.; Kirubakaran, R. Pentoxifylline and vitamin E alone or in combination for preventing and treating side effects of radiation therapy and concomitant chemoradiotherapy. *Cochrane Database Syst. Rev.* **2016**, *2016*, CD012117. [CrossRef]
7. Wen, W.X.; Lee, S.Y.; Siang, R.; Koh, R.Y. Repurposing pentoxifylline for the treatment of fibrosis: An overwiew. *Adv. Ther.* **2017**, *34*, 1245–1269. [CrossRef]
8. Okunieff, P.; Augustine, E.; Hicks, J.E.; Cornelison, T.L.; Altemus, R.M.; Naydich, B.G.; Ding, I.; Huser, A.K.; Abraham, E.H.; Smith, J.J.; et al. Pentoxifylline in the treatment of radiation-induced fibrosis. *J. Clin. Oncol.* **2004**, *22*, 2207–2213. [CrossRef] [PubMed]
9. Delanian, S.; Porcher, R.; Balla-Mekias, S.; Lefaix, J. Randomised, placebo-controlled trial of combined pentoxifylline and tocopherol for regression of superficial radiation-induced fibrosis. *J. Clin. Oncol.* **2003**, *21*, 2545–2550. [CrossRef]
10. Fakhria, S.; Abbaszadehb, F.; Dargahic, L.; Jorjania, M. Astaxanhin: A mechanistic review on its biological activities and health benefits. *Pharmacol. Res.* **2018**, *136*, 1–20. [CrossRef]
11. Binatti, E.; Zoccatelli, G.; Zanoni, F.; Donà, G.; Mainente, F.; Chignola, R. Phagocytosis of astaxanthin-loaded microparticles modulates TGFβ production and intracellular ROS levels in J774A.1 macrophages. *Mar. Drugs* **2021**, *19*, 163. [CrossRef] [PubMed]
12. Liu, R.M.; Desai, L.P. Reciprocal regulation of TGF-β and reactive oxygen species: A perverse cycle for fibrosis. *Redox. Biol.* **2015**, *6*, 565–577. [CrossRef]
13. Bessler, H.; Gilgal, R.; Djaldetti, M.; Zahavi, I. Effect of pentoxifylline on the phagocytic activity, cAMP levels, and superoxide anion production by monocytes and polymorphonuclear cells. *J. Leukoc Biol.* **1986**, *40*, 747–754. [CrossRef] [PubMed]
14. Fan, H.; Kim, S.M.; Cho, Y.J.; Eo, M.Y.; Lee, S.K.; Woo, K.M. New approach for the treatment of osteoradionecrosis with pentoxifylline and tocopherol. *Biomater. Res.* **2014**, *18*, 13. [CrossRef]
15. Slinker, B.K. The statistics of synergism. *J. Mol. Cell. Cardiol.* **1998**, *30*, 723–731. [CrossRef]
16. Kobayashi, M.; Takizono, T.; Nishio, N.; Nagai, S.; Kurimura, Y.; Tsuji, Y. Antioxidant role of astaxanthin in the green alga Haematococcus pluvialis. *Appl. Microbiol.* **1997**, *48*, 351–356. [CrossRef]
17. Mishra, K.; Ojha, H.; Chaudhury, N.K. Estimation of antiradical properties of antioxidants using DPPH assay: A critical review and results. *Food Chem.* **2012**, *130*, 1036–1043. [CrossRef]
18. Shahidi, F.; Ambigaipalan, P. Phenolics and polyphenolics in foods, beverages and spices: Antioxidant activity and health effects–A review. *J. Funct. Foods* **2015**, *18*, 820–897. [CrossRef]
19. Azzam, E.; Jay Gerin, J.P.; Pain, D. Ionizing radiation-induced metabolic oxidative stress and prolonged cell injury. *Cancer Lett.* **2012**, *327*, 48–60. [CrossRef]
20. Franken, N.A.P.; Rodermond, H.M.; Stap, J.; Haveman, J.; Van Bree, C. Clonogenic assay of cells in vitro. *Nat. Protoc.* **2006**, *1*, 2315–2319. [CrossRef] [PubMed]
21. Chignola, R.; Sega, M.; Molesini, B.; Baruzzi, A.; Stella, S.; Milotti, E. Collective radioresistance of T47D breast carcinoma cells is mediated by a Syncytin-1 homologous protein. *PLoS ONE* **2019**, *14*, e0206713. [CrossRef] [PubMed]
22. Pizzino, G.; Irrera, N.; Cucinotta, M.; Pallio, G.; Mannino, F.; Arcoraci, V.; Squadrito, V.; Altavilla, D.; Bitto, A. Oxidative stress: Harms and benefits for human health. *Oxid. Med. Cell. Longev.* **2017**, *2017*, 8416763. [CrossRef]
23. Pouget, J.; Georgakilas, A.G.; Ravanat, J. Targeted and off-target (bystander and abscopal) effects of radiation therapy: Redox mechanisms and risk/benefit analysis. *Antioxid. Redox Signal.* **2018**, *29*, 1447–1487. [CrossRef] [PubMed]

24. Chen, L.; Deng, H.; Cui, H.; Fang, J.; Zuo, Z.; Deng, J.; Li, Y.; Wang, X.; Zhao, L. Inflammatory responses and inflammation-associated diseases in organs. *Oncotarget* **2018**, *9*, 7204–7218. [CrossRef] [PubMed]
25. Ranjbar, A.; Ghahremani, M.H.; Sharifzadeh, M.; Golestani, A.; Ghazi-Khansari, M.; Baeeri, M.; Abdollahi, M. Protection by pentoxifylline of malathion-induced toxic stress and mithocondrial damage in rat brain. *Hum. Exp. Toxicol.* **2010**, *29*, 851–864. [CrossRef] [PubMed]

molecules

MDPI

Review

AQP3 and AQP5—Potential Regulators of Redox Status in Breast Cancer

Lidija Milković and **Ana Čipak Gašparović** *

Division of Molecular Medicine, Ruđer Bošković Institute, HR-10000 Zagreb, Croatia; lidija.milkovic@irb.hr
* Correspondence: acipak@irb.hr; Tel.: +385-1-457-1212

Abstract: Breast cancer is still one of the leading causes of mortality in the female population. Despite the campaigns for early detection, the improvement in procedures and treatment, drastic improvement in survival rate is omitted. Discovery of aquaporins, at first described as cellular plumbing system, opened new insights in processes which contribute to cancer cell motility and proliferation. As we discover new pathways activated by aquaporins, the more we realize the complexity of biological processes and the necessity to fully understand the pathways affected by specific aquaporin in order to gain the desired outcome–remission of the disease. Among the 13 human aquaporins, AQP3 and AQP5 were shown to be significantly upregulated in breast cancer indicating their role in the development of this malignancy. Therefore, these two aquaporins will be discussed for their involvement in breast cancer development, regulation of oxidative stress and redox signalling pathways leading to possibly targeting them for new therapies.

Keywords: AQP3; AQP5; oxidative stress

Citation: Milković, L.; Čipak Gašparović, A. AQP3 and AQP5—Potential Regulators of Redox Status in Breast Cancer. *Molecules* **2021**, *26*, 2613. https://doi.org/10.3390/molecules26092613

Academic Editor: Hiroyuki Kataoka

Received: 31 March 2021
Accepted: 27 April 2021
Published: 29 April 2021

1. Introduction

Despite the progress in research and treatment procedures, cancer still remains the leading cause of death. Today, cancer is targeted via different approaches which is determined by diagnosis, tumour marker expression, and specific mutations. One of the first approaches was targeting highly proliferating (cancer) cells and this is the basic mechanism of conventional chemotherapy. Conventional chemotherapy, such as anthracyclines, taxols, cisplatin or other platinum-containing drugs, targets either DNA replication or repair, or interferes with tubulins, disabling proper chromatid exchange in mitosis. Oxidative stress commonly accompanies these therapies, which damage cancer cells, but also normal cells, thereby causing side effects such as cardiotoxicity or ototoxicity [1–3].

Oxidative stress is both a cause and consequence of pathological states. On the other hand, numerous physiological processes depend on mild oxidative stress and certainly redox signalling is now recognized for its role in physiology [4]. Although ROS were at first thought to be by-products of metabolic pathways, now we are repurposing proteins that produce and control ROS movements thereby contributing to and controlling redox signalling [5]. Aquaporins are one of these molecules which were attributed as water channels making a breakthrough in our knowledge of water movement or transport across the cellular membrane [6]. In the light of accumulating evidence on the role of aquaporin 3 and aquaporin 5 in breast cancer, we will bring an overview of current knowledge of these two aquaporins and their role in the regulation of redox signalling in breast cancer.

2. Aquaporins

Water movement in and out of the cells is one of the most important biological processes since it is the base for cell movement, regulation of cytoplasmic viscosity, and consequently kinetics of signal transduction. Water was supposed to enter through the cell membrane via diffusion, but this process has slow kinetics, and the changes of volume that occur due to water movement are quite rapid indicating a certain level of active regulation.

The discovery of aquaporins, the first one described was CHIP28/AQP1 [7], opened new research perspective. At first, aquaporins were figuratively described as cell's plumbing system focusing on their role in water transport [8]. Today, we are distinguishing this family according to molecules channelled to aquaporins and aquaglyceroporins. Aquaporin number varies between the species, e.g., *Saccharomyces cerevisiae* has two orthodox (Aqy1 and Aqy2) and two glyceroaquaporins (YFL054Cp and Fps1), *Escherichia coli* has one orthodox, AqpZ, and one structurally similar to glyceroaquaporins, GlpF, while plants can have more than 35 isoforms [9].

Today we know 13 human aquaporins (AQP0-QP12), each one with a unique structure and cellular and subcellular location and role [10]. Aquaporins are grouped according to their permeability and primary structure as follows: orthodox aquaporins (AQP0, AQP1, AQP2, AQP4, AQP5, AQP6, and AQP8) which channel primarily water, glyceroaquaporins (AQP3, AQP7, AQP9, and AQP10) which channel primarily glycerol, and unorthodox or S-aquaporins (AQP11 and AQP12) [11,12]. The unorthodox S-aquaporins or super-aquaporins are located strictly inside the cell, on the membranes of organelles [13]. This group is found in animal cells but not in plants, bacteria no fungi, and has low homology with the other family members [14]. Interestingly, unorthodox aquaporins were initially attributed to channel water, but AQP11 was shown also to facilitate glycerol transport [14]. Still, their major role is the regulation of organelle volume and intra-vesicular transport [12, 15]. As stated, the initial grouping of aquaporins defined the water and glycerol channels, but both of these groups also channel other small molecules (ions and other polar molecules, as well as gases) [16,17]. In the light of new targets for transport through aquaporins, hydrogen peroxide emerged due to its relevance in biological processes and signalling. Therefore, Henzler and Steudle proposed a new term "peroxiporins" which refers to aquaporins that facilitate hydrogen peroxide flux [18,19]. Peroxiporins are members of all three previously mentioned groups and are as follows AQP0, AQP1, AQP3, AQP5, AQP8, AQP9, and AQP11 [19,20].

Aquaporins are highly conserved transmembrane channels built as tetramers, with each monomer consisting of about 320 amino acids, and of 28kDa molecular weight [10]. Each monomer has six transmembrane domains connected with five loops (A-E) [6]. Another motif, asparagine-proline-alanine (NPA) is present on loop B on the cytoplasmic side and loop E on extracellular side is common to both, orthodox and aquaglyceroporins. NPA motif serves as structure maintenance motif of each monomer, while loop D regulates gating of the pore [21]. In addition to structure maintenance, loop B and loop E, each form pseudo transmembrane segment [22]. Regulation sequences of aquaporins are located in loops, specifically loop E, which contains sites for inhibitors, Hg+ and tetraethylammonium, and loop D, which is sensitive to protonation and thereby regulates gating of the pore [21]. In contrast to ion channels where the channel is centrally situated, each monomer of aquaporin is a channel of its own, regulated independently of other monomers. The tetramer centre is a channel itself, mostly hydrophobic, and channels gasses, such as CO_2, nitric oxide, and ions; which of these species would be channelled depends specifically on each aquaporin [23–28].

Transcriptional regulation of aquaporins (Figure 1) is still not completely investigated, but AQP3 has several response elements and transcription factor binding sites in the promoter region, such as oestrogen response element (ERE), ROR/REV-ERB-response element (RORE), SP1 site, FOXA1 site [29]. In the salivary gland, FOXO1 was shown to be a direct regulator of AQP5 expression [30]. Additionally, insulin can upregulate the expression of AQP1, AQP5, and AQP8 in the submandibular glands of diabetic rats [31]. These data indicate that aquaporins can be regulated by different stimuli further exerting protective effects on the targeted cell.

Aquaporins are regulated post-translationally by protein trafficking from the intracellular deposits and transport to the plasma membrane when needed [32]. Trafficking may depend on the phosphorylation of monomers, AQP2 needs phosphorylation of at least three monomers to determine its position in the plasma membrane [33]. In prostate cancer cells, PC-3, AQP3 is translocated from the cytoplasm to the cell membrane after silencing of RAS like proto-oncogene A (RalA) [34]. Regulation of the flux through the channel is referred to as gating and is regulated by pH, phosphorylation, temperature, membrane tension, solvent gradient, and pressure [24,35,36]. Orthodox aquaporins are known for the quickest response to stimuli by changing the permeability which implicates sophisticated regulation. Regarding the link between structure and function, each monomer functions independently of others, therefore it is not clear why they form tetramers, especially as there is no evidence of cooperative interdependence in the quaternal structure [37]. Even more, aquaporins are considered to be homotetramer, but heterotetramers were found for AQP1 and AQP4 [38,39].

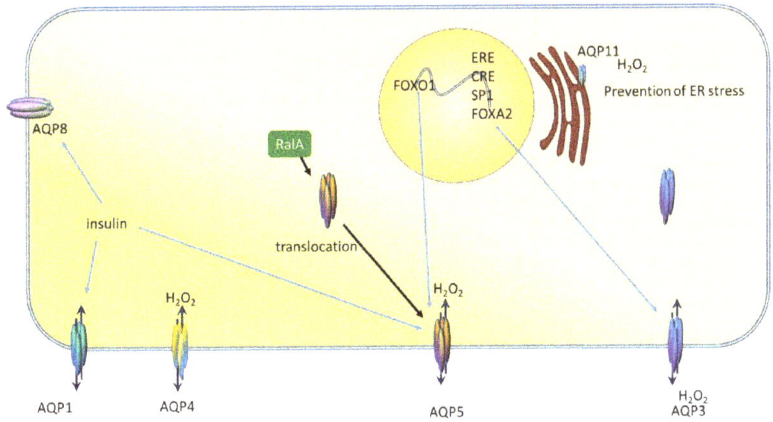

Figure 1. Regulation of different aquaporins-peroxiporins. Aquaporins are regulated by several transcription factors as well as hormones. Insulin increases the expression of AQP1, AQP5, and AQP8. AQP3 has several response elements in its promoter region (oestrogen response element, ERE, cAMP response element, CRE), as well as transcription binding sites (SP1, FOXA2). An additional level of regulation is translocation to the cell membrane, as observed for AQP5. AQP11 is a member of the S-aquaporin family which channels H_2O_2 and is located on the endoplasmic reticulum (ER). AQP11 channels H_2O_2 thereby preventing ER stress.

Aquaporins control water movement across the membrane, but also they control the osmotic pressure through the regulation of intracellular glycerol concentration [40]. Control of water movement and osmotic pressure is further linked to their involvement in cellular functions, such as migration, proliferation, and adhesion [41]. During migration, aquaporins selectively transport water due to osmotic gradient which is achieved by actin depolarisation. Water enters the cell at the leading site causing local membrane expansion which is then compensated by actin rearrangement for the maintenance of the membrane integrity [42]. The mechanism by which aquaporins contribute to cell proliferation is not simple and unambiguous for each aquaporin. Namely, the inhibition of AQP1 inhibits proliferation and migration of HT29 cells, which have a high basal expression of AQP1, while this inhibition does not affect HCT116 cells, with low expression of AQP1 [43]. In addition to AQP1, AQP3 is also related to cell proliferation giving correlation of AQP3 overexpression and increased proliferation of gastric cancer cells SGC7901 and MGC803, while its downregulation had the opposite effect [44]. Taken that in gastric cancer APQ3 is significantly higher than in normal gastric mucosa and was correlated to EMT proteins in gastric cancer tissues, overexpression of AQP3 is linked to poorer prognosis of these patients. Therefore, in gastric cancer, upregulation of these two aquaporins is linked to a more malignant phenotype, which is achieved by activation of ERK and Ras, as well as PI3K/AKT/Snail signalling pathway [44,45]. In lung cancer stem cells AQP3 silencing caused upregulation of Wnt/glycogen synthase kinase-3β (GSK-3β)/β-catenin pathway implying its role in reducing the activity of this signalling pathway and, thereby, inhibiting apoptosis and reducing differentiation of lung cancer cells keeping stemness of these cell [46]. AQP3 can, therefore, through inhibition of apoptosis and reduction in differentiation support stemness of these lung cancer cells, which implies the use of APQ3 for determining the malignant potential and recurrence of the primary disease [46]. These studies indicate the complexity of aquaporins in their cellular functions and involvement in signalling pathways.

3. Oxidative Stress and Antioxidative Defence

One of the important factors in the development of different pathologies is oxidative stress, an imbalance in cellular redox homeostasis [47]. This imbalance can occur due to decreased antioxidant defence or due to increased production of reactive oxygen species (ROS). At first, oxidative stress was considered to be a stressful and damaging condition [48,49]. Today, we know that this point is not so straight forward, and oxidative stress is considered as altered balance which can have either positive or negative consequences [50]. These two options are denoted as eustress and distress, where distress is damaging stress with possibly lethal consequences, and eustress is hormesis, adaptive biological response to stress condition [51,52]. Eustress (hormesis) came in the focus of the research by opening numerous questions on mechanisms by which this adaptation is achieved. There are several review papers that nicely provide an overview of this topic and a historical timeline of all the major points and discoveries leading to our current perception of oxidative stress [53–56].

The concept of eustress refers to ROS as a factors that contributes to physiological processes in the cell or organism and moving our understanding of their role away from exclusively damaging factors. Mild oxidative stress, a term that is referring to low levels of ROS, is very useful because it stimulates the defence mechanisms [57]. An example of this is the exercise: during exercise the ROS production increases which stimulates the cellular antioxidative system [57]. However, if vitamin C is consumed before the activity, it will neutralize ROS produced and they will fail to stimulate the cellular antioxidative system, diminishing the positive effect of the exercise [58,59]. Increasing oxidative stress and entering the "moderate" oxidative stress activates a whole different set of proteins, stimulating inflammation and changing the expression of cytokines and chemokines [60]. Of course, an additional increase in oxidative stress inevitably leads to cell death [61]. Therefore, oscillations of oxidative stress are very welcomed by means of adaptation to environmental stress building cellular antioxidant defence system and are called the "Goldilocks Zone" referring to the story of Goldilocks and her search for "just right" conditions [55].

As stress conditions disturb this balance and shift the cell to the so called "red zone", it is expected that the oxidative stress levels are related to the development of different pathologies, but making them an important mechanism in treating pathologies as well. Still, when treating pathologies, especially cancer, adaptation to oxidative stress is a big obstacle as cancer cells adapt and become resistant to therapy making it inefficient.

ROS and other electrophiles are highly reactive and cause oxidation of all cellular macromolecules, DNA, proteins, and lipids [62]. ROS introduce single- and double- strand breaks in DNA, and oxidize the bases which results in mutations. If not lethal, these DNA damages cause inactivation or overexpression of genes that regulate the cell cycle and lead to tumour development. Oxidation of protein disrupts their function. Lipids are vulnerable to peroxidation causing cell death due to disturbance of the physical barrier of the cell. Unlike DNA mutations and protein oxidation, lipid peroxidation is an autocatalytic process which is multiplying and expands the damage [47]. Especially vulnerable to lipid peroxidation are polyunsaturated fatty acids (PUFA) due to the double bond which is highly reactive with ROS. Lipid peroxidation is a self-multiplying process and has to be stopped by antioxidants, but it also results in the production of highly reactive aldehydes [47]. These aldehydes are stable enough to diffuse from the sight of the origin and further react with proteins, thereby modifying their reactivity and function finally affecting cellular processes [47,63,64].

As mentioned above, positive effects of mild stress are achieved by adaptation and include the nuclear factor erythroid 2 (NF-E2)-related factor 2-Kelch-like ECH-associated protein 1 (NRF2-KEAP1) pathway [52]. NRF2 is a transcription factor which is, in non-stimulating conditions, located in the cytoplasm bound to KEAP1 [65]. In the complex with KEAP1, NRF2 is ubiquitinated by Cul3-ubiquitin E2 ligase and is degraded in the proteasome [66]. If ROS or electrophiles are present in the cell, they oxidize the disulphide bond between two cysteines in KEAP1. The newly formed disulphide bond changes KEAP1 conformation which releases NRF2. NRF2 now translocates to the nucleus and binds to small MAF protein and induces transcription of antioxidant genes. NRF2 target genes are involved in glutathione synthesis (glutamate-cysteine ligase, both, catalytic and modifier subunits), in ROS detoxification (thioredoxin reductase 1, peroxiredoxin 1), xenobiotics detoxification (NQO1, NAD(P)H quinone dehydrogenase 1, glutathione-S-transferase), but also in drug transport (multidrug resistance-associated proteins, MRP) [65]. In addition, another protein family, Forkhead box O (FOXO) proteins, also contribute to antioxidative protection [67]. Among the antioxidant genes activated by the FOXO family are catalase, manganese-dependent superoxide dismutase (MnSOD), DNA damage binding protein 1 (DDB1), Fas ligand (FasL), cyclin-dependent kinase B1 (KIP1, p27), and protein for drug export, ABC1 [68].

Both, NRF2 and FOXO proteins are regulated by the PI3K/AKT signalling pathway [67,69]. Interestingly, NRF2 is activated directly by PI3K [69], but can be also activated indirectly, by 4-hydroxynonenal (HNE) activation of atypical protein kinase C (PKC) [70] and ERK [71]. FOXO family is negatively regulated by PI3K pathway, while JNK serves as a positive regulator by phosphorylating them upon the stress signal and thereby activating them [67]. Unlike NRF2, FOXO activation depends on the severity of stress, if stress levels are low, antioxidative defence mechanisms are activated, and if levels are high, apoptosis is activated [72].

4. Aquaporins, Oxidative Stress, and Cancer

As already described, one of the possible mechanisms by which aquaporins achieve their role in proliferation, differentiation, and apoptosis, is the regulation of small molecule transport such as hydrogen peroxide (H_2O_2), nitric oxide (NO), urea, and CO_2 [23–25]. As hydrogen peroxide and nitric oxide regulate and modulate redox signalling pathways thereby regulating proliferation, differentiation, and apoptosis, regulation of these molecules intake can directly or indirectly contribute to the modulation of these processes resulting in tumour growth and development.

As mentioned, tumour development and progression can be a result of oxidative stress. Therefore, factors that regulate oxidative stress highly influence tumour development as well as its fate. Interestingly enough, these protective factors in many cases are actually double edge sword. This situation especially refers to NRF2, which builds up the antioxidative defence of the cell. In normal cells this is a positive set of reactions that protects from malignant transformation, while in tumour cells NRF2 leads to protection from the therapy [65,73]. Additionally, the NRF2 pathway affects and promotes tumorigenesis not only directly by activation in tumour cells, but also indirectly through cancer associated fibroblasts (CAF). Tumour cells interact with normal cells in their microenvironment, changing normal fibroblasts to CAF which are reprogrammed to support tumour growth. The CAF reprogramming occurs by activation of p62, which then targets KEAP1 to lysosomal degradation. Degradation of KEAP1 activates NRF2 and enhances transcription of ATF6 finally mediating ER stress response [74]. Evidence also suggests that FOXO, a tumour suppressor, can activate tumour resistance mechanisms, but this can happen only in combination with other events in the cell [67].

In the light of antioxidative defence acting as a double edged sword, aquaporins introduce a new moment by regulating the flux of H_2O_2. Delicate regulation of intracellular levels of H_2O_2 opens possibilities in the stimulation of proliferation and survival mechanism for tumour cells, leading to resistance and increased mobility. Aquaporins were found to be upregulated in numerous tumours [28,75,76] opening possibilities to target them as a part of tumour therapy. The exact mechanisms and signalling pathways affected by aquaporins are still to be determined, but it is not a simple and straightforward interaction. The increased presence of particular aquaporin in the cell membrane controls H_2O_2 flux, as well as water flux, and other small molecules mentioned above. Import and export of these small molecules, in addition to water and H_2O_2, can also modify signalling pathways involved in proliferation, differentiation and migration. Additionally, glycerol intake could also be one of the factors by which aquaporins achieve stimulation of proliferation intake [75]. Therefore, we will further discuss two aquaporins AQP3 and AQP5, representing glyceroaquaporin and orthodox aquaporin in details for their role in breast cancer.

5. AQP3 and Breast Cancer

AQP3 is a member of aquaglyceroporin, and, as such, facilitates glycerol transport [77] in addition to water [78]. Although some authors state that AQP3 is a weak water channel [46], in the kidney, AQP3 is constitutively active together with AQP4 [78] thereby regulating water excretion. Beside the role in physiology, its role in cancer was recognized very early for skin cancer [79]. Soon, evidence of AQP3 overexpression in several other types of cancers, including breast cancer, accumulated [80–82].

Analysis of genomics data from The Cancer Genome Atlas (TCGA) project freely available from web-portal UALCAN [83] revealed pattern of AQP3 in normal (median 31.64 transcripts per million (TPM) (1.448–87.454 TPM)) vs. different subclasses of breast cancer (luminal–16.183 TPM (0.363–134.985 TPM), HER2-positive-36.481 TPM (0.521–329.342 TPM), TNBC-13.484 TPM (0.713–103.167 TPM) (Figure 2).

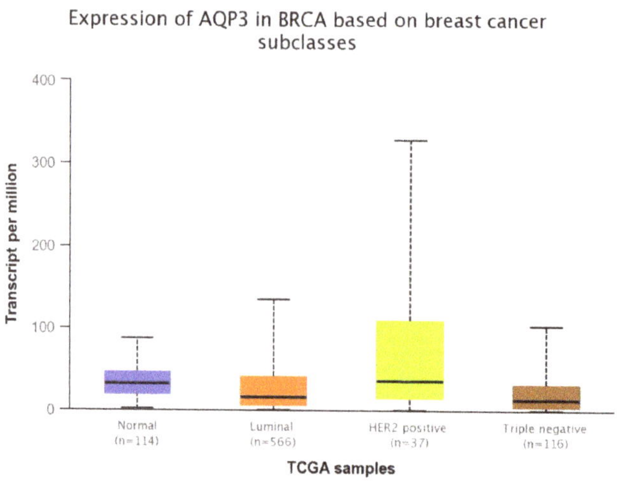

Comparison	Statistical significance
Normal-vs-Luminal	**1.593860 × 10^{-2}**
Normal-vs-HER2 Positive	**2.355900 × 10^{-2}**
Normal-vs-TNBC	1.427980 × 10^{-1}
Luminal-vs-HER2 Positive	**4.785800 × 10^{-2}**
Luminal-vs-TNBC	6.301800 × 10^{-1}
HER2 Positive-vs-TNBC	7.607900 × 10^{-2}

Figure 2. The expression of AQP3 in normal and breast cancer tissue subdivided according to expression of hormone and HER2 receptors retrieved from UALCAN [83].

Once recognized as overexpressed in cancer, AQP3 was getting in the focus as a possible prognostic marker for triple negative breast cancer together with AQP5 [84], as well as for HER2 positive early breast cancer [85]. In parallel, it was shown that the *AQP3* gene has an oestrogen-response element and responds to oestrogen stimuli by increasing its expression [86] suggesting a link between AQP3 and oestrogen receptor positive breast cancer. The possibility to use AQP3 as a prognostic marker in breast cancer can be attributed to its role in cell migration, which is facilitated by channelling both, water and glycerol, further resulting in lamellipodia formation and consequently, cell movement and migration [86,87]. Studies confirmed that AQP3 overexpression increased cell migration and invasion [86] for oestrogen-receptor positive breast cancer cells, as well as for keratinocytes [87]. Further, in keratinocytes, AQP3 facilitates glycerol transport into the cell as well, resulting in ATP generation and proliferation [88]. The fact that AQP3 facilitates the transport of both, water and glycerol, together with H_2O_2 puts AQP3 high on the list of potential targets for tumour therapy. At the same time, signalling pathways and cellular processes affected by changes in AQP3 levels need to be clarified. Water and glycerol transport affect migration and metabolic processes (especially lipid metabolism), while H_2O_2 is the one that affects signalling pathways. H_2O_2 fuels several signalling pathways in the cell and a channel that can facilitate H_2O_2 transport is a potential candidate that provides some level of control over those pathways. In support of this assumption is the study by Hara-Chikuma et al. [89] showing in keratinocytes that TNFα stimulus is facilitated through NADPH oxidase isoform 2 (NOX2) production of H_2O_2. AQP3 then transports H_2O_2 resulting in regulation (inhibition) of protein phosphatase 2A and activation of nuclear factor kappa B (NF-κB). Even more, C-X-C motif chemokine ligand 12 (CXCL12) stimulates H_2O_2 transport across the membrane by AQP3 in breast cancer cells MDA-MB-231 and DU4475 [90]. The oxidation of phosphatase and tensin homolog/protein tyrosine phosphatase 1B (PTEN/PTP1B) occurs due to H_2O_2 followed by activation of the AKT pathway and again, resulting in cell migration. Knockdown of AQP3 impairs this process thereby supporting the role of AQP3 in migration [90]. The signalling pathways are summarized in Figure 3.

The need to study pathways affected by AQP3 overexpression and mechanisms of action is reflected by the finding that Auphen, aquaporin gold-containing inhibitor, blocked glycerol transport quite efficiently (about 90% inhibition), while water transport was blocked only partially (20% inhibition) [91]. Having in mind that AQP3 also facilitate H_2O_2 transport across the plasma membrane [92,93] and structural similarities between H_2O_2 and water [94], inhibitors should be carefully examined for their ability to block AQP3 all three molecules channelled via AQP3. For these reasons, the fact that an inhibitor can block glycerol, but not water transport implies that H_2O_2 transport can also be unaffected or affected at smaller rate suggests activation and modification of cellular processes in an undesired direction, driving to progression rather than to regression of the tumour. Interestingly, although it is well known that AQP3 channels H_2O_2, there are several papers on the effect of aquaporins in general on the antioxidative defence system, specifically NRF2 transcription factor, regardless of the disease [95–97]. In breast cancer cell lines, MCF7, SUM159 and SkBr3, AQP3 was the most expressed aquaporin, and in HER2 positive cells it was upregulated together with NRF2 by H_2O_2 [95] implying for then need to study effects of AQP3 overexpression in relation to the parts of antioxidative system.

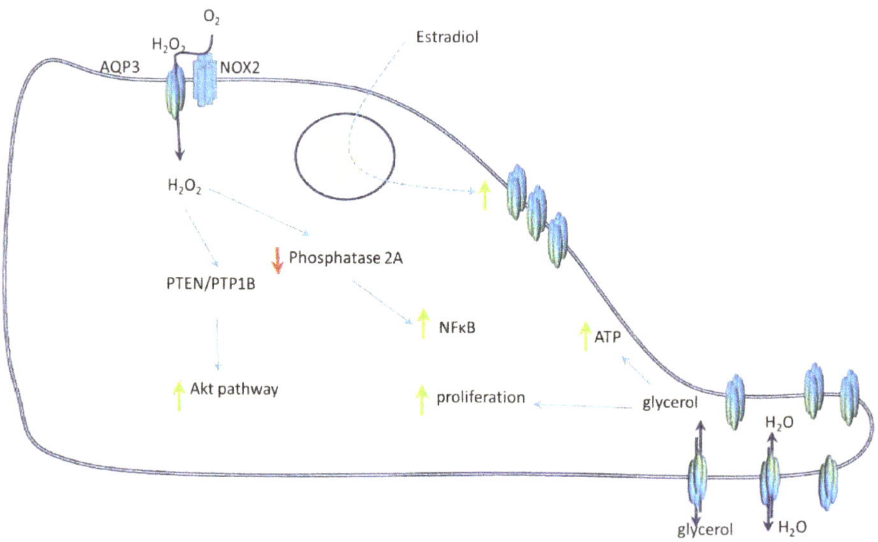

Figure 3. AQP3 involvement in cellular signalling pathways and processes. AQP3 respond to estradiol stimuli and increases expression. IT is also located in the membrane by the NOX2 and imports H_2O_2. H_2O_2 then oxidizes PTEN and activates AKT signalling pathways. AQP3 also channels extracellular H_2O_2 which inhibits phosphatase A and thereby activates NFκB. Increased positioning of AQP3 on the leading side regulates water and glycerol intake which leads to lamellipodia formation and cellular migration. Additionally, glycerol intake by AQP3 increases intracellular ATP and causes proliferation.

6. AQP5 and Breast Cancer

AQP5 is involved in normal mammary development and milk production, as well as in breast carcinogenesis [80,98]. Jung et al. showed that silencing of AQP5 or induction of hyperosmotic stress to MCF-7 cells decreases the expression of AQP5 and negatively affects cell proliferation and migration. Additionally, AQP5 expression in benign tumours and invasive ductal carcinoma showed different patterns, expression of AQP5 in apical domains of ductal epithelial cells vs. increased expression in cancer cells with the loss in ducts accompanied with the loss of apical polarity, thus suggesting its role in breast cancer progression [99]. In triple-negative breast cancer (TNBC) patients, markedly higher expression of AQP5 and AQP3 was observed in cancer tissue than in adjacent normal tissue. Overexpression of AQP5 was mainly observed within Ki-67 high TNBC samples and, together with the higher expression of AQP3, associated with the more progressive disease with poorer overall survival proposing their co-expression as an independent prognostic marker in TNBC patients [84]. Additionally, overexpression of AQP5 was associated with worse outcomes in early breast cancer patients regardless of tumour type and stage, suggesting it as an independent prognostic marker of survival, particularly in hormone receptor-positive patients who underwent curative surgery [100].

Analysis of genomics data from The TCGA project, freely available from web-portal UALCAN [83], revealed significantly different gene expression pattern of AQP5 in normal (median 7.366 transcripts per million (TPM) (0–44.252 TPM)) vs. different subclasses of breast cancer (luminal–0.34 TPM (0–5.848 TPM), HER2-positive-1.79 TPM (0–105.814 TPM), TNBC-8.469 TPM (0–169.313 TPM); Figure 4).

Therefore, in our previous study, we examined how oxidative challenge (known contributor to breast carcinogenesis and a mechanism of effective anticancer therapy too [101]) affects lipid profile, levels of oxidative stress mediators and NRF2, the expression pattern of AQP1, AQP3, AQP5, and sensitivity to H_2O_2 in three breast cancer cell lines (representing hormone-positive (MCF-7), HER2-positive (SkBr-3) and TNBC (SUM 159)) [95]. Levels of polyunsaturated fatty acids (PUFA) were cell-type dependent with the highest observed in triple-negative SUM 159 cell line, and along with lower NRF2 levels may explain their higher sensitivity to H_2O_2. AQPs' expression pattern was cell-type specific, also. While AQP3 was the most expressed isoform in all cell lines tested, exposure to H_2O_2 increased AQP3 expression in MCF-7 and SkBr-3 cells whereas in SUM 159 AQP3 was decreased. At the same time, the expression of AQP5 and AQP1 was similar in SUM 159 and SkBr-3, increased upon oxidative challenge, while decreased in MCF-7.

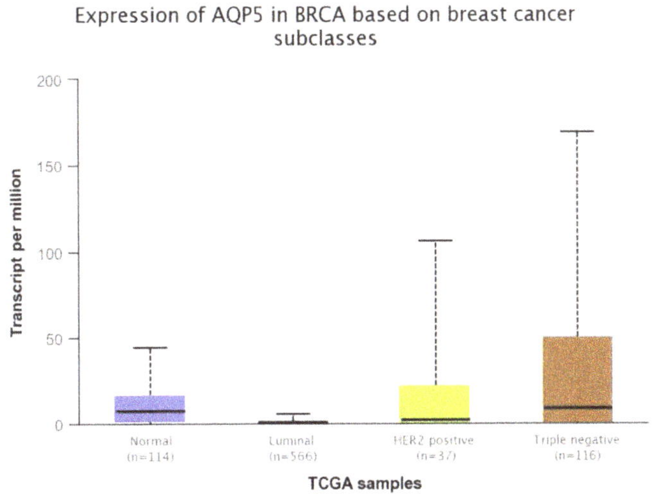

Comparison	Statistical significance
Normal-vs-Luminal	1.699820×10^{-3}
Normal-vs-HER2 Positive	3.596300×10^{-2}
Normal-vs-TNBC	$1.61221999999128 \times 10^{-5}$
Luminal-vs-HER2 Positive	2.156500×10^{-2}
Luminal-vs-TNBC	$1.89340000011917 \times 10^{-7}$
HER2 Positive-vs-TNBC	4.157800×10^{-1}

Figure 4. The expression of AQP5 in normal and breast cancer tissue subdivided according to expression of hormone and HER2 receptors retrieved from UALCAN [83].

Aside from being a water channel, a study by Rodrigues et al. has shown a highly efficient peroxiporin activity of AQP5, with external oxidative stress stimuli rescuing the suppression of cancer cells' migratory ability induced by AQP5 silencing. Hence, the authors highlighted the role of AQP5 in dynamic fine-tuning of the intracellular levels of H_2O_2 [102] that are important for redox signalling and regulation of cell fate [103]. Thus AQP5 might show promise in anticancer therapy. Indeed, the discovery of three AQP5-regulating miRNAs (miR-1226–3p, miR-19a-3p, and miR-19b-3p) that, by decreasing the translation of AQP5, reduce breast cancer cell migration, supports its further investigation as a possible therapeutic target [104].

The connection of ROS and AQP5 was also observed in a study by Oh et al. They studied how hypercholesterolemia and the inhibition of xanthine oxidase (ROS-generating enzyme) affects breast cancer progression in both, in vitro and mouse xenograft model. Hyperlipidemic conditions were shown to contribute to ROS production, breast cancer progression, and MAPK activation, whereas treatment with febuxostat, xanthine oxidase inhibitor, by diminishing ROS levels and AQP5 expression, mitigated proliferative and migratory ability of breast cancer cells, as well as pulmonary metastases [105].

Whether the involvement of AQP5 in breast carcinogenesis is causative or merely a consequence of breast cancer cell's need to grow, involving metabolic reprogramming and redox signalling including ROS, particularly H_2O_2, still needs to be elucidated.

7. Conclusions

Aquaporin 3 and aquaporin 5 are upregulated in breast cancer and certainly support processes leading to breast cancer growth and metastasis. Current knowledge indicates that these two aquaporins as potential biomarkers of breast cancer malignancy making them potential therapeutic targets. In order to define AQP3 and AQP5 as targets for cancer treatment, it is needed to thoroughly study all possible aspects and pathways affected, as inadequate inhibition or stimulation of each AQP could drive cancer cells to the more malignant phenotype. One of the currently neglected aspects is certainly, crosstalk with the antioxidative system, especially, as AQP3 and AQP5 channel H_2O_2 which then plays an active role in signalling pathways.

Author Contributions: Conceptualization, L.M. and A.Č.G.; investigation, L.M. and A.Č.G.; writing—original draft preparation L.M. and A.Č.G.; writing—review and editing, L.M. and A.Č.G. All authors have read and agreed to the published version of the manuscript.

Funding: This research was funded by HrZZ, grant number IP-2020-02-3617.

Institutional Review Board Statement: Not applicable.

Informed Consent Statement: Not applicable.

Conflicts of Interest: The authors declare no conflict of interest.

References

1. Chaiswing, L.; Cole, M.P.; St Clair, D.K.; Ittarat, W.; Szweda, L.I.; Oberley, T.D. Oxidative Damage Precedes Nitrative Damage in Adriamycin-Induced Cardiac Mitochondrial Injury. *Toxicol. Pathol.* **2004**, *32*, 536–547. [CrossRef] [PubMed]
2. Zhang, W.-B.; Lai, X.; Guo, X.-F. Activation of Nrf2 by miR-152 Inhibits Doxorubicin-Induced Cardiotoxicity via Attenuation of Oxidative Stress, Inflammation, and Apoptosis. *Oxid. Med. Cell. Longev.* **2021**, *2021*, 1–14.
3. Atalay, F.; Tatar, A.; Dincer, B.; Gundogdu, B.; Koycegiz, S. Protective Effect of Carvacrol against Paclitaxel-Induced Ototoxicity in Rat Model. *Turkish Arch. Otorhinolaryngol.* **2021**, *58*, 241–248. [CrossRef] [PubMed]
4. Finkel, T. Signal transduction by reactive oxygen species. *J. Cell Biol.* **2011**, *194*, 7–15. [CrossRef]
5. Heppner, D.E.; van der Vliet, A. Redox-dependent regulation of epidermal growth factor receptor signaling. *Redox Biol.* **2016**, *8*, 24–27. [CrossRef] [PubMed]
6. King, L.S.; Kozono, D.; Agre, P. From structure to disease: The evolving tale of aquaporin biology. *Nat. Rev. Mol. Cell Biol.* **2004**, *5*, 687–698. [CrossRef] [PubMed]
7. Preston, G.M.; Agre, P. Isolation of the cDNA for erythrocyte integral membrane protein of 28 kilodaltons: Member of an ancient channel family. *Proc. Natl. Acad. Sci.* **1991**, *88*, 11110–11114. [CrossRef]
8. Agre, P.; Bonhivers, M.; Borgnia, M.J. The aquaporins, blueprints for cellular plumbing systems. *J. Biol. Chem.* **1998**, *273*, 14659–14662. [CrossRef]
9. Madeira, A.; Moura, T.F.; Soveral, G. Detecting aquaporin function and regulation. *Front. Chem.* **2016**, *4*, 3. [CrossRef]
10. da Silva, I.V.; Soveral, G. Aquaporins in Obesity. In *Aquaporins. Advances in Experimental Medicine and Biology*; Yang, B., Ed.; Springer: Dordrecht, The Netherlands, 2017; Volume 969, pp. 227–238.
11. Verkman, A.S. Aquaporins: Translating bench research to human disease. *J. Exp. Biol.* **2009**, *212*, 1707–1715. [CrossRef]
12. Ishibashi, K.; Tanaka, Y.; Morishita, Y. The role of mammalian superaquaporins inside the cell. *Biochim. Biophys. Acta* **2014**, *1840*, 1507–1512. [CrossRef]
13. Yeste, M.; Morató, R.; Rodríguez-Gil, J.E.; Bonet, S.; Prieto-Martínez, N. Aquaporins in the male reproductive tract and sperm: Functional implications and cryobiology. *Reprod. Domest. Anim.* **2017**, *52*, 12–27. [CrossRef]
14. Madeira, A.; Fernández-Veledo, S.; Camps, M.; Zorzano, A.; Moura, T.F.; Ceperuelo-Mallafré, V.; Vendrell, J.; Soveral, G. Human Aquaporin-11 is a water and glycerol channel and localizes in the vicinity of lipid droplets in human adipocytes. *Obesity* **2014**, *22*, 2010–2017. [CrossRef]
15. Morishita, Y.; Sakube, Y.; Sasaki, S.; Ishibashi, K. Molecular mechanisms and drug development in aquaporin water channel diseases: Aquaporin superfamily (superaquaporins): Expansion of aquaporins restricted to multicellular organisms. *J. Pharmacol. Sci.* **2004**, *96*, 276–279. [CrossRef]

16. Madeira, A.; Moura, T.F.; Soveral, G. Aquaglyceroporins: Implications in adipose biology and obesity. *Cell. Mol. Life Sci.* **2015**, *72*, 759–771. [CrossRef] [PubMed]
17. Galán-Cobo, A.; Ramírez-Lorca, R.; Echevarría, M. Role of aquaporins in cell proliferation: What else beyond water permeability? *Channels* **2016**, *10*, 185–201. [CrossRef]
18. Henzler, T.; Steudle, E. Transport and metabolic degradation of hydrogen peroxide in chara corallina: Model calculations and measurements with the pressure probe suggest transport of h202 across water channels. *J. Exp. Bot.* **2000**, *51*, 2053–2066. [CrossRef]
19. Prata, C.; Hrelia, S.; Fiorentini, D. Peroxiporins in cancer. *Int. J. Mol. Sci.* **2019**, *20*, 1371. [CrossRef]
20. Varadaraj, K.; Kumari, S.S. Lens aquaporins function as peroxiporins to facilitate membrane transport of hydrogen peroxide. *Biochem. Biophys. Res. Commun.* **2020**, *524*, 1025–1029. [CrossRef] [PubMed]
21. Yool, A. Functional Domains of Aquaporin-1: Keys to Physiology, and Targets for Drug Discovery. *Curr. Pharm. Des.* **2007**, *13*, 3212–3221. [CrossRef]
22. Roche, J.V.; Törnroth-Horsefield, S. Aquaporin protein-protein interactions. *Int. J. Mol. Sci.* **2017**, *18*, 2255. [CrossRef] [PubMed]
23. Bienert, G.P.; Chaumont, F. Aquaporin-facilitated transmembrane diffusion of hydrogen peroxide. *Biochim. Biophys. Acta* **2014**, *1840*, 1596–1604. [CrossRef] [PubMed]
24. Rodrigues, C.; Mósca, A.; Martins, A.; Nobre, T.; Prista, C.; Antunes, F.; Cipak Gasparovic, A.; Soveral, G. Rat Aquaporin-5 Is pH-Gated Induced by Phosphorylation and Is Implicated in Oxidative Stress. *Int. J. Mol. Sci.* **2016**, *17*, 2090. [CrossRef] [PubMed]
25. Rodrigues, C.; Tartaro Bujak, I.; Mihaljević, B.; Soveral, G.; Cipak Gasparovic, A. Yeast aquaporin regulation by 4-hydroxynonenal is implicated in oxidative stress response. *IUBMB Life* **2017**, *69*, 355–362. [CrossRef] [PubMed]
26. Alishahi, M.; Kamali, R. A novel molecular dynamics study of CO2 permeation through aquaporin-5. *Eur. Phys. J. E.* **2019**, *42*, 151. [CrossRef] [PubMed]
27. Kourghi, M.; Pei, J.V.; De Ieso, M.L.; Nourmohammadi, S.; Chow, P.H.; Yool, A.J. Fundamental structural and functional properties of Aquaporin ion channels found across the kingdoms of life. *Clin. Exp. Pharmacol. Physiol.* **2018**, *45*, 401–409. [CrossRef]
28. De Ieso, M.L.; Yool, A.J. Mechanisms of aquaporin-facilitated cancer invasion and metastasis. *Front. Chem.* **2018**, *6*, 135. [CrossRef] [PubMed]
29. Yde, J.; Keely, S.J.; Moeller, H.B. Expression, regulation and function of Aquaporin-3 in colonic epithelial cells. *Biochim. Biophys. Acta-Biomembr.* **2021**, *1863*, 183619. [CrossRef]
30. Lee, S.M.; Lee, S.W.; Kang, M.; Choi, J.K.; Park, K.; Byun, J.S.; Kim, D.Y. FoxO1 as a Regulator of Aquaporin 5 Expression in the Salivary Gland. *J. Dent. Res.* **2021**, 002203452110034. [CrossRef]
31. Cui, F.; Hu, M.; Li, R.; Li, B.; Huang, D.; Ma, W.; Jia, X.; Lv, Z. Insulin on changes in expressions of aquaporin-1, aquaporin-5, and aquaporin-8 in submandibular salivary glands of rats with Streptozotocin-induced diabetes. *Int. J. Clin. Exp. Pathol.* **2021**, *14*, 221–229.
32. Kreida, S.; Törnroth-Horsefield, S. Structural insights into aquaporin selectivity and regulation. *Curr. Opin. Struct. Biol.* **2015**, *33*, 126–134. [CrossRef]
33. Kamsteeg, E.J.; Heijnen, I.; Van Os, C.H.; Deen, P.M.T. The subcellular localization of an aquaporin-2 tetramer depends on the stoichiometry of phosphorylated and nonphosphorylated monomers. *J. Cell Biol.* **2000**, *151*, 919–929. [CrossRef]
34. Chen, Q.; Zhu, L.; Zong, H.; Song, X.; Wang, L.; Wang, X.; Yang, D.; Wang, J. Subcellular localization of aquaporin 3 in prostate cancer is regulated by RalA. *Oncol. Rep.* **2018**, *39*, 2171–2177. [CrossRef]
35. Soveral, G.; Madeira, A.; Loureiro-Dias, M.C.; Moura, T.F. Membrane tension regulates water transport in yeast. *Biochim. Biophys. Acta* **2008**, *1778*, 2573–2579. [CrossRef]
36. Soveral, G.; Macey, R.I.; Moura, T.F. Membrane stress causes inhibition of water channels in brush border membrane vesicles from kidney proximal tubule. *Biol. Cell* **1997**, *89*, 275–282. [CrossRef]
37. Ozu, M.; Galizia, L.; Acuña, C.; Amodeo, G. Aquaporins: More Than Functional Monomers in a Tetrameric Arrangement. *Cells* **2018**, *7*, 209. [CrossRef]
38. Smith, A.J.; Jin, B.-J.; Ratelade, J.; Verkman, A.S. Aggregation state determines the localization and function of M1- and M23-aquaporin-4 in astrocytes. *J. Cell Biol.* **2014**, *204*, 559–573. [CrossRef]
39. Neely, J.D.; Christensen, B.M.; Nielsen, S.; Agre, P. Heterotetrameric composition of aquaporin-4 water channels. *Biochemistry* **1999**, *38*, 11156–11163. [CrossRef]
40. Ahmadpour, D.; Geijer, C.; Tamás, M.J.; Lindkvist-Petersson, K.; Hohmann, S. Yeast reveals unexpected roles and regulatory features of aquaporins and aquaglyceroporins. *Biochim. Biophys. Acta* **2014**, *1840*, 1482–1491. [CrossRef]
41. Papadopoulos, M.C.; Saadoun, S. Key roles of aquaporins in tumor biology. *Biochim. Biophys. Acta* **2015**, *1848*, 2576–2583. [CrossRef]
42. Tomita, Y.; Dorward, H.; Yool, A.J.; Smith, E.; Townsend, A.R.; Price, T.J.; Hardingham, J.E. Role of Aquaporin 1 Signalling in Cancer Development and Progression. *Int. J. Mol. Sci.* **2017**, *18*, 299. [CrossRef]
43. Dorward, H.S.; Du, A.; Bruhn, M.A.; Wrin, J.; Pei, J.V.; Evdokiou, A.; Price, T.J.; Yool, A.J.; Hardingham, J.E. Pharmacological blockade of aquaporin-1 water channel by AqB013 restricts migration and invasiveness of colon cancer cells and prevents endothelial tube formation in vitro. *J. Exp. Clin. Cancer Res.* **2016**, *35*, 36. [CrossRef]
44. Chen, J.; Wang, T.; Zhou, Y.-C.; Gao, F.; Zhang, Z.-H.; Xu, H.; Wang, S.-L.; Shen, L.-Z. Aquaporin 3 promotes epithelial-mesenchymal transition in gastric cancer. *J. Exp. Clin. Cancer Res.* **2014**, *33*, 38. [CrossRef]

45. Wang, Z.; Wang, Y.; He, Y.; Zhang, N.; Chang, W.; Niu, Y. Aquaporin-1 facilitates proliferation and invasion of gastric cancer cells via GRB7-mediated ERK and Ras activation. *Anim. Cells Syst.* **2020**, *24*, 253–259. [CrossRef]
46. Liu, C.; Liu, L.; Zhang, Y.; Jing, H. Molecular mechanism of AQP3 in regulating differentiation and apoptosis of lung cancer stem cells through Wnt/GSK-3β/β-Catenin pathway. *JBUON* **2020**, *25*, 1714–1720. [PubMed]
47. Guéraud, F.; Atalay, M.; Bresgen, N.; Cipak, A.; Eckl, P.M.M.; Huc, L.; Jouanin, I.; Siems, W.; Uchida, K.; Gueraud, F.; et al. Chemistry and biochemistry of lipid peroxidation products. *Free Radic. Res.* **2010**, *44*, 1098–1124. [CrossRef] [PubMed]
48. Cadenas, E.; Sies, H. Oxidative stress: Excited oxygen species and enzyme activity. *Adv. Enzyme Regul.* **1985**, *23*, 217–237. [CrossRef]
49. Sies, H.; Cadenas, E. Oxidative stress: Damage to intact cells and organs. *Philos. Trans. R. Soc. Lond. B. Biol. Sci.* **1985**, *311*, 617–631. [PubMed]
50. Niki, E. Oxidative stress and antioxidants: Distress or eustress? *Arch. Biochem. Biophys.* **2016**, *595*, 19–24. [CrossRef] [PubMed]
51. Calabrese, E.J.; Baldwin, L.A. Defining hormesis. *Hum. Exp. Toxicol.* **2002**, *21*, 91–97. [CrossRef]
52. Davies, K.J.A. Adaptive homeostasis. *Mol. Aspects Med.* **2016**, *49*, 1–7. [CrossRef]
53. Sies, H. Oxidative Eustress: On Constant Alert for Redox Homeostasis. *Redox Biol.* **2021**, *41*, 101867. [CrossRef]
54. Sies, H. Findings in redox biology: From H2O2 to oxidative stress. *J. Biol. Chem.* **2020**, *295*, 13458–13473. [CrossRef]
55. Alleman, R.J.; Katunga, L.A.; Nelson, M.A.M.; Brown, D.A.; Anderson, E.J. The "Goldilocks Zone" from a redox perspective-Adaptive vs. deleterious responses to oxidative stress in striated muscle. *Front. Physiol.* **2014**, *5*, 358. [CrossRef]
56. Ursini, F.; Maiorino, M.; Forman, H.J. Redox homeostasis: The Golden Mean of healthy living. *Redox Biol.* **2016**, *8*, 205–215. [CrossRef] [PubMed]
57. Antoncic-Svetina, M.; Sentija, D.; Cipak, A.; Milicic, D.; Meinitzer, A.; Tatzber, F.; Andrisic, L.; Zelzer, S.; Zarkovic, N. Ergometry induces systemic oxidative stress in healthy human subjects. *Tohoku J. Exp. Med.* **2010**, *221*, 43–48. [CrossRef]
58. Gomez-Cabrera, M.C.; Domenech, E.; Romagnoli, M.; Arduini, A.; Borras, C.; Pallardo, F.V.; Sastre, J.; Viña, J. Oral administration of vitamin C decreases muscle mitochondrial biogenesis and hampers training-induced adaptations in endurance performance. *Am. J. Clin. Nutr.* **2008**, *87*, 142–149. [CrossRef]
59. Morrison, D.; Hughes, J.; Della Gatta, P.A.; Mason, S.; Lamon, S.; Russell, A.P.; Wadley, G.D. Vitamin C and e supplementation prevents some of the cellular adaptations to endurance-training in humans. *Free Radic. Biol. Med.* **2015**, *89*, 852–862. [CrossRef]
60. Peluso, I.; Yarla, N.S.; Ambra, R.; Pastore, G.; Perry, G. MAPK signalling pathway in cancers: Olive products as cancer preventive and therapeutic agents. *Semin. Cancer Biol.* **2019**, *56*, 185–195. [CrossRef] [PubMed]
61. Milkovic, L.; Cipak Gasparovic, A.; Zarkovic, N.; Gasparovic, A.C.; Zarkovic, N.; Cipak Gasparovic, A.; Zarkovic, N. Overview on major lipid peroxidation bioactive factor 4-hydroxynonenal as pluripotent growth-regulating factor. *Free Radic. Res.* **2015**, *49*, 850–860. [CrossRef]
62. Esterbauer, H.; Schaur, R.J.; Zollner, H. Chemistry and biochemistry of 4-hydroxynonenal, malonaldehyde and related aldehydes. *Free Radic. Biol. Med.* **1991**, *11*, 81–128. [CrossRef]
63. Zarkovic, N.; Cipak, A.; Jaganjac, M.; Borovic, S.; Zarkovic, K. Pathophysiological relevance of aldehydic protein modifications. *J. Proteom.* **2013**, *92*, 239–247. [CrossRef]
64. Esterbauer, H.; Muskiet, F.; Horrobin, D.F. Cytotoxicity and genotoxicity of lipid-oxidation products. *Am. J. Clin. Nutr.* **1993**, *57*, 779S–785S. [CrossRef]
65. Milkovic, L.; Zarkovic, N.; Saso, L. Controversy about pharmacological modulation of Nrf2 for cancer therapy. *Redox Biol.* **2017**, *12*, 727–732. [CrossRef]
66. Taguchi, K.; Motohashi, H.; Yamamoto, M. Molecular mechanisms of the Keap1-Nrf2 pathway in stress response and cancer evolution. *Genes Cells* **2011**, *16*, 123–140. [CrossRef]
67. Beretta, G.L.; Corno, C.; Zaffaroni, N.; Perego, P. Role of FoxO Proteins in Cellular Response to Antitumor Agents. *Cancers* **2019**, *11*, 90. [CrossRef]
68. Laissue, P. The forkhead-box family of transcription factors: Key molecular players in colorectal cancer pathogenesis. *Mol. Cancer* **2019**, *18*, 5. [CrossRef]
69. Chen, J.; Wang, L.; Chen, Y.; Sternberg, P.; Cai, J. Phosphatidylinositol 3 Kinase Pathway and 4-Hydroxy-2-Nonenal-Induced Oxidative Injury in the RPE. *Investig. Opthalmology Vis. Sci.* **2009**, *50*, 936. [CrossRef]
70. Numazawa, S.; Ishikawa, M.; Yoshida, A.; Tanaka, S.; Yoshida, T. Atypical protein kinase C mediates activation of NF-E2-related factor 2 in response to oxidative stress. *Am. J. Physiol. Cell Physiol.* **2003**, *285*, C334–C342. [CrossRef]
71. Łuczaj, W.; Gęgotek, A.; Skrzydlewska, E. Antioxidants and HNE in redox homeostasis. *Free Radic. Biol. Med.* **2017**, *111*, 87–101. [CrossRef]
72. Carter, M.E.; Brunet, A. FOXO transcription factors. *Curr. Biol.* **2007**, *17*, R113–R114. [CrossRef] [PubMed]
73. Bhakta-Guha, D.; Efferth, T. Hormesis: Decoding Two Sides of the Same Coin. *Pharmaceuticals* **2015**, *8*, 865–883. [CrossRef] [PubMed]
74. Kang, J.I.; Kim, D.H.; Sung, K.W.; Shim, S.M.; Cha-molstad, H.; Soung, N.K.; Lee, K.H.; Hwang, J.; Lee, H.G.; Kwon, Y.T.; et al. P62-induced cancer-associated fibroblast activation via the Nrf2-ATF6 pathway promotes lung tumorigenesis. *Cancers* **2021**, *13*, 864. [CrossRef] [PubMed]
75. Aikman, B.; de Almeida, A.; Meier-Menches, S.M.; Casini, A. Aquaporins in cancer development: Opportunities for bioinorganic chemistry to contribute novel chemical probes and therapeutic agents. *Metallomics* **2018**, *10*, 696–712. [CrossRef]

76. Khan, S.; Ricciardelli, C.; Yool, A.J. Targeting Aquaporins in Novel Therapies for Male and Female Breast and Reproductive Cancers. *Cells* **2021**, *10*, 215. [CrossRef] [PubMed]
77. van Lieburg, A.F.; Knoers, N.V.A.M.; Deen, P.M.T. Discovery of aquaporins: A breakthrough in research on renal water transport. *Pediatr. Nephrol.* **1995**, *9*, 228–234. [CrossRef]
78. Verkman, A.S.; Anderson, M.O.; Papadopoulos, M.C. Aquaporins: Important but elusive drug targets. *Nat. Rev. Drug Discov.* **2014**, *13*, 259–277. [CrossRef]
79. Verkman, A.S.; Hara-Chikuma, M.; Papadopoulos, M.C. Aquaporins—New players in cancer biology. *J. Mol. Med.* **2008**, *86*, 523–529. [CrossRef]
80. Mobasheri, A.; Barrett-Jolley, R. Aquaporin water channels in the mammary gland: From physiology to pathophysiology and neoplasia. *J. Mammary Gland Biol. Neoplasia* **2014**, *19*, 91–102. [CrossRef]
81. Moosavi, M.S.; Elham, Y. Aquaporins 1, 3 and 5 in Different Tumors, their Expression, Prognosis Value and Role as New Therapeutic Targets. *Pathol. Oncol. Res.* **2020**, *26*, 615–625. [CrossRef]
82. Marlar, S.; Jensen, H.H.; Login, F.H.; Nejsum, L.N. Aquaporin-3 in cancer. *Int. J. Mol. Sci.* **2017**, *18*, 2106. [CrossRef] [PubMed]
83. Chandrashekar, D.S.; Bashel, B.; Balasubramanya, S.A.H.; Creighton, C.J.; Ponce-Rodriguez, I.; Chakravarthi, B.V.S.K.; Varambally, S. UALCAN: A Portal for Facilitating Tumor Subgroup Gene Expression and Survival Analyses. *Neoplasia* **2017**, *19*, 649–658. [CrossRef] [PubMed]
84. Zhu, Z.; Jiao, L.; Li, T.; Wang, H.; Wei, W.; Qian, H. Expression of AQP3 and AQP5 as a prognostic marker in triple-negative breast cancer. *Oncol. Lett.* **2018**, *16*, 2661–2667. [CrossRef] [PubMed]
85. Kang, S.; Chae, Y.S.; Lee, S.J.; Kang, B.W.; Kim, J.G.; Kim, W.W.; Jung, J.H.; Park, H.Y.; Jeong, J.-H.; Jeong, J.Y.; et al. Aquaporin 3 Expression Predicts Survival in Patients with HER2-positive Early Breast Cancer. *Anticancer Res.* **2015**, *35*, 2775–2782.
86. Huang, Y.T.; Zhou, J.; Shi, S.; Xu, H.Y.; Qu, F.; Zhang, D.; Chen, Y.D.; Yang, J.; Huang, H.F.; Sheng, J.Z. Identification of Estrogen Response Element in Aquaporin-3 Gene that Mediates Estrogen-induced Cell Migration and Invasion in Estrogen Receptor-positive Breast Cancer. *Sci. Rep.* **2015**, *5*, 12484. [CrossRef] [PubMed]
87. Hara-Chikuma, M.; Verkman, A.S. Aquaporin-3 facilitates epidermal cell migration and proliferation during wound healing. *J. Mol. Med.* **2008**, *86*, 221–231. [CrossRef]
88. Hara-Chikuma, M.; Verkman, A.S. Prevention of Skin Tumorigenesis and Impairment of Epidermal Cell Proliferation by Targeted Aquaporin-3 Gene Disruption. *Mol. Cell. Biol.* **2008**, *28*, 326–332. [CrossRef]
89. Hara-Chikuma, M.; Satooka, H.; Watanabe, S.; Honda, T.; Miyachi, Y.; Watanabe, T.; Verkman, A.S. Aquaporin-3-mediated hydrogen peroxide transport is required for NF-Î° B signalling in keratinocytes and development of psoriasis. *Nat. Commun.* **2015**, *6*, 7454. [CrossRef]
90. Satooka, H.; Hara-Chikuma, M. Aquaporin-3 Controls Breast Cancer Cell Migration by Regulating Hydrogen Peroxide Transport and Its Downstream Cell Signaling. *Mol. Cell. Biol.* **2016**, *36*, 1206–1218. [CrossRef]
91. Martins, A.P.; Marrone, A.; Ciancetta, A.; Galán Cobo, A.; Echevarría, M.; Moura, T.F.; Re, N.; Casini, A.; Soveral, G. Targeting Aquaporin Function: Potent Inhibition of Aquaglyceroporin-3 by a Gold-Based Compound. *PLoS ONE* **2012**, *7*, e37435. [CrossRef]
92. Miller, E.W.; Dickinson, B.C.; Chang, C.J. Aquaporin-3 mediates hydrogen peroxide uptake to regulate downstream intracellular signaling. *Proc. Natl. Acad. Sci.* **2010**, *107*, 15681–15686. [CrossRef]
93. Bienert, G.P.; Møller, A.L.B.; Kristiansen, K.A.; Schulz, A.; Møller, I.M.; Schjoerring, J.K.; Jahn, T.P. Specific aquaporins facilitate the diffusion of hydrogen peroxide across membranes. *J. Biol. Chem.* **2007**, *282*, 1183–1192. [CrossRef] [PubMed]
94. Bienert, G.P.; Schjoerring, J.K.; Jahn, T.P. Membrane transport of hydrogen peroxide. *Biochim. Biophys. Acta.* **2006**, *1758*, 994–1003. [CrossRef] [PubMed]
95. Rodrigues, C.; Milkovic, L.; Bujak, I.T.; Tomljanovic, M.; Soveral, G.; Cipak Gasparovic, A. Lipid Profile and Aquaporin Expression under Oxidative Stress in Breast Cancer Cells of Different Malignancies. *Oxid. Med. Cell. Longev.* **2019**, *2019*, 1–10. [CrossRef] [PubMed]
96. Ampawong, S.; Luplertlop, N. Experimental Scedosporiosis Induces Cerebral Oedema Associated with Abscess regarding Aquaporin-4 and Nrf-2 Depletions. *Biomed Res. Int.* **2019**, *2019*. [CrossRef] [PubMed]
97. Song, D.; Liu, X.; Diao, Y.; Sun, Y.; Gao, G.; Zhang, T.; Chen, K.; Pei, L. Hydrogen-rich solution against myocardial injury and aquaporin expression via the PI3K/Akt signaling pathway during cardiopulmonary bypass in rats. *Mol. Med. Rep.* **2018**, *18*, 1925–1938. [CrossRef] [PubMed]
98. Lu, C.; Ma, Z.; Cheng, X.; Wu, H.; Tuo, B.; Liu, X.; Li, T. Pathological role of ion channels and transporters in the development and progression of triple-negative breast cancer. *Cancer Cell Int.* **2020**, *20*, 1–11. [CrossRef]
99. Jung, H.J.; Park, J.Y.; Jeon, H.S.; Kwon, T.H. Aquaporin-5: A marker protein for proliferation and migration of human breast cancer cells. *PLoS ONE* **2011**, *6*, e28492. [CrossRef] [PubMed]
100. Lee, S.J.; Chae, Y.S.; Kim, J.G.; Kim, W.W.; Jung, J.H.; Park, H.Y.; Jeong, J.Y.; Park, J.Y.; Jung, H.J.; Kwon, T.H. AQP5 expression predicts survival in patients with early breast cancer. *Ann. Surg. Oncol.* **2014**, *21*, 375–383. [CrossRef]
101. Hecht, F.; Pessoa, C.F.; Gentile, L.B.; Rosenthal, D.; Carvalho, D.P.; Fortunato, R.S. The role of oxidative stress on breast cancer development and therapy. *Tumor Biol.* **2016**, *37*, 4281–4291. [CrossRef]
102. Rodrigues, C.; Pimpao, C.; Coxixo, A.; Lopes, D.; Pedersen, P.A.; Antunes, F.; Soveral, G. Human Aquaporin-5 Facilitates Hydrogen Peroxide and Cancer Cell Migration. *Cancers* **2019**, *11*, 932. [CrossRef] [PubMed]

103. Milkovic, L.; Cipak Gasparovic, A.; Cindric, M.; Mouthuy, P.A.; Zarkovic, N. Short Overview of ROS as Cell Function Regulators and Their Implications in Therapy Concepts. *Cells* **2019**, *8*, 793. [CrossRef] [PubMed]
104. Park, E.J.; Jung, H.J.; Choi, H.J.; Jang, H.J.; Park, H.J.; Nejsum, L.N.; Kwon, T.H. Exosomes co-expressing AQP5-targeting miRNAs and IL-4 receptor-binding peptide inhibit the migration of human breast cancer cells. *FASEB J.* **2020**, *34*, 3379–3398. [CrossRef] [PubMed]
105. Oh, S.H.; Choi, S.Y.; Choi, H.J.; Ryu, H.M.; Kim, Y.J.; Jung, H.Y.; Cho, J.H.; Kim, C.D.; Park, S.H.; Kwon, T.H.; et al. The emerging role of xanthine oxidase inhibition for suppression of breast cancer cell migration and metastasis associated with hypercholesterolemia. *FASEB J.* **2019**, *33*, 7301–7314. [CrossRef] [PubMed]

molecules

MDPI

Review

Involvement of NRF2 in Breast Cancer and Possible Therapeutical Role of Polyphenols and Melatonin

Alev Tascioglu Aliyev [1,†], Emiliano Panieri [2], Višnja Stepanić [3], Hande Gurer-Orhan [1,*] and Luciano Saso [2]

1 Department of Toxicology, Faculty of Pharmacy, Ege University, 35040 Izmir, Turkey; alev.aliyev@ibg.edu.tr
2 Department of Physiology and Pharmacology "Vittorio Erspamer", Sapienza University of Rome, P. le Aldo Moro 5, 00185 Rome, Italy; emiliano.panieri@hotmail.it (E.P.); luciano.saso@uniroma1.it (L.S.)
3 Laboratory for Machine Learning and Knowledge Representation, Ruđer Bošković Institute, Bijenička 54, 10000 Zagreb, Croatia; Visnja.Stepanic@irb.hr
* Correspondence: hande.gurer.orhan@ege.edu.tr; Tel.: +90-506-315-8241
† Current address: Izmir Biomedicine and Genome Center, 35340 Izmir, Turkey.

Abstract: Oxidative stress is defined as a disturbance in the prooxidant/antioxidant balance in favor of the former and a loss of control over redox signaling processes, leading to potential biomolecular damage. It is involved in the etiology of many diseases, varying from diabetes to neurodegenerative diseases and cancer. Nuclear factor erythroid 2-related factor 2 (NRF2) is a transcription factor and reported as one of the most important oxidative stress regulators. Due to its regulatory role in the expression of numerous cytoprotective genes involved in the antioxidant and anti-inflammatory responses, the modulation of NRF2 seems to be a promising approach in the prevention and treatment of cancer. Breast cancer is the prevalent type of tumor in women and is the leading cause of death among female cancers. Oxidative stress-related mechanisms are known to be involved in breast cancer, and therefore, NRF2 is considered to be beneficial in its prevention. However, its overactivation may lead to a negative clinical impact on breast cancer therapy by causing chemoresistance. Some known "oxidative stress modulators", such as melatonin and polyphenols, are suggested to play an important role in the prevention and treatment of cancer, where the activation of NRF2 is reported as a possible underlying mechanism. In the present review, the potential involvement of oxidative stress and NRF2 in breast cancer will be reviewed, and the role of the NRF2 modulators—namely, polyphenols and melatonin—in the treatment of breast cancer will be discussed.

Keywords: breast cancer; NRF2; oxidative stress; polyphenols; melatonin

Citation: Tascioglu Aliyev, A.; Panieri, E.; Stepanić, V.; Gurer-Orhan, H.; Saso, L. Involvement of NRF2 in Breast Cancer and Possible Therapeutical Role of Polyphenols and Melatonin. *Molecules* **2021**, *26*, 1853. https://doi.org/10.3390/molecules26071853

Academic Editor: Roberto Fabiani

Received: 4 March 2021
Accepted: 22 March 2021
Published: 25 March 2021

1. Introduction

According to the global cancer statistics data (GLOBOCAN 2020) from the International Agency for Research on Cancer, breast cancer has the highest incidence and mortality rates among women [1]. The mortality-to-incidence ratio in breast cancer has been decreasing over the years, which indicates that the survival rate is gradually increasing. However, our century still has the economical, physical and psychological burden of cancer therapy. Despite the remarkable advances in scientific knowledge, the risk factors and the mechanisms responsible for breast cancer are still under investigation.

Breast cancer is a heterogeneous disease and is classified as a luminal A (estrogen receptor (ER) and/or progesterone receptor (PR) (+) and human epidermal growth factor receptor 2 (HER2) (−), with low ki-67 levels); luminal B (ER and/or PR (+) and HER2 (−), with high ki-67 levels); basal-like (also called triple-negative due to the lack of ER, PR and HER2 expression) and HER2-enriched (ER and PR (−) and HER (+)) molecular subtypes [2]. The hormone receptor status is not the only prognostic marker for the disease, but other molecular markers such as the PD-L1 gene presence [3], tissue localization (ductal and lobular) and invasion status of the cancer cells are also used to define better treatment strategies. With a better understanding of the genetic and epigenetic alterations and estrogen and oxidative stress signaling pathways, breast cancer therapy is expected to take a decisive step towards the personalized and effective treatments of patients.

2. Mechanisms Underlying Breast Cancer

Hanahan and Weinberg defined the "hallmarks of cancer" to understand tumoral changes in normal cells for all cancer types. To explain the tumoral behavior of the cell, the authors addressed eight key features: unrestrained proliferation, avoided growth suppressor signaling, resistance to cell death, sustained replication, induction of angiogenesis, activation of invasion and metastasis and altered cellular metabolism, including redox homeostasis and evaded immune destruction [4,5]. More generally, the stages of carcinogenesis can be summarized as a multistep process wherein excessive and irreversible cellular DNA damage initiates cancer cell formation, while the subsequent alteration of the signaling pathways promotes the sustained proliferation and genetic instability of preneoplastic cells, leading to their clonal expansion. Afterwards, the occurrence of further genetic changes converts these cells into cancer cells that progressively acquire more aggressive features, culminating in tissue invasion and metastasis formation [6,7].

Multiple risk factors, such as aging, early menarche/late menopause, hormone replacement therapy, oral contraceptive usage, obesity, family history, benign lesions, radiation therapy, nutrition and other lifestyle habits, are involved in breast cancer development [8]. Estrogens are a driven factor for the promotion and progression of around 75–85% of breast cancer patients, which have hormone receptor-positive subtypes (luminal A and luminal B). However, patients with basal-like (triple-negative) breast cancer have inherited breast cancer gene (BRCA) mutations. The current section explains the risk factors and risk factor-related underlying mechanisms of breast cancer, with a particular emphasis on oxidative stress.

2.1. *Oxidative Stress-Independent Mechanisms*

Family history, race and ethnicity are related to the development of breast cancer. Inherited genetic mutations—in other words, germline mutations in specific genes—were found to be related to breast cancer development. At the mid-1990s, the role of BRCA1 and BRCA2 mutations in elevating breast cancer risk was considered [9]. A prospective study with 3886 breast cancer patients showed that a cumulative risk was 72% for BRCA1 and 69% for BRCA2 carriers [10]. The genetic analyses of patients revealed that tumor-suppressor genes—namely, PTEN, p53, STK11 and cell adhesion regulator CDH1 genes, are high-penetrance breast cancer-susceptibility genes and increase the breast cancer risk more than fourfold. Mutations in moderate-penetrance genes such as PALB2, ATM and CHEK2 were also reported to increase the risk of breast cancer occurrence two to four times [11].

Other conditions—specifically, hormone replacement therapy, early menarche/late menopause, a history of never being pregnant and/or breastfeeding and oral contraceptive usage—are stated as risk factors for the development of breast cancer, since elevated estrogen exposure was found to enhance the aberrant epithelial cell proliferation and to cause neoplastic formation [12]. Hilton et al., in a seminal paper, reviewed the relationship between steroid hormone receptors and breast cancer [13]. The authors reported an aberrant hormonal activation in steroid receptor-positive breast cancer results with a metabolic switch between autocrine signaling, which led to cell proliferation and prevented proapoptotic signaling. Estrogens and progesterone also regulate histone modifications, and enhanced levels of those hormones can cause epigenetic changes [13].

Nutritional and lifestyle habits such as the consumption of alcohol, high-calorie diets, obesity and physical inactivity contribute to breast cancer initiation and/or promotion [14]. For example, alcohol consumption increases the steroid hormone levels, which, in turn, alters the DNA methylation status by affecting one carbon metabolism [15]. A high-calorie intake with reduced physical activity led to an increase in the body mass index (BMI).

Being overweight ($25 \leq$ BMI < 30 kg/m^2) and obese (BMI $\geq$ 30 kg/m^2) was found to be significantly associated with breast cancer progression in postmenopausal women [16]. Mechanistically, adipose cells secrete two adipokines, antioncogenic adiponectin and pro-oncogenic leptin. Jarde et al. revealed that the leptin expression was higher than adiponectin expression in epithelial ductal breast cancers, while, in the normal tissue adjacent to cancer, the expression of adiponectin was prevalent [17]. Mechanistically, obesity-related alterations of adipose stem cells increase leptin secretion, which becomes a source for enhanced estrogen-related tumor growth by upregulating the estrogen receptor and aromatase [18].

An age and cancer incidence correlation was well-established, although the underlying mechanisms are still under investigation. Oxidative stress-independent age-related mechanisms were reviewed, such as accumulated somatic mutations, epigenetic changes in DNA methylation and chromatin remodeling and changes in breast tissue and the cell microenvironment [19].

2.2. *Oxidative Stress-Dependent Mechanisms*

Oxidative stress was defined as "a disturbance in prooxidant-antioxidant balance in favor of the former" by Sies at 1985 [20], while Dean Jones further expanded this definition, including also the disruption of redox signaling circuitries [21]. The production of reactive oxygen species (ROS) and reactive nitrogen species (RNS) is a normal consequence of the cell metabolism. Mitochondria produce ATP through oxidative phosphorylation, involving the passage of electrons through a series of complexes, part of the electron transport chain, promoting the reduction of molecular oxygen to water. During this process, electrons can leak through the inner membrane space and produce superoxide radical ion, which is the precursor of various ROS and RNS molecules. Cells such as macrophages, leukocytes and monocytes also produce superoxide radical anion ($O_2^{\bullet -}$) via NADPH oxidase activity. On the other hand, mammalian cells maintain the redox balance through the coordinated activity of antioxidant enzymes and antioxidant molecules. For example, the superoxide dismutase (SOD) converts $O_2^{\bullet -}$ to the less electrophilic hydrogen peroxide H_2O_2, which is rapidly detoxified into water H_2O and oxygen molecules O_2 by glutathione peroxidase (GPx), catalase (CAT) or peroxiredoxin (Prx). However, the excessive production of ROS and/or inadequate supply of antioxidant molecules (e.g., endogenous glutathione) may cause a further reaction between H_2O_2 and Fe^{2+} ions (Fenton reaction) and produce highly reactive hydroxyl radical $^{\bullet}OH$. Due to its electrophilic properties, the hydroxyl radical attacks nucleophilic cellular molecules such as proteins, lipids and DNA. In such a way, increased oxidative stress generates DNA adducts, protein damage and lipid peroxidation products such as the 4-hydroxynonenal (4-HNE) [6]. DNA adduct formation is mainly initiated by $^{\bullet}OH$. Hydroxyl radical attacks the 2′-deoxyribose of purine and pyrimidine bases. As a consequence, this reaction generates 8-hydroxy-2′-deoxyguanosine (8-OH-dG) and 8-oxo-7,8-dihydro-2′-deoxyguanosine (8-oxo-dG), two well-recognized markers of oxidative stress-related DNA adduct formation [22]. Insufficient DNA repair leads to the accumulation of DNA adducts in the cells. Moreover, the division of damaged cells promotes tumoral changes [23]. As a result, oxidative stress derived by different sources such as ionizing radiation, exposure to xenobiotics or their metabolites, aging, alcohol consumption, obesity and estrogens may trigger carcinogenesis.

Nour Eldin et al. showed that the plasmatic levels of 8-OH-dG exhibit a steady-state increase from normal breast tissue to benign and malignant lesions for different patient groups [24]. In the same study, the authors also showed that malignant breast tumors with highly invasive subtypes were also characterized by higher 8-OH-dG levels than tumors with less invasive behaviors. In contrast, Sova et al. did not find a positive association between the tumor stages and serum 8-OH-dG levels, while, conversely, the content of 8-OH-dG was found to be related with the tumor stage in the immunohistochemical analyses [25]. By comparing the serum levels and the immunohistochemical content of 8-OH-dG, the study revealed that both these features were strongly associated with tumor aggressiveness. Taken together, it can be concluded that breast cancer development partially occurs via DNA adduct formation by oxidative stress, and the disease progression may be monitored by the quantitative analysis of oxidative stress-related DNA products.

ROS Formation via Estrogen Metabolism

Breast cancer development by estrogens can derive from two different mechanisms: estrogen receptor activation and the alteration of estrogen metabolism-related pathways [12,26].

Generally, chemicals are metabolized through phase I reactions that include oxidation, reduction and hydroxylation of the substance and phase II reactions that promote enzymatic conjugation of the metabolite with endogenous molecules, such as glutathione (GSH), sulfate and glucuronic acid, and facilitate its elimination. Similarly, estrogens (estradiol and estrone) undergo phase I metabolism by cytochrome p-450 enzymes and are converted to 2-hydroxycatechol estrogen or to 4-hydroxycatechol metabolites. Catechol metabolites of estrogens are detoxified mainly by catechol-O-methyl transferase via conjugation. At the same time, catechol estrogens undergo autooxidation, which promotes the formation of a hydroxyl radical. Catechol metabolites can also be oxidized to highly reactive estrogen 3,4-semiquinone, estradiol-2,3-semiquinone and, additionally, to estrogen 3,4-quinone and estradiol-2,3-quinone. Commonly, the phase II detoxification of quinone metabolites are catalyzed by glutathione-S-transferase [12,27]. However, under certain circumstances, such as excessive exposure to estrogens by hormone replacement therapy, oral contraceptive usage and/or limited detoxification capacity, reactive estrogen metabolites can react with DNA and cause genomic instability, as extensively reviewed by Cavalieri et al. [27].

2.3. *Role of Nuclear Factor Erythroid 2-Related Factor 2 (NRF2) in Breast Cancer*

Compelling evidence from the last decade indicates that nuclear factor erythroid 2-related factor 2 (NRF2) is a master regulator of a cytoprotective response centered on the activation of detoxifying mechanisms in response to oxidative/electrophilic stress or xenobiotics. Importantly, it is well-recognized that sustained NRF2 signaling in cancer cells can be instrumental to the orchestration of a pro-oncogenic "program" that ultimately promotes the malignant progression and the development of therapy resistance, leading to poor clinical outcomes. An increased NRF2 expression in breast cancer patients resulted in a lower overall survival and disease-free survival [28]. Despite that NRF2 can participate in the regulation of oncogenic signaling and cancer-specific hallmarks, other data suggest that its activation in normal cells can exert a chemopreventive role by suppressing ROS-dependent DNA damage and carcinogenesis. In agreement with the still-unresolved scientific debate, this reflects the dual role of NRF2 in cancer and suggests that both pro-oncogenic and antioncogenic activities can be exerted by this transcription factor, depending on additional factors that require a case-by-case assessment (Figure 1). In the following section, we will describe in more detail the role of NRF2 in the regulation of breast cancer hallmarks.

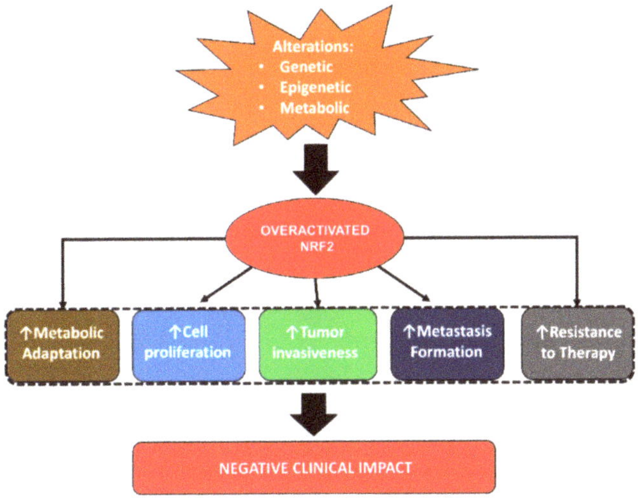

Figure 1. Roles of nuclear factor erythroid 2-related factor 2 (NRF2) in breast cancer.

2.3.1. NRF2 in Breast Cancer Cell Proliferation, Growth, Invasion and Metastasis

In a quite recent study, the genetic silencing of NRF2 was found to significantly impair cell proliferation and migration in the MDA-MB-231 and MCF-7 breast cancer cells by downregulating the small GTPase and transforming protein RhoA, while its reconstituted expression was able to promote an increased growth rate and invasiveness, also restoring the levels of RhoA. Mechanistically, NRF2 was shown to bind the promoter region of ERR1 (nuclear member receptor estrogen-related receptor α), a protein that promotes RhoA ubiquitination, suggesting that the transcriptional repression of ERR1 might prevent RhoA degradation and, therefore, facilitate the activation of its downstream pro-oncogenic signaling [29]. A positive role for NRF2 in controlling the proliferation of breast cancer cells was also described in a study wherein an mir-101 mimic significantly reduced the mRNA levels of NRF2, impairing both the proliferation and colony formation rate of MCF-7 cells, while opposite changes were produced by mir-101 inhibition [30]. Along similar lines of evidence, De Blasio and colleagues reported that NRF2 enhanced both the proliferation and antioxidant capacity of triple-negative MDA-MB-231 breast cancer cells through the downmodulation of miR-29b-1-5p expression [31]. In marked contrast, however, other studies support the notion that decreased NRF2 signaling, rather than its overactivation, can promote cancer growth and cell proliferation. In this regard, Xu et al. demonstrated that the genetic silencing of NRF2 in MCF-7 breast cancer cells resistant to Adriamycin was paralleled by a marked upregulation of the CDCA4 protein (cell division cycle-associated protein 4) that was ultimately responsible of the enhanced growth and proliferation of these cells in vitro [32]. Similarly, by using MDA-MB-231 breast cancer cells, NRF2 was recently shown to induce the expression of ferroportin (FPN), a transmembrane protein that regulates the intracellular iron content by promoting iron efflux [33]. Here, the genetic silencing of FPN accelerated MDA-MB-231 cell proliferation and growth in vitro and in vivo, while its forced expression produced the opposite changes. Of note, a further analysis revealed that the NRF2 mRNA and protein levels were significantly lower in tumor specimens from breast cancer patients compared to the corresponding adjacent tissues and paralleled by similar changes in the FPN content. The authors concluded that defective NRF2 signaling can induce alterations in the iron metabolism that might ultimately promote breast cancer growth. Taken together, these studies suggest that NRF2 might exert divergent effects on breast cancer cell proliferation and growth, most likely depending on the specific context and genetic background of its activation or repression.

Despite that sustained NRF2 activation is considered a driver of the malignant progression of several types of tumors, its role in breast cancer (migration, invasion and metastasis) still remains a matter of intense debate due to conflicting data. In this regard, the initial studies identified NRF2 as a negative regulator of RON (Recepteur d' origine nantais) gene expression, a tyrosine kinase receptor frequently overactivated in breast tumors and associated with tamoxifen resistance, as well as metastatic disease [34]. Here, from the comparative immunohistochemical analysis of the tissue microarray samples, a high NRF2 and low RON expression was observed in normal tissues, while a low or absent NRF2 content and high RON levels were found in breast tumors. A mechanistic investigation conducted on a panel of breast cancer cells (and other types of tumors) revealed that NRF2 was able to directly bind to the RON promoter and repress its genetic induction, while, accordingly, the reconstituted expression of NRF2 not only decreased the RON levels but, also, impaired breast cancer cell migration and invasion. Along similar lines of evidence, other research conducted on 4T1 and JC murine breast cancer cells has shown that the flavonoid fisetin, a known NRF2 activator, can exert a tumor-suppressive function by decreasing the expression of matrix metalloproteinases (MMP-2 and MMP-9), two key regulators of tumor invasion and metastatic spreading [35]. Mechanistically, fisetin was found to enhance the NRF2 nuclear accumulation and heme oxygenase-1 (HO-1) expression, causing a significant decrease in the mRNA levels and the activity of MMP-2 and MMP-9, while these changes were antagonized by NRF2 silencing. Therefore, the authors proposed that the NRF2-dependent activation of HO-1 can attenuate the metastatic potential of breast cancer cells by inhibiting the MMP-2 and MMP-9 expression and enzymatic activity but, also, cell motility through the NRF2–HO-1 axis, in agreement with the evidence from other types of tumors [36]. Additionally, a very recent study investigated the role of NRF2 in the transcriptional regulation of the chemokine CXCL13 and its receptor CXCR5, two known drivers of breast cancer cells migration and metastatic dissemination, from primary breast tumors to lymph nodes [37]. Here, NRF2 was found to directly recognize multiple sites within the CXCL13 gene promoter and to negatively modulate its transcription when overexpressed in MDA-MB-231 cells, also suppressing the activating effects normally induced by the transcription factor RelA. The authors concluded that NRF2 is a negative regulator of CXCL13 and a potential tumor suppressor in breast cancer, as also suggested by the presence of higher NRF2 levels in ER (−) breast cancer cells and the increased frequency of CXCL13/CXCR5 co-expression in ER (+) breast cancer cells with lower NRF2 contents. In a marked contrast, however, a number of recent studies have indicated that NRF2 plays a pro-oncogenic role in breast cancer, promoting tumor invasion and metastasis spreading. For instance, Zhou et al. showed that the oncoprotein HBXIP (mammalian hepatitis B X-interacting protein) plays a critical role in modulating cancer malignancy and tumor progression and is involved in breast carcinoma progression [38]. More specifically, an immunohistochemistry assay, qPCR on the mRNA and immunofluorescence on the protein levels confirmed that HBXIP expression was positively correlated with NRF2 expression in clinical samples from breast cancer tissues, suggesting their implication in breast cancer development. The mechanistic insights on MCF-7 breast cancer cells showed that HBXIP reduced the ROS levels by promoting nuclear NRF2 accumulation and the subsequent transactivation of NRF2-dependent target genes such as NAD(P)H dehydrogenase-1 (NQO1), glutamate-cysteine ligase catalytic and modifier subunits (GCLC and GCLM) and AKR1C1. Of note, HBXIP silencing attenuated the expression of NRF2 and its nuclear accumulation markedly enhancing the intracellular ROS levels. Furthermore, by comparing the impact of HBXIP-induced NRF2 activation in MDA-MB-436 cells with the stable knockdown or reconstituted expression of HBXIP, the authors revealed that the proliferation, migration and invasion abilities of these cells were strongly impaired by the genetic depletion or site-specific mutation of HBXIP, while, conversely, were rescued by the re-expression of the HBXIP-wt form. These observations were further confirmed in vivo, since HBXIP knockdown led to reduced nodule formation in mouse lungs and impaired tumor growth in metastatic and orthotopic xenograft models. Lastly, both these phenotypic changes were reverted by HBXIP-wt reconstitution. In summary, the authors concluded that HBXIP promotes the malignancy of breast cancer by modulating abnormal redox regulation in vitro and in vivo through the overactivation of NRF2 signaling.

2.3.2. NRF2 in the Regulation of Breast Cancer Cell Stemness and Therapy Resistance

In another work, the group of Mi-Kyoung investigated the role of NRF2 in the modulation of the cancer stem cell (CSC) phenotype, establishing a breast CSC-like model by isolating the subpopulations of MCF-7 (doxorubicin sensitive), MCF-7/ADR (doxorubicin-resistant) and MDA-MB-231 cells with high expression levels from cluster of differentiation 44 (CD44 high), a common CSC marker associated with drug resistance, tumor recurrence and metastasis formation [39]. Notably, a detailed analysis revealed that the expression of CD44, along with typical CSC markers such as SOX2, OCT-4 and MDR-1, was significantly higher in MCF7/ADR than in MCF7 cells and that NRF2 signaling was also more active in CD44 high cells compared to the CD44 low counterparts. Mechanistic studies on the stable CD44 high cell line (ADR44P) revealed that the increased NRF2 content was caused by CD44-p62 signaling, while the genetic silencing of NRF2 was able to markedly suppress the aggressive phenotype of CSC, including drug resistance, colony/sphere formation and cell migration in vitro, but, also, tumor growth in vivo [40]. Therefore, this indicates that the CD44–NRF2 axis might be an effective therapeutic target to impair the stress resistance and survival of the CD44 high CSC population in breast cancer, in agreement with the previous findings.

Regarding the clinical impact of NRF2, a consistent body of evidence supports the notion that its activation is a major determinant of therapy resistance in breast cancer cells. For instance, Wei et al. showed that both NRF2 and p62 were overexpressed in breast cancer samples compared to normal tissues but, also, in MCF-7/ADR breast cancer cells compared to MCF-7 cells [41]. Interestingly, the genetic silencing of either NRF2 or p62 revealed the existence of a reciprocal regulation between these two proteins and led to the impaired cell proliferation and increased sensitivity of MCF-7/ADR but not MCF-7 cells to doxorubicin. Of note, these data were also confirmed in vivo, since the administration of the anticancer drug, PA-MSHA (*Pseudomonas aeruginosa* mannose-sensitive hemagglutinin) to MCF-7/ADR-xenografted mice was able to significantly impair the tumor growth by downregulating the NRF2 and p62 levels.

A study from Del Vecchio et al. tried to elucidate the role of NRF2 in the acquisition of multidrug resistance (MDR) caused by cellular dedifferentiation by comparing normally differentiated or dedifferentiated isogenic human breast epithelial cells by the induction of epithelial-to-mesenchymal transition (EMT) due to the expression of the transcription factor TWIST [42]. Here, in differentiated cells, NRF2 activation was mediated by its oxidation, while in dedifferentiated cells, it was caused by PERK-dependent phosphorylation. These findings were further substantiated in therapy-resistant basal breast cancer cells and animal models, wherein the inhibition of the PERK–NRF2 axis reverted the MDR phenotype and sensitized the drug-resistant cancer cells to chemotherapy, decreasing their intracellular GSH content. In addition, the analysis of the patient tumor datasets revealed that the PERK gene expression signature positively correlated with a basal breast cancer gene signature, tumor grade and chemotherapy resistance while it was negatively correlated with the differentiation status and the overall survival of the patients. Based on these data, the authors proposed that dedifferentiated breast cancer cells upregulate MDR-related genes through PERK–NRF2 signaling, while targeting this pathway might increase the sensitivity of poorly differentiated tumors refractory to anticancer drugs treatment.

Another experimental work elucidated the functional interrelation between NRF2 and heat shock factor-1 (HSF1), a protein that is frequently overexpressed and implicated in the survival and proliferation of cancer cells, also correlated with a poor prognosis of cancer patients [43]. Here, by using different cancer cell lines, including MCF-7, the authors showed that NRF2 could interact with two distinct AREs sites within the HSF1 promoter, inducing a marked increase in its mRNA and protein levels in response to oxidative or proteotoxic stress. Importantly, the NRF2–HSF1 axis was found to play an important role in MCF-7 malignancy, since interfering with NRF2-mediated HSF1 activation also suppressed the survival, migration and the expression of E-cadherin and N-cadherin, two markers of EMT in MCF7 breast cancer cells. These results demonstrated that NRF2 can transcriptionally regulate HSF1 and that this event plays an important role in the progression of breast cancers, influencing the growth, migration and survival of malignant cells. The importance of NRF2 in chemoresistance of breast cancer cells was further substantiated in a recent study wherein NRF2 activation in response to hypoxia-induced ROS accumulation conferred an insensitivity to cisplatin in MCF-7 cells [44]. Here, the enhanced NRF2 nuclear accumulation was found to induce the expression of antioxidant enzymes, GCLC and GCLM, leading to increased GSH biosynthesis, under hypoxic conditions. Importantly, the genetic or pharmacologic inhibition of NRF2 activation restored the sensitivity to cisplatin in vitro, while the concomitant use of the NRF2 inhibitor trigonelline was able to potentiate the efficacy of cisplatin in a xenograft mouse model. Finally, Carlisi et al. focused on the role of NRF2 in the development of chemoresistance using a model of triple-negative breast cancer [45]. Here, prolonged incubation of MDA-MB-231 cells with sublethal doses of doxorubicin or mitoxantrone led to an increased expression of NRF2 and the acquisition of a resistant phenotype. Of note, the use of parthenolide, a sesquiterpene lactone known for its anti-inflammatory and anticancer effects, was able to partially restore the sensitivity to both doxorubicin and mitoxantrone by preventing the overexpression of NRF2 and its target proteins and promoting intracellular ROS accumulation. Taken together, these data indicate that, in most cases, NRF2 mediates a therapy resistance in breast cancer cells in response to the altered redox balance caused by anticancer drug administration, but this does not exclude the possibility that other types of mechanisms, ROS-independent, might also account for enhanced NRF2 signaling under different biological contexts.

2.3.3. NRF2 in Metabolic Adaptation of Breast Cancer Cells

It is well-recognized that metabolic reprogramming is a common hallmark of cancer cells and that NRF2 participates in the regulation of metabolic pathways supporting tumor progression [46]. In a recent study, it was demonstrated that the overexpression of NRF2 can promote the proliferation and migration of breast cancer cells by upregulating the expression of glucose-6-phosphate dehydrogenase (G6PD), a key enzyme in the pentose phosphate pathway (PPP) [47]. More in detail, by using MCF-7 and MDA-MB-231 breast cancer cells, the authors showed that the inhibition of NRF2 and overexpression of Kelch-like ECH-associated protein-1 (KEAP1) reduced the expression of G6PD, while NRF2 overexpression or KEAP1 knockdown had the opposite effect. Additionally, the dissection of the molecular mechanism revealed that NRF2 promoted the expression of Notch1 through the activation of the G6PD/HIF-1α (hypoxia-inducing factor 1α) pathway in both MCF-7 and MDA-MB-231 cells. Importantly, the NRF2 genetic depletion was significantly abrogated, while its overexpression or KEAP1 knockdown markedly enhanced breast cancer cell proliferation, migration and invasion [47]. Therefore, the authors concluded that NRF2 plays an essential function in the regulation of breast cancer malignancy by influencing the Notch1 signaling pathway through the upregulation of G6PD, a rate-limiting enzyme of the PPP. Consistently, other experimental works have provided evidence supporting the notion that the NRF2-dependent modulation of HIF-1α downstream signaling might represent a conserved mechanism through which breast cancer cells undergo metabolic adaptation and reprogramming [48,49].

A study from Walker et al. showed another pathway for the metabolic adaptation of breast cancer cells [50]. According to their experimental work, a glucose deprivation in cancer cells led to the induction of autophagy, which supports cellular survival. This metabolic adaptation under the metabolic stress condition also decreased the p62 levels and eventually upregulated the NRF2 levels. The increased NRF2 levels in cancer cells also maintained the regular ROS levels. This data supports the role of NRF2 in the metabolic adaptation of cancer cells to a nutrient-deprived environment.

In another study, the analysis of human breast cancer datasets revealed that the expression of caveolin-1 (CAV-1) was inversely correlated to that of NRF2 or superoxide dismutase-2 (SOD-2), and this could predict the development of more aggressive forms of cancer [51–53]. Interestingly, the reconstitution of CAV-1 expression in MCF-7 breast cancer cells that are normally defective for this protein was sufficient to suppress the NRF2 activity by promoting its interaction with KEAP1 and, consequently, its faster degradation, an event ultimately leading to a decreased SOD-2 expression. Of note, these changes were sufficient to induce a metabolic switch characterized by impaired glycolytic activity and increased mitochondria-dependent ATP production. The mechanistic insights revealed that CAV-1 loss induces SOD-2 upregulation and H_2O_2 accumulation, which, in turn, promotes AMPK activation and, ultimately, enhances the glycolytic rate of the MCF-7 cells. Consistently, the rescued expression of CAV-1 led to a marked inhibition of anchorage-independent growth and suppressed the AMPK-dependent activation of the glycolytic switch through the inhibition of H_2O_2 production derived from the SOD-2-dependent conversion of the superoxide anion. Furthermore, by using a mouse model of genetically induced breast cancer, the reduced expression of CAV-1, associated with elevated SOD-2 and enhanced AMPK activation, was also confirmed in tissue sections from mammary tumors but not in their healthy counterparts. Therefore, the authors concluded that the progressive loss of CAV-1 during breast cancer progression induces the activation of NRF2 and the subsequent upregulation of SOD-2, promoting an AMPK-dependent glycolytic switch that is permissive to the acquisition of a highly aggressive phenotype that negatively impacts the overall survival and prognosis of patients [54].

2.3.4. NRF2 in Breast Cancer Prognosis

Accumulating evidence indicates that NRF2 is frequently overexpressed in different types of malignant tumors and associated with a poor prognosis. However, the pathological and clinical significance of NRF2 in breast cancers has revealed contrasting results. For example, an early study showed that NRF2 was frequently depleted in breast cancer biopsies and breast cancer cell lines due to its augmented proteasomal degradation caused by the concomitant overexpression of the E3 ubiquitin ligase CUL3 [55]. Here, the use of siRNA against CUL3 in MCF-7 breast cancer cells increased the levels of NRF2-regulated proteins, including GCL, NQO1, AKR1C1, UGDH, TXN and the drug transporter ABCC1, ultimately conferring a resistance to the oxidants and conventional anticancer agents. Despite the limited number of samples analyzed, and the lack of clear correlations between NRF2 and tumor grading or the survival of patients, the authors proposed that the molecular signature characterized by a high CUL3 and low NRF2 content might identify a cohort of patients more sensitive to chemical carcinogenesis but, also, to the use of chemotherapeutic drugs.

The first study demonstrating an association between nuclear NRF2 immunoreactivity and the adverse clinical outcome of breast cancer patients was conducted by Onodera et al. [56]. Here, by analyzing around 100 specimens of invasive breast carcinoma, nuclear NRF2 immunoreactivity was detected in 44% of the carcinoma cases, while the NRF2 status was significantly associated with the NQO1 and p62 protein contents, the Ki-67 index and the histological grade. Of note, the multivariate analysis revealed that the NRF2 status was an independent adverse prognostic factor for both the recurrence and disease-free survival of the patients, while in vitro studies confirmed that NRF2 was able to control the proliferation and migration of the MCF-7 and SK-BR-3 cells. These results indicate that the nuclear NRF2 status plays an important role in controlling the progression of breast cancer and might be considered a robust prognostic marker in breast cancer patients. In another study, Hartikainen et al. investigated the significance of NRF2 expression and its target gene sulfiredoxin (SRXN1) by using tissue microarrays representative of invasive breast carcinomas [57]. Here, nine single-nucleotide polymorphisms of the NRF2 gene were analyzed in 452 patients with breast cancer and 370 controls while the subsequent protein expression analysis revealed high cytoplasmic NRF2 positivity in 66% (237 of 361) and nuclear positivity in 26% (96 of 365) of the cases. More in detail, the authors showed that the NRF2 polymorphism rs6721961 was associated with breast cancer risk, while the NRF2 rs2886162 AA genotype variant independently predicted a poorer survival among patients who received chemotherapy or radiotherapy. These data indicated that the NRF2 pathway can influence both the predisposition toward developing breast cancer and the overall clinical outcome of breast cancer patients, confirming its importance in cancer progression. In a later study from the same group, it was consistently demonstrated that genetic polymorphisms in KEAP1 also affect the breast cancer risk and clinical outcome, modifying the effects of radiotherapy and tamoxifen treatments on patient survival [58]. Interestingly, among the five genetic variants, the KEAP1 rs11085735 minor allele A was significantly associated with a lower KEAP1 protein expression and high NRF2 nuclear expression. Additionally, when the treatment data were included, a multivariate survival analysis revealed that this SNP was associated with a poorer relapse-free survival and breast cancer-specific survival among all the invasive cases and with a shorter relapse-free survival among the tamoxifen-treated cases. In conclusion, the authors confirmed that the genetic variants in the KEAP1 gene were associated with the outcomes of patients with breast cancer and that these SNPs may cause defects in the antioxidant defense mechanisms, underscoring the importance of the NRF2/KEAP1 signaling pathway.

An indirect confirmation that high levels of NRF2 adversely impact the prognosis of breast cancer patients comes from a quite recent study wherein the authors analyzed the expression data and clinical data from The Cancer Genome Atlas (TCGA) [59]. Here, by using univariate and multivariate survival analyses, it was shown that the median survival time of patients with a low SETD7 expression (18.1 years) was twice as much that of patients expressing high levels of SETD7 (9.5 years), a SET domain-containing lysine methyltransferase 7 that monomethylates histone and nonhistone proteins. Additionally, the SETD7 expression was found to positively correlate with the expression of NRF2 and a bunch of its target genes—namely, ME1, TXNRD1, GCLC and GCLM. The mechanistic insights revealed that the stable knockdown of SETD7 significantly impaired the cell proliferation and viability in MCF-7 and MDA-MB-231 breast cancer cells, also leading to an increased intracellular ROS content and a decreased GSH/GSSG ratio due to the repression of NRF2-dependent antioxidant genes expression. Therefore, the authors concluded that SETD7 is a prognostic marker in breast cancer patients and an upstream transcriptional regulator of antioxidant proteins in breast cancer cells dependent on the KEAP1-NRF2 pathway.

In another study from Lu et al., it was shown that dipeptidyl-peptidase 3 (DPP3), a KEAP1-binding protein that promotes NRF2 accumulation by competitively binding and sequestering KEAP1, was able to influence NRF2 signaling in breast cancer cells [60]. Here, MCF-7 ER (+) breast cancer cells were used to demonstrate that the DPP3 interaction with KEAP1 was dose-dependently reinforced by increasing the amount of oxidative stressors. Mechanistically, DPP3 promoted NRF2 nuclear accumulation and activity through a competitive binding with KEAP1 independently from its enzymatic activity, while its overexpression enhanced the KEAP1 levels and its genetic knockdown prevented H_2O_2-dependent NRF2 nuclear accumulation. Finally, by analyzing the data from the TCGA database, DPP3 was found to be overexpressed in human breast cancer and to correlate with increased NRF2 target gene expression and poor prognosis, especially in ER (+) breast cancer. Based on these data, the authors proposed that DPP3 overexpression promotes breast cancer progression, metastasis and drug resistance, stabilizing NRF2, while the signature characterized by high levels of NRF2 and DPP3 might represent a potential biomarker for breast cancer prognosis and treatment.

In a marked contrast, other evidence has shown that the NRF2 mRNA levels analyzed in two independent breast cancer patient cohorts were inversely correlated with the clinical outcome of the disease [61]. Indeed, not only the NRF2 mRNA levels were higher in normal breast tissue than in breast tumor tissue of the same patient, but patients with high NRF2 mRNA levels had a better disease-specific survival and overall survival compared to those with low NRF2 mRNA contents. Interestingly, this prognostically relevant association was even more pronounced in the subgroup of patients with ER (+), but not in patients with ER (−), tumors. These data support the notion that NRF2 can also act as a tumor suppressor in breast cancer, underlying the complexity of NRF2/KEAP1 signaling.

In summary, the available data indicate that NRF2 can have both pro- and antioncogenic effects, most likely depending on additional factors that are context-specific. Among the others, the genetic background, the subcellular location, the presence of genetic polymorphisms, the interactions with upstream and downstream regulators and the KEAP1 status represent the major determinants of whether NRF2 will exert tumor-promoting or tumor-suppressive functions. These findings underscore the intricate complexity of NRF2 signaling and pave the way to further investigations aimed at elucidating the clinical impact of NRF2 in breast cancer.

3. NRF2-Related Mechanisms as a Target in Breast Cancer

NRF2 activation shows important activities against oxidative stress and xenobiotic detoxification, which assure NRF2's preventive role against cancer. However, recent studies have shown that excessive NRF2 levels in cancer can cause cell metabolism reprogramming, resulting in chemo- and radiotherapy resistance. NRF2 targeting in cancer therapy mainly relies on two different mechanisms: NRF2 activation to prevent cancer development and NRF2 inhibition to improve cancer therapy sensitivity [62]. In this section, the possible roles of natural molecules, polyphenols and melatonin in breast cancer prevention and therapy via NRF2-related mechanisms are reviewed.

3.1. Therapeutical Role of Polyphenols

Curcumin, which is a polyphenolic metabolite of *Curcuma* spp., induces NRF2; facilitates the upregulation of antioxidative enzymes such as NQO1, HO-1, GST and glutathione reductase (GR) and enables cellular senescence [63]. The mechanism under NRF2 activation by curcumin relies on the modulation of KEAP1 thiol [64]. A previous study from Rushworth et al. provided evidence of another mechanism, since curcumin mediates NRF2 phosphorylation via stimulating PKC and enhances the ARE-mediated expression of HO-1 and GCLM in human monocytes [65]. In a study that set out to determine the NRF2-mediated HO-1 activation by polyphenols, curcumin was shown to induce the NRF2 and HO-1 protein levels in MDA-MB-468 and HBL100 breast cancer cells [66]. The researchers also investigated wild type (NRF2 +/+) and null (NRF2-/-) mouse embryo fibroblasts and revealed that curcumin induces HO-1 through NRF2 activation. In another study, the oxidative DNA damage induced by benzo-a-pyrene (BaP), a well-known human carcinogen, was shown to be prevented by a curcumin treatment in Swiss albino rats [67]. A curcumin treatment also decreased the CYP1A catalyzed bioactivation of BaP and induced the NRF2, GST and NQO1 protein levels. In contrast with this study, Jain et al. showed that curcumin downregulated the 2-amino-1-methyl-6-phenylimidazo(4,5-b) pyridine (PhIP; the heterocyclic amin is possibly carcinogenic to humans)-induced NQO1 expression while inhibiting the DNA adduct and ROS formation in MCF-10A cells. The authors concluded that the increased NRF2 activation ensures cells reduce the oxidative stress and DNA adduct formation [68]. However, the authors did not explain that the NQO1 expression is regulated via NRF2, which has been shown to be reduced by curcumin. NRF2 regulates not only ARE but, also, Flap endonuclease I (FENI), which has a role in DNA repair, as well as cell proliferation. Curcumin inhibits MCF-7 cell proliferation while inducing the NRF2 levels in a dose-dependent manner [69]. Here, the NRF2 protein downregulated the FENI expression. Since tumor aggressiveness was found related to the increased FENI activity, this study revealed that NRF2 activation by curcumin may also be effective in cancer therapy. Curcumin's inhibitory effect on breast cancer cell proliferation also mediated with the inhibition of oncogenic miR-19. MCF-7 cell transfection with miR-19a and -19b represented the decreased PTEN, AKT and p53 pathway-associated protein levels [70]. Moreover, curcumin inhibited bisphenol A-induced cell proliferation; downregulated miR-19a and

-19b and modulated the PTEN, AKT and p53 expression. Taken together, both in vitro and in vivo studies have provided important insights into curcumin's preventive role in chemically induced DNA damage and oxidative stress. Additionally, curcumin-mediated NRF2 activation may have an impact on cancer therapy via its downstream gene-related tumor growth inhibition.

Epigallocatechin-3-gallate (EGCG) is most abundantly found in green tea. This polyphenol induces NRF2 and downstream genes—mainly, phase II detoxification enzymes, similarly to curcumin [71]. NRF2 induction by EGCG was investigated with the Western blot analysis in MCF-7 and MDA-MB-231 breast cancer cells. According to the study, EGCG increased the NRF2 levels, and consequently, both cell lines became resistant to the growth inhibitory effects of doxorubicin or paclitaxel [72]. In contrast, the EGCG cotreatment reduced the apoptotic response of cisplatin in MDA-MB-231 cells [71]. The authors also studied a triple-negative breast cancer model in vivo with 4T1 murine breast cancer cells, showing that a cotreatment of cisplatin with EGCG suppressed tumor growth more potently than cisplatin alone. Increased oxidative stress in the cancer cells resulted in enhanced tumor growth and chemoresistance [73]. NRF2 induction by polyphenols such as curcumin and/or EGCG can cope with oxidative stress and improve the sensitivity to chemotherapy. However, these controversial findings can be explained by the fact that polyphenols also induce the NRF2 downstream phase II detoxification pathway, which metabolizes chemotherapeutics easily and causes chemoresistance.

The grapefruit polyphenol resveratrol is well-known for its biphasic effects [74]. At low concentrations, the resveratrol treatment induces cell proliferation in breast cancer cells, while its increased concentration causes cytotoxicity. Similarly, the antioxidant features of resveratrol have been seen in low concentrations, while it represents the prooxidant profile at higher dosages. A study from Rai et al. treated MCF-7 and MDA-MB-231 cells with resveratrol between the 50–400-μM concentration range, which had strong cytotoxicity in a dose-dependent manner [75]. Furthermore, in MCF-7 cells, a resveratrol combination with doxorubicin acted synergistically to inhibit cell proliferation, colony formation, cell migration and to promote apoptosis while, at the protein level, downregulate the regulatory genes of inflammation (NF-kB and COX-2), oxidative stress (NRF2) and autophagy (LC3B and Beclin-1) and increase the proapoptotic protein ratios (BAX/BCL2). Resveratrol alone increases NRF2 at the protein level. The concentration of doxorubicin was decreased to obtain a similar therapeutic efficacy with resveratrol. Resveratrol also showed a promising suppressing effect on estrogen-induced breast cancer [76]. The resveratrol treatment for DMBA-induced ER (+) breast cancer in Sprague–Dawley rats resulted with a decreased tumor volume. The study revealed that the increased expression of NRF2 with a resveratrol-treated group promoted NQO1, HO1 and UGT1A8 expression. The estradiol downregulation of NRF2 and its downstream antioxidant (NQO1 and SOD3) genes, detoxification (FMO1 and AOX1) genes and DNA repair (OGG1) gene were reinduced by resveratrol in August Copenhagen Irish rats. Similarly, an estradiol-induced mammary tumor incidence was decreased with resveratrol. In the same study, estradiol-induced DNA damage was evaluated in MCF-10A cells with both estradiol and estradiol plus resveratrol-treated groups. These studies have revealed that resveratrol induces apoptosis, decreases oxidative stress-related DNA adduct production and inhibits colony and mammosphere formation in estradiol-treated MCF-10A cells. The authors concluded that resveratrol exerted a preventive role against estradiol-induced breast tumor development via the NRF2-mediated pathway. The study was consistent with resveratrol's protective role against BaP-induced oxidative stress in BRCA1-defective MCF-10A cells [77]. Overall, these studies confirmed that resveratrol mediated the induction of NRF2, resulting in a chemoprevention against xenobiotic-induced cancer development in contrast to the inhibition of NRF2, which resulted in chemosensitivity.

A limited mechanistical study was conducted to evaluate the inhibition of NRF2 by polyphenols. Luteolin-loaded nanoparticles with increased luteolin bioavailability were shown to decrease the NRF2, HO1 and MDR1 expression at the mRNA level in

MDA-MB-231 cells [78]. Additionally, luteolin nanoparticles enhanced the doxorubicin sensitivity of MDA-MB-231 cells. A study from Zhong et al. showed that wogonin treatment remodulates increased the NRF2, HO-1 and NQO1 protein levels in doxorubicin-resistant MCF-7 cells [79]. These results suggest that the NRF2 modulation by polyphenols differs between the molecules, while fisetin, curcumin, EGCG and resveratrol induce NRF2, luteolin and wogonin and inhibit NRF2 expression (Figure 2).

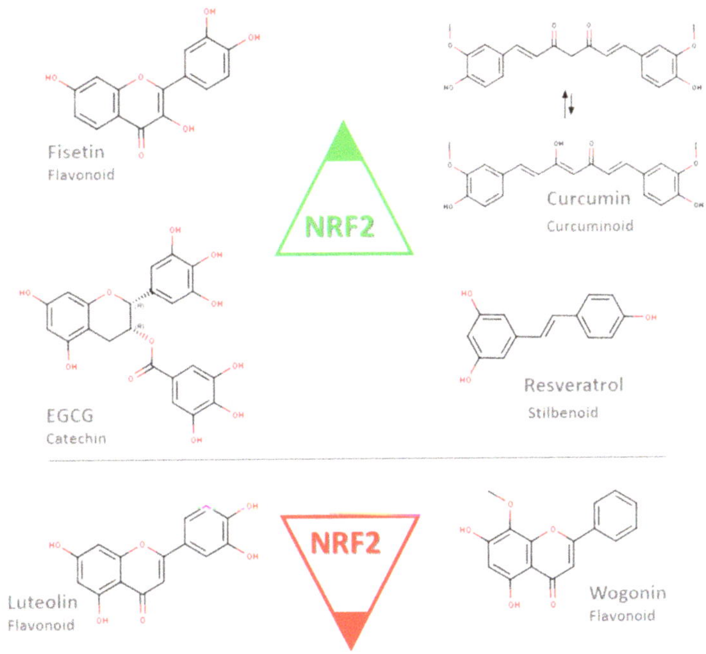

Figure 2. Polyphenols of various subgroups modulating the expression of NRF2 at the mRNA or/and protein level. Fisetin, epigallocatechin-3-gallate (EGCG), curcumin and resveratrol induce NRF2 in contrast to luteolin and wogonin (Marvin was used for drawing chemical structures, Marvin 20.20.0 (2020), ChemAxon [80]).

3.2. Therapeutical Role of Melatonin

Melatonin is an indolic pineal hormone. The impaired circadian synthesis and secretion of melatonin is shown to be an important risk factor for the development and progression of breast cancer [81,82]. A plethora of research was conducted to evaluate the role of melatonin in the prevention and treatment of breast cancer, including the modulation of oxidative stress [8] and the regulation of miRNAs that are associated with apoptotic, cellular senescence and proliferation genes [83]. Melatonin also plays a crucial role in sustaining the mitochondrial functions, which was reviewed by Almeida Chuffa et al. [84]. The authors underlined that melatonin favors the mitochondria by protecting the mitochondrial proteins and mtDNA against oxidative damage and promoting electron chain activity against mitochondrial dysfunction in various cancers, including breast cancer.

Melatonin activates NRF2 via the upregulation of cellular mediators such as PKC [85], SIRT1 [86] and PI3K/AKT [87], according to the oxidative stress-mediated cellular responses. Furthermore, Janjetovic et al. showed that the melatonin-mediated activation of NRF2 protects melanocytes against ultraviolet light B-induced oxidative stress and DNA adduct formation [88]. Increased protein levels of NRF2, NQO1 and HO-1 were also found in a melatonin-treated colon cancer model [89]. The authors showed that melatonin reverses the downregulation of inflammation (NF-kB, STAT3 and COX2) and oxidative stress

(NRF2, NQO1 and HO-1) regulatory proteins by 1, 2-dimethylhydrazine dihydrochloride, a colon cancer inducer, consequently leading to reduced inflammatory (MPO, IL-17, IL-6 and TNF-a) and oxidative stress (TBARS and GSH) responses while protecting against DNA damage. Another study confirmed the NRF2 activation by melatonin, which is mediated by melatonin receptors and resulted with increased HO-1, NQO1 and GCLC levels [90]. Contrarily, melatonin is reported to inhibit the serum and glucocorticoid-induced kinase 1 (SGK1)-mediated NRF2 upregulation in cervical cancer cells. As a result, the increased oxidative stress caused cellular damage and inhibited the tumor growth in the cervical cancer model [91]. According to these current findings, melatonin seems to controversially modulate NRF2 in tumorigenic and nontumorigenic cells, which is possibly related to the different mitochondrial functions and ROS levels in the two cell types (Figure 3).

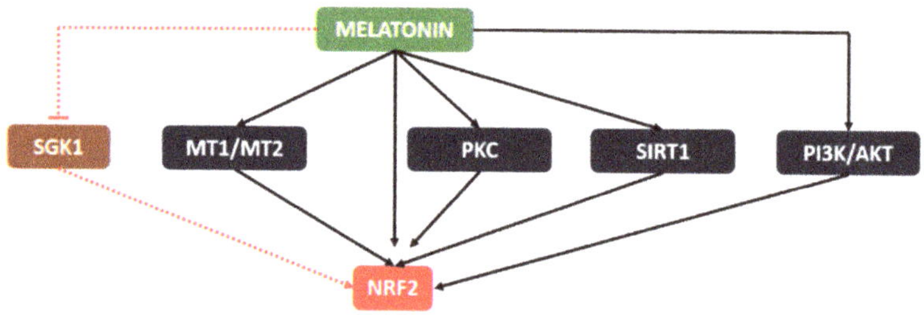

Figure 3. Melatonin stimulates NRF2 via the melatonin receptors (MT1 and MT2), SIRT1 and PI3K/AKT pathways in nontumorigenic cells while inhibiting the SGK1-mediated upregulation of NRF2 in tumorigenic cells.

A study from Paroni et al. evaluated melatonin's NRF2-mediated inhibition of prostate cancer in vivo. Here, melatonin induced NRF2, HIF1a and the VEGF protein while reducing the microvessel occurrence and CD31, Ki67-positive tumor cell number and inhibited tumor growth [92]. Melatonin was reviewed extensively as a "full-service anticancer agent against breast cancer" in a recent paper [93]. However, the NRF2-mediated actions of melatonin in breast cancer development and therapy have not been investigated in detail yet.

4. Conclusions

Oxidative stress-related mechanisms are known to be involved in the initiation, promotion and progression stages of breast cancer. NRF2 as an antioxidant response transcription factor might have a role in cancer prevention and cancer treatment. In the present review, both polyphenols and melatonin, well-known "oxidative stress modulators", are reported to have therapeutic potential in breast cancer via the activation of NRF2. Future mechanistic studies will shed light on the NRF2-related potential of melatonin in inhibiting breast cancer initiation, promotion and/or progression.

Funding: This research received no external funding.

Acknowledgments: The authors are thankful to Ege University for supporting the doctoral research of Alev Tascioglu Aliyev with grant #18ECZ008.

Conflicts of Interest: The authors declare no conflict of interest.

References

1. WHO. GLOBOCAN 2018 Database. Issued by World Health Organization (WHO). Available online: http://gco.iarc.fr/today (accessed on 26 February 2021).
2. Sims, A.H.; Howell, A.; Howell, S.J.; Clarke, R.B. Origins of breast cancer subtypes and therapeutic implications. *Nat. Clin. Pr. Oncol.* **2007**, *4*, 516–525. [CrossRef] [PubMed]
3. Planes-Laine, G.; Rochigneux, P.; Bertucci, F.; Chrétien, A.-S.; Viens, P.; Sabatier, R.; Gonçalves, A. PD-1/PD-L1 Targeting in Breast Cancer: The First Clinical Evidences Are Emerging. A Literature Review. *Cancers* **2019**, *11*, 1033. [CrossRef]

4. Hanahan, D.; Weinberg, R.A. Hallmarks of Cancer: The Next Generation. *Cell* **2011**, *144*, 646–674. [CrossRef]
5. Hanahan, D.; Robert, A.W. Biological Hallmarks of cancer. *Holl. Frei Cancer Med.* **2017**, *1*, 1–10. [CrossRef]
6. Halliwell, B.; Gutteridge, J.M.C. *Free Radicals in Biology and Medicine*; Oxford University Press: Oxford, UK, 2015.
7. Klaunig, J.E. Chemical Carcinogenesis. In *Casarette and Doull's Toxicology*; Klaassen, C.D., Ed.; Mc Graw Hill Educations: New York, NY, USA, 2013; pp. 393–444.
8. Gurer-Orhan, H.; Ince, E.; Konyar, D.; Saso, L.; Suzen, S. The Role of Oxidative Stress Modulators in Breast Cancer. *Curr. Med. Chem.* **2018**, *25*, 4084–4101. [CrossRef]
9. Thorlacius, S.; Struewing, J.P.; Hartage, P.; Olafsdottir, G.H.; Sigvaldason, H.; Tryggvadottir, L.; Wacholder, S.; Tulinius, H.; Eyfjörd, J.E. Population-based study of risk of breast cancer in carriers of BRCA2 mutation. *Lancet* **1998**, *352*, 1337–1339. [CrossRef]
10. Kuchenbaecker, K.B.; Hopper, J.L.; Barnes, D.R.; Phillips, K.-A.; Mooij, T.M.; Roos-Blom, M.-J.; Jervis, S.; van Leeuwen, F.E.; Milne, R.L.; Andrieu, N.; et al. Risks of Breast, Ovarian, and Contralateral Breast Cancer for BRCA1 and BRCA2 Mutation Carriers. *JAMA* **2017**, *317*, 2402–2416. [CrossRef] [PubMed]
11. Kleibl, Z.; Kristensen, V.N. Women at high risk of breast cancer: Molecular characteristics, clinical presentation and management. *Breast* **2016**, *28*, 136–144. [CrossRef]
12. Yager, J.D.; Davidson, N.E. Estrogen Carcinogenesis in Breast Cancer. *N. Engl. J. Med.* **2006**, *354*, 270–282. [CrossRef] [PubMed]
13. Hilton, H.N.; Clarke, C.L.; Graham, J.D. Estrogen and progesterone signalling in the normal breast and its implications for cancer development. *Mol. Cell. Endocrinol.* **2018**, *466*, 2–14. [CrossRef]
14. Barrios-Rodríguez, R.; Toledo, E.; Martinez-Gonzalez, M.A.; Aguilera-Buenosvinos, I.; Romanos-Nanclares, A.; Jiménez-Moleón, J.J. Adherence to the 2018 World Cancer Research Fund/American Institute for Cancer Research Recommendations and Breast Cancer in the SUN Project. *Nutrients* **2020**, *12*, 2076. [CrossRef]
15. Freudenheim, J.L. Alcohol's Effects on Breast Cancer in Women. *Alcohol Res. Curr. Rev.* **2020**, *40*, 11. [CrossRef]
16. Rock, C.L.; Thomson, C.; Gansler, T.; Gapstur, S.M.; McCullough, M.L.; Patel, A.V.; Ba, K.S.A.; Bandera, E.V.; Spees, C.K.; Robien, K.; et al. American Cancer Society guideline for diet and physical activity for cancer prevention. *CA Cancer J. Clin.* **2020**, *70*, 245–271. [CrossRef]
17. Jardé, T.; Caldefie-Chézet, F.; Damez, M.; Mishellany, F.; Perrone, D.; Penault-Llorca, F.; Guillot, J.; Vasson, M.P. Adi-ponectin and leptin expression in primary ductal breast cancer and in adjacent healthy epithelial and myoepithelial tissue. *Histopathology* **2008**, *53*, 484–487. [CrossRef] [PubMed]
18. Sabol, R.A.; Beighley, A.; Giacomelli, P.; Wise, R.M.; Harrison, M.A.A.; O'Donnnell, B.A.; Sullivan, B.N.; Lampenfeld, J.D.; Matossian, M.D.; Bratton, M.R.; et al. Obesity-Altered Adipose Stem Cells Promote ER^+ Breast Cancer Metastasis through Estrogen Independent Pathways. *Int. J. Mol. Sci.* **2019**, *20*, 1419. [CrossRef]
19. La Barge, M.A.; Mora-Blanco, E.L.; Samson, S.; Miyano, M. Breast Cancer beyond the Age of Mutation. *Gerontology* **2016**, *62*, 434–442. [CrossRef]
20. Sies, H. Oxidative Stress: Introductory Remarks. In *Oxidative Stress*; Academic Press: Cambridge, MA, USA, 1985; pp. 1–5.
21. Jones, D.P. Redefining Oxidative Stress. *Antioxid. Redox Signal.* **2006**, *8*, 1865–1879. [CrossRef] [PubMed]
22. Valavanidis, A.; Vlachogianni, T.; Fiotakis, C. 8-hydroxy-2′ -deoxyguanosine (8-OHdG): A Critical Biomarker of Oxidative Stress and Carcinogenesis. *J. Environ. Sci. Heal. Part C* **2009**, *27*, 120–139. [CrossRef]
23. Barnes, J.L.; Zubair, M.; John, K.; Poirier, M.C.; Martin, F.L. Carcinogens and DNA damage. *Biochem. Soc. Trans.* **2018**, *46*, 1213–1224. [CrossRef] [PubMed]
24. Eldin, E.E.M.N.; El-Readi, M.Z.; Eldein, M.M.N.; Alfalki, A.A.; Althubiti, M.A.; Kamel, H.F.M.; Eid, S.Y.; Al-Amodi, H.S.; Mirza, A.A. 8-Hydroxy-2′-deoxyguanosine as a Discriminatory Biomarker for Early Detection of Breast Cancer. *Clin. Breast Cancer* **2019**, *19*, e385–e393. [CrossRef] [PubMed]
25. Sova, H.; Jukkolavuorinen, A.; Puistola, U.; Kauppila, S.; Karihtala, P. 8-Hydroxydeoxyguanosine: A new potential independent prognostic factor in breast cancer. *Br. J. Cancer* **2010**, *102*, 1018–1023. [CrossRef]
26. Chuffa, L.G.D.A.; Lupi-Júnior, L.A.; Costa, A.B.; Amorim, J.P.D.A.; Seiva, F.R.F. The role of sex hormones and steroid receptors on female reproductive cancers. *Steroids* **2017**, *118*, 93–108. [CrossRef] [PubMed]
27. Cavalieri, E.; Frenkel, K.; Liehr, J.G.; Rogan, E.; Roy, D. Chapter 4: Estrogens as Endogenous Genotoxic Agents—DNA Adducts and Mutations. *J. Natl. Cancer Inst. Monogr.* **2000**, *2000*, 75–94. [CrossRef] [PubMed]
28. Almeida, M.; Soares, M.; Ramalhinho, A.C.; Moutinho, J.F.; Breitenfeld, L.; Pereira, L. The prognostic value of NRF2 in breast cancer patients: A systematic review with meta-analysis. *Breast Cancer Res. Treat.* **2019**, *179*, 523–532. [CrossRef]
29. Zhang, C.; Wang, H.-J.; Bao, Q.-C.; Wang, L.; Guo, T.-K.; Chen, W.-L.; Xu, L.-L.; Zhou, H.-S.; Bian, J.-L.; Yang, Y.-R.; et al. NRF2 promotes breast cancer cell proliferation and metastasis by increasing RhoA/ROCK pathway signal transduction. *Oncotarget* **2016**, *7*, 73593–73606. [CrossRef]
30. Yi, J.; Huang, W.-Z.; Wen, Y.-Q.; Yi, Y.-C. Effect of miR-101 on proliferation and oxidative stress-induced apoptosis of breast cancer cells via Nrf2 signaling pathway. *Eur. Rev. Med. Pharmacol. Sci.* **2019**, *23*, 8931–8939. [PubMed]
31. De Blasio, A.; di Fiore, R.; Pratelli, G.; Drago-Ferrante, R.; Saliba, C.; Baldacchino, S.; Grech, G.; Scerri, C.; Vento, R.; Tesoriere, G. A loop involving NRF2, miR-29b-1-5p and AKT, regulates cell fate of MDA-MB-231 triple-negative breast cancer cells. *J. Cell. Physiol.* **2020**, *235*, 629–637. [CrossRef] [PubMed]
32. Xu, Y.; Wu, X.; Li, F.; Huang, D.; Zhu, W. CDCA4, a downstream gene of the Nrf2 signaling pathway, regulates cell proliferation and apoptosis in the MCF-7/ADM human breast cancer cell line. *Mol. Med. Rep.* **2017**, *17*, 1507–1512. [CrossRef]

33. Chen, Y.; Zhang, S.; Wang, X.; Guo, W.; Wang, L.; Zhang, D.; Yuan, L.; Zhang, Z.; Xu, Y.; Liu, S. Disordered signaling governing ferroportin transcription favors breast cancer growth. *Cell. Signal.* **2015**, *27*, 168–176. [CrossRef]
34. Thangasamy, A.; Rogge, J.; Krishnegowda, N.K.; Freeman, J.W.; Ammanamanchi, S. Novel Function of Transcription Factor Nrf2 as an Inhibitor of RON Tyrosine Kinase Receptor-mediated Cancer Cell Invasion. *J. Biol. Chem.* **2011**, *286*, 32115–32122. [CrossRef]
35. Tsai, C.-F.; Chen, J.-H.; Chang, C.-N.; Lu, D.-Y.; Chang, P.-C.; Wang, S.-L.; Yeh, W.-L. Fisetin inhibits cell migration via inducing HO-1 and reducing MMPs expression in breast cancer cell lines. *Food Chem. Toxicol.* **2018**, *120*, 528–535. [CrossRef]
36. Tertil, M.; Golda, S.; Skrzypek, K.; Florczyk, U.; Weglarczyk, K.; Kotlinowski, J.; Maleszewska, M.; Czauderna, S.; Pichon, C.; Kieda, C.; et al. Nrf2-heme oxygenase-1 axis in mucoepidermoid carcinoma of the lung: Antitumoral effects associated with down-regulation of matrix metalloproteinases. *Free. Radic. Biol. Med.* **2015**, *89*, 147–157. [CrossRef]
37. Biswas, S.; Chowdhury, S.R.; Mandal, G.; Purohit, S.; Gupta, A.; Bhattacharyya, A. RelA driven co-expression of CXCL13 and CXCR5 is governed by a multifaceted transcriptional program regulating breast cancer progression. *Biochim. Biophys. Acta Mol. Basis Dis.* **2019**, *1865*, 502–511. [CrossRef] [PubMed]
38. Zhou, X.-L.; Zhu, C.-Y.; Wu, Z.-G.; Guo, X.; Zou, W. The oncoprotein HBXIP competitively binds KEAP1 to activate NRF2 and enhance breast cancer cell growth and metastasis. *Oncogene* **2019**, *38*, 4028–4046. [CrossRef] [PubMed]
39. Ryoo, I.-G.; Choi, B.-H.; Ku, S.-K.; Kwak, M.-K. High CD44 expression mediates p62-associated NFE2L2/NRF2 activation in breast cancer stem cell-like cells: Implications for cancer stem cell resistance. *Redox Biol.* **2018**, *17*, 246–258. [CrossRef] [PubMed]
40. Wu, T.; Harder, B.G.; Wong, P.K.; Lang, J.E.; Zhang, D.D. Oxidative stress, mammospheres and Nrf2-new implication for breast cancer therapy? *Mol. Carcinog.* **2014**, *54*, 1494–1502. [CrossRef]
41. Wei, Y.; Liu, D.; Jin, X.; Gao, P.; Wang, Q.; Zhang, J.; Zhang, N. PA-MSHA inhibits the growth of doxorubicin-resistant MCF-7/ADR human breast cancer cells by downregulating Nrf2/p62. *Cancer Med.* **2016**, *5*, 3520–3531. [CrossRef]
42. Del Vecchio, C.A.; Feng, Y.; Sokol, E.S.; Tillman, E.J.; Sanduja, S.; Reinhardt, F.; Gupta, P.B. De-Differentiation Confers Multidrug Resistance Via Noncanonical PERK-Nrf2 Signaling. *PLoS Biol.* **2014**, *12*, e1001945. [CrossRef] [PubMed]
43. Paul, S.; Ghosh, S.; Mandal, S.; Sau, S.; Pal, M. NRF2 transcriptionally activates the heat shock factor 1 promoter under oxidative stress and affects survival and migration potential of MCF7 cells. *J. Biol. Chem.* **2018**, *293*, 19303–19316. [CrossRef]
44. Syu, J.-P.; Chi, J.-T.; Kung, H.-N. Nrf2 is the key to chemotherapy resistance in MCF7 breast cancer cells under hypoxia. *Oncotarget* **2016**, *7*, 14659–14672. [CrossRef] [PubMed]
45. Carlisi, D.; de Blasio, A.; Drago-Ferrante, R.; di Fiore, R.; Buttitta, G.; Morreale, M.; Scerri, C.; Vento, R.; Tesoriere, G. Parthenolide prevents resistance of MDA-MB231 cells to doxorubicin and mitoxantrone: The role of Nrf2. *Cell Death Discov.* **2017**, *3*, 17078. [CrossRef] [PubMed]
46. Wang, Y.-Y.; Chen, J.; Liu, X.-M.; Zhao, R.; Zhe, H. Nrf2-Mediated Metabolic Reprogramming in Cancer. *Oxidative Med. Cell. Longev.* **2018**, *2018*, 1–7. [CrossRef]
47. Zhang, H.; Zhang, Z.; Du, G.; Sun, H.; Liu, H.; Zhou, Z.; Gou, X.; Wu, X.; Yu, X.; Huang, Y. Nrf2 promotes breast cancer cell migration via up-regulation of G6PD/HIF-1α/Notch1 axis. *J. Cell. Mol. Med.* **2019**, *23*, 3451–3463. [CrossRef] [PubMed]
48. Zhang, H.-S.; Du, G.-Y.; Zhang, Z.-G.; Zhou, Z.; Sun, H.-L.; Yu, X.-Y.; Shi, Y.-T.; Xiong, D.-N.; Li, H.; Huang, Y.-H. NRF2 facilitates breast cancer cell growth via HIF1α-mediated metabolic reprogramming. *Int. J. Biochem. Cell Biol.* **2018**, *95*, 85–92. [CrossRef] [PubMed]
49. Lee, S.; Hallis, S.P.; Jung, K.-A.; Ryu, D.; Kwak, M.-K. Impairment of HIF-1α-mediated metabolic adaption by NRF2-silencing in breast cancer cells. *Redox Biol.* **2019**, *24*, 101210. [CrossRef] [PubMed]
50. Walker, A.; Singh, A.; Tully, E.; Woo, J.; Le, A.; Nguyen, T.; Biswal, S.; Sharma, D.; Gabrielson, E. Nrf2 signaling and autophagy are complementary in protecting breast cancer cells during glucose deprivation. *Free Radic. Biol. Med.* **2018**, *120*, 407–413. [CrossRef] [PubMed]
51. Sorlie, T.; Perou, C.M.; Tibshirani, R.; Aas, T.; Geisler, S.; Johnsen, H.; Hastie, T.; Eisen, M.B.; van de Rijn, M.; Jeffrey, S.S.; et al. Gene expression patterns of breast carcinomas distinguish tumor subclasses with clinical implications. *Proc. Natl. Acad. Sci. USA* **2001**, *98*, 10869–10874. [CrossRef]
52. Curtis, C.; Shah, S.P.; Chin, S.-F.; Turashvili, G.; Rueda, O.M.; Dunning, M.J.; Speed, D.; Lynch, A.G.; Samarajiwa, S.A.; Yuan, Y.; et al. The genomic and transcriptomic architecture of 2000 breast tumours reveals novel subgroups. *Nature* **2012**, *486*, 346–352. [CrossRef]
53. Kao, K.-J.; Chang, K.-M.; Hsu, H.-C.; Huang, A.T. Correlation of microarray-based breast cancer molecular subtypes and clinical outcomes: Implications for treatment optimization. *BMC Cancer* **2011**, *11*, 143. [CrossRef]
54. Hart, P.C.; Ratti, B.A.; Mao, M.; Ansenberger-Fricano, K.; Shajahan-Haq, A.N.; Tyner, A.L.; Minshall, R.D.; Bonini, M.G. Caveolin-1 regulates cancer cell metabolism via scavenging Nrf2 and suppressing MnSOD-driven glycolysis. *Oncotarget* **2015**, *7*, 308–322. [CrossRef]
55. Loignon, M.; Miao, W.; Hu, L.; Bier, A.; Bismar, T.A.; Scrivens, P.J.; Mann, K.; Basik, M.; Bouchard, A.; Fiset, P.O.; et al. Cul3 overexpression depletes Nrf2 in breast cancer and is associated with sensitivity to carcinogens, to oxidative stress, and to chemotherapy. *Mol. Cancer Ther.* **2009**, *8*, 2432–2440. [CrossRef] [PubMed]
56. Onodera, Y.; Motohashi, H.; Takagi, K.; Miki, Y.; Shibahara, Y.; Watanabe, M.; Ishida, T.; Hirakawa, H.; Sasano, H.; Yamamoto, M.; et al. NRF2 immunolocalization in human breast cancer patients as a prognostic factor. *Endocr. Relat. Cancer* **2013**, *21*, 241–252. [CrossRef] [PubMed]

57. Hartikainen, J.M.; Tengström, M.; Kosma, V.-M.; Kinnula, V.L.; Mannermaa, A.; Soini, Y. Genetic Polymorphisms and Protein Expression of NRF2 and Sulfiredoxin Predict Survival Outcomes in Breast Cancer. *Cancer Res.* **2012**, *72*, 5537–5546. [CrossRef] [PubMed]
58. Hartikainen, J.M.; Tengström, M.; Winqvist, R.; Jukkola-Vuorinen, A.; Pylkäs, K.; Kosma, V.-M.; Soini, Y.; Mannermaa, A. KEAP1 Genetic Polymorphisms Associate with Breast Cancer Risk and Survival Outcomes. *Clin. Cancer Res.* **2015**, *21*, 1591–1601. [CrossRef]
59. Huang, R.; Li, X.; Yu, Y.; Ma, L.; Liu, S.; Zong, X.; Zheng, Q. SETD7 is a prognosis predicting factor of breast cancer and regulates redox homeostasis. *Oncotarget* **2017**, *8*, 94080–94090. [CrossRef]
60. Lu, K.; Alcivar, A.L.; Ma, J.; Foo, T.K.; Zywea, S.; Mahdi, A.; Huo, Y.; Kensler, T.W.; Gatza, M.L.; Xia, B. NRF2 Induction Supporting Breast Cancer Cell Survival Is Enabled by Oxidative Stress–Induced DPP3–KEAP1 Interaction. *Cancer Res.* **2017**, *77*, 2881–2892. [CrossRef]
61. Wolf, B.; Goebel, G.; Hackl, H.; Fiegl, H. Reduced mRNA expression levels of NFE2L2 are associated with poor outcome in breast cancer patients. *BMC Cancer* **2016**, *16*, 821. [CrossRef]
62. Catanzaro, E.; Calcabrini, C.; Turrini, E.; Sestili, P.; Fimognari, C. Nrf2: A potential therapeutic target for naturally occurring anticancer drugs? *Expert Opin. Ther. Targets* **2017**, *21*, 781–793. [CrossRef] [PubMed]
63. Das, L.; Vinayak, M. Long Term Effect of Curcumin in Restoration of Tumour Suppressor p53 and Phase-II Antioxidant Enzymes via Activation of Nrf2 Signalling and Modulation of Inflammation in Prevention of Cancer. *PLoS ONE* **2015**, *10*, e0124000. [CrossRef]
64. Shin, J.W.; Chun, K.-S.; Kim, D.-H.; Kim, S.-J.; Kim, S.H.; Cho, N.-C.; Na, H.-K.; Surh, Y.-J. Curcumin induces stabilization of Nrf2 protein through Keap1 cysteine modification. *Biochem. Pharmacol.* **2020**, *173*, 113820. [CrossRef]
65. Rushworth, S.A.; Ogborne, R.M.; Charalambos, C.A.; O'Connell, M.A. Role of protein kinase C δ in curcumin-induced antioxidant response element-mediated gene expression in human monocytes. *Biochem. Biophys. Res. Commun.* **2006**, *341*, 1007–1016. [CrossRef] [PubMed]
66. Andreadi, C.K.; Howells, L.M.; Atherfold, P.A.; Manson, M.M. Involvement of Nrf2, p38, B-Raf, and Nuclear Factor-κB, but Not Phosphatidylinositol 3-Kinase, in Induction of Hemeoxygenase-1 by Dietary Polyphenols. *Mol. Pharmacol.* **2005**, *69*, 1033–1040. [CrossRef] [PubMed]
67. Garg, R.; Gupta, S.; Maru, G.B. Dietary curcumin modulates transcriptional regulators of phase I and phase II enzymes in benzo[a]pyrene-treated mice: Mechanism of its anti-initiating action. *Carcinogenesis* **2008**, *29*, 1022–1032. [CrossRef]
68. Jain, A.; Samykutty, A.; Jackson, C.; Browning, D.D.; Bollag, W.B.; Thangaraju, M.; Takahashi, S.; Singh, S.R. Curcumin inhibits PhIP induced cytotoxicity in breast epithelial cells through multiple molecular targets. *Cancer Lett.* **2015**, *365*, 122–131. [CrossRef]
69. Chen, B.; Zhang, Y.; Wang, Y.; Rao, J.; Jiang, X.; Xu, Z. Curcumin inhibits proliferation of breast cancer cells through Nrf2-mediated down-regulation of Fen1 expression. *J. Steroid Biochem. Mol. Biol.* **2014**, *143*, 11–18. [CrossRef]
70. Li, X.; Xie, W.; Xie, C.; Huang, C.; Zhu, J.; Liang, Z.; Deng, F.; Zhu, M.; Zhu, W.; Wu, R.; et al. Curcumin Modulates miR-19/PTEN/AKT/p53 Axis to Suppress Bisphenol A-induced MCF-7 Breast Cancer Cell Proliferation. *Phytother. Res.* **2014**, *28*, 1553–1560. [CrossRef] [PubMed]
71. Foygel, K.; Sekar, T.V.; Paulmurugan, R. Monitoring the Antioxidant Mediated Chemosensitization and ARE-Signaling in Triple Negative Breast Cancer Therapy. *PLoS ONE* **2015**, *10*, e0141913. [CrossRef]
72. Hu, L.; Miao, W.; Loignon, M.; Kandouz, M.; Batist, G. Putative chemopreventive molecules can increase Nrf2-regulated cell defense in some human cancer cell lines, resulting in resistance to common cytotoxic therapies. *Cancer Chemother. Pharmacol.* **2009**, *66*, 467–474. [CrossRef] [PubMed]
73. Galadari, S.; Rahman, A.; Pallichankandy, S.; Thayyullathil, F. Reactive oxygen species and cancer paradox: To promote or to suppress? *Free Radic. Biol. Med.* **2017**, *104*, 144–164. [CrossRef]
74. Calabrese, E.J.; Mattson, M.P.; Calabrese, V. Resveratrol commonly displays hormesis: Occurrence and biomedical significance. *Hum. Exp. Toxicol.* **2010**, *29*, 980–1015. [CrossRef] [PubMed]
75. Rai, G.; Mishra, S.; Suman, S.; Shukla, Y. Resveratrol improves the anticancer effects of doxorubicin in vitro and in vivo models: A mechanistic insight. *Phytomedicine* **2016**, *23*, 233–242. [CrossRef]
76. Zhou, X.; Zhao, Y.; Wang, J.; Wang, X.; Chen, C.; Yin, D.; Zhao, F.; Yin, J.; Guo, M.; Zhang, L.; et al. Resveratrol represses estrogen-induced mammary carcinogenesis through NRF2-UGT1A8-estrogen metabolic axis activation. *Biochem. Pharmacol.* **2018**, *155*, 252–263. [CrossRef] [PubMed]
77. Singh, B.; Shoulson, R.; Chatterjee, A.; Ronghe, A.; Bhat, N.K.; Dim, D.C.; Bhat, H.K. Resveratrol inhibits estrogen-induced breast carcinogenesis through induction of NRF2-mediated protective pathways. *Carcinogenesis* **2014**, *35*, 1872–1880. [CrossRef]
78. Sabzichi, M.; Hamishehkar, H.; Ramezani, F.; Sharifi, S.; Tabasinezhad, M.; Pirouzpanah, M.; Ghanbari, P.; Samadi, N. Luteolin-loaded phytosomes sensitize human breast carcinoma MDA-MB 231 cells to doxorubicin by suppressing Nrf2 mediated signalling. *Asian Pac. J. Cancer Prev.* **2014**, *15*, 5311–5316. [CrossRef] [PubMed]
79. Zhong, Y.; Zhang, F.; Sun, Z.; Zhou, W.; Li, Z.-Y.; You, Q.-D.; Guo, Q.-L.; Hu, R. Drug resistance associates with activation of Nrf2 in MCF-7/DOX cells, and wogonin reverses it by down-regulating Nrf2-mediated cellular defense response. *Mol. Carcinog.* **2013**, *52*, 824–834. [CrossRef]
80. Marvin 20.20.0. ChemAxon. 2020. Available online: https://www.chemaxon.com (accessed on 26 February 2021).

81. Jasser, S.A.; Blask, D.E.; Brainard, G.C. Light During Darkness and Cancer: Relationships in Circadian Photoreception and Tumor Biology. *Cancer Causes Control* **2006**, *17*, 515–523. [CrossRef]
82. Hill, S.M.; Belancio, V.P.; Dauchy, R.T.; Xiang, S.; Brimer, S.; Mao, L.; Hauch, A.; Lundberg, P.W.; Summers, W.; Yuan, L.; et al. Melatonin: An inhibitor of breast cancer. *Endocr. Relat. Cancer* **2015**, *22*, R183–R204. [CrossRef] [PubMed]
83. Chuffa, L.G.D.A.; Carvalho, R.F.; Justulin, L.A.; Cury, S.S.; Seiva, F.R.F.; Jardim-Perassi, B.V.; Zuccari, D.A.P.D.C.; Reiter, R.J. A meta-analysis of microRNA networks regulated by melatonin in cancer: Portrait of potential candidates for breast cancer treatment. *J. Pineal Res.* **2020**, *69*, 12693. [CrossRef]
84. Chuffa, L.G.D.A.; Seiva, F.R.F.; Cucielo, M.S.; Silveira, H.S.; Reiter, R.J.; Lupi, L.A. Mitochondrial functions and melatonin: A tour of the reproductive cancers. *Cell. Mol. Life Sci.* **2019**, *76*, 837–863. [CrossRef]
85. Li, B.; Feng, X.J.; Hu, X.Y.; Chen, Y.P.; Sha, J.C.; Zhang, H.Y.; Fan, H.-G. Effect of melatonin on attenuating the isoflurane-induced oxidative damage is related to PKCα/Nrf2 signaling pathway in developing rats. *Brain Res. Bull.* **2018**, *143*, 9–18. [CrossRef]
86. Shi, S.; Lei, S.; Tang, C.; Wang, K.; Xia, Z. Melatonin attenuates acute kidney ischemia/reperfusion injury in diabetic rats by activation of the SIRT1/Nrf2/HO-1 signaling pathway. *Biosci. Rep.* **2019**, *39*. [CrossRef] [PubMed]
87. Zhang, Y.; Wei, Z.; Liu, W.; Wang, J.; He, X.; Huang, H.; Zhang, J.; Yang, Z. Melatonin protects against arsenic trioxide-induced liver injury by the upregulation of Nrf2 expression through the activation of PI3K/AKT pathway. *Oncotarget* **2016**, *8*, 3773–3780. [CrossRef]
88. Janjetovic, Z.; Jarrett, S.G.; Lee, E.F.; Duprey, C.; Reiter, R.J.; Slominski, A.T. Melatonin and its metabolites protect human melanocytes against UVB-induced damage: Involvement of NRF2-mediated pathways. *Sci. Rep.* **2017**, *7*, 1–13. [CrossRef] [PubMed]
89. Trivedi, P.; Jena, G.; Tikoo, K.; Kumar, V. Melatonin modulated autophagy and Nrf2 signaling pathways in mice with colitis-associated colon carcinogenesis. *Mol. Carcinog.* **2016**, *55*, 255–267. [CrossRef] [PubMed]
90. Fang, J.; Yan, Y.; Teng, X.; Wen, X.; Li, N.; Peng, S.; Liu, W.; Donadeu, F.X.; Zhao, S.; Hua, J. Melatonin prevents senescence of canine adipose-derived mesenchymal stem cells through activating NRF2 and inhibiting ER stress. *Aging* **2018**, *10*, 2954–2972. [CrossRef]
91. Wang, M.; Xue, Y.; Shen, L.; Qin, P.; Sang, X.; Tao, Z.; Yi, J.; Wang, J.; Liu, P.; Cheng, H. Inhibition of SGK1 confers vulnerability to redox dysregulation in cervical cancer. *Redox Biol.* **2019**, *24*, 101225. [CrossRef]
92. Paroni, R.; Terraneo, L.; Bonomini, F.; Finati, E.; Virgili, E.; Bianciardi, P.; Favero, G.; Fraschini, F.; Reiter, R.J.; Rezzani, R.; et al. Antitumour activity of melatonin in a mouse model of human prostate cancer: Relationship with hypoxia signalling. *J. Pineal Res.* **2014**, *57*, 43–52. [CrossRef]
93. Reiter, R.J.; Rosales-Corral, S.A.; Tan, D.-X.; Acuna-Castroviejo, D.; Qin, L.; Yang, S.-F.; Xu, K. Melatonin, a Full Service Anti-Cancer Agent: Inhibition of Initiation, Progression and Metastasis. *Int. J. Mol. Sci.* **2017**, *18*, 843. [CrossRef] [PubMed]

molecules

MDPI

Review

Review and Chemoinformatic Analysis of Ferroptosis Modulators with a Focus on Natural Plant Products

Višnja Stepanić [1,*] and Marta Kučerová-Chlupáčová [2,*]

1 Laboratory for Machine Learning and Knowledge Representation, Ruđer Bošković Institute, Bijenička 54, 10000 Zagreb, Croatia
2 Department of Pharmaceutical Chemistry and Pharmaceutical Analysis, Faculty of Pharmacy in Hradec Králové, Charles University, Ak. Heyrovského 1203/8, 500 05 Hradec Králové, Czech Republic
* Correspondence: visnja.stepanic@irb.hr (V.S.); kucerom@faf.cuni.cz (M.K.-C.); Tel.: +385-1-457-1356 (V.S.); +420-495-067-372 (M.K.-C.)

Abstract: Ferroptosis is a regular cell death pathway that has been proposed as a suitable therapeutic target in cancer and neurodegenerative diseases. Since its definition in 2012, a few hundred ferroptosis modulators have been reported. Based on a literature search, we collected a set of diverse ferroptosis modulators and analyzed them in terms of their structural features and physicochemical and drug-likeness properties. Ferroptosis modulators are mostly natural products or semisynthetic derivatives. In this review, we focused on the abundant subgroup of polyphenolic modulators, primarily phenylpropanoids. Many natural polyphenolic antioxidants have antiferroptotic activities acting through at least one of the following effects: ROS scavenging and/or iron chelation activities, increased GPX4 and NRF2 expression, and LOX inhibition. Some polyphenols are described as ferroptosis inducers acting through the generation of ROS, intracellular accumulation of iron (II), or the inhibition of GPX4. However, some molecules have a dual mode of action depending on the cell type (cancer versus neural cells) and the (micro)environment. The latter enables their successful use (e.g., apigenin, resveratrol, curcumin, and EGCG) in rationally designed, multifunctional nanoparticles that selectively target cancer cells through ferroptosis induction.

Keywords: ferroptosis; inducers; inhibitors; drug-likeness; cancer; neurodegenerative; polyphenol

Citation: Stepanić, V.; Kučerová-Chlupáčová, M. Review and Chemoinformatic Analysis of Ferroptosis Modulators with a Focus on Natural Plant Products. *Molecules* **2023**, *28*, 475. https://doi.org/10.3390/molecules28020475

Academic Editor: Andrea Trabocchi

Received: 3 December 2022
Revised: 23 December 2022
Accepted: 24 December 2022
Published: 4 January 2023

1. Introduction

Cancer is one of the most common causes of death in humans worldwide [1]. Prevention programs and early cancer detection through regular medical check-ups and with the use of specific biomarkers, as well as the development of novel therapeutics (such as the development of various protein kinase inhibitors and immunotherapeutics), reduce cancer mortality [2,3]. Cancer therapy (chemotherapy, targeted therapy, radiotherapy, or immunotherapy) generally aims to destroy cancer cells without too many harmful effects on healthy cells. The induction of natural, programmed cell death pathways through the use of low-molecular-weight (MW) compounds has been widely explored as a way to combat death-escaping cancer cells. In cancer chemoprevention and chemotherapy, the induction and promotion of cancer cell apoptosis by small molecule agents have been extensively studied [4,5]. However, the main limitation of this approach is that cancer acquires resistance to such drugs, including targeted therapies, leading to their failure [4]. Dysregulated mechanisms that sustain cancer resistance to various other types of cell death pathways have also been studied in solid tumors and hematological malignancies [6].

Cancer initiation and promotion are generally linked with oxidative stress [7]. Oxidative stress causes DNA mutations, cell component damage, and pro-oncogenic signaling and, thus, triggers and sustains carcinogenesis [8]. Sustained overproduction of reactive oxygen species (ROS) may lead to persistent, chronic oxidative stress and injury through nonlethal modifications of normal cellular growth control mechanisms such as modified intercellular communication, protein kinase activity, membrane structure and function, and gene expression. Most conventional chemotherapeutic agents increase ROS production and, thus, cause cell and tissue damage and activate an inflammatory response.

Ferroptosis is a regulated cell death mechanism caused directly by (over)production and accumulation of specific kinds of ROS, which may enable the selective killing of cancer cells without causing significant toxicity to normal cells. The druggability of ferroptosis by low-MW compounds has already been demonstrated. In fact, ferroptosis was discovered to be a programmed cell death pathway by using low-MW compounds [9]. The treatment of an NRAS oncogene mutant containing HT-1080 fibrosarcoma cells with the compound erastin (10 μM) induced a time-dependent, continuous increase in cytosolic and lipid ROS, which resulted in cell death with a distinct non-apoptotic phenotype. Cell death was suppressed by each of the following low-MW agents: iron chelators deferoxamine (100 μM) and ciclopirox olamine (5 μM), the glutathione peroxidase mimetic organoselenium compound ebselen (5 μM), the mitogen-activated protein kinase (MEK) inhibitor U0126 (5 μM), and the antioxidants trolox (100 μM) and ferrostatin-1 (EC_{50} = 60 nM). Since ferroptosis was described as a distinct regulated cell death pathway in 2012, more than a few hundred low-MW inducers and inhibitors of ferroptosis have been reported.

Recently, many reviews on ferroptosis relating to various biological aspects have been published [10–12]. Herein, we collect sets of 30 representative inducers (Tables 1 and S1) and 48 suppressors/inhibitors (Tables 2 and S2) of ferroptotic cell death with a MW of less than 800 and analyze them in relation to structural, physicochemical/drug-likeness, and biological/pharmacological aspects. Thereafter, the review focuses on describing the biological activities/effects of the subset of (poly)phenolic ferroptosis modulators since, to our knowledge, there has been no such comprehensive review of polyphenols as ferroptosis modulators [13–17]. We focus only on the activities of (poly)phenolic compounds (many of which are already known) in conjunction with their reported (anti)ferroptotic effects. We review their influence on ferroptosis through activities affecting the three major components of ferroptosis. Tables 1 and 2 list these activities for natural plant molecules and put them in the context with other ferroptosis modulators with analogous effects.

Table 1. Ferroptosis inducer representatives—chemical structures with the main mode of action [1] and predicted lipophilicity coefficient and potential for crossing the blood-to-brain barrier (BBB) by the program ADMET Predictor™ [18].

Inducer Name	2D Structure	Mode of Action	log*P*	BBB
(1*S*,3*R*)-RSL3	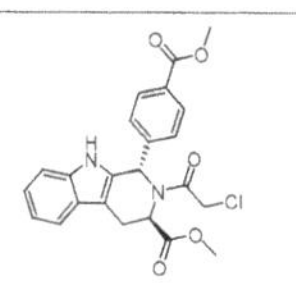	GPX4 ↓	3.6	low
2-phenethylisothiocyanate	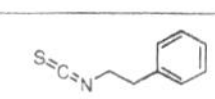	GPX4 ↓	3.4	high
gallic acid	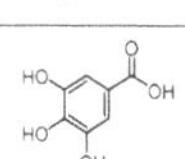	GPX4 ↓	0.7	low
FIN56	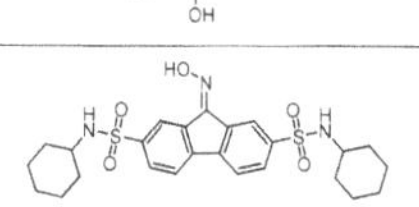	GPX4 ↓	4.7	low
albiziabioside A	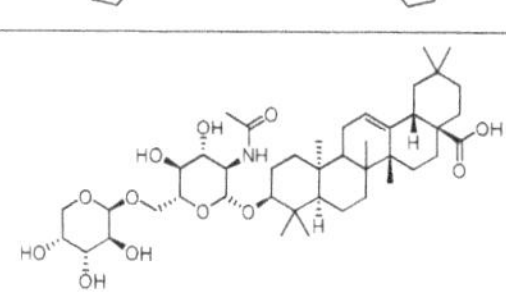	GPX4 ↓	4.5	low

Table 1. *Cont.*

Inducer Name	2D Structure	Mode of Action	log*P*	BBB
dihydroisotanshinone I	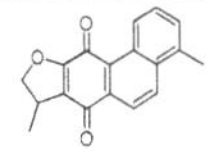	GPX4 ↓	3.7	high
withaferin A	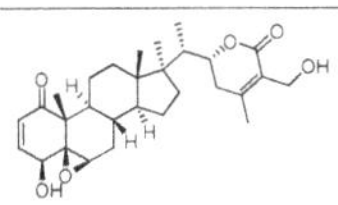	GPX4 ↓	3.2	high
ML162	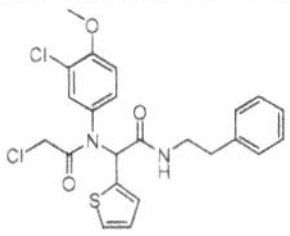	GPX4 ↓	4.4	low
altretamine	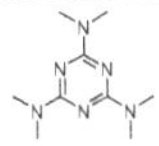	GPX4 ↓	2.5	high
typhaneoside	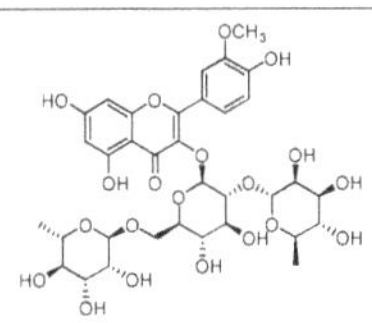	GPX4 ↓	−0.8	low
artemisinin	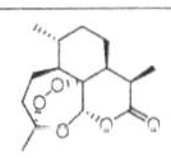	iron ↑; GPX4 ↓	2.4	high
dihydroartemisinin	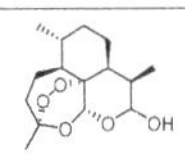	iron ↑; GPX4 ↓	2.2	high
artesunate	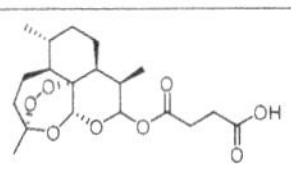	iron ↑	2.3	low
ruscogenin	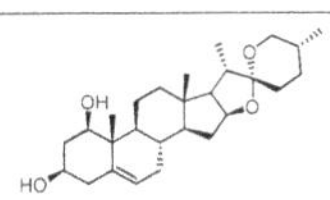	iron ↑	4.5	high
salinomycin	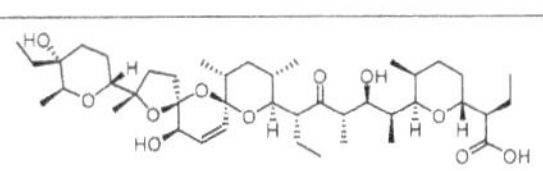	iron ↑	5.4	low
amentoflavone	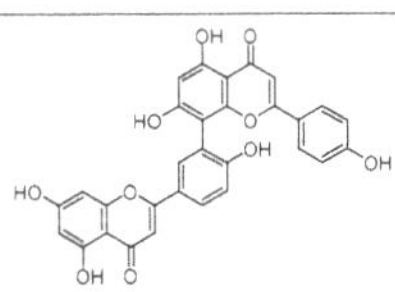	iron ↑	4.5	low
BSO	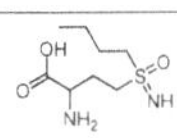	ROS ↑	−2.1	low
FINO2	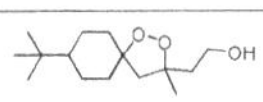	ROS ↑	4.0	high

Table 1. *Cont.*

Inducer Name	2D Structure	Mode of Action	log*P*	BBB
epunctanone	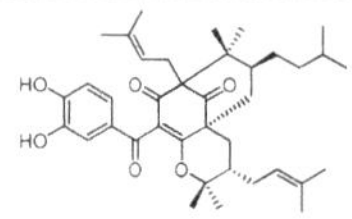	ROS ↑	7.7	low
piperlongumine	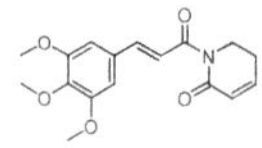	ROS ↑	1.9	high
robustaflavone A	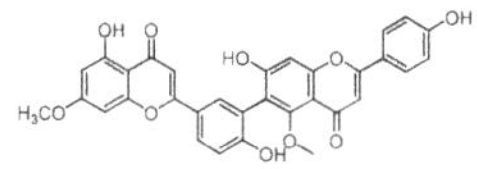	ROS ↑	4.6	low
sorafenib	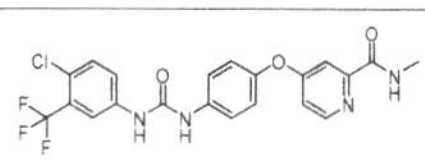	X_c^- system ↓	5.1	low
erastin	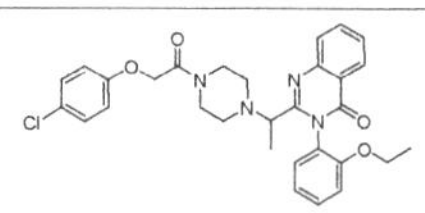	X_c^- system ↓	3.8	low
sulfasalazine	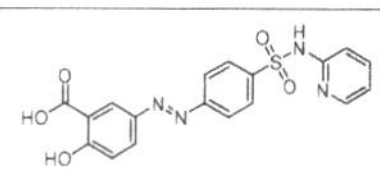	X_c^- system ↓	3.1	low
ferroptocide	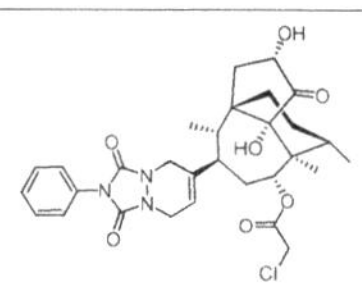	Other	2.1	low
maneb	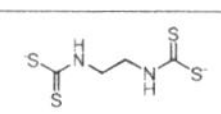	other	0.2	low
gambogic acid	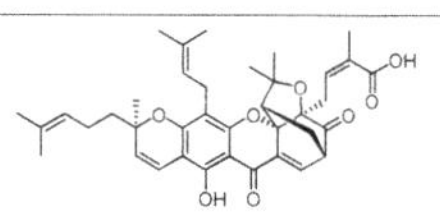	other	7.4	low
paraquat	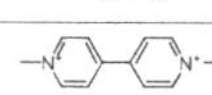	other	−6.3	low
soyauxinium	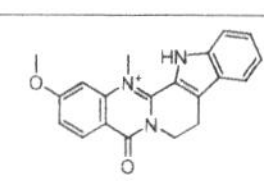	other	0.4	high
erianin	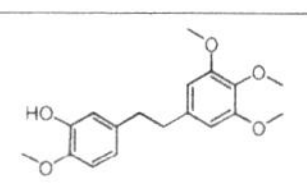	other	3.4	high

[1] Mode of action: GPX4 ↓ inhibition of GPX4; iron ↑/ROS ↑ intracellular accumulation of iron/ROS; other mechanisms.

Table 2. Ferroptosis inhibitor representatives—chemical structures with the main mode of action [1] and predicted lipophilicity coefficient and potential for crossing the BBB.

Inhibitor Name	Structure	Mode of Action	log*P*	BBB
butylatedhydroxytoluene	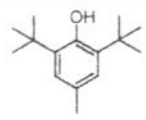	antioxidative	5.5	high
ferrostatin-1	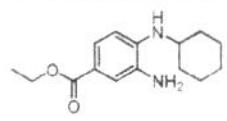	antioxidative	3.7	high
α-tocopherol	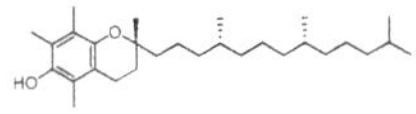	antioxidative	11.5	high
β-carotene	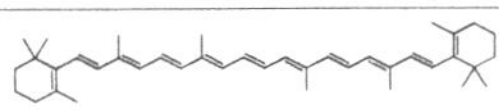	antioxidative	11.6	high
glutathione	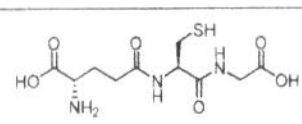	antioxidative	−3.4	low
N-acetylcysteine	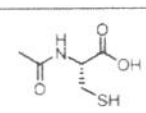	antioxidative	−0.6	low
ascorbic acid	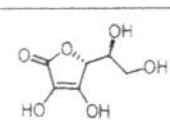	antioxidative	−1.6	low
edaravone	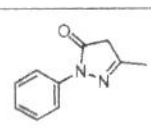	antioxidative	1.3	high
GSK2334470	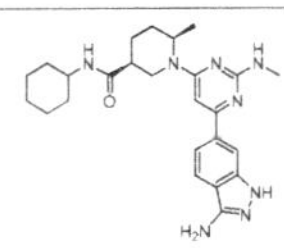	antioxidative	4.3	low
liproxstatin-1	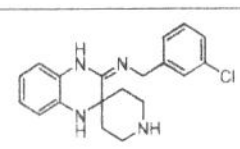	antioxidative	3.3	high
trolox	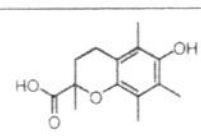	antioxidative	2.9	low
α-lipoic acid	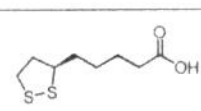	antioxidative; NRF2 ↑	2.7	low
7-*O*-cinnamoyltaxifolin	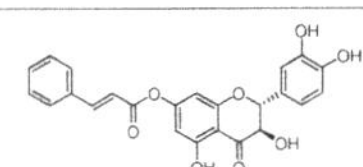	antioxidative	3.7	low
7-*O*-feruloyltaxifolin	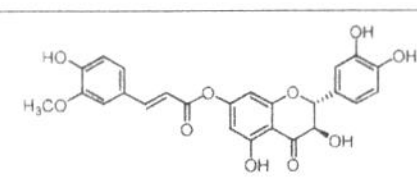	antioxidative	3.1	low
butein	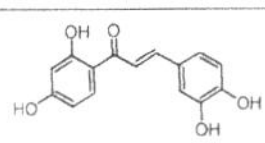	antioxidative	2.8	low
butin	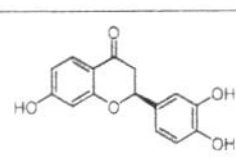	antioxidative	1.9	low
isorhapontigenin	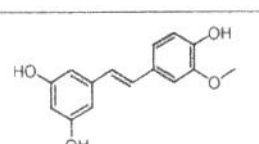	antioxidative	3.0	high

Table 2. *Cont.*

Inhibitor Name	Structure	Mode of Action	log*P*	BBB
morachalcone D	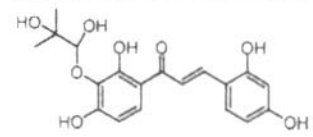	antioxidative	1.7	low
piceatannol-3′-*O*-glucoside	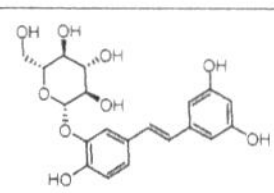	antioxidative	0.4	low
rhapontigenin	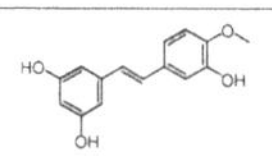	antioxidative	3.1	low
rhapontin	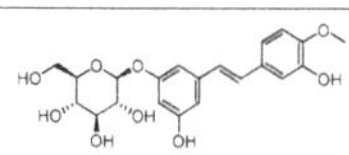	antioxidative	0.6	low
synthetic chalcone 1 [2]	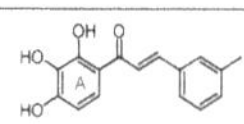	antioxidative	3.8	low
Baicalein [2]	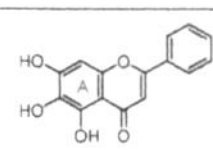	antioxidative; 15-LOX ↓	3.0	low
dopamine	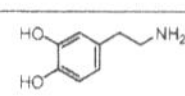	GPX4 ↑	−0.3	low
galangin	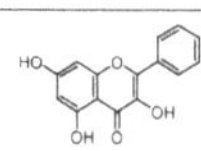	GPX4 ↑	2.7	low
apigenin	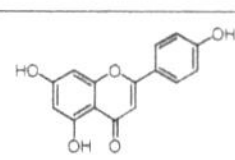	GPX4 ↑	2.9	low
silibinin	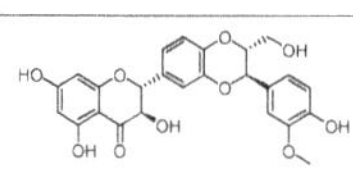	iron chelation	1.8	low
deferoxamine	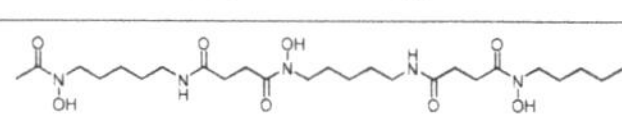	iron chelation	−1.3	low
phytic acid	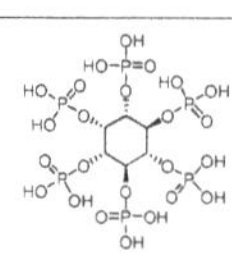	iron chelation	−11.2	high
8-hydroxyquinoline	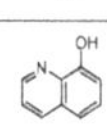	iron chelation	2.1	high
ciclopirox olamine	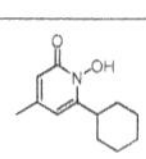	iron chelation	2.5	high
deferasirox	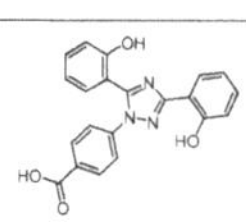	iron chelation	3.8	low
deferiprone	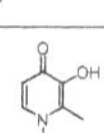	iron chelation	−0.6	high
maltol	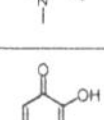	iron chelation	0.0	high

Table 2. *Cont.*

Inhibitor Name	Structure	Mode of Action	log*P*	BBB
quercetin	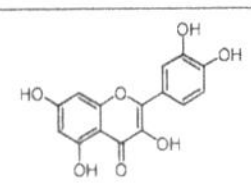	iron chelation	2.0	low
curcumin	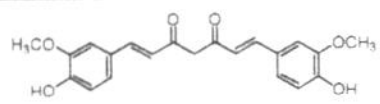	iron chelation	3.0	high
EGCG	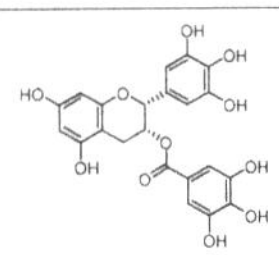	iron chelation	2.2	low
ellagic acid	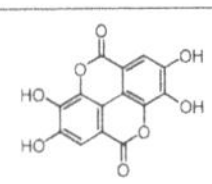	iron chelation	1.9	low
gossypitrin	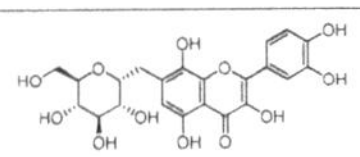	iron chelation	−0.7	low
sterubin	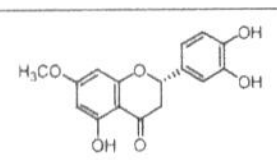	iron chelation	2.0	low
fisetin	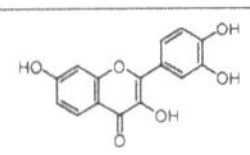	iron chelation; NRF2 ↑	2.2	low
carnosic acid	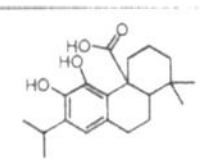	NRF2 ↑	4.5	low
melatonin	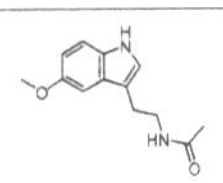	NRF2 ↑	1.7	high
zileuton	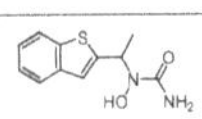	5-LOX ↓	1.9	high
bakuchiol	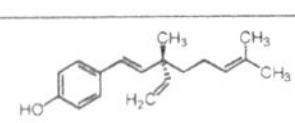	other	5.7	high
psoralidin	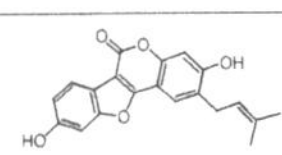	other	4.1	low

[1] Mode of action: GPX4 ↑/NRF2 ↑ activation of GPX4/NRF2; 5-LOX ↓/15-LOX ↓ inhibition of 5-LOX/15-LOX; antioxidative—radical scavenging; other mechanisms. [2] Conventional notation of rings used within text for discussion.

2. Ferroptosis—Three Main Factors

Ferroptosis was described in 2012 as a non-apoptotic cell death mechanism. It is a regulated form of cell death that is triggered and driven by the (over)production and accumulation of lipid and phospholipid (hydroxy)peroxides, the formation of which is specifically mediated by Fe^{2+} ferrous ions. It is named after Fe^{2+} ferrous ions since cytosolic and mitochondrial Fe^{2+} ions are essential factors in ferroptosis. Once ignited, ferroptosis can remain local or rapidly spread to surrounding cells, depending on the ignition mechanism [19]. It is morphologically, biochemically, and genetically distinct from apoptosis, necrosis, and autophagy but similar to oxytosis. Ferroptosis is closely associated with mitochondria, and its primary morphological markers are aberrant mitochondria characterized by a reduced number of mitochondrial cristae, inner membrane condensation, outer membrane rupture, and size shrinkage [20,21]. Physiologically, ferroptosis may contribute to embryonic development, erythropoiesis, aging, and antiviral and anticancer defense mechanisms [11]. Pathologically, it is associated with neurological diseases, myocardial infarction, atherosclerosis, renal and liver diseases, and cancer.

Three factors are necessary for the induction of and maintaining ferroptosis. These are: (i) the intracellular accumulation of Fe^{2+} ferrous ions, (ii) the accumulation of lipid peroxides generated from polyunsaturated fatty acids (PUFA) with bis-allylic fragments localized mainly in membranes, and (iii) the deficient repair of lipid peroxides. It is possible to induce or inhibit ferroptotic cell death in a relatively straightforward manner by directly or indirectly targeting one of the three necessary factors. They are considered markers for ferroptosis.

Iron is an essential metal in the human body, and its uptake, distribution, storage, and retrieval are coordinated at cellular and systemic levels by a complex and finely balanced network of regulatory pathways. Iron is involved in a variety of physiological functions and processes, including DNA replication, the tricarboxylic acid cycle, ATP production via the electron transport chain, and signal transduction. It is a cofactor in 6.5% of all human enzymes, localized mainly in the endoplasmic reticulum and mitochondria in the form of iron ions, heme, or iron–sulfur (FeS) clusters [22]. Its aberrant metabolism, leading to excessive Fenton reactions and/or impairment of mitochondrial function and energy metabolism, induces ferroptosis. Free, non-protein-bound Fe^{2+} ions can be quickly released from labile iron pools that are available within living cells and serve as a transient hub of the cellular iron metabolism. Intracellular accumulation of free iron can lead to high production of ROS which can override the antioxidant defense of a cell.

Other forms of regulated cell death mechanisms such as apoptosis, necroptosis, and pyroptosis may also depend on iron-induced ROS and oxidative stress. However, ferroptosis is dependent on lipid peroxidation. Ferroptosis is characterized by excessive peroxidation of PUFA bound in certain phospholipids such as phosphatidylethanolamines (e.g., peroxides of phosphoethanolamines (PEs) with arachidonic acid (AA) (18:0/20:4(5Z,8Z,11Z,14Z) PE-AA) or adrenic acid (AdA) (18:0/22:4 (7Z,10Z,13Z,16Z) PE-AdA), phosphatidylcholines, and other types of phospholipids [23]. These phospholipids are synthesized mainly in the membranes of mitochondria and the endoplasmic reticulum. Lipids are peroxidized through non-enzymatic and enzymatic mechanisms involving Fe^{2+} ions. In the non-enzymatic Fenton reaction of Fe^{2+} ions with hydrogen peroxide (H_2O_2), hydroxyl radicals (HO•) are produced, which can cause oxidative damage to cellular components such as lipids and induce cell death. Lipoxygenase (LOX) enzymes are non-heme, iron-containing dioxygenases that catalyze the stereospecific oxygenation of free and esterified PUFAs, generating a spectrum of bioactive lipid mediators that can initiate autocatalytic lipid autoxidation.

Since lipid peroxidation fuels the spread of ferroptosis, regulation of the activity and expression of proteins involved in lipid metabolic pathways has a major impact on ferroptosis [24]. These are the enzymes responsible for the formation of phospholipids and their incorporation into various cell membranes (ACSL4 (acyl-CoA synthetase long-chain family member 4/long-chain fatty-acid-CoA ligase 4) and LPCAT3 (lysophospholipid acyltransferase 5)), as well as enzymes regulating lipid peroxidation such as LOXs and GPX4 (glutathione peroxidase 4). Lipid peroxide radicals are neutralized non-enzymatically by exogenous, lipophilic free radical scavengers such as α-tocopherol and β-carotene and enzymatically by the selenoprotein GPX4. GPX4 is unique among the eight GPX isoforms in that it is the only enzyme capable of reducing oxidized, esterified fatty acids and cholesterol hydroperoxides, thus, protecting against lipid peroxidation in cells and structural function in mature sperm cells [25,26]. It reduces fatty acids/phospholipid hydroperoxides to lipid alcohols with the help of glutathione (GSH) as a cofactor, even if they are incorporated in lipoproteins and membranes. There are three isoforms of GPX4: mGPX4 distributed in mitochondria, nGPX4 distributed in nucleoli, and cGPX4 distributed in the nucleus and cytosol and also strongly associated with membranes. GPX4 activity decreases under GSH depletion and can also be directly inhibited via a covalent interaction with selenocysteine in the active site. Deficiency of GPX4 in quantity and/or activity leads to ferroptosis. Its cofactor, GSH, is synthesized from cysteine and glutamate, and cystine is a precursor of cysteine. GPX4 activity can be reduced by depletion of GSH through inhibition of the cystine/glutamate antiporter X_c^- system, which exchanges intracellular L-glutamate for extracellular L-cystine across the cellular plasma membrane. The X_c^-/GSH/GPX4 axis crucially controls ferroptosis. The heterodimeric X_c^- system consists of two transmembrane amino acid transporters: SLC3A2 (solute carrier family 3 member 2) and cystine/glutamate transporter SLC7A11 (solute carrier family 7 member 11). Glutamate itself can be replenished by importing glutamine via transporter SLC1A5. Cystine is reduced inside cells to cysteine. The enzyme ACSL4 enriches cellular membranes with long ω-6 PUFAs, determining the ferroptosis sensitivity of cells [27]. It was found to be preferentially expressed in a panel of basal-like breast cancer cell lines.

3. Ferroptosis as a Potential Therapeutical Target

Cell death is impaired in cancer, and the common method of cancer therapy is to induce death mechanisms in immortal cancer cells. Cancer cells are characterized by a rapid proliferation rate that requires a high iron load, and they are well adapted to acquire iron and prevent its loss [28]. Iron plays an important role in modulating the tumor microenvironment and metastasis, maintaining genomic stability, and controlling epigenetics. To meet the high iron demand, neoplastic cells remodel iron metabolic pathways, including iron uptake, storage, and efflux, making the manipulation of iron homeostasis an important approach for cancer therapy [29]. Metabolic reprogramming of cancer cells also involves mitochondrial dysfunction and dysregulated p53 expression, which has been implicated in the regulation of ferroptosis [30,31]. To facilitate iron uptake, TfR1 (transfer iron protein receptor 1) is highly expressed on the surface of cancer cells, and iron is accumulated within cells transformed with the oncogene RAS because of the upregulation of TfR1. Ferritin, an intracellular protein that stores iron, is also elevated in many cancers, including breast cancer, and can be used as a prognostic marker for breast cancer progression.

The process of ferroptosis involves signaling pathways that play a role in cancer biology, such as the AMP-activated protein kinase (AMPK)-RAS/MAPK and AMPK/mTOR/p70S6k pathways and the NRF2-KEAP1 pathway. NRF2 (nuclear factor erythroid 2-related factor 2) is an important transcription factor in the regulation of oxidative stress and plays a major role in the induction of drug insensitivity or resistance in cancer cells. Its activity affects the expression of antiferroptosis genes encoding for GPX4, SLC7A11, and iron storage protein ferritin subunits—FTH1 (ferritin heavy chain 1) and FTL (ferritin light chain) [32]. Cancer chemoresistance may be caused by activation of the NRF2 and downstream NRF2/FTH1 or NRF2/SLC7A11 pathways, resulting in the lowering of free intracellular iron (TfR1 downregulation and ferritin upregulation) or enhanced neutralization of lipid peroxides (GPX4 and FSP1 (ferroptosis suppressor protein 1) upregulation) compared to those in drug-sensitive cells. Thus, ferroptosis is a specific weakness of cancer suitable for use in the treatment of certain therapy-resistant cancers. Inducing ferroptosis in synergy with classic chemotherapeutic agents can sensitize cancer cells to treatments.

Several dozen compounds have been reported to induce ferroptosis in cancer cells by direct modulation of the ferroptosis targets (Table 1). Sensitivity profiling of 177 cancer cell lines to 12 ferroptosis-inducing small molecules (including erastin, RSL3, and their analogs) revealed that diffuse large B-cell lymphomas and renal cell carcinomas are particularly susceptible to ferroptosis through GPX4 inhibition [26]. Some natural inducers promote ferroptosis by regulating the ROS/AMPK/mTOR signaling pathways to inhibit cancer cell viability and proliferation, such as dihydroartemisinin (DHA) and amentoflavone.

Since ferroptosis is associated with neurodegeneration, numerous modulators have been discovered through studies of neurological, pathological conditions. It was found that ferroptosis induced by erastin is similar in cancer cells and primary neurons [33]. Therefore, we collected compounds discovered to act as ferroptosis modulators in cancer and/or neuronal cells and analyzed them in relation to chemical/structural and physicochemical/drug-likeness molecular aspects.

In the context of ferroptosis, it should be stressed that an additional, important common factor in cancer, neuronal networks, and ferroptosis is hydrogen peroxide (H_2O_2) (Figure 1). Cancer cells including melanomas, neuroblastoma, colon carcinoma, ovarian carcinoma, and cancer-associated fibroblasts, as well as myofibroblasts, macrophages, and neutrophils, are the major producers of H_2O_2 [34]. H_2O_2 is not only used as an effective biological weapon but is also an important signaling molecule in cancer and neuronal networks [35,36]. The localized and concentrated production of H_2O_2 and ROS is enabled by the packaging of active ingredients including ferroptotic modulators in multifunctional nanosystems.

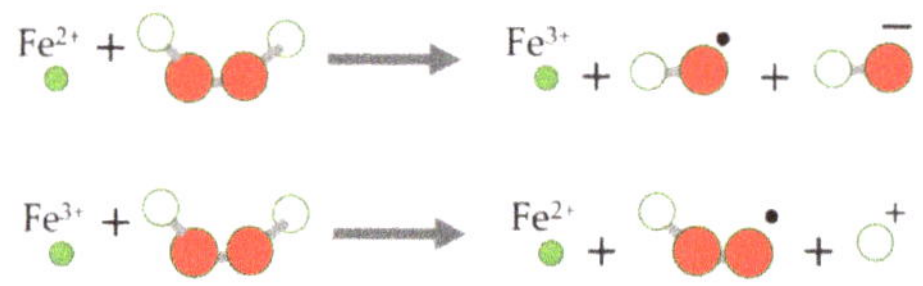

Figure 1. Fenton reaction of iron ions and H_2O_2 for production of powerful ROS. Under the condition of accumulated reactants, it can become uncontrolled, causing extensive oxidative damage and death.

The iron-seeking phenotype of cancers can be exploited in two ways: (i) to restrict iron availability and (ii) to exploit the redox properties of excess iron to promote cytotoxic oxidative stress in cancer cells. To foster ROS generation, nanoparticles provide rationally designed strategies to preferentially deliver drugs/active ingredients into cancer cells [37]. Recently, different ferroptotic inducers have also been applied in the form of rationally designed, multifunctional nanoparticles on a variety of cancer cells. Their efficacy has been enhanced by the delivery of nanoparticles to a target place in chemotherapy, radiotherapy, and immunotherapy [38]. The development of nanotherapeutics offers a route to overcome the toxic, off-target effects of ferroptosis on normal cells and the shortcomings of ferroptosis-driven therapeutics due to their low bioavailability caused by their low aqueous solubility and membrane permeability. Some ferroptosis-driven nanotherapeutics contain the two basic elements of the Fenton response: Fe^{2+} ions and/or H_2O_2 to trigger and promote a Fenton reaction in cancer cells. The lower pH caused by hypoxia in the tumor microenvironment facilitates the release of iron ions from nanomaterials, which triggers the Fenton reaction and leads to the ferroptotic death of cancer cells. Some nanotherapeutics enhance the uptake of ferroptosis inducers into cancer cells, while others provide exogenous regulation of lipid peroxidation to cancer cells through, e.g., PUFA supplementation.

4. Chemical Aspects of Ferroptosis Modulators

The induction and inhibition of ferroptosis by small compounds have been well characterized so far. They alter the concentration of ROS and, thus, of (per)oxidized lipid species through general mechanisms such as free radical scavenging and iron chelation and/or by modulation of specific biochemical pathways.

Ferroptosis inducers are compounds that stimulate iron accumulation and/or inhibit GPX4 expression and/or activity, thereby promoting lipid ROS production and accumulation. GPX4-regulated ferroptosis can be induced in two ways. Class I inducers such as erastin inhibit GPX4 by causing depletion of its cofactor, intracellular GSH. Class II inducers such as RSL3 directly inhibit GPX4 by binding to it. Ferroptosis inhibitors, such as iron chelators and lipophilic antioxidants, have the opposite effect of reducing lipid ROS concentration.

In this analysis, a set of 78 organic ferroptosis modulators (Tables 1 and 2) with a MW up to 800 is described and compared in relation to the aspects of structural diversity and drug-likeness. The collected ferroptosis modulators represent different chemical scaffolds that have been employed as privileged scaffolds for designing novel chemical libraries of ferroptosis inducers or inhibitors and for structure–activity relationship (SAR) analysis. The dendrograms in Figure 2a,b depict the structural diversity of the collected inducers and inhibitors, respectively. The collected set of ferroptosis inducers is a group of chemically diverse compounds. Employing a TC value of 0.85 (Jaccard = 1 − TC of 0.15; cyan line in Figure 2a,b) as a lower limit of chemical similarity, only artemisinin and its dihydrogenated derivate DHA are mutually structurally similar, and they are placed within the same group. Other inducers are structurally unique compounds forming one-member clusters. In comparison, ferroptosis inhibitors are more mutually similar molecules (Figure 2b). The 23 polyphenolic antioxidants are grouped into three clusters as they possess the same core scaffolds with various numbers and positions of OH substituents and/or sugars. The inducers and the inhibitors have a few structural similarities (Figure 2c).

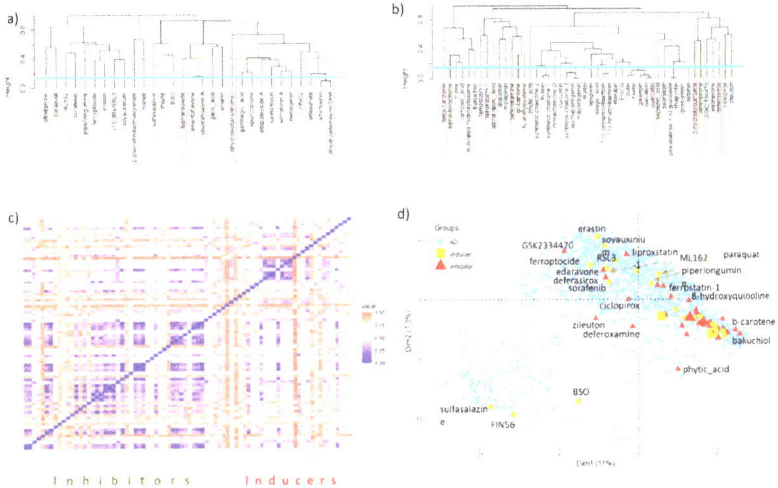

Figure 2. Analysis of structural diversity of ferroptosis modulators described with MACCS fingerprints and using the Jaccard index (1 − TC) as a dissimilarity measure (y-axis). Cluster dendrograms for (**a**) 30 inducers and (**b**) 48 inhibitors. Clusters are denoted with borders drawn at a TC level of 0.6, while the cyan line corresponds to a TC of 0.85. (**c**) Structural (dis)similarity of ferroptosis modulators. More blue/red values denote greater/less structural similarity. (**d**) The ferroptosis inducers and inhibitors fall within the chemical space of approved drugs (AD) which is represented by the first two PCA components, explaining 18.3% of the variance in MACCS fingerprints.

The structures of ferroptosis modulators are compared to those of drugs available on the market(s) by principal component analysis (PCA). Most of the modulators (53/78 in total; 16/30 inducers; 37/48 inhibitors) are grouped in the bottom-right part of the PCA score plot (Figure 2d), following their proximity already shown in the dendrograms (Figure 2a,b). They are inhibitors with antioxidative (20/25), iron chelation (11/15), and GPX4 activation (3/3) modes of action plus inducers with iron accumulation (6/6) or GPX deactivation (5/10) effects. They are clustered together with around two-thirds of the drugs from the ATC groups D (dermatologicals), G (genito-urinary system and sex hormones), and H (systemic hormonal preparations, excluding sex hormones and insulins). The rest of the ferroptosis modulators are scattered in the score plot.

The placement of ferroptosis modulators within the structural space of drugs also implies their drug-like character (Figure 2d). The general, whole-molecule physicochemical properties of compounds are known to influence their ADME (absorption, distribution, metabolism, and excretion) profile and drug-likeness [39]. Although research on the significance of ferroptosis in anticancer therapy has progressed, there are still limitations to its clinical application because of the low solubility and poor membrane permeability of ferroptosis modulators (Tables S1 and S2). For instance, the iron-chelating agent deferoxamine is a classic ferroptosis inhibitor used to avoid excessive ferroptosis and injury to normal cells and tissues. However, its low solubility limits its application. The poor ADME profiles of compounds not only decrease target efficiency but also increase the likelihood of undesirable, off-target effects on normal cells.

The majority of low-MW ferroptosis modulators are natural or natural-based molecules that fit the drug-likeness profile in terms of log*P* values and numbers of H-bond donors, which are considered the two most important parameters in predicting oral bioavailability and drug-likeness [39]. The 10 ferroptosis modulators do not obey the well-known Lipinski rule of five, mainly due to molecular size and the number of H-bond donors greater than 500 and 5, respectively. The 75 molecular physicochemical features important for biological activities were calculated and used for uncovering the (di)similarity of ferroptosis modulators by applying PCA. Five statistically significant molecular features were found to be important for the differentiation of ferroptosis modulators in a physicochemical space. They are: molecular weight (MW), diffusion coefficient in water (Hayduk–Laudie formula) ($\log(\text{Diff} \times 10^{-5}/(\text{cm}^2/\text{s}))$, the number of hydrogen-bond-donating atoms (HBD), lipophilicity coefficient (log*P*), and the number of amide groups (Amide) (Figure 3). These descriptors are known to influence aqueous solubility and distribution properties such as passing biological membranes [40].

Ferroptosis modulators with different modes of action have different physicochemical profiles (Figure 3b). Ferroptosis inhibitors with antioxidant activity are diverse molecules in terms of their structures and mechanisms of action, as well as their physicochemical profiles. Small, hydrophilic antioxidants such as vitamin C, *N*-acetylcysteine, and GSH acting in the aqueous environment are placed in the lower-left part of the PCA plot. Relatively large lipophilic antioxidants such as α-tocopherol and β-carotene scavenge lipid ROS in the hydrophobic environment in membranes. Iron-chelating inhibitors are scattered in the PCA plot, which illustrates that they have various H-bond/H-atom-donating capacities and molecular size, but almost all lie under the *x*-axis, showing that they are, in general, hydrophilic compounds. X_c^- system inhibitors and iron accumulation inducers are among the largest and most lipophilic ferroptosis modulators.

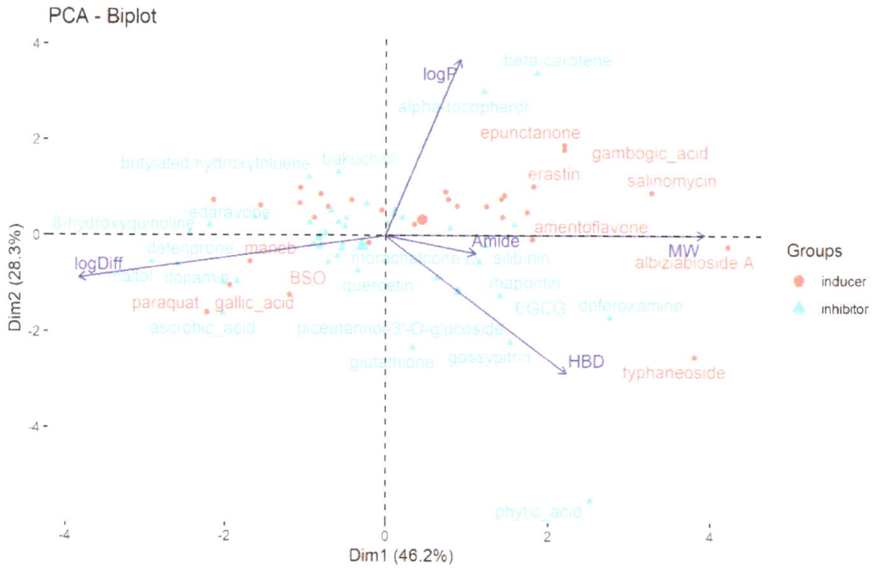

(**a**)

Figure 3. *Cont.*

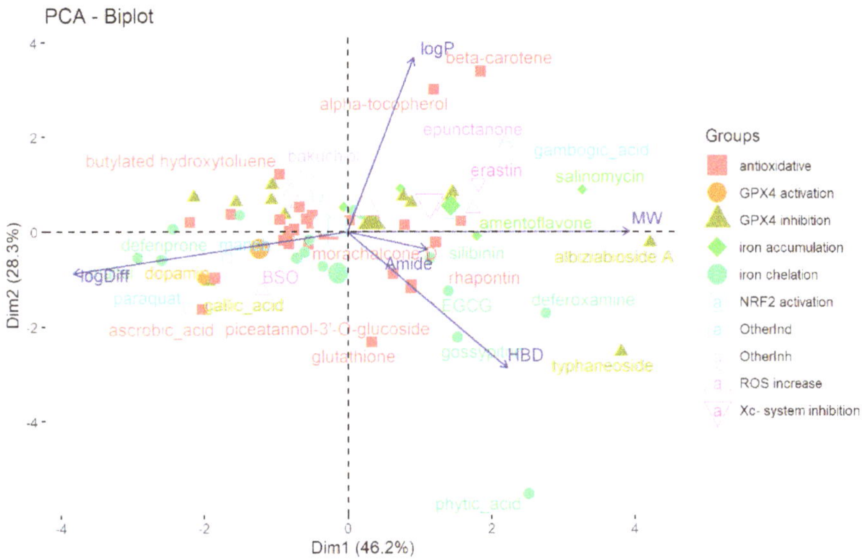

(**b**)

Figure 3. The PCA biplot generated in terms of molecular traits (blue arrows) that describe similarity/difference in the physicochemical space of 78 ferroptosis modulators colored according to (**a**) the class of inhibitors/inducers and (**b**) the mode of action.

The ferroptosis-activating or -suppressing effects of low-MW compounds were mainly detected in terms of their in vitro anticancer and neuroprotective effects on cancer cell lines and neuronal cells, respectively. Hence, the estimation of the capacity of a compound for crossing the BBB and retention in the brain once taken up (Tables 1, 2, S1 and S2) was used to assess whether or not a compound is likely to be active in the central nervous system (CNS). CNS activity may be a(n) (un)desirable side effect of a cancer chemotherapeutic. Dozens of inducers (e.g., FINO2, dihydroisotanshinone I, artemisinin, DHA, 2-phenethylisothiocyanate, altretamine, piperlongumine) and inhibitors (e.g., α-tocopherol, bakuchiol, butylated hydroxytoluene, liproxstatin-1, ferrostatin-1, 8-hydroxyquinoline) are estimated to cross the BBB and be retained in the brain (Tables S1 and S2) [33,41].

5. Biological Aspects of Ferroptosis Modulators

We further analyzed the SAR for ferroptosis modulators. We used the ClassyFire approach to assign chemical (super)classes (Tables S1 and S2) [42]. The class was determined by the core scaffold in a compound. Most ferroptosis modulators are natural products or semisynthetic derivatives belonging to the superclasses of "phenylpropanoids and polyketides" (29 compounds) or "lipids and lipid-like molecules" (13 compounds) according to their biosynthetic origin (Tables 1 and 2). The majority of the remaining modulators are derivatives of organic acids or compounds with structurally diverse, heterocyclic scaffolds.

Almost one-third of the inducers and some inhibitors belonging to the "lipids and lipid-like molecules" are prenol lipids synthesized by the condensation of isoprene subunits. This class includes the inducers artesunate, ruscogenin, and salinomycin, which increase intracellular iron(II) concentration and lipid peroxidation, and lipophilic antioxidant inhibitors α-tocopherol and β-carotene. Of the 29 ferroptosis modulators from "phenylpropanoids and polyketides", six molecules are ferroptosis inducers, and most of the remaining 23 molecules are polyphenolic ferroptosis inhibitors belonging to the flavonoid stilbene and linear 1,3-diarylpropanoid chalcone types. Other ferroptosis inhibitors are lipophilic (vitamin A and α-lipoic acid) and hydrophilic (GSH, vitamin C, dopamine) compounds (Table 2). In this article, we focus mainly on (poly)phenols that act as ferroptosis inducers and inhibitors, or both, and which belong mainly to the superclass of "phenylpropanoids and polyketides".

5.1. Ferroptosis Inducers

The first ferroptosis modulators discovered were the inducers erastin and RAS-selective lethal 3 RSL3 ((1*S*,3*R*)-RSL3 given the stereochemistry). Erastin is an inhibitor of the X_c^- system, while RSL3 is a covalent inhibitor of GPX4. These synthetic compounds were discovered as ferroptosis inducers by the screening of diverse compounds. In 2003, erastin was detected to induce a new form of non-apoptotic cell death in RAS-overexpressing cancer cells [43]. In 2007, it was found to bind directly to voltage-dependent anion channel protein VDAC2 in the mitochondrial outer membrane, leading to mitochondrial production of ROS, mitochondrial dysfunction, and non-apoptotic cell death of RAS-overexpressing cancer cells. In 2008, RSL3 was revealed to induce cell death by involving labile iron [44]. Erastin, glutamate, sorafenib, and sulfasalazine are type I inducers that block the X_c^- system, resulting in a decrease in GSH levels. RSL3, altretamine, ML162, and FIN56 are type II inducers that affect GPX4 either by directly inhibiting the enzyme or by reducing its expression level. Erastin and RSL3 are not readily bioavailable molecules. However, erastin analogs (e.g., PRLX 93936) and other X_c^- system inhibitors, such as multikinase inhibitors sorafenib and sulfasalazine, as well as GPX4 inhibitors, e.g., altretamine and withaferin A, are under clinical anticancer investigation [45].

Some currently used drugs have been found to induce ferroptosis in addition to their already known mechanisms of action [46]. For example, the cancer chemotherapeutic agent cisplatin not only prevents DNA repair but also induces ferroptosis by depleting intracellular GSH in lung A549 and colon HCT116 cancer cells [47]. The most potent anti-malarial drug isolated from traditional Chinese herb *Artemisia annua* L., artemisinin, and its derivative DHA, selectively targets KRAS-reprogrammed pancreatic cancer cells that are resistant to apoptosis. They accumulate in lysosomes and increase ferritin degradation and, thus, iron release, which promotes the accumulation of cellular ROS and leads to ferroptotic cancer cell death. DHA was found to accelerate the degradation of ferritin through downregulation of the activity of the mTOR/p70S6k signaling pathway by activating the phosphorylation of AMPK. DHA also induced ferroptosis of acute myeloid leukemia (AML) cells through autophagy by regulating the activity of the AMPK/mTOR/p70S6k signaling pathway. The water-soluble artemisinin derivative artesunate selectively killed head and neck cancer (HNC) cells but not normal cells by decreasing cellular GSH level and increasing lipid ROS levels [48]. Inhibition of the NRF2 pathway (by NRF2 genetic silencing or by trigonelline) increased the artesunate sensitivity of ferroptosis-resistant HNC cells.

Polyphenols are natural products known for a plethora of bioactivities, including their diverse effects on ferroptosis cell death [13–17,49]. They have shown significant anticancer activity in vitro, even against an aggressive type of cancer, and their chemopreventive role has been summarized elsewhere [50,51]. Grounded in many in vitro and in vivo (mostly rodent) preclinical studies, clinical trials with polyphenols alone or in a combination with anticancer drugs have been carried out on hormone-dependent prostate and breast cancers, bladder and renal cancers, and colorectal and lung cancers, as well as on leukemia. They can trigger several cell death mechanisms simultaneously. They can act as ferroptosis inducers or inhibitors or both. It is known that the dominant target activity of polyphenols depends on the dose, treatment duration, and cell/tissue specificity [52].

Polyphenols have already been used in dermatology and are known to exert antiproliferative, proapoptotic, and antimetastatic activities on melanoma cell lines, whereas they have no cytotoxic effect in healthy cells [53,54]. Since ferroptotic pathways contribute to the regulation of the differentiation state of melanoma cells and their resistance to certain therapeutic agents, ferroptosis inducers are expected to have strong therapeutic potential in melanoma [55]. The majority of collected polyphenolic ferroptosis modulator representatives were found to be structurally most similar to drugs from related ATC groups D, G, and H (Figure 2d). The local, topical application of polyphenols may have advantages over their oral administration because these molecules are generally rapidly metabolized and weakly bioavailable, in addition to having multitargeting effects. The observed effects on cancer cells of ferroptosis inducers from the "phenylpropanoids and polyketides" superclasses (except gallic acid) are described further.

Typhaneoside, a major flavonoid in the extract of *Typha* spp. (Typhaceae) pollen, was found to strongly inhibit the proliferation and growth of AML cells by interacting with multiple targets simultaneously [56]. Typhaneoside (40 μM) induced apoptosis, autophagy, and ferroptosis in AML cells. The induction of ferroptosis was iron-dependent and attended by mitochondrial dysfunction and reduced GSH and GPX4 levels. It triggered autophagy in AML cells by promoting AMPK signaling, which contributes to ROS accumulation, ferritin degradation, and ferroptotic cell death. The anticancer activity of typhaneoside was also confirmed in vivo using BALB/c mice xenografts bearing HL60 cells.

A biflavonoid obtained by the oxidative coupling of two molecules of apigenin robustaflavone A isolated from *Selaginella trichoclada* (Selaginellaceae) induced ferroptosis in breast cancer cells MCF-7 with a cytotoxic IC_{50} value of 11.89 μM (doxorubicin 12.62 μM, MTT test) [57]. It (5 and 10 μM) decreased the expression of E3 ubiquitin ligase Nedd4 (neuronal precursor cell-expressed developmentally downregulated 4), thereby reducing the ubiquitination and proteasomal degradation of VDAC2 proteins. In response, VDAC2 protein expression was enhanced, leading to lipid peroxidation and ROS production in mitochondria and MCF-7 cell death. After the addition of the ferroptosis inhibitors, the antioxidant ferrostatin-1, or the iron chelator deferoxamine, the MCF-7 viability was significantly increased. Recently, by RNA sequencing and KEGG functional enrichment analysis, robustaflavone A was shown to reduce the expression of ferroptosis-related genes including *ACSL4*, *NOXO1*, *NOXA1*, *ACSL5*, *STEAP3*, *LPCAT3*, *ATG7*, and *TP53* in MCF-7 cells [58].

Another biflavonoid of apigenin, widely distributed in the *Selaginella* species, amentoflavone was reported to have a multitarget anticancer ability. This polyphenol (10 and 20 μM) was found to suppress growth and induce death of glioblastoma cells U251 and U373 by triggering ferroptosis in an autophagy-dependent manner [59,60]. Its inhibitory effect on cell proliferation was suppressed by deferoxamine and ferrostatin-1 as well as by the upregulation of FTH1. In cells U251 and U373, amentoflavone increased intracellular levels of iron, malondialdehyde (MDA), and lipid ROS and decreased GSH level and mitochondrial membrane potential. Selectively, in glioma cells, and not in normal human astrocytes, amentoflavone induced ferroptosis by modulating iron homeostasis through suppression of FTH1 levels. It induced autophagy via the AMPK/mTOR/p70S6K pathway, which resulted in the downregulation of FTH1 expression. Its effect on inhibition of tumor growth by inducing autophagy-dependent ferroptosis was additionally demonstrated in vivo in a xenograft mouse model.

Erianin is a natural polyphenol found in *Dendrobium chrysotoxum Lindl* (Orchidaceae), and it has anticancer activity against various cancers—osteosarcoma, nasopharyngeal carcinoma, bladder, and lung cancer. Erianin inhibited the proliferation and metastasis of lung cancer via calcium/calmodulin-dependent ferroptosis in vitro and in vivo [61]. In lung cancer cells H460 and H1299, erianin (at concentrations ranging from 12.5 to 100 nM) induced G2/M-phase arrest, ROS accumulation, lipid peroxidation, GSH depletion, and downregulation of the expression of the negative regulatory proteins for ferroptosis, GPX4, CHAC2, SLC40A1, SLC7A11, and glutaminase. The ferroptosis inhibitors ferrostatin-1 and liproxstatin-1, but not the pan-caspase inhibitor Z-VAD-FMK, the potent inhibitor of autophagy chloroquine, or the potent inhibitor of necroptosis necrostatin-1, rescued cells from erianin-induced cell death. Calcium/calmodulin signaling is a critical mediator of erianin-induced ferroptosis. The blockade of this signaling with ruthenium red and antagonist calmidazolium, significantly rescued cell death induced by erianin treatment by suppressing ferroptosis. Inhibition of calcium/calmodulin signaling significantly reduced the expression of transferrin and increased the expression of GPX4 and SLC7A11.

Ferroptosis was also found to contribute largely to the erianin-induced cell death of bladder cancer in vitro and in vivo [62]. Erianin (100 μg/mL, i.e., 314 μM) suppressed the growth of bladder cancer cell lines KU-19–19 and RT4. It induced G2/M-phase arrest, ROS, and MDA accumulation, GSH depletion, and downregulation of ferroptosis-related proteins FTH1, GPX4, HO-1, glutaminase, and X_c^-/SLC7A11 and inactivated NRF2. The compound *tert*-butylhydroquinone, an NRF2 activator, suppressed erianin-induced ferroptosis, whereas NRF2 inhibition by shRNA augmented the ferroptosis response induced by erianin treatment.

Epunctanone, a polyprenylated benzophenone isolated from two African plants *Garcinia epunctata* (Clusiaceae) and *Ptycholobium contortum* (Fabaceae), has been identified as a promising cytotoxic molecule against nine cancer cell lines including multidrug-resistant ones [63]. It was shown to induce ferroptosis in addition to apoptosis in leukemia cells CCRF-CEM (at a concentration of 11.8 μM).

Piperlongumine, an electrophilic, natural compound isolated from the long pepper *Piper longum* L. (Piperaceae), was selectively toxic to cancer cells in vitro and in vivo [64,65]. It induced the death of human pancreatic cancer cells PANC-1, in part by ferroptosis [66]. The combined treatment of piperlongumine (5–10 μM) with the plant growth regulator cotylenin A (24.1 μM) and/or the clinically approved ferroptosis inducer sulfasalazine (200 μM) enhanced PANC-1 cell death. Its cancer-cell-killing activity was inhibited by the antioxidant *N*-acetylcysteine (3 mM) and the ferroptosis inhibitors ferrostatin-1 (1 μM) and liproxstatin-1 (1 μM), as well as the iron chelator deferoxamine (200 μM), but not by the apoptosis inhibitor Z-VAD-FMK or the necrosis inhibitor necrostatin-1. Piperlongumine in the form of a nanosystem was found to induce the non-apoptotic cell death of mouse breast cancer cell line TH1 via ferroptosis and pyroptosis (an inherently inflammatory kind of programmed cell death activated by caspases 1, 4, and 5). The nanosystem Tf-LipoMof@PL consists of piperlongumine (PL) encapsulated in an iron-containing metal–organic framework (MOF) coated with a transferrin-decorated, pH-sensitive lipid layer (Tf-Lipo). In this multifunctional nanosystem, piperlongumine contributed to ferroptotic cell death by providing H_2O_2 to increase ROS generation through the Fenton reaction [64]. Transferrin facilitated the accumulation of intracellular iron levels, while a pH-sensitive DOPE (dioleoylphosphatidylethanolamine) lipid layer enhanced cellular uptake, prevented early drug leakage, and enabled pH-responsive piperlongumine release in response to low pH at the tumor site. The anticancer effect of the dual-inductive nanosystem was demonstrated in vivo on the xenograft mice model.

The anticancer effect of classical chemotherapeutics, targeted drugs, or radiotherapy can be enhanced by combination with ferroptosis inducers [47]. Gallic acid (3,4,5-trihydroxy benzoic acid, GA) is a phenolic compound with anticancer and antioxidant properties [67]. It was found to cause the death of cervical cancer cells HeLa by early-stage ferroptosis via inhibition of GPX4 activity, mid-stage apoptosis, and late-stage necroptosis at 50 μg/mL concentration (i.e., 294 μM) [68,69]. In another study, while low-level laser irradiation (red) was unable to cause death in breast (MDA-MB-231) and melanoma (A375) cancer cell lines, the pre-irradiation followed by treatment with gallic acid (IC_{50} of 25 and 50 μg/mL, i.e., 147 μM and 294 μM, respectively) reduced the cancer cell survival significantly more than gallic acid alone [68]. The viability of the human skin fibroblasts was not altered, and the effect was greater than that caused by the first treatment with gallic acid followed by low-level laser irradiation. Irradiation of the cells promoted penetration of gallic acid and caused (in addition to apoptosis) ferroptosis via decreasing GPX4 activity and increasing lipid peroxidation. The biomimetic nanoreactor was composed of 4,4′-azonzenecarboxylic acid (Azo)–BSA functionalized hybrid zeolitic imidazolate framework (ZIF), which encapsulated Fe(III)–gallic acid and glucose oxidase (GOx, for sustained oxygen consumption resulting in hypoxia microenvironment) [70]. This Fe(III) –GA/GOx@ZIF–Azo nanoplatform was applied to the MCF-7 breast cell line. It enabled hypoxia-activated positive feedback of cellular uptake and more efficient ferroptotic therapy. The Fenton reaction was accelerated not only by the sustained supply of Fe^{2+} and H_2O_2 but also by the low pH and photothermal stimulation of the reduction of Fe^{3+} to Fe^{2+} ions. Gallic acid has also been used for the synthesis of nanoprobes applicable for magnetic resonance imaging and photothermal cancer therapy [71]. The nanoprobes were constructed using luminescence nanoparticles (UCNP) as the core and the Fe^{3+}/gallic acid complex as the shell, allowing the release of Fe^{3+} ions in the tumor microenvironment in response to the slightly acidic pH. UCNP@GA-FeIII probes acquire specific tumor-targeting ability by absorbing the unsaturated transferrin from serum that recognizes the overexpressed TfR1 on the surface of various solid, malignant tumor cells, including colorectal cancer cells LS180.

5.2. Ferroptosis Inhibitors

Ferroptosis inhibitors can act as GPX4 activators, free radical scavengers, iron chelators, and/or NRF2 activators. The most commonly used ferroptosis inhibitors for in vitro experiments, ferrostatin-1 and liproxstatin-1, act as antioxidants. Many polyphenols are well characterized as radical scavengers, iron chelators, and/or NRF2 activators [51,72], which contributes to their suppressive effect on ferroptosis. The mechanism of their antioxidant effects depends on the number and position of hydroxy groups on the core benzo-γ-pyrone fragment, as well as on the dose and cell type [7].

The food flavonoid apigenin (20 μM) inhibited ferroptosis induced by kainic acid in human neuroblastoma cells SH-SY5Y [73]. Apigenin also rescued mouse brain in vivo from myeloperoxidase (MPO)-mediated oxidative stress. Kainic-acid-induced upregulation of MPO, and, thus, HClO generation, was accompanied by reduced activities of SIRT1 and GPX4, while apigenin treatment decreased expression of MPO and upregulated expression of SIRT1 and the intracellular antioxidants GPX4, TrxR (thioredoxin reductase), and GSH. Conversely, in multiple myeloma cell line NCl-H929, apigenin (5–40 μM) induced cell death by ferroptosis even in the presence of ferrostatin-1 and deferoxamine [74]. Chloroform extract from *Fumaria officinalis* (Fumariaceae), which contains two flavonoids, apigenin and isoquercetin, in addition to isoquinoline alkaloids, also stimulated iron-dependent death in the cell line NCI-H929 [75]. Apigenin also induced the death of lung cancer cells A549 partly through ferroptosis (in addition to apoptosis) [76]. The ferroptotic effect of apigenin (API) on A549 cells was synergistically enhanced by its incorporation into magnetic iron oxide nanoparticles modified on the surface with polysaccharide hyaluronic acid Fe_2O_3/Fe_3O_4@mSiO_2-HA. Such a nanocomposite enabled the sustained release of poorly water-soluble apigenin and the specific targeting of cancer cells with expressed channel protein CD44 on the cell membrane. The API-Fe_2O_3/Fe_3O_4@mSiO_2-HA nanosystem was found to significantly increase ROS and cell lipid peroxidation levels, as well as downregulate GPX4 and FTH1 in A549 cells, compared to pure apigenin.

Two other 3-hydroxy flavones, or so-called flavonols, galangin and kaempferol (Figure 4), protect neurons from ferroptosis. Galangin, a flavonol from the Chinese medicinal herb *Alpinia officinarum* (Zingiberaceae), protected hippocampal neurons in gerbils after ischemia reperfusion by inhibiting ferroptosis via activation of the SLC7A11/GPX4 axis. A deficiency of GSH indirectly caused suppression of GPX4 activity [77]. Kaempferol has been used to treat neuronal cells after oxygen-glucose deprivation/reoxygenation associated with ischemic stroke. It protected cells by activation of the NRF2/SLC7A11/GPX4 signaling pathway [78].

Naringenin, a flavanone from fruits and herbs, was studied in cardiomyocytes H9C2 [79]. Hypoxia reperfusion induced with erastin in cardiomyocytes was alleviated by naringenin, as it increased the expression of proteins NRF2, SLC7A11, GPX4, FTH1, and the iron export protein ferroportin 1 (FPN1) and decreased the expression of NADPH oxidase NOX1. Iron uptake mediated the activation of NOX1 signaling, which induced the release of ROS and mitochondrial damage. All four polyphenols shared resorcinol moiety in ring A (Figure 4) and suppressed ferroptosis through regulation of the NRF2/SLC7A11/GPX4 axis.

Another polyphenolic antioxidant with a *para*-OH group in B-ring puerarin, 8C-glucoside of isoflavone daidzein (Figure 4), was shown to inhibit ferroptosis and lipopolysaccharides (LPS)-induced inflammatory response in A549 cells. Total iron and divalent iron levels, lipid peroxidation, and the NOX1 expression were decreased significantly upon addition of puerarin in LPS-modified A549 cells. In contrast, the expression of SLC7A11, GPX4, and FTH1 was increased [80].

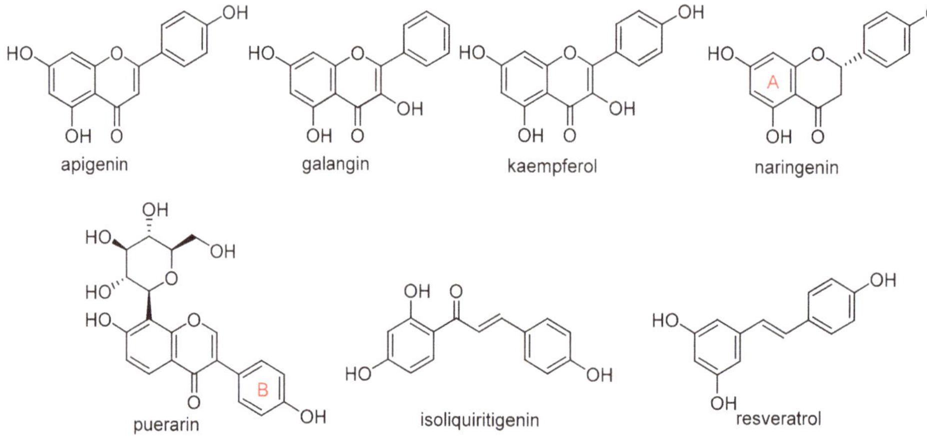

Figure 4. Flavonoids and resveratrol with resorcinol and/or *para*-OH phenyl moieties showing antiferroptotic activity.

There are plenty of structurally diverse polyphenols containing catechol (1,2-dihydroxy benzene) or galloyl (1,2,3-trihydroxy benzene) moieties, which are responsible for their strong antioxidant activity. In addition to their high potential to directly scavenge free radicals, they can form complexes with Fe^{2+} ions, which can further stimulate the oxidation of Fe^{2+} to Fe^{3+} and prevent the recycling of Fe^{2+} ions [81].

Quercetin, named after *Quercus* sp. (oak tree) [82], is a catecholic flavonol. Its antioxidant and chelating effects, as well as its anticancer activity, have been reviewed many times [6,83–86]. Its metal-chelating activity has been linked to antioxidant activity, and its metal complexes also exert antioxidant activity [82,87]. Quercetin could chelate Fe^{2+} ion by catechol functionality at a stoichiometric ratio of 1:2 metal/ligand and attenuated lipid peroxidation, as well as protein oxidation, in the liver, kidneys, and hearts of mice overloaded by iron–dextran complex [88]. Quercetin is more toxic to cancer cells than to normal cells [83]. While quercetin has antioxidant and chemopreventive effects at low concentrations, it exhibits pro-oxidant effects at high doses [86,89]. The oxidized product quercetin quinone is responsible for its pro-oxidant activity, and the pro-oxidant effects of quercetin also depend on the intracellular GSH levels [51]. Quercetin can also exist in plants as quercetin Diels–Alder anti-dimer (QDAD, Figure 5). The antiferroptotic activity of quercetin and its biflavonoid QDAD was investigated in erastin-induced ferroptosis in a bone-marrow-derived mesenchymal stem cell model [90]. Quercetin had better antiferroptotic activity than QDAD, which may be due to the stronger antioxidant action of quercetin. The IC_{50} value for the antioxidant effect against lipid peroxidation for quercetin (2.2 μM) was better than for QDAD (15.5 μM) and referent antioxidants trolox (136 μM) and ascorbic acid (3.0 μM). Quercetin alleviated ferroptosis in pancreatic β-cells in diabetic mice in vivo [91]. It also inhibited ferroptosis and subsequent inflammation in renal proximal tubular epithelial cells, which contributed to the mitigation of acute kidney injury in vivo [92]. Quercetin reduced MDA and lipid ROS levels and increased GSH level. It was found to inhibit ferroptosis by blocking the expression of activation transcription factor 3 (ATF3), the activation of which plays an important role in cell ferroptosis. It did not affect the ATF3 level in normal mice.

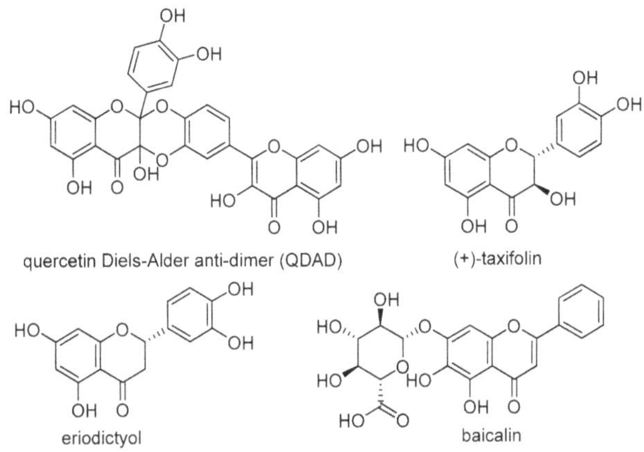

Figure 5. Antiferroptotic flavonoids with catecholic moiety not included in Table 2.

Taxifolin (Figure 5) is a dihydrogenated quercetin derivative belonging to the flavanonol subclass of flavonoids. Its synthetic 7-*O*-esters with cinnamic and ferulic acids (Table 2) showed antiferroptotic activity in the mouse hippocampal neuron cell model HT22, in a way that was different from parent taxifolin and phenolic acids [93]. The HT22 cell line represents immortalized mouse hippocampal neuronal cells that do not express cholinergic and glutamate receptors such as mature hippocampal neurons in vivo. They are commonly used in in vitro studies of the neuronal differentiation and neurotoxicity implicated in brain injuries or neurological diseases.

Taxifolin's analog, without the group 3-OH and with the methylated 7-OH group, the flavanone sterubin from Yerba santa (*Eriodictyon californicum*, Boraginaceae), was studied in association with neuroprotective compounds that can be used against Alzheimer's disease [94]. It was also found to have dose-dependent protection against ferroptosis inducers erastin and RSL3 in hippocampal cells HT22. It has a chelating ability that has been determined in a ferrozine-based assay.

The demethylated derivative of sterubin, eriodictyol (Figure 5), found in the peel of citrus fruits and some Chinese herbal medicines, also showed an anti-Alzheimer's-disease effect through antiferroptotic activity in vitro in HT22 cells and in vivo in the APP/PS1 mouse model of Alzheimer's disease [95]. Treatment with eriodictyol (50 mg/kg) inhibited ferroptosis in the brains of APP/PS1 mice by reduction of the levels of Fe^{2+} ions and total iron, as well as the reduction of the expression levels of TfR1 and FTH1, and increase of the GPX4 and vitamin D receptor (VDR) levels in the cortex and hippocampus of APP/PS1 mice. The iron export protein FPN1 was also upregulated upon eriodictyol treatment. This suggests that eriodictyol might maintain iron balance in cells by reducing iron intake and increasing iron output. The mechanism of antiferroptotic action investigated in HT22 cells was shown to be related to the activation of the NRF2/heme oxygenase-1 (HO-1) signaling pathway mediated by VDR.

By screening a natural product library, a flavon baicalein with the galloyl ring A, was found to be a ferroptosis inhibitor. It markedly inhibited erastin-induced ferroptosis in pancreatic cancer cells compared to known ferroptosis inhibitors. It limited erastin-induced Fe^{2+} iron accumulation, GSH depletion, and lipid peroxidation by suppressing erastin-mediated degradation of GPX4 [96]. Another mechanism that may contribute to the antiferroptotic effect of baicalein is the significant and selective inhibition of 12/15-lipoxygenase (12/15-LOX), which was demonstrated in an experiment with RSL3-stimulated lipid peroxidation in acute lymphoblastic leukemia (ALL) model cell lines Molt-4 and Jurkat [97]. In a study focused on Alzheimer's disease, baicalein decreased ferroptosis markers lipid ROS, 4-hydroxynonenal, and COX2 (cyclooxygenase-2) and inhibited the expression of 12/15-LOX in a HT22 cell model of iron-induced injury, which was also validated in a mouse model of posttraumatic epilepsy [98]. The antiferroptotic effect of baicalein was also confirmed in melanocytes in vitiligo. Baicalein upregulated GPX4 and reduced TfR1 levels in melanocytes treated with RSL3 and ferric ammonium citrate [99]. 7-*O*-glycoside baicalin (Figure 5), found in the roots of the traditional Chinese medicinal plant *Scutellaria baicalensis* (Lamiaceae), suppressed autophagy-dependent ferroptosis in a study which focused on early brain injury following subarachnoid hemorrhage. The study was based on in vitro and in vivo models, and the evaluated parameters were Fe^{2+}, MDA, ROS, and GSH levels [100]. However, in contrast, baicalin exerted anticancer activity by triggering FTH1-dependent ferroptosis in bladder cancer cells in vitro (in cell lines 5637 (50–60 μg/mL, i.e., 112–134 μM) and KU-19-19 (100–120 μg/mL, i.e., 224–269 μM)) and in vivo (in mice, 200 mg/kg), accompanied by intracellular ROS and iron accumulation [101]. The ferroptosis inhibitor deferoxamine rescued baicalin-induced cell death in both 5637 and KU-19-19 cell lines.

A change in iron concentration may influence the dominant mechanism of action of polyphenols. While an excess of iron does not influence the beneficial effects of flavanone sterubin, this is not the case with flavonol fisetin. Differently to Cu^{2+} ions, Fe^{2+} ions have a significant influence on its anti-inflammatory effect [102]. The antioxidant activity of fisetin, suppression of ROS production, and maintenance of GSH levels are altered differently by metals. Fisetin (5 μM) has been tested in several assays in association with its neuroprotective and anti-inflammatory activities. It has direct antioxidant activity and can also chelate Fe^{2+} ions. Fisetin induced NRF2 in hippocampal HT22 cells and microglial BV-2 cells, but not in the presence of Fe^{2+}. Iron ions blocked NRF2 induction by fisetin in cells of both types. Fisetin additionally reduced glutamate-induced ROS production, but the presence of Fe^{2+} also blocked this effect. In HT22 cells, fisetin completely blocked the ROS production induced by RSL3. While ROS production was not significantly increased by Fe^{2+} ions, it was greatly potentiated by the combination of RSL3 and Fe^{2+} ions with fisetin. This was attributed to the ability of iron to oxidize fisetin and, thereby, change fisetin's effect on the induction of antioxidant transcription factor NRF2. The oxidized form of fisetin acts as a strong pro-oxidant. In contrast, although sterubin also binds iron, the metal does not affect the ability of sterubin to induce NRF2.

Gossypitrin is a hydrophilic 7-*O*-glucoside of flavonol gossypetin, isolated from *Talipariti elatum* Sw. (Majagua azul from the Malvaceae family), which has antioxidant activity [103]. It was tested on iron-induced oxidative damage in HT22 cells and mitochondria isolated from rat brains [104]. It was able to rescue HT22 cells from damage induced by 100 μM Fe(II)-citrate, with an EC_{50} of 8.6 μM. The effect was associated with the prevention of iron-induced mitochondrial membrane potential dissipation and ATP depletion. This substance also prevented Fe(II)-citrate-induced mitochondrial lipid peroxidation with an IC_{50} value of 12.45 μM, which was about nine times more efficient than the prevention of *tert*-butylhydroperoxide-induced peroxidation. It also decreased Fe^{2+} concentration with time, while increasing the O_2 consumption rate and impairing Fe^{3+} reduction by ascorbate. Gossypitrin forms a complex with ferrous Fe^{2+} ions in a 2:1 ratio, accelerates its oxidation to a more stable complex with iron in the ferric Fe^{3+} state, and, thus, impedes iron recycling back to the pro-oxidant Fe^{2+} state required for the ROS production and, thus, suppresses the propagation phase of lipid peroxidation.

The flavonoid group also includes catechins (Figure 6), which belong to the subgroup of flavan-3-ols. These compounds are present in fruits, vegetables, various beverages, wine, juice, cocoa, and chocolate. They are known to be favorable components of green tea [105]. They have cardiac and neurological beneficial effects [106]. The anticancer activities of catechins through various mechanisms have been summarized, and controlled cell death has also been implicated in the mechanisms [107]. (−)-Epigallocatechin (EGC) and (−)-epigallocatechin gallate (EGCG), as well as other catechins, are potent scavengers of superoxide radicals, but they may also act as pro-oxidants [7,108]. The ability to chelate Fe^{3+} decreases in the order EGCG > epicatechin gallate (ECG) > EGC > (−)-epicatechin [106]. Brain-permeable (−)-epicatechin (15 mg/kg) reduced lesion volume and ameliorated neurologic deficits in a collagenase model of intracerebral hemorrhage in mice. It downregulated ferroptosis-related gene expression and acted through NRF2-dependent and independent pathways [109]. The neuroprotective function of EGCG was investigated in cerebellar granule neurons as a simulation of spinal cord injury. EGCG (50 μM) increased the survival rate, inhibited ferroptosis, and upregulated phosphorylation of protein kinase D1 under ferroptotic conditions. The effect was verified in rats [110]. Pretreatment of pancreatic β-cell line MIN6 with EGCG or curcumin (20 μM) inhibited the iron accumulation and ferroptotic cell death triggered by erastin. Both polyphenols attenuated GSH depletion, GPX4 inactivation, and lipid peroxidation [111].

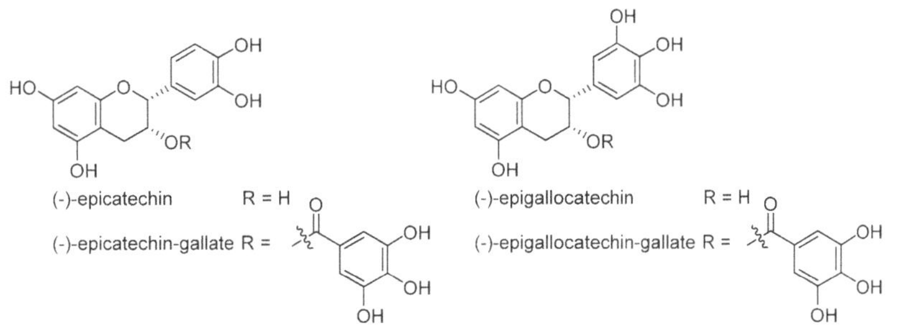

Figure 6. Structure of catechin derivatives with observed antiferroptotic activity.

In contrast, EGCG has been used in the preparation of nanocarriers aimed at inducing ferroptosis selectively in cancer cells. Since EGCG can reduce Fe^{3+} to Fe^{2+}, it has been used in the preparation of nanoparticles with the trio doxorubicin/Fe^{3+}/EGCG, able to induce cancer cell death via ferroptosis and apoptosis. Fe^{3+} ions are released after uptake by cancer cells because cancer cells are acidic and have a high level of glutathione. EGCG is responsible for intracellular iron reduction and the consequent production of hydroxyl radicals through the Fenton reaction and induction of ferroptosis, which enhanced doxorubicin-induced apoptosis in mouse lung carcinoma cell line LL2 [112].

Chalcones (1,3-diphenylprop-2-en-1-ones) are compounds of natural origin serving as starting components for flavonoid biosynthesis in plants [113,114]. Due to the interesting range of biological activities of chalcone derivatives and analogs, the chalcone scaffold is considered to be an important synthetic moiety in medicinal chemistry [115,116]. The antiferroptotic activity of synthetic, hydroxylated chalcones ((2*E*)-3-(3-methylphenyl)-1-(2,3,4-trihydroxyphenyl)prop-2-en-1-one (synthetic chalcone 1 in Table 2), (2*E*)-3-(4-chlorophenyl)-1-(2,3,4-trihydroxyphenyl)prop-2-en-1-one, and (2*E*)-3-(4-methoxyphenyl)-1-(2,3,4-trihydroxyphenyl)prop-2-en-1-one), each containing the galloyl ring A, was tested for lipid peroxidation inhibition in cellular assays. All three compounds were included in the set of inhibitors for chemoinformatic analysis (Table S2). Pretreatment of cells with each of the chalcones (25 μM) was shown to inhibit amyloid-β peptide (Aβ) aggregation in human neuroblastoma SH-SY5Y cells as well as ferroptosis in human embryonic kidney HEK-293 cells. They were able to reduce lipid peroxidation stimulated by RSL3 or erastin in HEK-293 cells with an IC_{50} value of 0.45–1.77 μM and 3.15–3.88 μM, respectively, whereas ECGC did not show any effect up to the concentration of 20 μM [117].

The natural chalcone butein was studied to inhibit ferroptosis, and its antioxidant effect was compared to that of its cyclized product flavanone butin at a dose of 30 μM. Butein inhibited ferroptosis more effectively in erastin-treated, bone-marrow-derived mesenchymal stem cells and showed a stronger antioxidant effect in five different antioxidant assays than butin. The authors concluded that butein exerts an anti-apoptotic effect due to antioxidant action based on the hydrogen-atom transfer pathway. The difference in action between butein and butin was attributed to the decrease in π–π conjugation in butein due to the saturation of the α,β double bond and loss of the 2-hydroxy group upon biocatalytical isomerization [118].

Morachalcone D (Table 2) and morachalcone E are prenylated chalcones isolated from mulberry leaves. In HT22 cells, morachalcone D attenuated erastin-induced ferroptosis with an EC_{50} value of 33.7 ± 0.89 μM. Its effect was compared to quercetin EC_{50} 9.55 ± 0.43 μM. Morachalcone E was only slightly active, which might have been influenced by the elimination of the active hydroxy group and different prenyl pattern. Morachalcone D inhibited the iron accumulation triggered by erastin, which was confirmed by FeRhoNox™-1, an activatable probe detecting labile Fe^{2+} ions in living cells via orange fluorescence. It also upregulated the expression of genes involved in antioxidant defense, including *GPX4, CAT, SOD2, NRF2, HMOX1,* and *SLC7A11,* in erastin-treated HT22 cells in a dose-dependent manner [119].

The antiferroptotic effect of isoliquiritigenin (Figure 4), a component of licorice root, was detected in human kidney epithelial tubular cell line HK2. Pretreatment of the cells with this chalcone inhibited Fe^{2+} ions accumulation and lipid peroxidation in LPS-stimulated HK2 cells. Isoliquiritigenin increased the expression of GPX4 and the chain subunit of the cystine/glutamate transporter SLC7A11 and attenuated mitochondria injury in renal tubular following LPS injection in mice [120].

In a study on constituents of traditional Chinese medicine, coumarins and coumestans were isolated from the fruits of *Psoralea corylifolia* (syn. *Cullen corylifolium,* Fabaceae). Their activity was explored in erastin-exposed HT22 cells. Among the isolated phytochemicals, the coumestan psoralidin was the most active one, with an IC_{50} of 5.21 μM (for comparison, IC_{50} for the standard compound ferrostatin-1 was 0.45 μM). The coumarins psoralen and isopsoralen showed no significant activity [121].

The stilbenoid resveratrol protects cells against oxidative agents at low doses but can promote the production of ROS at high concentrations. It stimulates the KEAP1/NRF2 pathway by activating NRF2 and increasing its expression [122,123]. Its precise mechanism of action is the modification of the amino acid residue Cys151 in KEAP1, which causes newly formed NRF2 to escape from ubiquitination [124]. In mouse pancreatic β cells MIN6, resveratrol inhibited ferroptosis induced by acrolein, a food and environmental pollutant and a risk factor for diabetes. The inhibition was determined by analysis of the biomolecules associated with ferroptosis GPX4, COX2, ACSL4, MDA, GSH, and 5-hydroxyeicosatetraenoic acid (HETE) [125]. The ferroptosis inhibitory effect of resveratrol was also observed in an oxygen-glucose deprivation/reoxygenation model of myocardial ischemia–reperfusion injury in H9C2 cells. Resveratrol (10 μM) alleviated induced oxidative stress and inhibited ferroptosis. It decreased TfR1 expression and increased the expressions of FTH1 and GPX4. It was found to inhibit ferroptosis via the regulation of ubiquity-specific peptidase 19 (USP19)-Beclin1-induced autophagy. The attenuating effect on ferroptosis was also confirmed in vivo in rats [126]. Resveratrol is also known for its neuroprotective effects, but its poor oral bioavailability limits its clinical application. The poor bioavailability of resveratrol has been improved by incorporation into MPEG-PLGA nanoparticles. MPEG-PLGA nanoparticles containing resveratrol accumulated in the endoplasmic reticulum and lysosomes in Madin–Darby canine kidney (MDCK) cells and passed across physiological barriers in a zebrafish model. They inhibited erastin-induced ferroptosis in mouse hippocampal HT22 cells and an intracerebral hemorrhage injury mouse model [127].

However, in head and neck cancer cells HN3 and HN4, resveratrol (20 μM) increased ferroptosis when used in combination with RSL3. Resveratrol is a histone deacetylase SIRT1 inducer. The pharmacological activation of SIRT1 and associated epithelial–mesenchymal transition (EMT) epigenetic reprogramming induced by resveratrol lead to increased sensitivity to the ferroptosis inducer RSL3. The cell viability was significantly decreased after the addition of resveratrol due to a weakened antioxidant system [128].

Another non-flavonoid, polyphenol curcumin (Table 2), is the main component of turmeric, a dietary spice extracted from the root of *Curcuma longa* (Zingiberaceae). This curcuminoid exerts plenty of biological activities, including anticancer and neuroprotective ones. It possesses strong antioxidant activity at low concentrations, whereas, at higher concentrations, it behaves as a potent pro-oxidant. In association with ferroptosis, iron chelation by curcumin is worth mentioning. Its chelating ability was demonstrated at a concentration of 25 μM in liver cancer cell line Huh7 with iron overload obtained by fe-nitriloacetic acid or ferric ammonium citrate [129]. Curcumin is known to have renoprotective properties. It has been studied in an acute kidney injury model in mice caused by rhabdomyolysis. Pre- and posttreatment with curcumin (at a dose of 1 g/kg intraperitoneally) mediated HO-1 induction to prevent oxidative stress and inflammation in vivo. Analogous to the in vivo results, an in vitro mechanistic study conducted in proximal murine tubular epithelial cells showed that pretreatment with curcumin (10 μM) also reduced TLR4/NF-kB and ERK1/2 activation. Within this study, it was demonstrated that ferroptosis is involved in rhabdomyolysis-associated renal damage [130]. The disadvantages of curcumin, such as its poor water solubility, limited oral bioavailability, and inability to efficiently transit across physiological barriers, can be overcome by its encapsulation into polymer-based nanoparticles (Cur-NPs). Cur-NPs (PEG-PTMC) were shown to attenuate the severity of intracerebral hemorrhage injury in a mice model and to suppress erastin-induced ferroptosis in HT22 cells. The absorption, distribution, and elimination properties of Cur-NPs were explored in vitro in MDCK cells and a zebrafish model and in vivo in the brain and plasma of treated mice. Cur-NPs were accumulated in lysosomes, the endoplasmic reticulum, and mitochondria. In HT22 cells, Cur-NPs inhibited ROS production by regulating the NRF2/HO-1 pathway [131]. Due to the presence of two enone moieties in the structure, it can react with Cys151 in KEAP1 to permit NRF2 dissociation and stabilization [128].

However, in cancer cells, curcumin has been shown to induce ferroptosis. Curcumin has anticancer properties that operate through a variety of mechanisms, including inhibition of cancer cell proliferation, invasion and metastasis, regulation of apoptosis, and autophagy. It has been shown to inhibit glioblastoma, breast, and non-small-cell lung cancer (NSCLC) cells via the regulation of ferroptosis. It induced characteristic changes of ferroptosis in vivo in tumor tissues and in vitro in cancer cell lines. Curcumin significantly triggered the cytological and molecular characteristics of ferroptotic cell death in LLC (Lewis lung cancer)-bearing mice (dose 100 mg/kg/day intraperitoneally) and in A549 and H1299 cells (at a 30 μM dose), including depletion of GSH, lipid peroxidation, and accumulation of ROS and iron. In the tumor tissue of mice, the protein level of ACSL4 was upregulated, while the protein levels of SLC7A11 and GPX4 were significantly downregulated by curcumin [132]. In breast cancer cell lines MCF-7 and MDA-MB-231, curcumin (with cell viability IC_{50} values after 48 h of 41.90 μM and 53.51 μM, respectively) caused marked accumulation of intracellular iron, ROS, lipid peroxides, and MDA, while it downregulated GSH levels significantly. It was found to upregulate a variety of ferroptosis genes related to redox regulation, including HO-1, but to downregulate the expression of GPX4 [133]. Curcumin also induced ferroptosis in clear-cell renal cell carcinoma (ccRCC) cells resistant to sunitinib, a tyrosine kinase inhibitor which blocks angiogenesis. Curcumin reversed resistance and enhanced the sensitivity of 786-O-DR (drug-resistant) cells to sunitinib. It reduced the iron content, upregulated the expression of the *ADAMTS18* gene, and significantly reduced expression levels of the ferritin autophagic cargo receptor NCOA4 (nuclear receptor coactivator 4) and the proteins FTH1 and p53 in the cells [134].

6. Data Set and Methods

The sets of diverse inducers and inhibitors of ferroptosis were collected from the literature. The collected ferroptosis modulators were compared mutually and with drugs in terms of their structural and physicochemical/drug-likeness properties. The set of 1390 approved drugs was downloaded from the open-access drug discovery resource ChEMBL (https://www.ebi.ac.uk/chembl, 29 August 2021) [135]. The Anatomical Therapeutic Chemical (ATC) first-level categories of drugs were collected from databases ChEMBL and KEGG [136].

Ferroptosis modulators and drugs are represented with SMILES and corresponding MACCS keys calculated by the R package *rcdk* [137]. The physicochemical molecular features important for biological activities were calculated by the programs ADMET Predictor™ 10.0 (Simulations Plus Inc., Lancaster, CA, USA) (43 descriptors) [18] and DataWarrior (32 features) [138]. The ionization state of molecules was estimated by the program ADMET Predictor™ (Supplementary Table S1).

Structural similarity of the compounds was estimated by applying clustering in terms of MACCS fingerprints with Jaccard distance as a dissimilarity measure and principal component analysis (PCA). The Jaccard distance equals the difference 1−Tanimoto coefficient (TC). TC is the ratio of the number of common features to the number of different features present in two compared molecules. The clustering and PCA were performed by the R functions *hclust* using the complete linkage method and princomp, respectively [137]. The biplot visualization was carried out with the R package factoextra.

The chemical classes of ferroptosis modulators were determined by the ClassyFire algorithm [42].

All chemical structures have been drawn using the program CS ChemDraw Professional version 20.0.0.41 (PerkinElmer, Waltham, MA, USA).

Information on the ferroptosis-related activities of the selected natural and semisynthetic derivatives was gathered through a literature search.

7. Conclusions

In the review, we aimed to present the results of our in silico analysis of the collected structurally diverse representatives of ferroptosis modulators in terms of their structural and physicochemical/drug-likeness properties and to summarize in vitro and in vivo results observed mainly for the large subgroup of natural ferroptosis modulators plant (poly)phenols, primarily phenylpropanoids (Figure 7). Most polyphenols (at the micromolar range of concentrations) have antiferroptotic activity, which may contribute to their neuroprotective capacity. The efficient anticancer activity of typhaneoside, robustaflavone A, amentoflavone, and erianin (at μM levels) in vitro and in vivo is ascribed partly to their capacity to induce ferroptosis. However, the effects of polyphenols considerably depend upon their (micro)environment, for example, on the amount and type of free iron. Some polyphenols (apigenin, baicalin, resveratrol, curcumin) can have an inducing or inhibitory effect on ferroptosis depending on the cell type or composition of multifunctional nanoformulation. Such a dependence enables the construction of different kinds of multifunctional, nanoformulated drug delivery systems of active pharmacological ingredients, including polyphenols, which allow selective and specific induction of ferroptosis in pathological cancer cells.

Phenylpropanoid ferroptosis modulators

Ferroptosis induction for anticancer activity	Dual action polyphenols	Ferroptosis inhibition for neuroprotective activity
Typhaneoside	Apigenin	Galangin
Robustaflavone A	Baicalin	Kaempferol
Amentoflavone	Resveratrol	Naringenin
Erianin	Curcumin	Puerarin
Epunctanone		Quercetin
Piperlongumine		Taxifolin
		Eriodictyol
		Balcalein
		Fisetin
		Gossypitrin
		Green tea catechins
		Butein
		Morachalcone A
		Isoliquiritigenin

Figure 7. Neuroprotective and anticancer activities of plant ferroptosis modulators from phenylpropanoid biochemical origin and represented in the review.

In addition, the performed in silico comparison with approved drugs placed (poly)phenols in the same chemical space as around two-thirds of the drugs from the ATC groups D (dermatologicals), G (genito urinary system and sex hormones), and H (systemic hormonal preparations, excluding sex hormones and insulins). This observation is in accordance with already performed studies on the anticancer activities of polyphenols on pancreatic, prostate, breast, bladder, and renal cancers and melanoma.

Our results can be used to quickly gain insight into the chemical scaffolds and druglike properties of ferroptosis inducers and inhibitors and to motivate relatively new, targeted use of polyphenols.

Supplementary Materials: The following are available online at https://www.mdpi.com/article/10.3390/molecules28020475/s1, Table S1: ClassyFire chemical description, simple molecular descriptors, and physicochemical parameters calculated for the set of 30 ferroptosis inducers, Table S2: ClassyFire chemical description, simple molecular descriptors, and physicochemical parameters calculated for the set of 48 ferroptosis inhibitors.

Author Contributions: Conceptualization, V.S. and M.K.-C.; methodology, V.S.; software, V.S.; formal analysis, V.S.; investigation, V.S. and M.K.-C.; writing—original draft preparation, V.S. and M.K.-C.; writing—review and editing, V.S. and M.K.-C.; visualization, M.K.-C. and V.S. All authors have read and agreed to the published version of the manuscript.

Funding: This research was funded by the Croatian Ministry of Science and Education and the Ministry of Health of the Czech Republic, grant no. NU21-05-00482.

Data Availability Statement: The data used for analysis in this study are available in the Supplementary Material.

Acknowledgments: V.S. would like to acknowledge the use of the software provided by the project Bioprospecting of the Adriatic Sea (KK.01.1.1.01.0002).

Conflicts of Interest: The authors declare no conflict of interest.

References

1. Sung, H.; Ferlay, J.; Siegel, R.L.; Laversanne, M.; Soerjomataram, I.; Jemal, A.; Bray, F. Global cancer statistics 2020: GLOBOCAN estimates of incidence and mortality worldwide for 36 cancers in 185 countries. *CA Cancer J. Clin.* **2021**, *71*, 209–249. [CrossRef] [PubMed]
2. Attwood, M.M.; Fabbro, D.; Sokolov, A.V.; Knapp, S.; Schioth, H.B. Trends in kinase drug discovery: Targets, indications and inhibitor design. *Nat. Rev. Drug Discov.* **2021**, *20*, 839–861. [CrossRef] [PubMed]
3. Carioli, G.; Bertuccio, P.; Boffetta, P.; Levi, F.; La Vecchia, C.; Negri, E.; Malvezzi, M. European cancer mortality predictions for the year 2020 with a focus on prostate cancer. *Ann. Oncol.* **2020**, *31*, 650–658. [CrossRef] [PubMed]
4. Pistritto, G.; Trisciuoglio, D.; Ceci, C.; Garufi, A.; D'Orazi, G. Apoptosis as anticancer mechanism: Function and dysfunction of its modulators and targeted therapeutic strategies. *Aging* **2016**, *8*, 603–619. [CrossRef] [PubMed]
5. Aggarwal, B.B.; Bhardwaj, A.; Aggarwal, R.S.; Seeram, N.P.; Shishodia, S.; Takada, Y. Role of resveratrol in prevention and therapy of cancer: Preclinical and clinical studies. *Anticancer Res.* **2004**, *24*, 2783–2840. [PubMed]
6. Khan, F.; Niaz, K.; Maqbool, F.; Hassan, F.I.; Abdollahi, M.; Venkata, K.C.N.; Nabavi, S.M.; Bishayee, A. Molecular Targets Underlying the Anticancer Effects of Quercetin: An Update. *Nutrients* **2016**, *8*, 529. [CrossRef] [PubMed]
7. Stepanic, V.; Gasparovic, A.C.; Troselj, K.G.; Amic, D.; Zarkovic, N. Selected Attributes of Polyphenols in Targeting Oxidative Stress in Cancer. *Curr. Top. Med. Chem.* **2015**, *15*, 496–509. [CrossRef]
8. Klaunig, J.E.; Kamendulis, L.M. The role of oxidative stress in carcinogenesis. *Annu. Rev. Pharmacol. Toxicol.* **2004**, *44*, 239–267. [CrossRef]
9. Dixon, S.J.; Lemberg, K.M.; Lamprecht, M.R.; Skouta, R.; Zaitsev, E.M.; Gleason, C.E.; Patel, D.N.; Bauer, A.J.; Cantley, A.M.; Yang, W.S.; et al. Ferroptosis: An Iron-Dependent Form of Nonapoptotic Cell Death. *Cell* **2012**, *149*, 1060–1072. [CrossRef]
10. Greco, G.; Catanzaro, E.; Fimognari, C. Natural Products as Inducers of Non-Canonical Cell Death: A Weapon against Cancer. *Cancers* **2021**, *13*, 304. [CrossRef]
11. Yan, H.F.; Zou, T.; Tuo, Q.Z.; Xu, S.; Li, H.; Belaidi, A.A.; Lei, P. Ferroptosis: Mechanisms and links with diseases. *Signal Transduct. Target. Ther.* **2021**, *6*, 49. [CrossRef] [PubMed]
12. Xie, Y.; Hou, W.; Song, X.; Yu, Y.; Huang, J.; Sun, X.; Kang, R.; Tang, D. Ferroptosis: Process and function. *Cell Death Differ.* **2016**, *23*, 369–379. [CrossRef] [PubMed]
13. Ulrich-Merzenich, G.; Zeitler, H.; Vetter, H.; Kraft, K. Synergy research: Vitamins and secondary plant components in the maintenance of the redox-homeostasis and in cell signaling. *Phytomedicine* **2009**, *16*, 2–16. [CrossRef]

14. Firuzi, O.; Miri, R.; Tavakkoli, M.; Saso, L. Antioxidant Therapy: Current Status and Future Prospects. *Curr. Med. Chem.* **2011**, *18*, 3871–3888. [CrossRef]
15. Imam, M.U.; Zhang, S.S.; Ma, J.F.; Wang, H.; Wang, F.D. Antioxidants Mediate Both Iron Homeostasis and Oxidative Stress. *Nutrients* **2017**, *9*, 671. [CrossRef] [PubMed]
16. Zhao, X.Y.; Wang, X.N.; Pang, Y.Z. Phytochemicals Targeting Ferroptosis: Therapeutic Opportunities and Prospects for Treating Breast Cancer. *Pharmaceuticals* **2022**, *15*, 1360. [CrossRef]
17. de Souza, I.; Ramalho, M.C.C.; Guedes, C.B.; Osawa, I.Y.A.; Monteiro, L.K.S.; Gomes, L.R.; Rocha, C.R.R. Ferroptosis Modulation: Potential Therapeutic Target for Glioblastoma Treatment. *Int. J. Mol. Sci.* **2022**, *23*, 6879. [CrossRef]
18. Lawless, M.S.; Waldman, M.; Fraczkiewicz, R.; Clark, R.D. Using Cheminformatics in Drug Discovery. In *New Approaches to Drug Discovery*; Nielsch, U., Fuhrmann, U., Jaroch, S., Eds.; Springer International Publishing: Cham, Switzerland, 2016; pp. 139–168.
19. Davidson, A.J.; Wood, W. Igniting the spread of ferroptotic cell death. *Nat. Cell Biol.* **2020**, *22*, 1027–1029. [CrossRef]
20. Stockwell, B.R.; Angeli, J.P.F.; Bayir, H.; Bush, A.I.; Conrad, M.; Dixon, S.J.; Fulda, S.; Gascon, S.; Hatzios, S.K.; Kagan, V.E.; et al. Ferroptosis: A Regulated Cell Death Nexus Linking Metabolism, Redox Biology, and Disease. *Cell* **2017**, *171*, 273–285. [CrossRef]
21. Gao, M.H.; Yi, J.M.; Zhu, J.J.; Minikes, A.M.; Monian, P.; Thompson, C.B.; Jiang, X.J. Role of Mitochondria in Ferroptosis. *Mol. Cell* **2019**, *73*, 354–363.e3. [CrossRef]
22. Andreini, C.; Putignano, V.; Rosato, A.; Banci, L. The human iron-proteome. *Metallomics* **2018**, *10*, 1223–1231. [CrossRef] [PubMed]
23. Wiernicki, B.; Dubois, H.; Tyurina, Y.Y.; Hassannia, B.; Bayir, H.; Kagan, V.E.; Vandenabeele, P.; Wullaert, A.; Vanden Berghe, T. Excessive phospholipid peroxidation distinguishes ferroptosis from other cell death modes including pyroptosis. *Cell Death Dis.* **2020**, *11*, 11. [CrossRef] [PubMed]
24. Lee, J.Y.; Kim, W.K.; Bae, K.H.; Lee, S.C.; Lee, E.W. Lipid Metabolism and Ferroptosis. *Biology* **2021**, *10*, 184. [CrossRef]
25. Imai, H.; Matsuoka, M.; Kumagai, T.; Sakamoto, T.; Koumura, T. Lipid peroxidation-dependent cell death regulated by GPx4 and ferroptosis. *Curr. Top. Microbiol. Immunol.* **2017**, *403*, 143–170. [CrossRef] [PubMed]
26. Yang, W.S.; SriRamaratnam, R.; Welsch, M.E.; Shimada, K.; Skouta, R.; Viswanathan, V.S.; Cheah, J.H.; Clemons, P.A.; Shamji, A.F.; Clish, C.B.; et al. Regulation of Ferroptotic Cancer Cell Death by GPX4. *Cell* **2014**, *156*, 317–331. [CrossRef] [PubMed]
27. Doll, S.; Proneth, B.; Tyurina, Y.Y.; Panzilius, E.; Kobayashi, S.; Ingold, I.; Irmler, M.; Beckers, J.; Aichler, M.; Walch, A.; et al. ACSL4 dictates ferroptosis sensitivity by shaping cellular lipid composition. *Nat. Chem. Biol.* **2017**, *13*, 91–98. [CrossRef]
28. Steegmann-Olmedillas, J.L. The role of iron in tumour cell proliferation. *Clin. Transl. Oncol.* **2011**, *13*, 71–76. [CrossRef]
29. Wang, Y.F.; Yu, L.; Ding, J.; Chen, Y. Iron Metabolism in Cancer. *Int. J. Mol. Sci.* **2019**, *20*, 95. [CrossRef]
30. Liu, Y.Q.; Gu, W. p53 in ferroptosis regulation: The new weapon for the old guardian. *Cell Death Differ.* **2022**, *29*, 895–910. [CrossRef]
31. Kang, R.; Kroemer, G.; Tang, D.L. The tumor suppressor protein p53 and the ferroptosis network. *Free. Radic. Biol. Med.* **2019**, *133*, 162–168. [CrossRef]
32. Kerins, M.J.; Ooi, A. The Roles of NRF2 in Modulating Cellular Iron Homeostasis. *Antioxid. Redox Signal.* **2018**, *29*, 1756–1773. [CrossRef]
33. Ratan, R.R. The Chemical Biology of Ferroptosis in the Central Nervous System. *Cell Chem. Biol.* **2020**, *27*, 479–498. [CrossRef] [PubMed]
34. Szatrowski, T.P.; Nathan, C.F. Production of large amounts of hydrogen-peroxide by human tumor-cells. *Cancer Res.* **1991**, *51*, 794–798. [PubMed]
35. Lisanti, M.P.; Martinez-Outschoorn, U.E.; Lin, Z.; Pavlides, S.; Whitaker-Menezes, D.; Pestell, R.G.; Howell, A.; Sotgia, F. Hydrogen peroxide fuels aging, inflammation, cancer metabolism and metastasis the seed and soil also needs "fertilizer". *Cell Cycle* **2011**, *10*, 2440–2449. [CrossRef] [PubMed]
36. Rice, M.E. H_2O_2: A Dynamic Neuromodulator. *Neuroscientist* **2011**, *17*, 389–406. [CrossRef]
37. Fujihara, K.M.; Zhang, B.N.Z.; Clemons, N.J. Opportunities for Ferroptosis in Cancer Therapy. *Antioxidants* **2021**, *10*, 986. [CrossRef]
38. Luo, L.X.; Wang, H.; Tian, W.; Li, X.L.; Zhu, Z.; Huang, R.M.; Luo, H. Targeting ferroptosis-based cancer therapy using nanomaterials: And. *Theranostics* **2021**, *11*, 9937–9952. [CrossRef]
39. Ganesan, A. The impact of natural products upon modern drug discovery. *Curr. Opin. Chem. Biol.* **2008**, *12*, 306–317. [CrossRef]
40. Klein, V.G.; Bond, A.G.; Craigon, C.; Lokey, R.S.; Ciulli, A. Amide-to-Ester Substitution as a Strategy for Optimizing PROTAC Permeability and Cellular Activity. *J. Med. Chem.* **2021**, *64*, 18082–18101. [CrossRef]
41. Weiland, A.; Wang, Y.M.; Wu, W.H.; Lan, X.; Han, X.N.; Li, Q.; Wang, J. Ferroptosis and Its Role in Diverse Brain Diseases. *Mol. Neurobiol.* **2019**, *56*, 4880–4893. [CrossRef]
42. Feunang, Y.D.; Eisner, R.; Knox, C.; Chepelev, L.; Hastings, J.; Owen, G.; Fahy, E.; Steinbeck, C.; Subramanian, S.; Bolton, E.; et al. ClassyFire: Automated chemical classification with a comprehensive, computable taxonomy. *J. Cheminform.* **2016**, *8*, 20. [CrossRef]
43. Dolma, S.; Lessnick, S.L.; Hahn, W.C.; Stockwell, B.R. Identification of genotype-selective antitumor agents using synthetic lethal chemical screening in engineered human tumor cells. *Cancer Cell* **2003**, *3*, 285–296. [CrossRef] [PubMed]
44. Yang, W.S.; Stockwell, B.R. Synthetic lethal screening identifies compounds activating iron-dependent, nonapoptotic cell death in oncogenic-RAS-harboring cancer cells. *Chem. Biol.* **2008**, *15*, 234–245. [CrossRef] [PubMed]

45. Xia, X.J.; Fan, X.P.; Zhao, M.Y.; Zhu, P. The Relationship between Ferroptosis and Tumors: A Novel Landscape for Therapeutic Approach. *Curr. Gene Ther.* **2019**, *19*, 117–124. [CrossRef]
46. Wu, Y.N.; Yu, C.C.; Luo, M.; Cen, C.; Qiu, J.L.; Zhang, S.Z.; Hu, K.M. Ferroptosis in Cancer Treatment: Another Way to Rome. *Front. Oncol.* **2020**, *10*, 16. [CrossRef]
47. Guo, J.P.; Xu, B.F.; Han, Q.; Zhou, H.X.; Xia, Y.; Gong, C.W.; Dai, X.F.; Li, Z.Y.; Wu, G. Ferroptosis: A Novel Anti-tumor Action for Cisplatin. *Cancer Res. Treat.* **2018**, *50*, 445–460. [CrossRef]
48. Roh, J.L.; Kim, E.H.; Jang, H.; Shin, D. Nrf2 inhibition reverses the resistance of cisplatin-resistant head and neck cancer cells to artesunate-induced ferroptosis. *Redox Biol.* **2017**, *11*, 254–262. [CrossRef]
49. Kajarabille, N.; Latunde-Dada, G.O. Programmed Cell-Death by Ferroptosis: Antioxidants as Mitigators. *Int. J. Mol. Sci.* **2019**, *20*, 4968. [CrossRef]
50. Bhosale, P.B.; Ha, S.E.; Vetrivel, P.; Kim, H.H.; Kim, S.M.; Kim, G.S. Functions of polyphenols and its anticancer properties in biomedical research: A narrative review. *Transl. Cancer Res.* **2020**, *9*, 7619–7631. [CrossRef]
51. Gibellini, L.; Pinti, M.; Nasi, M.; Montagna, J.P.; De Biasi, S.; Roat, E.; Bertoncelli, L.; Cooper, E.L.; Cossarizza, A. Quercetin and Cancer Chemoprevention. *Evid.-Based Complement. Altern. Med.* **2011**, *2011*, 591356. [CrossRef]
52. Stepanić, V.; Kujundžić, R.N.; Trošelj, K.G. Epigenome, Cancer Prevention and Flavonoids and Curcumin. In *Epigenetics and Epigenomics*; Payne, C.J., Ed.; IntechOpen: London, UK, 2014; p. 220.
53. Pop, T.D.; Diaconeasa, Z. Recent Advances in Phenolic Metabolites and Skin Cancer. *Int. J. Mol. Sci.* **2021**, *22*, 9707. [CrossRef] [PubMed]
54. Costa, A.; Bonner, M.Y.; Arbiser, J.L. Use of Polyphenolic Compounds in Dermatologic Oncology. *Am. J. Clin. Dermatol.* **2016**, *17*, 369–385. [CrossRef] [PubMed]
55. Leu, J.I.; Murphy, M.E.; George, D.L. Targeting ErbB3 and Cellular NADPH/NADP(+)Abundance Sensitizes Cutaneous Melanomas to Ferroptosis Inducers. *ACS Chem. Biol.* **2022**, *17*, 1038–1044. [CrossRef] [PubMed]
56. Zhu, H.Y.; Huang, Z.X.; Chen, G.Q.; Sheng, F.; Zheng, Y.S. Typhaneoside prevents acute myeloid leukemia (AML) through suppressing proliferation and inducing ferroptosis associated with autophagy. *Biochem. Biophys. Res. Commun.* **2019**, *516*, 1265–1271. [CrossRef] [PubMed]
57. Xie, Y.; Zhou, X.; Li, J.; Yao, X.C.; Liu, W.L.; Kang, F.H.; Zou, Z.X.; Xu, K.P.; Xu, P.S.; Tan, G.S. Identification of a new natural biflavonoids against breast cancer cells induced ferroptosis via the mitochondrial pathway. *Bioorg. Chem.* **2021**, *109*, 11. [CrossRef] [PubMed]
58. Xie, Y.; Zhou, X.; Li, J.; Yao, X.C.; Liu, W.L.; Xu, P.S.; Tan, G.S. Cytotoxic effects of the biflavonoids isolated from Selaginella trichoclada on MCF-7 cells and its potential mechanism. *Bioorg. Med. Chem. Lett.* **2022**, *56*, 5. [CrossRef]
59. Xiong, X.F.; Tang, N.; Lai, X.D.; Zhang, J.L.; Wen, W.L.; Li, X.J.; Li, A.G.; Wu, Y.H.; Liu, Z.H. Insights Into Amentoflavone: A Natural Multifunctional Biflavonoid. *Front. Pharmacol.* **2021**, *12*, 24. [CrossRef]
60. Chen, Y.; Li, N.; Wang, H.J.; Wang, N.N.; Peng, H.; Wang, J.; Li, Y.H.; Liu, M.D.; Li, H.; Zhang, Y.; et al. Amentoflavone suppresses cell proliferation and induces cell death through triggering autophagy-dependent ferroptosis in human glioma. *Life Sci.* **2020**, *247*, 10. [CrossRef]
61. Chen, P.; Wu, Q.B.; Feng, J.; Yan, L.L.; Sun, Y.T.; Liu, S.P.; Xiang, Y.; Zhang, M.M.; Pan, T.; Chen, X.Y.; et al. Erianin, a novel dibenzyl compound in Dendrobium extract, inhibits lung cancer cell growth and migration via calcium/calmodulin-dependent ferroptosis. *Signal Transduct. Target. Ther.* **2020**, *5*, 11. [CrossRef]
62. Xiang, Y.; Chen, X.; Wang, W.; Zhai, L.; Sun, X.; Feng, J.; Duan, T.; Zhang, M.; Pan, T.; Yan, L.; et al. Natural Product Erianin Inhibits Bladder Cancer Cell Growth by Inducing Ferroptosis via NRF2 Inactivation. *Front. Pharmacol.* **2021**, *12*, 12. [CrossRef]
63. Mbaveng, A.T.; Fotso, G.W.; Ngnintedo, D.; Kuete, V.; Ngadjui, B.T.; Keumedjio, F.; Andrae-Marobela, K.; Efferth, T. Cytotoxicity of epunctanone and four other phytochemicals isolated from the medicinal plants Garcinia epunctata and Ptycholobium contortum towards multi-factorial drug resistant cancer cells. *Phytomedicine* **2018**, *48*, 112–119. [CrossRef] [PubMed]
64. Huang, B.K.; Langford, T.F.; Sikes, H.D. Using Sensors and Generators of H2O2 to Elucidate the Toxicity Mechanism of Piperlongumine and Phenethyl Isothiocyanate. *Antioxid. Redox Signal.* **2016**, *24*, 924–938. [CrossRef] [PubMed]
65. Roh, J.L.; Kim, E.H.; Park, J.Y.; Kim, J.W.; Kwon, M.; Lee, B.H. Piperlongumine selectively kills cancer cells and increases cisplatin antitumor activity in head and neck cancer. *Oncotarget* **2014**, *5*, 9227–9238. [CrossRef] [PubMed]
66. Yamaguchi, Y.; Kasukabe, T.; Kumakura, S. Piperlongumine rapidly induces the death of human pancreatic cancer cells mainly through the induction of ferroptosis. *Int. J. Oncol.* **2018**, *52*, 1011–1022. [CrossRef] [PubMed]
67. Tang, H.M.; Cheung, P.C.K. Gene expression profile analysis of gallic acid-induced cell death process. *Sci. Rep.* **2021**, *11*, 17. [CrossRef]
68. Khorsandi, K.; Kianmehr, Z.; Hosseinmardi, Z.; Hosseinzadeh, R. Anti-cancer effect of gallic acid in presence of low level laser irradiation: ROS production and induction of apoptosis and ferroptosis. *Cancer Cell Int.* **2020**, *20*, 14. [CrossRef]
69. Tang, H.M.; Cheung, P.C.K. Gallic Acid Triggers Iron-Dependent Cell Death with Apoptotic, Ferroptotic, and Necroptotic Features. *Toxins* **2019**, *11*, 492. [CrossRef]
70. An, P.J.; Gu, D.H.; Gao, Z.G.; Fan, F.Y.; Jiang, Y.; Sun, B.W. Hypoxia-augmented and photothermally-enhanced ferroptotic therapy with high specificity and efficiency. *J. Mat. Chem. B* **2020**, *8*, 78–87. [CrossRef]
71. Zhang, P.S.; Hou, Y.; Zeng, J.F.; Li, Y.Y.; Wang, Z.H.; Zhu, R.; Ma, T.C.; Gao, M.Y. Coordinatively Unsaturated Fe3+ Based Activatable Probes for Enhanced MRI and Therapy of Tumors. *Angew. Chem.-Int. Edit.* **2019**, *58*, 11088–11096. [CrossRef]

72. Ullah, A.; Munir, S.; Badshah, S.L.; Khan, N.; Ghani, L.; Poulson, B.G.; Emwas, A.H.; Jaremko, M. Important Flavonoids and Their Role as a Therapeutic Agent. *Molecules* **2020**, *25*, 5243. [CrossRef]
73. Shao, C.W.; Yuan, J.W.; Liu, Y.N.; Qin, Y.J.; Wang, X.A.; Gu, J.; Chen, G.Q.; Zhang, B.; Liu, H.K.; Zhao, J.; et al. Epileptic brain fluorescent imaging reveals apigenin can relieve the myeloperoxidase-mediated oxidative stress and inhibit ferroptosis. *Proc. Natl. Acad. Sci. USA* **2020**, *117*, 10155–10164. [CrossRef] [PubMed]
74. Adham, A.N.; Abdelfatah, S.; Naqishbandi, A.M.; Mahmoud, N.; Efferth, T. Cytotoxicity of apigenin toward multiple myeloma cell lines and suppression of iNOS and COX-2 expression in STAT1-transfected HEK293 cells. *Phytomedicine* **2021**, *80*, 15. [CrossRef] [PubMed]
75. Adham, A.N.; Naqishbandi, A.M.; Efferth, T. Cytotoxicity and apoptosis induction by Fumaria officinalis extracts in leukemia and multiple myeloma cell lines. *J. Ethnopharmacol.* **2021**, *266*, 10. [CrossRef] [PubMed]
76. Liu, R.; Rong, G.; Liu, Y.; Huang, W.; He, D.; Lu, R. Delivery of apigenin-loaded magnetic $Fe_2O_3/Fe_3O_4@mSiO_2$ nanocomposites to A549 cells and their antitumor mechanism. *Mater. Sci. Eng. C* **2021**, *120*, 111719. [CrossRef] [PubMed]
77. Guan, X.; Li, Z.H.; Zhu, S.; Cheng, M.J.; Ju, Y.T.; Ren, L.; Yang, G.L.; Min, D.Y. Galangin attenuated cerebral ischemia-reperfusion injury by inhibition of ferroptosis through activating the SLC7A11/GPX4 axis in gerbils. *Life Sci.* **2021**, *264*, 11. [CrossRef] [PubMed]
78. Yuan, Y.; Zhai, Y.Y.; Chen, J.J.; Xu, X.F.; Wang, H.M. Kaempferol Ameliorates Oxygen-Glucose Deprivation/Reoxygenation-Induced Neuronal Ferroptosis by Activating Nrf2/SLC7A11/GPX4 Axis. *Biomolecules* **2021**, *11*, 923. [CrossRef]
79. Xu, S.J.; Wu, B.X.; Zhong, B.Y.; Lin, L.Q.; Ding, Y.N.; Jin, X.; Huang, Z.W.; Lin, M.Y.; Wu, H.L.; Xu, D.P. Naringenin alleviates myocardial ischemia/reperfusion injury by regulating the nuclear factor-erythroid factor 2-related factor 2 (Nrf2)/System xc-/glutathione peroxidase 4 (GPX4) axis to inhibit ferroptosis. *Bioengineered* **2021**, *12*, 10924–10934. [CrossRef]
80. Xu, B.Y.; Wang, H.D.; Chen, Z. Puerarin Inhibits Ferroptosis and Inflammation of Lung Injury Caused by Sepsis in LPS Induced Lung Epithelial Cells. *Front. Pediatr.* **2021**, *9*, 7. [CrossRef]
81. Andjelkovic, M.; Van Camp, J.; De Meulenaer, B.; Depaemelaere, G.; Socaciu, C.; Verloo, M.; Verhe, R. Iron-chelation properties of phenolic acids bearing catechol and galloyl groups. *Food Chem.* **2006**, *98*, 23–31. [CrossRef]
82. Yang, D.Y.; Wang, T.C.; Long, M.; Li, P. Quercetin: Its Main Pharmacological Activity and Potential Application in Clinical Medicine. *Oxidative Med. Cell. Longev.* **2020**, *2020*, 13. [CrossRef]
83. Kashyap, D.; Mittal, S.; Sak, K.; Singhal, P.; Tuli, H.S. Molecular mechanisms of action of quercetin in cancer: Recent advances. *Tumor Biol.* **2016**, *37*, 12927–12939. [CrossRef] [PubMed]
84. Rauf, A.; Imran, M.; Khan, I.A.; Ur-Rehman, M.; Gilani, S.A.; Mehmood, Z.; Mubarak, M.S. Anticancer potential of quercetin: A comprehensive review. *Phytother. Res.* **2018**, *32*, 2109–2130. [CrossRef] [PubMed]
85. Reyes-Farias, M.; Carrasco-Pozo, C. The Anti-Cancer Effect of Quercetin: Molecular Implications in Cancer Metabolism. *Int. J. Mol. Sci.* **2019**, *20*, 3177. [CrossRef] [PubMed]
86. Ulusoy, H.G.; Sanlier, N. A minireview of quercetin: From its metabolism to possible mechanisms of its biological activities. *Crit. Rev. Food Sci. Nutr.* **2020**, *60*, 3290–3303. [CrossRef]
87. Xu, D.; Hu, M.J.; Wang, Y.Q.; Cui, Y.L. Antioxidant Activities of Quercetin and Its Complexes for Medicinal Application. *Molecules* **2019**, *24*, 1123. [CrossRef]
88. Xiao, L.; Luo, G.; Tang, Y.H.; Yao, P. Quercetin and iron metabolism: What we know and what we need to know. *Food Chem. Toxicol.* **2018**, *114*, 190–203. [CrossRef]
89. Batiha, G.E.; Beshbishy, A.M.; Ikram, M.; Mulla, Z.S.; Abd El-Hack, M.E.; Taha, A.E.; Algammal, A.M.; Elewa, Y.H.A. The Pharmacological Activity, Biochemical Properties, and Pharmacokinetics of the Major Natural Polyphenolic Flavonoid: Quercetin. *Foods* **2020**, *9*, 374. [CrossRef]
90. Li, X.C.; Zeng, J.Y.; Liu, Y.P.; Liang, M.S.; Liu, Q.R.; Li, Z.; Zhao, X.J.; Chen, D.F. Inhibitory Effect and Mechanism of Action of Quercetin and Quercetin Diels-Alder anti-Dimer on Erastin-Induced Ferroptosis in Bone Marrow-Derived Mesenchymal Stem Cells. *Antioxidants* **2020**, *9*, 205. [CrossRef]
91. Li, D.; Jiang, C.J.; Mei, G.B.; Zhao, Y.; Chen, L.; Liu, J.J.; Tang, Y.H.; Gao, C.; Yao, P. Quercetin Alleviates Ferroptosis of Pancreatic beta Cells in Type 2 Diabetes. *Nutrients* **2020**, *12*, 2954. [CrossRef]
92. Wang, Y.; Quan, F.; Cao, Q.; Lin, Y.; Yue, C.; Bi, R.; Cui, X.; Yang, H.; Yang, Y.; Birnbaumer, L.; et al. Quercetin alleviates acute kidney injury by inhibiting ferroptosis. *J. Adv. Res.* **2021**, *28*, 231–243. [CrossRef]
93. Gunesch, S.; Hoffmann, M.; Kiermeier, C.; Fischer, W.; Pinto, A.F.M.; Maurice, T.; Maher, P.; Decker, M. 7-O-Esters of taxifolin with pronounced and overadditive effects in neuroprotection, anti-neuroinflammation, and amelioration of short-term memory impairment in vivo. *Redox Biol.* **2020**, *29*, 14. [CrossRef] [PubMed]
94. Fischer, W.; Currais, A.; Liang, Z.B.; Pinto, A.; Maher, P. Old age-associated phenotypic screening for Alzheimer's disease drug candidates identifies sterubin as a potent neuroprotective compound from Yerba santa. *Redox Biol.* **2019**, *21*, 12. [CrossRef] [PubMed]
95. Li, L.; Li, W.J.; Zheng, X.R.; Liu, Q.L.; Du, Q.; Lai, Y.J.; Liu, S.Q. Eriodictyol ameliorates cognitive dysfunction in APP/PS1 mice by inhibiting ferroptosis via vitamin D receptor-mediated Nrf2 activation. *Mol. Med.* **2022**, *28*, 20. [CrossRef] [PubMed]
96. Xie, Y.C.; Song, X.X.; Sun, X.F.; Huang, J.; Zhong, M.Z.; Lotze, M.T.; Zeh, H.J.; Kang, R.; Tang, D.L. Identification of baicalein as a ferroptosis inhibitor by natural product library screening. *Biochem. Biophys. Res. Commun.* **2016**, *473*, 775–780. [CrossRef]

97. Probst, L.; Dächert, J.; Schenk, B.; Fulda, S. Lipoxygenase inhibitors protect acute lymphoblastic leukemia cells from ferroptotic cell death. *Biochem. Pharmacol.* **2017**, *140*, 41–52. [CrossRef]
98. Li, Q.; Li, Q.Q.; Jia, J.N.; Sun, Q.Y.; Zhou, H.H.; Jin, W.L.; Mao, X.Y. Baicalein Exerts Neuroprotective Effects in $FeCl_3$-Induced Posttraumatic Epileptic Seizures via Suppressing Ferroptosis. *Front. Pharmacol.* **2019**, *10*, 13. [CrossRef]
99. Yang, M.; Li, X.Y.; Li, H.J.; Zhang, X.X.; Liu, X.L.; Song, Y.Q. Baicalein inhibits RL S3-induced ferroptosis in melanocytes. *Biochem. Biophys. Res. Commun.* **2021**, *561*, 65–72. [CrossRef]
100. Zheng, B.; Zhou, X.W.; Pang, L.J.; Che, Y.J.; Qi, X. Baicalin suppresses autophagy-dependent ferroptosis in early brain injury after subarachnoid hemorrhage. *Bioengineered* **2021**, *12*, 7794–7804. [CrossRef]
101. Kong, N.; Chen, X.Y.; Feng, J.; Duan, T.; Liu, S.P.; Sun, X.N.; Chen, P.; Pan, T.; Yan, L.L.; Jin, T.; et al. Baicalin induces ferroptosis in bladder cancer cells by downregulating FTH1. *Acta Pharm. Sin. B* **2021**, *11*, 4045–4054. [CrossRef]
102. Maher, P. Modulation of the Neuroprotective and Anti-inflammatory Activities of the Flavonol Fisetin by the Transition Metals Iron and Copper. *Antioxidants* **2020**, *9*, 1113. [CrossRef]
103. Becquer-Viart, M.A.; Gonzalez-Yaque, J.; Fonseca-Fonseca, L.A.; Nunez-Figueredo, Y.; Andreu, G.L.P. Antioxidant and neuroprotective effects of gossypitrin, a flavonoid from Talipariti elatum, against chemical hypoxia-induced PC12 cell death. *J. Pharm. Pharmacogn. Res.* **2018**, *6*, 72–80.
104. Becquer-Viart, M.A.; Armentero-Lopez, A.; Alvarez-Alminaque, D.; Fernandez-Acosta, R.; Matos-Peralta, Y.; D'Vries, R.F.; Marin-Prida, J.; Pardo-Andreu, G.L. Gossypitrin, A Naturally Occurring Flavonoid, Attenuates Iron-Induced Neuronal and Mitochondrial Damage. *Molecules* **2021**, *26*, 3364. [CrossRef] [PubMed]
105. Matic, S.; Takac, M.J.M.; Barbaric, M.; Lucic, B.; Troselj, K.G.; Stepanic, V. The Influence of In Vivo Metabolic Modifications on ADMET Properties of Green Tea Catechins-In Silico Analysis. *J. Pharm. Sci.* **2018**, *107*, 2957–2964. [CrossRef] [PubMed]
106. Bernatova, I. Biological activities of (-)-epicatechin and (-)-epicatechin-containing foods: Focus on cardiovascular and neuropsychological health. *Biotechnol. Adv.* **2018**, *36*, 666–681. [CrossRef] [PubMed]
107. Butt, M.S.; Ahmad, R.S.; Sultan, M.T.; Qayyum, M.M.N.; Naz, A. Green Tea and Anticancer Perspectives: Updates from Last Decade. *Crit. Rev. Food Sci. Nutr.* **2015**, *55*, 792–805. [CrossRef]
108. Nakagawa, T.; Yokozawa, T. Direct scavenging of nitric oxide and superoxide by green tea. *Food Chem. Toxicol.* **2002**, *40*, 1745–1750. [CrossRef]
109. Chang, C.F.; Cho, S.; Wang, J. (-)-Epicatechin protects hemorrhagic brain via synergistic Nrf2 pathways. *Ann. Clin. Transl. Neurol.* **2014**, *1*, 258–271. [CrossRef]
110. Wang, J.J.; Chen, Y.; Chen, L.; Duan, Y.Z.; Kuang, X.J.; Peng, Z.; Li, C.H.; Li, Y.H.; Xiao, Y.; Jin, H.; et al. EGCG modulates PKD1 and ferroptosis to promote recovery in ST rats. *Transl. Neurosci.* **2020**, *11*, 173–181. [CrossRef]
111. Kose, T.; Vera-Aviles, M.; Sharp, P.A.; Latunde-Dada, G.O. Curcumin and (-)- Epigallocatechin-3-Gallate Protect Murine MIN6 Pancreatic Beta-Cells against Iron Toxicity and Erastin-Induced Ferroptosis. *Pharmaceuticals* **2019**, *12*, 26. [CrossRef]
112. Mu, M.; Wang, Y.L.; Zhao, S.S.; Li, X.L.; Fan, R.R.; Mei, L.; Wu, M.; Zou, B.W.; Zhao, N.; Han, B.; et al. Engineering a pH/Glutathione-Responsive Tea Polyphenol Nanodevice as an Apoptosis/Ferroptosis-Inducing Agent. *ACS Appl. Bio Mater.* **2020**, *3*, 4128–4138. [CrossRef]
113. Banoth, R.K.; Thatikonda, A. A Review on Natural Chalcones an Update. *Int. J. Pharm. Sci. Res.* **2020**, *11*, 546–555. [CrossRef]
114. Rammohan, A.; Reddy, J.S.; Sravya, G.; Rao, C.N.; Zyryanov, G.V. Chalcone synthesis, properties and medicinal applications: A review. *Environ. Chem. Lett.* **2020**, *18*, 433–458. [CrossRef]
115. Matos, M.J.; Vazquez-Rodriguez, S.; Uriarte, E.; Santana, L. Potential pharmacological uses of chalcones: A patent review (from June 2011-2014). *Expert Opin. Ther. Patents* **2015**, *25*, 351–366. [CrossRef]
116. Zhuang, C.; Zhang, W.; Sheng, C.; Xing, C.; Miao, Z. Chalcone: A Privileged Structure in Medicinal Chemistry. *Chem. Rev.* **2017**, *117*, 7762–7810. [CrossRef] [PubMed]
117. Cong, L.; Dong, X.Y.; Wang, Y.; Deng, Y.L.; Li, B.; Dai, R.J. On the role of synthesized hydroxylated chalcones as dual functional amyloid-beta aggregation and ferroptosis inhibitors for potential treatment of Alzheimer's disease. *Eur. J. Med. Chem.* **2019**, *166*, 11–21. [CrossRef]
118. Liu, J.; Li, X.C.; Cai, R.X.; Ren, Z.W.; Zhang, A.Z.; Deng, F.D.; Chen, D.F. Simultaneous Study of Anti-Ferroptosis and Antioxidant Mechanisms of Butein and (S)-Butin. *Molecules* **2020**, *25*, 674. [CrossRef]
119. Wen, L.R.; Shi, D.D.; Zhou, T.; Tu, J.M.; He, M.; Jiang, Y.M.; Yang, B. Identification of two novel prenylated flavonoids in mulberry leaf and their bioactivities. *Food Chem.* **2020**, *315*, 11. [CrossRef]
120. Tang, Y.; Luo, H.J.; Xiao, Q.; Li, L.; Zhong, X.; Zhang, J.; Wang, F.; Li, G.S.; Wang, L.; Li, Y. Isoliquiritigenin attenuates septic acute kidney injury by regulating ferritinophagy-mediated ferroptosis. *Ren. Fail.* **2021**, *43*, 1551–1560. [CrossRef]
121. Yaseen, A.; Yang, F.; Zhang, X.; Li, F.; Chen, B.; Faraag, A.H.I.; Wang, M.K.; Shen, X.F.; Wang, L. Ferroptosis inhibitory constituents from the fruits of Cullen corylifolium. *Nat. Prod. Res.* **2021**, *35*, 5364–5368. [CrossRef]
122. Xia, N.; Daiber, A.; Forstermann, U.; Li, H.G. Antioxidant effects of resveratrol in the cardiovascular system. *Br. J. Pharmacol.* **2017**, *174*, 1633–1646. [CrossRef]
123. Yi, M.; Li, J.J.; Chen, S.N.; Cai, J.; Ban, Y.Y.; Peng, Q.; Zhou, Y.; Zeng, Z.Y.; Peng, S.P.; Li, X.L.; et al. Emerging role of lipid metabolism alterations in Cancer stem cells. *J. Exp. Clin. Cancer Res.* **2018**, *37*, 18. [CrossRef]
124. Qu, Z.; Sun, J.C.; Zhang, W.N.; Yu, J.Q.; Zhuang, C.L. Transcription factor NRF2 as a promising therapeutic target for Alzheimer's disease. *Free. Radic. Biol. Med.* **2020**, *159*, 87–102. [CrossRef] [PubMed]

125. Zhang, X.H.; Jiang, L.P.; Chen, H.B.; Wei, S.; Yao, K.; Sun, X.C.; Yang, G.; Jiang, L.J.; Zhang, C.; Wang, N.N.; et al. Resveratrol protected acrolein-induced ferroptosis and insulin secretion dysfunction via ER-stress- related PERK pathway in MIN6 cells. *Toxicology* **2022**, *465*, 12. [CrossRef] [PubMed]
126. Li, T.; Tan, Y.; Ouyang, S.; He, J.; Liu, L.L. Resveratrol protects against myocardial ischemia-reperfusion injury via attenuating ferroptosis. *Gene* **2022**, *808*, 9. [CrossRef] [PubMed]
127. Mo, Y.S.; Duan, L.N.; Yang, Y.N.; Liu, W.; Zhang, Y.; Zhou, L.G.; Su, S.Y.; Lo, P.C.; Cai, J.Y.; Gao, L.Q.; et al. Nanoparticles improved resveratrol brain delivery and its therapeutic efficacy against intracerebral hemorrhage. *Nanoscale* **2021**, *13*, 3827–3840. [CrossRef]
128. Lee, J.; You, J.H.; Kim, M.S.; Roh, J.L. Epigenetic reprogramming of epithelial-mesenchymal transition promotes ferroptosis of head and neck cancer. *Redox Biol.* **2020**, *37*, 12. [CrossRef]
129. Rainey, N.E.; Moustapha, A.; Saric, A.; Nicolas, G.; Sureau, F.; Petit, P.X. Iron chelation by curcumin suppresses both curcumin-induced autophagy and cell death together with iron overload neoplastic transformation. *Cell Death Discov.* **2019**, *5*, 15. [CrossRef]
130. Guerrero-Hue, M.; Garcia-Caballero, C.; Palomino-Antolin, A.; Rubio-Navarro, A.; Vazquez-Carballo, C.; Herencia, C.; Martin-Sanchez, D.; Farre-Alins, V.; Egea, J.; Cannata, P.; et al. Curcumin reduces renal damage associated with rhabdomyolysis by decreasing ferroptosis-mediated cell death. *Faseb J.* **2019**, *33*, 8961–8975. [CrossRef]
131. Yang, C.; Han, M.M.; Li, R.Y.; Zhou, L.G.; Zhang, Y.; Duan, L.N.; Su, S.Y.; Li, M.; Wang, Q.; Chen, T.K.; et al. Curcumin Nanoparticles Inhibiting Ferroptosis for the Enhanced Treatment of Intracerebral Hemorrhage. *Int. J. Nanomed.* **2021**, *16*, 8049–8065. [CrossRef]
132. Tang, X.; Ding, H.; Liang, M.L.; Chen, X.; Yan, Y.X.; Wan, N.S.; Chen, Q.Q.; Zhang, J.; Cao, J. Curcumin induces ferroptosis in non-small-cell lung cancer via activating autophagy. *Thorac. Cancer* **2021**, *12*, 1219–1230. [CrossRef]
133. Li, R.H.; Zhang, J.; Zhou, Y.F.; Gao, Q.; Wang, R.; Fu, Y.R.; Zheng, L.W.; Yu, H. Transcriptome Investigation and In Vitro Verification of Curcumin-Induced HO-1 as a Feature of Ferroptosis in Breast Cancer Cells. *Oxidative Med. Cell. Longev.* **2020**, *2020*, 18. [CrossRef] [PubMed]
134. Xu, B.; Zhu, W.J.; Peng, Y.J.; Cheng, S.D. Curcumin reverses the sunitinib resistance in clear cell renal cell carcinoma (ccRCC) through the induction of ferroptosis via the ADAMTS18 gene. *Transl. Cancer Res.* **2021**, *10*, 3158–3167. [CrossRef] [PubMed]
135. Gaulton, A.; Hersey, A.; Nowotka, M.; Bento, A.P.; Chambers, J.; Mendez, D.; Mutowo, P.; Atkinson, F.; Bellis, L.J.; Cibrian-Uhalte, E.; et al. The ChEMBL database in 2017. *Nucleic Acids Res.* **2017**, *45*, D945–D954. [CrossRef]
136. Kanehisa, M.; Sato, Y.; Kawashima, M.; Furumichi, M.; Tanabe, M. KEGG as a reference resource for gene and protein annotation. *Nucleic Acids Res.* **2016**, *44*, D457–D462. [CrossRef] [PubMed]
137. R Core Team. R: A Language and Environment for Statistical Computing. R Foundation for Statistical Computing. Available online: http://www.R-project.org/ (accessed on 1 January 2021).
138. Sander, T.; Freyss, J.; von Korff, M.; Rufener, C. Data Warrior: An Open-Source Program For Chemistry Aware Data Visualization And Analysis. *J. Chem Inf. Model.* **2015**, *55*, 460–473. [CrossRef] [PubMed]

MDPI
St. Alban-Anlage 66
4052 Basel
Switzerland
Tel. +41 61 683 77 34
Fax +41 61 302 89 18
www.mdpi.com

Molecules Editorial Office
E-mail: molecules@mdpi.com
www.mdpi.com/journal/molecules